Spatially Resolved Single-Cell and Multiple Omics in Human Diseases

Jen-Tsung Chen
https://orcid.org/0000-0002-3540-4449
National University of Kaohsiung, Taiwan

IGI Global
Scientific Publishing
Publishing Tomorrow's Research Today

Vice President of Editorial Melissa Wagner
Director of Acquisitions Mikaela Felty
Director of Book Development Jocelynn Hessler
Production Manager Mike Brehm
Cover Design Jose Rosado

Published in the United States of America by
IGI Global Scientific Publishing
701 East Chocolate Avenue
Hershey, PA, 17033, USA
Tel: 717-533-8845 | Fax: 717-533-7115
Website: https://www.igi-global.com E-mail: cust@igi-global.com

Library of Congress Cataloging-in-Publication Data

LCCN: 2025950950 (CIP Data Pending)
ISBN13: 979-8-3373-4849-0
Isbn13Softcover: 979-8-3373-4850-6
EISBN13: 979-8-3373-4851-3

British Cataloguing in Publication Data
A Cataloguing in Publication record for this book is available from the British Library.

Table of Contents

Detailed Table of Contents

The term "Precision medicine" indicates treatment customization for a particular subset of a population that displays contrast vulnerability towards a specific disease or medication. The understanding of disease mechanisms and responses to curative approaches has been hugely revolutionized by the current technological evolutions, which have facilitated enhanced quality profiling of tissues and cells. An overview of spatial and single-cell multi-omics technologies blending into the workflow of precision medicine is essential. It covers platforms of spatial and single-cell multi-omics technologies, computational perspectives, prime biological understandings achieved to date, challenges faced in data integration, and future prospects. Further it highlights application of these technologies for precision healthcare of several ailment aspects like oncology, immunology, neurology, infectious disease, and many. The aim lies in setting an assist to the interest of researchers and clinicians in the convergence of leading-edge multi-omics technologies and individualized tailored healthcare

Precision genomics is a new paradigm, a combination of high-throughput genomic and multi-omics technologies, which facilitates proactive, personalized prediction, prevention, diagnostics and treatment of disease. These include; incomplete variant identification with short-read sequencing, low statistical power to analyze high-dimensional omics data, high computation and storage needs, data misalignments across platforms and complex issues in privacy, ownership and consent of genomic data. Future directions are focused on precision genomics combined with digital twin technology, diagnostics via nanotechnology and nano-genomics, and quantum computing on intractable genome problems, protein folding, and drug discovery. These innovations indicate the direction towards more predictive and preventive, as well as personalized healthcare eco systems.

Spatial multi-omics (SMO) is revolutionizing drug discovery by integrating molecular profiling with spatial tissue context to provide deep insights into cellular heterogeneity, molecular interactions, and disease microenvironments. Combining spatial transcriptomics, proteomics, metabolomics, and epigenomics, SMO maintains tissue architecture while mapping genes, proteins, metabolites, and regulatory elements at single-cell or subcellular resolution. This approach enables accurate

Neurodegenerative diseases represent a significant global health burden, characterized by the progressive deterioration of neuronal dysfunction and complex molecular pathologies. Traditional omics approaches have provided valuable insights into the genetic, transcriptomic, and proteomic alterations involved; however, they often lack spatial resolution, obscuring the tissue-specific and cellular-contextual dynamics of disease progression. Spatial multi-omics technologies integrate spatial transcriptomics, proteomics, epigenomics, and metabolomics to enable high-resolution mapping of molecular features within intact tissues. This chapter explores recent advancements in spatial multi-omics technologies and also the role of spatial multi-omics involving metabolomics, epigenomics, and their application to neurodegenerative disease models, as well as the challenges and future directions in translating these insights into clinical impact.

Fungal infections are an increasing global health concern, especially in individuals with weakened immune systems. Their complexity stems from intricate interactions between fungal pathogens and host cells, making diagnosis and treatment difficult. Although advances have been made, key aspects of host–pathogen interactions, such as invasion mechanisms, immune evasion, and variable host responses, remain unclear. Systems biology provides a holistic way to study these processes by integrating large-scale data on genes, proteins, metabolites, and immune pathways. Multiomics approaches, including genomics, transcriptomics, proteomics,

Spatial multi-omics has become a transformative frontier of cardiovascular research, offering integrative understandings of cellular heterogeneity and spatially organized cardiac tissues. Through integration of omics by high-resolution spatial profiling, these technologies empower mapping of interactions underlying cardiac development, disease, and remodelling. Developments for atherosclerosis, myocardial infarction, and cardiomyopathies have revealed dynamic networks of communication, metabolic reprogramming, and evolution of the fibrotic niche. Next-generation computational platforms, including graph neural networks, variational autoencoders, and digital pathology guided by AI, support multi-layered data integration and predictive modelling. Novel human heart spatial atlases additionally connect molecular architecture and clinical phenotypes, advancing precision diagnostics and targeted therapeutics. Although data harmonization and standardization are key challenges, spatial multi-omics represents untold promise for transformative precision cardiology and regenerative medicine.

Kidney diseases encompass a heterogeneous group of disorders with complex molecular underpinnings and diverse spatial manifestations within renal tissue. This chapter explores the transformative potential of spatial multi-omics technologies in unraveling the cellular and molecular architecture of kidney diseases. By integrating genomics, transcriptomics, proteomics, and metabolomics with spatial resolution, spatial multi-omics enables high-resolution mapping of disease-associated alterations within distinct renal compartments. The chapter discusses recent advances in spatial transcriptomics, and single-cell omics that have deepened our understanding of kidney pathophysiology. Additionally, the chapter addresses the challenges in data integration, computational analysis, and clinical translation. Ultimately, spatial multi-omics holds promise for redefining kidney disease diagnostics, identifying novel biomarkers, and guiding targeted therapies through a more precise, tissue-contextual understanding of renal pathology.

Spatial multi-omics is transforming veterinary infectious disease research by integrating genomic, transcriptomic, proteomic, and metabolomic data with spatial and histological context. This approach reveals host–pathogen interactions, cellular heterogeneity, and localized immune or metabolic responses that shape disease outcomes. Emerging technologies such as SM-Omics, SpatialGlue, and MOSAIK enable multi-layered data integration across imaging and molecular scales. In veterinary medicine, spatial multi-omics provides insights into pathogen localization, immune microenvironments, and species-specific mechanisms in bacterial, viral, and parasitic infections. Combined with artificial intelligence (AI) and digital pathology, it enhances diagnostic accuracy, biomarker discovery, and outbreak prediction. Despite challenges in cost and data standardization, the convergence of spatial omics, AI, and One Health strategies is paving the way for precision veterinary medicine and improved surveillance of zoonotic diseases.

The integration of AI in transcriptomics has revolutionized the analysis of complex biological data. This chapter highlights AI's role in digital transcriptomics, especially in spatial transcriptomics, pathology, and toxicity studies. AI techniques like ML and DL tackle high-dimensional data through preprocessing, feature extraction, and modeling. Tools such as HE2RNA predict RNA-Seq data from histology slides, aiding virtual gene expression mapping and tumor analysis. GANs, like TransTox, translate organ-specific profiles to support drug safety. In cancer immunotherapy, AI helps detect immune signatures and tumor heterogeneity. AI also enables multi-omics integration, improving signal clarity and accelerating precision medicine.

Preface

INTRODUCTION

The post-genomics era has ushered in a transformative phase in biomedical research, characterized by the rapid expansion and convergence of multi-omics technologies. Over the past decades, advances in genomics, transcriptomics, proteomics, and related disciplines have moved beyond bulk analyses toward spatially resolved, single-cell–level interrogation of biological systems. These developments have fundamentally reshaped our ability to understand the cellular heterogeneity, spatial organization, and molecular complexity that underlie human development, physiology, and disease.

At the same time, the emergence and maturation of artificial intelligence (AI) and advanced computational methods have created unprecedented opportunities for the efficient integration and interpretation of high-dimensional, multi-omics data. By bridging experimental innovation with computational intelligence, spatial and single-cell multi-omics approaches now offer powerful frameworks for uncovering disease mechanisms, identifying biomarkers, and advancing precision medicine. The application of these technologies to human diseases is therefore not only timely but also highly anticipated.

This edited volume, *Spatially Resolved Single-Cell and Multiple Omics in Human Diseases*, aims to provide a comprehensive overview of current technological developments and recent applications in this rapidly evolving field. Particular emphasis is placed on conceptual advances, methodological innovations, and practical examples of how artificial intelligence and computational strategies are being integrated with spatial and single-cell multi-omics to enhance human health management and disease research. Through contributions from experts across disciplines, the book highlights both the promise and the challenges of translating these cutting-edge approaches into biomedical and clinical insights.

The intended readership of this volume includes students, educators, researchers, and professionals working in human disease management, medical and biomedical research, drug discovery, artificial intelligence, bioinformatics, computational biology, and systems biology. It is our hope that this book will serve not only as a reference for current knowledge, but also as a source of inspiration for future interdisciplinary research and innovation at the interface of biology, medicine, and data science.

As Editor, I am deeply grateful to all contributors for sharing their expertise and perspectives, and I trust that this volume will contribute meaningfully to the continued advancement of spatial and single-cell multi-omics in the understanding and treatment of human diseases.

ORGANIZATION OF THE BOOK

Chapter 1: Spatial and Single-Cell Multi-Omics for Precision Medicine: An Overview

In this opening chapter, the contributors establish the conceptual and technological foundations of precision medicine through the lens of spatial and single-cell multi-omics. The chapter introduces how recent advances have transformed our ability to profile tissues and cells with unprecedented resolution, enabling a deeper understanding of disease heterogeneity and therapeutic response. By surveying major spatial and single-cell platforms, computational frameworks, and integration challenges, the chapter provides a cohesive overview of the field. Importantly, it contextualizes these technologies within real-world applications across oncology, immunology, neurology, infectious diseases, and beyond, setting the stage for the chapters that follow and underscoring the convergence of cutting-edge multi-omics with individualized healthcare.

Chapter 2: Precision Genomics Integrates Advanced Genomic Technologies

This chapter expands the precision medicine narrative by focusing on precision genomics as a unifying paradigm that integrates high-throughput genomics with broader multi-omics data. The authors critically examine current limitations, including incomplete variant detection, analytical power constraints, computational demands, and ethical concerns related to data governance. Looking ahead, the chapter highlights emerging directions such as digital twin models, nano-genomics, and quantum computing, illustrating how these innovations may overcome existing barriers. Together, these perspectives outline a forward-looking vision of predictive, preventive, and personalized healthcare ecosystems driven by advanced genomic technologies.

Chapter 3: Spatial Multi-Omics in Drug Discovery

Chapter 3 explores how spatial multi-omics is reshaping drug discovery by embedding molecular information within its native tissue context. By integrating spatial transcriptomics, proteomics, metabolomics, and epigenomics, this approach enables precise mapping of cellular heterogeneity and molecular interactions at single-cell and subcellular resolution. The chapter demonstrates how these insights accelerate target identification, clarify drug mechanisms, and reveal spatial biomarkers of response or resistance. Applications across oncology, neuroscience, autoimmune, and infectious diseases highlight the translational value of spatial multi-omics, particularly when coupled with AI-driven modeling and visualization tools that bridge preclinical and clinical research.

Chapter 4: Advancing Personalized Medicine with AI-Driven Microbiome and Nutrigenomics

This chapter examines the rapidly evolving intersection of artificial intelligence, microbiome science, and nutrigenomics in advancing personalized medicine. The authors illustrate how AI-driven analytics uncover complex relationships among host genetics, diet, metabolism, and gut microbial communities, enabling more precise diagnostics and tailored interventions. Alongside technological progress, the chapter thoughtfully addresses challenges related to data bias, standardization, ethics, and privacy. By highlighting future developments such as portable biosensors, cloud-based platforms, and digital health tools, the chapter emphasizes the growing potential for scalable, real-time personalized care.

Chapter 5: Spatial Multi-Omics in Neurodegenerative Diseases

Focusing on neurodegenerative disorders, this chapter highlights the critical importance of spatial context in understanding complex brain pathologies. While traditional omics have yielded important molecular insights, the authors demonstrate how spatial multi-omics overcomes prior limitations by mapping transcriptomic, proteomic, epigenomic, and metabolomic features within intact neural tissues. The chapter reviews recent technological advances and their application to disease models, revealing tissue- and cell-specific dynamics of neurodegeneration. It also discusses key challenges and future directions for translating spatial multi-omics discoveries into clinically meaningful interventions.

Chapter 6: Spatial Multi-Omics in Infectious Diseases

In this chapter, the focus shifts to infectious diseases, with particular emphasis on fungal infections and their complex host–pathogen interactions. The authors adopt a systems biology perspective, illustrating how integrated multi-omics approaches can elucidate mechanisms of invasion, immune evasion, and host response. By combining genomics, transcriptomics, proteomics, metabolomics, and immunomics, spatially informed analyses reveal pathways that govern infection progression and immune defense. The chapter underscores how these insights can guide the development of improved diagnostics and therapeutic strategies for challenging infectious diseases.

Chapter 7: Spatial Multi-Omics in Cancers: Integrating Molecular Landscapes for Precision Oncology

Chapter 7 addresses one of the most impactful applications of spatial multi-omics: cancer research. The contributors demonstrate how simultaneous measurement of multiple molecular layers within the tumor microenvironment reveals critical spatial relationships among cancer, immune, and stromal cells. These insights are essential for understanding tumor heterogeneity, metastasis, and treatment response. The chapter highlights the growing role of artificial intelligence in analyzing large-scale spatial datasets and discusses expanding clinical applications, from patient stratification to trial design. It concludes by emphasizing the need for interpretable models that can directly inform precision oncology decisions.

Chapter 8: Spatial Multi-Omics in Cardiovascular Diseases

This chapter presents spatial multi-omics as a transformative approach in cardiovascular research, enabling detailed mapping of cellular heterogeneity and tissue organization in the heart and vasculature. The authors review advances in atherosclerosis, myocardial infarction, and cardiomyopathies, revealing dynamic molecular networks, metabolic reprogramming, and fibrotic remodeling. The integration of AI-driven computational frameworks and emerging human heart spatial atlases connects molecular architecture with clinical phenotypes. Despite ongoing challenges in data harmonization, the chapter underscores the promise of spatial multi-omics for precision cardiology and regenerative medicine.

Chapter 9: Spatial Multi-Omics in Kidney Diseases

Chapter 9 focuses on the application of spatial multi-omics to kidney diseases, a diverse group of disorders characterized by complex and compartment-specific pathology. By integrating multiple omics layers with spatial resolution, the chapter illustrates how disease-associated alterations can be mapped across distinct renal structures. Recent advances in spatial transcriptomics and single-cell approaches are discussed in the context of kidney pathophysiology, biomarker discovery, and therapeutic targeting. The authors also address computational and translational challenges, emphasizing the potential of spatial multi-omics to redefine renal diagnostics and precision treatment strategies.

Chapter 10: Spatial Multi-Omics in Veterinary Infectious Diseases

Extending spatial multi-omics into veterinary medicine, this chapter highlights its growing impact on the study of infectious diseases across animal species. By integrating molecular data with spatial and histological context, the approach reveals host–pathogen interactions, immune microenvironments, and species-specific disease mechanisms. The authors discuss emerging technologies and AI-enhanced digital pathology as tools for improving diagnostics, biomarker discovery, and outbreak prediction. Framed within a One Health perspective, the chapter illustrates how spatial multi-omics can advance precision veterinary medicine and strengthen surveillance of zoonotic diseases.

Chapter 11: AI Approaches for Constructing Digital Transcriptomics

The concluding chapter centers on artificial intelligence as a catalyst for innovation in transcriptomics, particularly in spatial and digital contexts. The authors review machine learning and deep learning approaches that address the challenges of high-dimensional data, from preprocessing to predictive modeling. Case studies such as virtual gene expression mapping from histology slides and generative models for toxicity prediction demonstrate the practical impact of AI. By enabling robust multi-omics integration and clearer biological signal extraction, this chapter reinforces AI's pivotal role in accelerating discovery and advancing precision medicine across diverse biomedical applications.

CONCLUSION

As the chapters of this volume collectively demonstrate, spatially resolved single-cell and multi-omics technologies are redefining the way we investigate human disease and approach precision health. The convergence of high-resolution molecular profiling with spatial context has moved the field beyond descriptive biology, enabling mechanistic insights into cellular heterogeneity, tissue organization, and dynamic disease processes that were previously inaccessible. When coupled with advances in artificial intelligence and computational modeling, these approaches provide a powerful framework for integrating complexity into actionable knowledge.

This edited reference book brings together diverse yet complementary perspectives spanning precision medicine, genomics, drug discovery, infectious and chronic diseases, veterinary science, and AI-driven analytics. Each contribution underscores a shared theme: meaningful progress in human and animal health increasingly depends on interdisciplinary integration—across omics layers, spatial scales, and analytical paradigms. While challenges remain in data standardization, computational scalability, ethical governance, and clinical translation, the collective work presented here highlights both the feasibility and the necessity of addressing these barriers collaboratively.

As Editor, it is my hope that this volume serves not only as a comprehensive reference for current methodologies and applications, but also as a catalyst for future innovation. By bridging experimental biology, medicine, and data science, this book aims to inspire researchers, clinicians, and trainees to further advance spatial and single-cell multi-omics toward predictive, preventive, and personalized healthcare. The rapid evolution of this field assures that the insights captured here represent both a milestone and a foundation for the next generation of discoveries in human disease research and precision medicine.

Chapter 1
Spatial and Single-Cell Multi-Omics for Precision Medicine:
An Overview

Kangkana Kataki
https://orcid.org/0009-0008-5191-3228
Mahapurusha Srimanta Sankaradeva Viswavidyalaya, India

Jayasree Talukdar
https://orcid.org/0009-0000-6803-9268
Pandu College, India

Subhash Medhi
https://orcid.org/0000-0002-2733-8293
Gauhati University, India

ABSTRACT

The term "Precision medicine" indicates treatment customization for a particular subset of a population that displays contrast vulnerability towards a specific disease or medication. The understanding of disease mechanisms and responses to curative approaches has been hugely revolutionized by the current technological evolutions, which have facilitated enhanced quality profiling of tissues and cells. An overview of spatial and single-cell multi-omics technologies blending into the workflow of precision medicine is essential. It covers platforms of spatial and single-cell multi-omics technologies, computational perspectives, prime biological understandings achieved to date, challenges faced in data integration, and future prospects. Further it highlights application of these technologies for precision healthcare of several ailment aspects like oncology, immunology, neurology, infectious disease, and

DOI: 10.4018/979-8-3373-4849-0.ch001

many. The aim lies in setting an assist to the interest of researchers and clinicians in the convergence of leading-edge multi-omics technologies and individualized tailored healthcare

1. INTRODUCTION

Figure 1. Graphical abstract

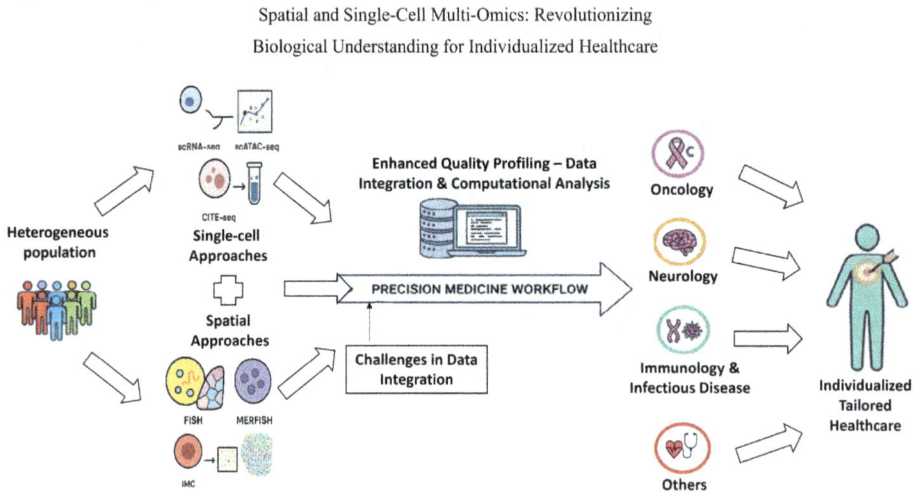

The 21st century has observed an exceptional improvement in disease diagnosis and treatment approaches in the influence of substantiation-based paradigm (Carmo-Fonseca, 2021). For decades, clinical diagnosis and therapy were dependent on the generalized structure of conventional bulk omics, which has a perspective on cellular heterogeneity with no availability of information on spatial factors. The term "Precision medicine" indicates treatment customization for a particular subset of a population that displays contrast vulnerability towards a specific disease or medication. However, in a conventional, one-size-fits-all model, individuals with similar health conditions are prescribed similar drugs regardless of the underlying molecular variations. Although this approach is effective in certain cases, but has limitations in complex diseases like autoimmune disorders, neurological conditions, and cancer. The perception of regulated treatment or precision medicine has arisen on the basis of an individual's inimitable biological composition (Shorter et al., 2024).

Precision medicine had its origin with Human Genome Project, which successfully generated the complete human genome sequence in the year 2003. This significant achievement ushered in a new era whereby genetic diversity could be systematically identified and linked to treatment response, prognosis, and vulnerability to diseases.

Although genomics represented a crucial foundation, very soon it was realized that disease diversity was unable to be understood merely through decoding the DNA sequence. Disease pathogenicity and treatment outcome can be influenced by several factors, like interaction between proteins, alteration in metabolism, epigenetic alterations, adaptive transcriptomic phases, and the surrounding microenvironment (Ahmed, 2024; Waldman & Terzic, 2011). This perspective provided a way for the establishment of multi-omics approaches, which evaluate multiple molecular levels in tandem to serve as a complete glance of the functioning of the biological system (Derks et al., 2015; Ivanisevic & Sewduth, 2023).

The understanding of the non-homogeneous nature of biological systems is perhaps the most significant conceptual advancement. Conventional bulk assessment obscures the variation that lies at the individual cellular level as it targets the measurement of aggregated signal from millions of cells (Walling & Shepard, 2011). Heterogeneity can be observed in several instances, like in a tumor, among the majority of chemotherapy-sensitive cells, there may exist a small number of drug-resistant cells which may survive the treatment and lead to relapse. Furthermore, in autoimmune condition, generally a subtype of immune cells plays the role of tissue destruction. These infrequent but clinically relevant subpopulation cannot be detected by averaging across diverse population approach of traditional bulk measurements (Pribluda et al., 2015; He, 2016; Shimagami et al., 2025).

This challenge was alleviated by the advancement of single-cell technologies, which rendered cellular-level molecular profiling. Owing to the advancements in droplet-based techniques, microfluidics, and advanced sequencing, understanding of several aspects like the genome, transcriptome, epigenome, proteome, and metabolome of distinct cells can be achieved (Lareau et al. 2019; Datlinger et al. 2021; Shang et al., 2024). These understandings have identified previously unexplored remarkable heterogeneity of tissues that lies in dynamic cellular levels and subpopulations, contributing to health and illness.

However, the single-cell approach also has limitations. Spatial insights of cell's position across its original microenvironment are compromised when tissues are separated into individual cells in single cell technologies. Knowledge of a cell's residence position is also as important as metabolic identity. Cells maintain domains which influences cell's behavior, communication with surroundings, and response to a range of signaling molecules. As an instance, in cancer, immune cells across the tumor edge may function very differently compared to those enmeshed deep within the tumor mass. In accordance to this, researchers have established a spatial-omics approach enabling precise monitoring of gene expression, protein availability and metabolite dispersion while still maintaining the structural framework of the intact tissue (Rawat, 2023; Suo et al., 2025).

Currently, one of the fascinating frontiers in biomedical studies is the confluence of single-cell omics and spatial omics. This advantageous integration provides precise molecular images of individual cells by a single cell approach, while the spatial approach integrates this image in the convoluted tissue architecture. The combination of single-cell and spatial omics is known as multi-omics, which accelerates the advancements of precision medicine, revolutionizing our knowledge regarding illness (Vandereyken et al., 2023).

Forthcoming succeeding discussion provides a thorough overview of implementation of spatial and single-cell multi-omics for precision healthcare. It includes the technological foundations that enabled these approaches, the computational strategies required for data integration and evaluation, and also the biological advancements the multi omics have generated in the domains of neuroscience, infectious disease, tumor biology, reconstructive medicine and immunology. Furthermore, the obstacles and restrictions that currently inhibit their widespread implementation will also be addressed as well as potential routes therapeutic applications in the future. Thus, with the orientation of these above-mentioned interconnecting themes this interlocution will provide students, researchers and clinicians an understanding of the convergence of customized medicine and modern omics technologies.

2. SINGLE-CELL OMICS TECHNOLOGIES

The single-cell biology has emerged as a revolutionary domain in recent biomedical research. As mentioned, prior to establishment of single-cell strategies, most of the molecular studies were dependent on bulk assessment of cell or tissue (Chen et al., 2020; Lee, 2025).

2.1. Single-Cell Genomics

The process of analysing of DNA sequences from individual cells that leads to tracking chromosomal transformation, structural differences and mutational patterns is known as single-cell genomics (scDNA-seq). Efficient retrieval of genomic material from an individual nucleus became feasible by the development whole-genome amplification techniques like Multiple Annealing and Looping-Based Amplification Cycles (MALBAC), Multiple Displacement Amplification (MDA) and Primary Template-Directed Amplification (PTA) (Kaur et al., 2019; Jäger, 2023). MDA employs high fidelity polymerase for obtaining extensive genome coverages; however, chances of amplification distortion may occur (Huang et al., 2015; Lu et al., 2023). MALBAC, on the contrary has reduced possibilities of amplification distortion and decreases biasness, consequently, it is useful for determining single nucleotide and copy number variations with reduced false positive frequency (Huang et al., 2015; Zhou et al., 2020). PTA is a more modern method with improved reproducibility and accuracy. PTA creates virtually an entire genome from specific bacterial cells, yielding a median accuracy of 81% in contrast to a lesser than 30% by normal conventional approaches. This method captures over 95% of single-cell genomes uniformly, facilitating more accurate variant detection These methods have rendered investigators to gain insights into mutational polymorphism in microbial populations, analyse mosaicism in disorders of development and monitor clonal development in cancer (Kalisky & Quake, 2011; Kaur et al., 2019; Bowers et al., 2024).

scDNA-seq has been particularly advantageous in cancer research. Tumor proliferation during numerous rounds of mutations and selections generate subclonal populations with specific treatment vulnerabilities. scDNA-seq has indicated that within the same tumor, genetically distinctive subpopulations exist, several which might lead to metastasis or therapeutic resistance. These understanding plays a prime role for establishment of beneficial individualized treatment regimens (Sharma & DasGupta, 2019; Li & Hou, 2018; Bai et al., 2022).

2.2. Single-Cell Transcriptomics

Single-cell transcriptomics, also referred as single-cell RNA sequencing (scRNA-seq), has established as extremely prominent omics technique. scRNA-seq identifies developmental progression, unusual populations, and dynamic cellular condition by assessing gene expression at the individual cell unit. This shift has been propelled by several technical infrastructure (Qu et al., 2023). SMART-seq is an efficient scRNA-seq process appropriate for comprehensive investigations of small cell populations due to its exceptional sensitivity and complete transcript range. Massive atlasing initiatives are now possible because to Drop-seq and 10x Genomics Chromium, which utilize droplet microfluidics to facilitate highly concurrent characterization of several thousands of cells (Otto, 2019; Zhang et al., 2023).

scRNA-seq has revolutionized the way researchers consider biology. It has made it feasible to distinguish between immune and neurological system in variety of cell types which was previously unidentifiable, recreate developmental differentiation pathways, and break down the tumor microenvironment into its constituent cells (Vats et al., 2024). The Human Cell Atlas project, particularly uses scRNA-seq as a fundamental method to classify all cell forms in the human being. It has been one of the most ambitious endeavors (Liu et al., 2020; Vats et al., 2024).

CITE-seq (Cellular Indexing of Transcriptomes and Epitopes by Sequencing) is an upgradation to scRNA-seq approach which offers for the comprehensive evaluation of cell properties by integrating transcriptomic monitoring with antibody mediated identification of protein. In contrast to scRNA sequencing this strategy bring together protein and RNA findings, generating better knowledge about the studied cell population (Stoeckius et al., 2017; Shi et al., 2023). Approaches like TEA-seq (trimodal single-cell measurement of transcripts, epitopes, and chromatin accessibility) further improves the potential of CITE-seq by evaluating scRNA sequence, epitopes and chromatin connectivity (scATAC-seq) in thousands of single cells concurrently. TEA-seq reveals an expanded comprehension of regulations of gene (Swanson et al., 2021). An additional improvement is marked by icCITE-seq (intracellular cellular indexing of transcriptomes and epitopes) which provide concurrent assessment of both surface and internal proteins together with transcription of genes. This multi-omic strategy strengthens the knowledge of cellular integrity and pathways associated with disease (Chen et al., 2025).

2.3. Single-Cell Epigenomics

Epigenomics reveals the mechanism of regulatory processes that govern transcription of gene, whereas genomics and transcriptomics illustrate what is encoded in DNA and being transcribed of the DNA respectively. Single-cell epigenomics encompasses several techniques. Single-cell whole-genome bisulfite sequencing (WGBS), it profiles DNA methylation which is essential for comprehension of the way genes are controlled and the way their responses impact the fate of cells (Cheng & McCarrey, 2023). Chromatin immunoprecipitation-sequencing (ChIP-seq) to analyze the genomic configurations of attaching transcription variables or certain histone alterations (Cheng & McCarrey, 2023). Single-cell assay for transposase-accessible chromatin sequencing (ATAC-seq) that tracks chromosome accessibility, revealing knowledge about the components of the genome that contribute in transcriptional modulation (Lo & Zhou, 2018; Dawson, 2024). Another innovative approach namely scNMT-seq (single-cell nucleosome, methylation, and transcription sequencing), combines several levels of molecules provides insight into networks of gene regulatory mechanisms by synchronously assessing gene transcription, genome methylation, and accessibility to chromatin (Clark et al., 2017; Myers, 2022). These methods provide information on how cellular characteristics and adaptability are shaped by epigenetic variability.

Single-cell epigenomic profiling has provided understanding regarding the way immune cells transit between activation and fatigue phases, drug resistance development in cancer cells by means of epigenetic modifications along with the process undertaken by stem cells to preserve multipotency. Strategies comprising multiomics analyses offer information on molecular processes controlling immune responses and variation among cells. The emergence of methodologies as scGET-seq (single-cell genome and epigenome by transposases sequencing) provides detailed chromatin configuration assessment, exposing the biological processes underpinning cancer cell resistance to therapies. Single-cell epigenomics contributes a vital component to our comprehension of cellular function by connecting transcriptional products and regulatory surroundings (Tedesco et al., 2021; Bond et al., 2020; LaFave et al., 2022).

The progression of epigenetic transformations throughout stem cell rejuvenation, which serve as significant for sustaining multipotency, are being revealed using single-cell epigenomics. For the purpose of to determine the progressions associated with these transformations and to enhance the knowledge of stem cell biology, chromatin velocities measures have shown indispensable. (Schulte-Schrepping et al., 2020; Li et al., 2020)

2.4. Single-Cell Proteomics and Metabolomics

As the essential mediators of processes within cells, proteins and metabolites need to be assessed at the level of individual cells in order to connect molecular characteristics to phenotypic outcomes. According to the advancements like single-cell western blotting and mass cytometry (CyTOF), it has become feasible to measure several of proteins concurrently from a single cell, raising understanding of the variety and purpose of cells. Correspondingly, single-cell metabolomics utilizes imaging-based methodologies or capillary electrophoresis–mass spectrometry has begun to identify metabolic diversity that affects the development of disease. Highly accurate and precise evaluation of proteins has been rendered accessible by approaches like SCOPE (Single Cell ProtEomics) and nanoPOTS (Nanodroplet Processing in One pot for Trace Samples), which are important to grasping sophisticated systems of biology (Singh, 2025).

Single-cell proteomics serves to be a crucial utility in immunology, response to therapy is potentially determined by the varied function of natural killer cells or T lymphocytes. Metabolomic characterization at single-cell precision in the field of oncology has shed light to the processes in which cancer cells use specific metabolic routes to endure in a situation of nutrient deficiency scenario. These understanding provide opportunity for treatment approaches that focus on metabolic processes of cell. In an effort to investigate metabolites at the single-cell level, methodologies which include capillary electrophoresis-mass spectrometry (CE-MS) as well as different mass spectrometry ionization approaches (e.g., nanoESI, MALDI) have proved significant. Rapid metabolic changes are documented by single-cell metabolomics, which makes it important in interpreting the way cells behave to external triggers (Duncan et al., 2019; Qin et al., 2023; Liu et al., 2022).

Single-cell technologies, including scRNA-seq, circumvent the vital disadvantage posed by conventional bulk investigation, which delivers a mean estimation within a population that is diverse, by offering the formerly unusual effectiveness of characterizing molecular information at the single cell scale. In domains which include developmental biology, immunology, and cancer research, this technology makes it achievable for researchers to confidently reveal variability in cells, recognize rare groups of cells, and observe dynamic cell changes in states and progenitors.

3. SPATIAL OMICS TECHNOLOGIES

As a way to present an extensive understanding of biological mechanisms, researchers have established spatial omics technologies that incorporate histological surroundings with molecular characterization.

Technologies associated with spatial omics can be classified into two primary groups namely 1. NGS-based techniques such as spatial barcoding and laser capture microdissection and 2. imaging-based techniques such as in situ sequencing and in situ hybridization. Each of them has distinctive advantages and disadvantages considering spatial accuracy, sensitivity, range, workload, emphasis on specific technology and data implementation. Currently none of the approaches satisfies all of these criteria. Consequently, the type of specimen and the targeted query determine the best suited approach to proceed (Conrad & Altmüller, 2022).

A revolutionary transformation has been made attainable by spatial omics approach, enabling investigators to understand pattern of cellular organization and the way these arrangements impact pathological and physiological processes.

3.1. Spatial Genomics

Spatial genomics is a growing field which intends to record genome wide characteristics precisely in an undamaged tissue segment. On the contrary, tissue differentiation is essential for both bulk and single cell genomics. Tumor variability in context to the structure of the tissue architecture can be identified by approaches which includes Targeted Imaging of Copy Number Variations (CNVs) or In Situ Sequencing of DNA (issDNA). This information is crucial for coupling proximal microenvironmental cells to specific oncogenic modifications. issDNA alters the sequencing procedure by combining hybridization, ligation and visualization. It is extremely beneficial for assessing of the spatial locations of single nucleotide variations (SNVs) or unique gene sequences. By recognizing structural differences within tumors, focused imaging of CNVs may provide illumination on clonal progression and microenvironmental consequences (Kamsalem et al., 2025; Zhao et al., n.d; Lomakin et al., 2022).

Widely, multiplexed kinds of non-sequencing approaches like Fluorescence In Situ Hybridization (FISH) utilize fluorescent probes customized for distinct genomic loci. However, investigators can determine CNVs by quantifying the number of signals for a specific probe embedded in the nucleus of the cell. The communication involving tumor cells as well as their environs, particularly impacts strategies for resistance and therapeutic reactions can be witnessed by spatial genomics. Diverse tumor populations can be tracked through methodologies like slide-DNA-seq, that demonstrate their spatial constriction and related microenvironments (Morse et al., 2023; Iafrate et al., 2020).

3.2. Spatial Transcriptomics

Spatial transcriptomics ranks as one of the earliest and most prevalent approaches in spatial omics, which evaluates the expression of genes precisely within tissue fragment. Conventionally, histology gives significant architectural knowledge, while immunohistochemistry facilitates the localization of specific proteins. Spatial transcriptomics, on the contrary, enables genome-wide accuracy, gathering thousands of transcripts in vivo by conserving tissue morphology. (Saul & Kosinsky, 2023)

Innovations related to this field have been propelled by several kinds of technical frameworks. Perhaps the most prominent and conveniently approachable is 10x Genomics' Visium system of spatial Gene Expression. Entire transcriptome assessment of tissue samples is rendered feasible by the Visium platform, which facilitates high-throughput mRNA sequencing with intact spatial details. Assessing tumor microenvironment (TME) necessitates a knack to determine the margins of malignant tissue and to monitor patterns in the expression of genes. For superior therapeutic results, the Visium system can be utilized in combination with additional methods, including immunotherapy and anti-angiogenic therapy. Creative strategies for instance Slide-tags merge tissue mounted methodologies alongside barcoded oligonucleotides for improved spatial precision and effectiveness (Nadaf & Rayan, 2025; Kuta Alam, 2022). Visium HD an advance approach, facilitates accurate spatial assessment of elements in cells within tissues by obtaining whole-transcriptome profiling at a resolution of 2 μm and wide transcriptome coverage (about 18,085 genes). Thus, it performs well in exploratory research however it remains less sensitive than alternatives for sparse transcripts (Lee et al., 2025; Luo et al., 2025). This approach provides a remarkable equilibrium between productivity and spatial accuracy.

Image-supported methods, such as seqFISH (Sequential Fluorescence In Situ Hybridization) and MERFISH (Multiplexed Error-Robust Fluorescence In Situ Hybridization), are currently been designed for greater clarity. These technologies utilize numerous cycles of fluorescent labelling and scanning to identify several thousand RNA sequences in undamaged portions of tissues. Augmented Multiplexing by the combinatorial barcodes have been utilized to encode gene transcripts, circumventing the downsides of traditional FISH, which only allows visualization of a restricted amount of genes concurrently. It has significantly outperformed conventional techniques by allowing simultaneously recognizing of hundreds to thousands of RNA variants. The MERFISH approach provides outstanding diagnosis effectiveness in thicker tissues (up to 200 μm) by recognizing and measuring RNA molecules employing error-robust barcodes. By conservation of spatial context methods it sustains the inherent spatial layout of cells thus, representation of cellular heterogeneity throughout tissues has been rendered considerably improved by MERFISH, that provides helpful knowledge on cell forms and phases that are necessary for comprehending disorders and designing remedies (Lerch, 2024; Emanuel & He, 2021; Moffitt et al., 2016; Wang et al., 2017). Akin to MERFISH, seqFISH gives it feasible for researchers to observe numerous RNA species concurrently, allowing increase appreciation of cellular heterogeneity and spatial structuring. In an analogous manner, techniques like Slide-seq, a scalable technology for measuring genome-wide expression at high spatial resolution, employ barcoded beads to record mRNA in substantial quantity, permitting transcriptomic visualization at almost cellular accuracy. Another to be mentioned platform is Xenium platform which enables subcellular clarity, permitting precision positioning of RNA targets, in each of the frozen and formalin-fixed paraffin-embedded (FFPE) tissue slices. It exhibits unparalleled sensitiveness and spatial resolution, which renders it appropriate for tailored cellular research and identification of rarer cell populations. CosMx works well for comprehensive architecture of tissues analysis since the technique achieves an ideal balance between spatial resolution and multiplexing capabilities (Lee et al., 2025; Cervilla et al., 2024; Houston, 2024; Cheung et al., 2024; Soja et al., 2025).

Spatial transcriptomics applications are transforming the discipline. By associating gene expression with specific anatomical positions, such approaches are being employed in neuroscience to determine the molecular framework of the mouse brain. By contributing to the compilation of extensive brain atlases, spatial transcriptomics enriches current knowledge of connections between neurons and activity. By analyzing the brain's cellular deficiencies and associations, it presents illumination on disorders like Alzheimer's disease (Wang et al., 2025).

Similarly, spatial transcriptomics is currently been utilized in oncology as well to discover immune-compromised tumor domains, discovering variations in hypoxia-associated expression of genes, and determine cancerous stem cell habitats that promote to recurrence. Furthermore, these methods reveal physiological understandings of the role of cellular surroundings in the progression of the disease. Effective therapeutic innovation involves the spatial transcriptomics capabilities to recognize immune suppressed tumor regions. Courses of action are influenced by modifications in gene expression correlated with deprivation and the natural environment of cancer stem cells (Ku et al., 2023; Jung & Kim, 2023).

3.3. Spatial Proteomics and Metabolomics

Acknowledging the potency of transcriptomics, the levels of proteins, which eventually regulate cellular function, rarely corelates with mRNA concentration. Hence, spatial proteomics has evolved as an important addition. By integrating immunohistochemistry alongside advanced dimensional sensing methodology, spatial proteomic approaches provide concurrent observation of numerous proteins within whole solid tissues (Dhar et al., 2025).

One of the most recognized techniques is Imaging Mass Cytometry (IMC), it essentially utilizes metal isotopes coupled antibodies in lieu of fluorophores. A laser serves cutting of tissue sections, while mass cytometry serves quantification of the isotopes. Since mass spectrometry can accommodate detection of several isotypes with no overlapping of spectral bands, IMC reads above 40 proteins in an individual tissue region excluding spectral overlap through the combination of mass spectrometry and flow cytometry. Since IMC data exhibits particular features that distinguish it from normal imaging approaches, it mandates the implementation of particular evaluation devices. It provides broad understanding regarding the communications between cells and disease mechanisms, which renders it especially valuable for investigating cancer biology including the TME. Likewise, multiplexed protein imaging is carried out by Multiplexed Ion Beam Imaging (MIBI), which utilizes secondary ion mass spectrometry. MIBI improves intracellular resolution by visualizing over 40 proteins concurrently employing time-of-flight secondary ion mass spectrometry. Modern innovations like the MIBIscope greatly enhance productivity and sensitiveness, facilitating speedy imaging and evaluation of intricate tissue specimens. MIBI aids in categorizing numerous cell groups as well as is helpful for immune assessment and for comprehending TME (Milošević, 2023; Schlecht et al., 2021; Ptacek et al., 2020).

Single-cell spatial proteomics—powered by approaches like MS imaging and mIF alters studies in biology. Incredibly high-dimensional precision is granted by mass spectrometry (MS) imaging, resulting in delivering profound knowledge on the position and magnitude of intracellular proteins. Multiplexed immunofluorescence (mIF) facilitates quick as well as high-throughput protein profiling. It plays an essential to single-cell spatial proteomics by generating novel details about cellular processes, The association of MS and mIF with additional system biology, has uplifted contribute to significant breakthroughs in both fundamental knowledge and application in medicine, overcoming current obstacles (Liyanage & Guo, 2024).

Significant factor multiplexing has been rendered achievable by CODEX, which makes use of antibodies bound to DNA barcodes that are constantly continuously labeled and eliminated through imaging phases. Numerous passes of fluorescent tags are implemented in the method, facilitating for the recording of enormous quantities of data from the same tissue specimen without encountering complications with spectrum overlap. By examining in excess of 40 features concurrently, CODEX is currently efficiently adopted for assessing immune cell penetration in tumors, boosting comprehension of the way cancer advances along with the way therapies respond. Deep tissue architecture assessment has been substantially feasible by the methodology, which gave researchers important new information regarding cell-to-cell communication in both normal and abnormal circumstances. The differentiation of immunological "hot" and "cold" malignancies depending upon the concretion and distribution of T cells, macrophages, and stromal cells is now achievable by spatial proteomics, resulting in new understanding into concerning TME. These approaches have rendered it apparent whether the arrangement of cells impacts pathogenesis in autoimmune ailments by analysing the structural makeup of immune penetration within inflammatory locales. Spatial proteomics has aided in the determination of synaptic proteins and their arrangement within the brain (Goltsev et al., 2018; Kennedy-Darling et al., 2016; Black et al., 2021; Cavarischia-Rega et al., 2025; Matsubayashi & Takano, 2025).

Along with proteins and gene sequences, metabolites constituting cellular by-products also offers an easy insight into physiological conditions. To understand the arrangement of metabolic compounds in tissues, spatial metabolomics integrates mass spectrometry and imaging strategies. Two of the main technologies are Desorption Electrospray Ionization (DESI) imaging and Matrix-Assisted Laser Desorption/Ionization (MALDI) imaging. MALDI imaging serves as a framework that aids facilitate the desorption and ionization of analytical substances. It permits for the imaging of metabolites in tissue fragments with outstanding spatial resolution. Typically employed to plant metabolomics to investigate pathways of metabolism and effects of stress. DESI Imaging is a label-free strategy that swiftly examines tissue interfaces void of substantial preparation of the sample. Responsible for tracking an extensive array of metabolites, particularly lipids and peptides, throughout distinct biological samples. Especially efficient for real-time assessment in clinical environments, such as tumor characterization (Wei et al., 2025; Ning & Xu, 2022; Gao et al., 2022; Myers et al., 2025; Walker, 2021).

The potential of spatial omics to bring together genetic and histological evidence provides one of its key benefits. These techniques enable an integrative view of biology which would not be feasible with dissociation-centered methodologies as this advanced approach detect genes, proteins, and metabolites directly within unbroken tissues. Through the implementation of spatial omics, pathologists and biological scientists can evaluate findings within the widely recognized paradigm of tissue structure, effectively spanning the disparity between the two distinct fields. Personalized medicine strategies can be assisted by the understandings generated by spatial omics, that may culminate in better diagnostic and prognostic aids. Despite the currently persistent challenges with incorporating information and computational evaluation as this discipline progresses, enhancements in multi-omics and computational intelligence are supposed to strengthen the medicinal value of spatial omics (Rawat, 2023; Liu et al., 2024; Bressan et al., 2023; Border et al., 2024; Fan & Lin, 2023).

4. COMPUTATIONAL APPROACHES IN SINGLE-CELL AND SPATIAL MULTI-OMICS

The enormous information created by single cell and spatial multi-omics provides a breakthrough but given rise to obstacles as well. A large collection of information, each reflecting several thousand genes or proteins recorded across numerous cells, is generated in a single analysis. The data effortlessly acquires incompatible levels that defy the use of traditional analytic approaches. As a consequence, computational biology emerges as an essential discipline offering a platform for data processing, incorporation, imagining, and interpretation. Cell type grouping and trajectory interpretation have become achievable by the growing importance of innovative computer methodologies, especially machine learning, in the handling and evaluation of single-cell knowledge. By eliminating divergent signals in numerous samples, such methods strengthen the clarity of the variability in cells (Raimundo et al., 2021; Yuan et al., 2017; Vandereyken et al., 2023).

4.1. Data Preprocessing and Quality Control

The unprocessed raw information requires to be processed prior to being employed in beneficial evaluation. Demultiplexing, read alignment, transcript quantification, and screening out low-quality cells are essential components of single-cell RNA sequencing. The approach of segregating intermingled collections of cells employing discrete barcodes is known as demultiplexing. The process has significance considering it enhances productivity as well as data accuracy by facilitating investigators for figuring out particular cells from mixed samples. The approach is simplified by programs like scPipe, that amalgamate barcode demultiplexing alongside various preprocessing phases. Aligning processed reads to a baseline transcriptome or genome is recognized as read alignment. For measuring gene expression levels, correct alignment is important as it help for multiple sequencing techniques. scPipe ensures ideal read alignment, being vital to future studies. Comprehensive transcript estimation is fundamental to determining the changing pattern of gene expression. Incorporating UMI-aware gene-level estimation, scPipe provides a count matrix that constitutes the framework of additional statistical examination. To forestall inaccurate outcomes, it is imperative to locate and eradicate low-quality cells. By confirming that solely cells that survive are considered in studies, strategies that evaluate biological and technological attributes may raise the efficiency of classification by over thirty percent (Tian et al., 2018; Ilicic et al., 2016).

Preprocessing for spatial omics encompasses barcode decoding, background correction, and picture alignment. In order to determine the spatial arrangement of RNA molecules within tissues, barcode decoding is crucial. It strengthens perception of tissue architecture by facilitating exact localization of transcript expression information for certain cellular conditions (Ospina et al., 2023). Accurate biological impulses in spatial omics data are frequently reduced by noise and distortions and are addressed through background remediation. SEraster is a certain methodology that efficiently suppresses noise whereas retaining the accuracy of data by consolidating cellular details into spatial pixels (Aihara et al., 2024). Precisely overlapping spatial data spanning numerous modalities or time points is achieved through picture alignment. Comprehensive assessment over multiple spatial omics datasets is simplified and feasible by the Spatial Data framework's offering for mapping to standard coordinate systems (Marconato et al., 2023). For ensuring accuracy of data, quality control measures are vital that includes excluding cells with low transcript counts, high mitochondrial gene expression, or doublet artifacts (Hippen et al., 2021).

4.2. Dimensionality Reduction and Visualization

Heterogeneity diminution techniques are significant considering the extremely dense nature of omics data. Cellular diversity is most commonly displayed in two- or three-dimensions utilizing methodologies like principal component analysis (PCA), t-distributed stochastic neighbour embedding (t-SNE), and Uniform Manifold Approximation and Projection (UMAP). PCA consistently incorporates deviation and provides value for primary data organization assessments. T-SNE highlights retaining localised connection, rendering it excellent for illustrating complicated data sets. UMAP corresponds to t-SNE, but is considerably more efficient and accessible (Chang, 2025; Jyothsna et al., 2024). Visual observation is widely integrated to tissue images imposing cellular signals on histological architecture. Dimensionality reduction promotes the investigation of several kinds of omics data, boosting predictive modeling in sectors including illness analysis (Kaur et al., 2021; Bécavin & Benecke, 2011).

4.3. Integration Across Modalities

Machine learning approaches that align datasets from multiple domains and investigations are crucial for integration. In this discipline, platforms like MOFA

(Multi-Omics Factor Analysis), Harmony, Scanpy, and Seurat have become ubiquitous. MOFA combines multi-modal single-cell data with a Bayesian Factor Analysis structure, offering collaborating simulation of alteration among numerous sample categories as well as data modalities. Considering assistance of the SeuratIntegrate package, Seurat gives a variety of integration strategies, including R and Python-based tools, consequently, users may choose the most suitable approach for their datasets. Whereas, Scanpy delivers a robust framework for evaluating and interpreting single-cell information, allowing convergence spanning multiple investigations and contexts, Equilibrium concentrates on aligning datasets by considering batch impacts. Deep learning methods, such as graph neural networks and autoencoders, are becoming more prevalent for establishing overlapping latent domains that integrate multidimensional datasets. These methods provide for the restoration of missing modalities, which reliably forecasts whatever couldn't be precisely recorded, in along with harmonizing different datasets (Specque et al., 2025; Argelaguet et al., 2020; Nunes et al., 2025).

4.4. Spatially Aware Computational Models

Localization is an additional dimension of depth that comes from spatial omics. Histological arrangement, neighbourhood impacts, and spatial autocorrelation ought to have require being taken into accounts in computational procedures. Cells are symbolized as nodes interlinked by lines that imply the distance between them in the graphing strategies. Quantifying the spatial arrangement of molecular signals can be assisted by spatial statistics like Ripley's K and Moran's I. These evaluations assist in assessing molecular communication clumping and spatial autocorrelation, making them vital to interpreting tissue microenvironments. By merging omics and pathology, machine learning algorithms developed on labelled sections of tissues can further identify tissue patches depending on molecular and cellular characteristics (Emons et al., 2025; Hu et al., 2021).

4.5. Clinical Translation of Computational Insights

Studies from computational biology need to be discernible and practically applicable to be beneficial to precision medicine, leading to the need for strategies that associate DNA fingerprints with disease symptoms, pharmaceutical interventions, and prognostic consequences. Initiatives are being undertaken to develop extensive patient-level databases by combining single-cell and spatial information with imaging modalities, digital medical records, and evidence from clinical trials, with a primary objective of establishing models of prediction that can facilitate real-time assessment, prognosis, and treatment decisions (Wang & He, 2024). Investigators can gather significant physiological knowledge from complicated databases by employing tools like SpaceSequest, which render simpler to interpret data from different platforms (Sun et al., 2025).

5. BIOLOGICAL INSIGHTS FROM SINGLE-CELL AND SPATIAL MULTI-OMICS

The authentic relevance of multi-omics has its roots in the physiological breakthroughs that it has rendered achievable, beyond the progress of technologies. Single-cell and spatial methodologies have completely altered the comprehension of disease phenomena, tissue arrangement, and cellular range in the last decade.

The importance of intra-tumoral polymorphism in malignancy has recently been illustrated by single-cell sequencing, which additionally revealed atypical subclones that carry mutations in resistance leading to recurrence. The environmental dimension of tumors has been emphasized by the identification of these specific clones' associations with immunological and stromal cell populations by spatial transcriptomics. Therapeutic procedures, like the emergence of mixed treatments which attack malignant cells as well as the microenvironments which nourish their growth, have been significantly impacted by these recent revelations.

Multi-omics has analysed in detail the wide range of immune cell behaviors, encompassing young, vibrant, and depleted phenotypic characteristics in immunology. Although spatial transcriptomics discovered chemokine patterns that control immune invasion, spatial proteomics has shown how damaged T cells concentrate at tumor border regions. These outcomes endorse attempts that employ inhibitors of checkpoints alongside other immunological treatments to enhance responses from the immune system (Yeo et al., 2024; Zhou et al., 2025).

In neuroscience, single-cell transcriptomics has generated diversified atlases of different types of cells of the brain, while spatial transcriptomics has linked the different types of cells within the brain to the anatomical framework. This research has rendered it apparent that the way distinct neuronal subgroups impact cognitive abilities, memory, and neurodegenerative illnesses. Spatial omics have additionally been employed to illuminate pathogen habitats in host histology in infectious illnesses, displaying how organisms such as parasites, viruses, and bacteria exploit particular microenvironments to defeat immunity (Jung & Kim, 2023; Close et al., 2021; Fernández-Moya et al., 2024; Jung & Kim, 2023).

Recent developments in single-cell and spatial omics technology have a significant role in understanding gene expression, signaling pathways, and cell-cell interactions enhancing comprehension of liver metabolism. These insights go beyond traditional metabolic zonation models. Immune zonation and its essential role in liver metabolism, outlines the geographical placement of immune cells—has attracted prominence. These achievements have boosted our awareness of the shifts in metabolism noticed in ailments like diabetes and fatty liver disease (Fujimoto & Tanaka, 2025).

By illustrating embryonic development, tumor immunology, and complex tissue functions, particularly those found in the kidney and brain, single cell and spatial approaches broaden translational application. Further, they demonstrate cellular interactions and oncogenic mechanisms, which facilitate biological and clinical applications in health and disease (Conrad & Altmüller, 2022).

Tumor microenvironment relationships become apparent by spatial technologies, while single-cell profiling increases awareness of cellular heterogeneity. In the final stages, these span the disparity between scientific findings and practical applications through offering knowledge regarding immune evasion, resistance to therapy, and clinically essential features which means perineural invasion (Gong et al., 2024).

Through permitting combined genome and transcriptome assessment in tumor analysis, especially in triple-negative breast cancer, and by implementing mass spectrometry imaging for spatial metabolome annotation, generating perspectives on the mechanisms of illness and patient prognosis, single-cell and spatial approaches augment translational relevance (Vandereyken et al., 2023).

Such instances contribute to emphasize an expanded concept: the concept of multi-omics gives us the ability to advance from observational inventories of cell kinds to operational portrayals that explain how cells operate within tissues. Single-cell and spatial omics provide a perspective on fundamental concepts underpinning both wellness and illness by simultaneously identifying the surrounding environment.

6. MULTI-OMICS TECHNOLOGIES TO DRIVE PRECISION MEDICINE AND CLINICAL INNOVATIONS

Since both single-cell and spatial technologies has evolved individually, a challenge of incorporating both these distinct categories of molecular data has surfaced to generate an accurate representation of biological systems. Thus, necessities convergence of multi-omics. Multi-omics implies progressive assessment of an array of molecular domains, notably genomics, transcriptomics, epigenomics, proteomics, and metabolomics, accompanied by computational approaches for integration of these datasets into reasonable biological inferences. Incorporation of multi-omics in the establishment of tailored therapies not only is in academics but has but has become an essential framework for research goals (Loscalzo, 2024; Vandereyken et al., 2023).

A partial understanding of cellular condition can be acquired by employing a single omics stratum. While somatic changes and transmissible mutations are observed by genomics, scientists are presently incapable to ascertain whether these modifications influence the expression of gene without the aid of transcriptomics. Similarly, though proteomics offers knowledge about functionality of the molecules, scientists are not capable to determine the regulatory factors controlling protein production without the aid of epigenomics. While metabolomics monitors the biochemical outcomes of cellular metabolism, but it could not readily understand the triggering factors of the metabolic behaviour without proteomic information. Therefore, this conveys that integration is obvious and essential (Haider & Pal, 2013; Dong et al., 2025; Manzoni et al., 2018).

Improvements in technological approaches have rendered single-cell multi-omics conceivable. Several applications like CITE-seq (simultaneous transcriptome and surface protein profiling), SHARE-seq (chromatin accessibility and expression of gene in a cell), and scNMT-seq (simultaneous profiling of nucleosome occupancy, DNA methylation, and transcriptomes) are illustrations of this integration. Developing techniques for spatially multi-omics research currently facilitate concurrent acquisition of proteins and RNA in undamaged sections of tissue, demonstrating a direct association between transcriptional fingerprints and in situ proteomic environments (Vandereyken et al., 2023; Conrad & Altmüller, 2022; Wu et al., 2024; Dimitriu et al., 2022).

The computational component is additionally essential. Multi-omics datasets are extremely wide, minimal, and extremely large. To coordinate and integrate various modalities, algorithms ranging from deep learning-based integration techniques, non-negative matrix factorization (NMF), and canonical correlation analysis (CCA) are currently being designed. Through utilizing such approaches, investigators may efficiently develop uniform schematics of cellular conditions by recognizing subtle variables accounted for heterogeneity across numerous omics strata (Computational Methods for Single-Cell Multi-Omics Integration and Alignment, 2022).

Recognition of essential trends and associations in multi-omics collections are rendered simpler by methods for machine learning, which involve supervised and unsupervised algorithms. Compared with standard methodologies, deep learning has displayed superior accuracy in incorporating multi-omics information, thereby permitting the computation of convoluted datasets with an extensive range of modalities. Recent investigation strengthens the awareness of single-cell physiological processes by categorizing deep learning methodologies for integration depending on data modality and layout (Athaya et al., 2023; Firdous et al., 2024).

Collaboration between multi-omics provides relevant knowledge for precision medicine. Oncology requires recognition of not merely the accumulating variants in the tumor but additionally the triggered pathways, proteins generated, and the metabolic deficiencies could potentially be targeted therapeutically. The integration also supports evaluation of the immune cells activity that combine cytokine production patterns, transcriptional priming, and epigenetic regulation in autoimmune disorders. It renders accessible understanding of pathogen-host interactions at different cellular phases of infectious diseases, indicating prospective therapeutic routes (Martinelli & Foreman, 2015; Raufaste-Cazavieille et al., 2022; Akhtar et al., 2025).

The prospects of multi-omics in precision medicine finally rely in its capabilities to translate biological convolution into understanding that is relevant for therapeutic contexts. This coherence assures that molecular observations are not merely explanatory but also applicable, spanning the gap between foundational molecular understanding and patient-tailored treatments.

Application of single cell and spatial technologies on patient samples continue to expand and the outcome have immediate impact on diagnosis, prognosis and therapeutic regimens (Gao et al., 2024; Zhang, 2022). There are several major domains like oncology, immunology, neurology, autoimmune and infectious disease wherein multi-omics strategies are contributing significant advancements.

7.1. Oncology

In cancer, prior to initiation of treatment, single cell sequencing is being used to assess tumor biopsies in an effort to identify resistant clones. Implementation of single cell multi-omics approaches facilitates recognition of variety of cells within tumor microenvironment and also reveals resistance to therapy strategies (Sabit et al., 2025). By modelling immune invasion on surgical tissue through spatial transciptomics, oncologist can evaluate the likelihood of beneficial effect of immunotherapy in the treatment (Boe et al., 2024). Integration of single cell RNA sequencing and spatial data, spatial transcriptomics enables understanding into cellular interactions and the structural properties of the tumor microenvironment (Yan et al., 2022). Clinician can associate transcriptional stages and proteomic findings with genomic variations by multi-omics integration permitting personalized therapeutics that are distinctive to individual tumor's molecular constitution. Understanding of spatial locations of single nucleotide variations (SNVs) or unique gene sequences by spatial genomic approach like issDNA helps investigators to understand clonal development and physical positions by occurrence of a validated oncogenic mutation with the adjacent tumor microenvironment (TME). Spatial genomics, in solid tumor, can determine the degree wherein a clone of tumor has expanded in the tissue by identifying the spatial confines based on certain amplification instances like for e.g. HER2 amplification. Regarding development of personalized therapeutics, spatial genomics delivers a rooted knowledge of tumor diversity (Le et al., 2025; Kamsalem et al., 2025). Research investigations assessing RNA and protein transcription display both regional transcriptomics and proteomics can be seamlessly merged to demonstrate imbalances that emphasize the crucial role of multi-omics strategies for grasping tumor biology (Cervilla et al., 2024).

Multi-omics integration enables an extensive understanding of biomarker surroundings, meanwhile the quick development of artificial intelligence (AI)-driven strategies boosts the assessment of complicated, multidimensional facts that possibly improve predicting value and therapeutic significance. Standardization, data integration, and real-time monitoring persist to raise an array of obstacles regardless of significant improvements. Still, integrating the combination of spatial and single-cell omics into biomarker investigation provides enormous prospects to establish individualized cancer immunotherapy (Zhang et al., 2025).

7.2. Immunology and Autoimmune Diseases

Single-cell omics has identified pathogenic immunological subtypes which are responsible for chronic inflammation in autoimmune conditions such as rheumatoid arthritis (RA) and lupus. Single-cell technologies are being utilized to characterizing several ensembles of cells and their respective functions in triggering the growth of the illness, to evaluate the immune microenvironment in disease like psoriasis. By strengthening comprehension of immune cell connections to dermatological cells, this strategy reveals the physiological processes that underlie the formation and recurrence of psoriasis (Cui et al., 2024). Single-cell RNA sequencing demonstrated diverse immune cell topologies and interferon characteristics, establishing that various autoimmune disorders possess distinctive pathogenic avenues. Personalized immunotherapeutic procedures are generated by developing an understanding certain immune types of cells and their roles in the body. As an illustration, focusing on TPH cells which express GPR56 in RA could suggest innovative approaches to therapy. Meanwhile, single-cell omics emphasizes the immunological deregulation influencing the wellness of bones in RA, highlighting prospective therapies for lowering the erosion of bones (Ding et al., 2022; Hu et al., 2023). MERSCOPE is an advanced imaging-derived methodology that advances awareness of tumor heterogeneity by accurately retrieving cell-type-specific transcriptome variations (Rademacher et al., 2025). Spatial proteomics aided visualization of immune invasion in affected tissues has brought insights into the structural makeup of inflammatory disorders. Immune cell subsets in numerous tissues have recently been comprehensively described owing to novel advances in spatial proteomics strategies like luminescence and radiant cell sorting. These strategies produced noteworthy proteome discrepancies amongst specific populations of dendritic cells, demonstrating multiple function in inflammation alongside potential opportunities for treatment. These understandings influence treatment approaches focused at specific diseased cells while sustaining protective immunity (Bhamidipati & Wei, 2022; Xu et al., 2025). Spatial single-cell omics technologies offer considerable possibilities for usage in therapy, for instance, improving patient stratification, and are critical in finding fresh perspectives regarding the intricate pathways of liver disorders. The implementation of advanced bioinformatics-powered spatial framework-based immunological categorization in immuno-oncology, can be validated to boost patient stratification, is a current example of clinical translation. This highlights the method by which better-defined patient groups for medical treatment could emerge from a knowledge of the spatial architecture of immune cells (Suo et al., 2025).

7.3. Neurology

Single-cell omics are currently utilized to a multitude of neurodegenerative illnesses, especially multiple sclerosis and Huntington's disease, yielding explanations regarding their special pathophysiologies (Pozojevic & Spielmann, 2023). Numerous types of cell compositions in Parkinson's disease have been discovered by single-cell RNA sequencing, resulting in enhanced awareness of the illness's biochemical pathophysiology (Ma & Lim, 2021). Single cell epigenomics has detected mechanism of regulation triggering neuronal fragility, helping in understanding the epigenetic changes. This understanding holds assurance of treatment developments with disease alteration possibilities (He et al., 2024; Rahimzadeh et al., 2024). Over one million plausible cis-regulatory regions have recently been uncovered by single-cell epigenomic assessments which have been connected to cognitive impairment and perseverance in Alzheimer's patients (Liu et al., 2025). Spatial omics is also progressively boosting the understanding of neurogenerative disorders in the field of neurology. The visualization of molecular dispersion in brain tissues rendered accessible by spatial omics methodologies highlights the spatial connections involving multiple kinds of cells and pathophysiological attributes corresponding to neurodegenerative conditions (Marshall et al., 2023). For instance, spatial transcriptomics has defined the phases of microglial activation illustrating the relationship between neurons, astrocytes, and microglia in the circumstances of Alzheimer's disease pathogenesis. The arrangement of amyloid plaques and neurofibrillary tangles, which serve as crucial for grasping the advancement of Alzheimer's illness, was recently established by means of spatial omics (Ma et al., 2024). A full comprehension of the cellular events triggering neurodegeneration could be achieved through integrating spatial omics alongside other omics technologies (genomic, transcriptomic, etc.). This multi-omic strategy enhances awareness of disease physiology and prospective strategies for treatment by facilitating an investigation of complicated mechanisms inside the brain (Ayaz et al., 2024).

Comprehending the health and susceptibility of neurons, specifically dopaminergic neurons involved in illnesses like Parkinson's disease, relies on proteomic investigations that have identified brain area-specific protein signals and the dynamic renewal of proteins. Protein networks in particular synapse types and synaptic regions can finally be evaluated beneath between physiological and diseased conditions due to closeness tagging methodologies like BioID and APEX. This sheds illumination on the molecular foundations of synaptic activation and diversity (Matsubayashi & Takano, 2025; Cavarischia-Rega, 2025).

7.4. Infectious Diseases

By exposing cellular heterogeneity, host-pathogen interactions, and immune responses, single-cell and spatial technologies improve knowledge of infectious illnesses. These render achievable to accurately define phenotypic and genetic biomarkers, fostering effective diagnostics and treatment advances in combating infections.

Additionally, these approaches facilitate quicker identification of particular cell kinds and infrequent subpopulations, which is necessary to generate tailored regimens and specialized diagnostics in precision medicine (Lin et al., 2020; Lee, 2025). The real time capabilities of single-cell and spatial technologies were demonstrated during the era of COVID-19 pandemic of viral infection. As spatial omics revealed the precise position of the virus in the lungs. Viral concentrations in lung tissues have been discovered by spatial transcriptomics, revealing locations of active SARS-CoV-2 replication, specifically at sites with significant immune cell invasion. Substantial enhancement of immune cell signaling routes and inflammatory indicators, which have been connected to tissue inflammation, was identified. Single-cell RNA sequencing showed enhanced T cell and macrophage populations in COVID-19 patients, particularly among advanced stages. By managing the special immune system reactions noticed in COVID-19, the expertise acquired through these breakthroughs could direct the establishment of customized treatments and patient care programs. These findings demonstrated the practical importance of these advancements in infectious diseases and consequently affect the development of therapies and care for patient strategies (Houseini et al., 2025; Park et al., 2022; Rendeiro et al., 2021; Zheng, 2023).

Figure 2. Evolution of treatment and therapeutic approach: From one-size-fits-all method to Individualized healthcare

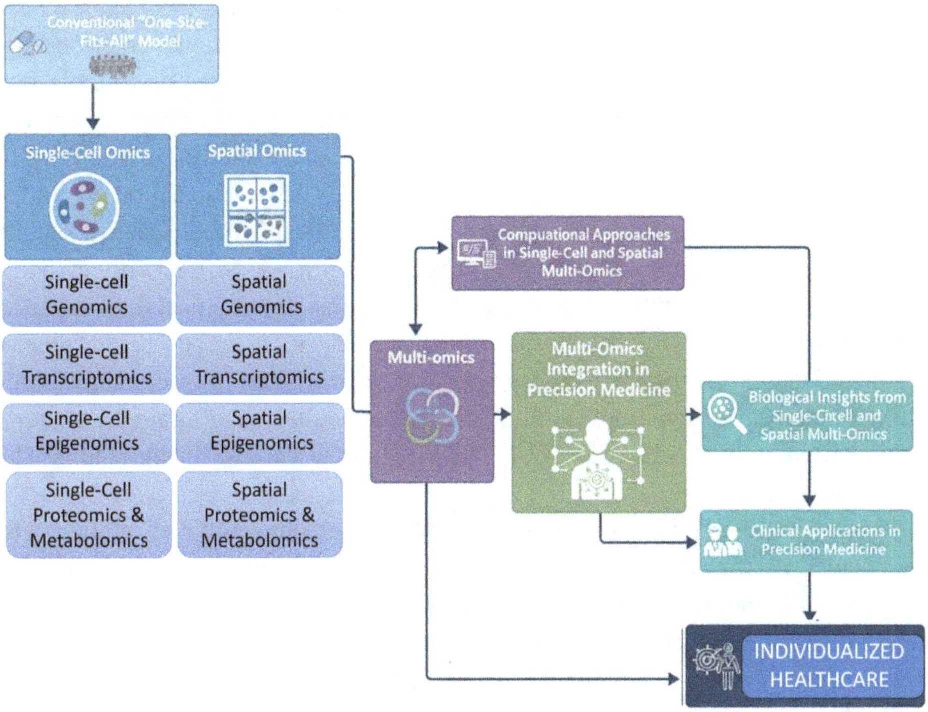

8. LIMITATIONS

Regardless of their revolutionary potential, single-cell and spatial multi-omics face numerous challenges from being used in healthcare adoption. The comprehension of biological phenomena in both wellness and sickness can get influenced by methodological constraints of single cell and spatial methodologies. Inadequate resolution, imprecise geographical details, diminished scalability, data complexity, and the prospect of preconceptions in the selection of cells and assessment constitute certain methodological constraints of single-cell and spatial approaches like transcriptomics. Such challenges prohibit their broad utilization in broad biological researches, prompting deeper investigation to maximize the effectiveness and integration with other omics analysis (Conrad & Altmüller, 2022; Chen et al., 2023; Greenstreet et al., 2023).

The capacity of humanity to evaluate the diversity of cells has recently been improved by single-cell omics approaches, but they frequently encounter one significant disadvantage i.e. the omission of spatial context. Preceding to molecular assessment, tissues are often physically or enzymatically detached into suspensions of distinct cells in single-cell approaches. This approach excludes essential details concerning the cell's original placement in the tissue and its association with adjacent cells, despite offering previously unavailable detail about each cell's molecular condition. For instance, single-cell RNA sequencing (scRNA-seq) lacks direct spatial information, and spatial transcriptomics holds cellular geographical data, however spatial transcriptomics might continue to have challenges like in consistent of precise characterization of various cell types in complicated tissues, low-content staining, steric hindrance from multiple antibodies, and difficulties in assigning pixel-level signals to individual cells (Chen et al., 2023; Hu et al., 2023; Huang et al., 2024; Lee et al., 2023).

Single-cell omics has become restricted mainly technically and intellectually by obstacles with integrating information, multidimensional evaluation, and the demand for higher sensitivity and resolution. Likewise, the widespread implementation of these technologies in clinical settings is impeded by issues of unresolved standardization and reproducibility (Le et al., 2025). Despite the fact that spatial technologies are currently extensively utilized, certain challenges persist that influence the dependability and efficacy of both single-cell and spatial methods in research. These concerns encompass standardization and benchmarking, sample compatibility, throughput, resolution, and overall necessity of training (Czarnewski et al., 2022; Kulasinghe et al., 2025; Li et al., 2025). Elevated transcriptomic platform expenses, probable pre-analytical elements impacting RNA purity, and spatial transcriptomics' shortcomings to achieve single-cell detail—requiring reduction methodologies for forecasting certain types of cells within transcriptome spots—are several methodological constraints (Rocque et al., 2024).

Furthermore, a lot of technologies only capture a small percentage of the molecules that are present in each cell due to their low sensitivity causing requirement of significant amounts of enhancement, which may trigger complications like dispersed genome distribution or allelic abandonment, alongside methodological error and amplifying bias. Likewise, several spatial transcriptomic methodologies presently acquire lesser transcripts per cell along with yield decreased productivity or less sensitivity as contrasted with single-cell sequencing. Considering relatively outstanding precision, imaging-based technologies commonly impose a limitation on the quantity of species of molecules they can assess concurrently. FixNCut eliminates artefacts from tissue processing while retaining intact RNA and cellular component to cope with specimen preparation concerns. It increases robustness in spatial and single-cell omics, permitting pinpoint gathering of information despite conserving sensitive cell forms and the patterns of gene expression (Jiménez-Gracia et al., 2023; Dezem et al., 2024; Williams et al., 2022)

Expense and manageability are significant obstacles from the perspective of healthcare especially when dealing with massive investigations. Several spatial and single-cell techniques demand a lot of resources, thereby rendering these impractical for regular diagnosis. Substantial verification, reproducibility, and clinical efficacy demonstration are also essential for rigorous validation (Kim et al., 2023; Xu et al., 2025).

Investigations are rendered more challenging by technical challenges like signal-to-noise ratio, autofluorescence, and tissue stability. Furthermore, the enormous, high-dimensional databases created through spatial omics demand powerful computational platforms for unification and assessment (Lee et al., 2022; Asp et al., 2020; Dezem et al., 2024).

Similarly, in both single cell omics and spatial omics evaluation of data is a computationally cost demanding process that requires infrastructure and a specialized understanding. There lies inadequate pipeline standardization, which causes variation in study outcome. Such issues can be alleviated by utilizing cloud computing and containerization, resulting in flexible, repeatable analysis remedies for numerous kinds of omics databases. Moreover, incorporating multi-omics datasets continues to be challenging while other approaches are constantly being developed. (Augustyn et al., 2021; Czarnewski et al., 2022).

Spatial omics technologies are recently developing substantially in conjunction with these limitations. With the goal to provide authentic multi-modal characterization in situ, recently developed methods are aiming to combine transcriptomic, proteomic, and metabolomic information throughout a single tissue region. Such techniques will eventually grow into more beneficial equipment in precision medicine, furnishing physicians with information on disease-causing cells' identities, specific positions inside tissues, and connections to cells within close proximity (Rawat, 2023; Bressan et al., 2023; Fan & Lin, 2023).

9. FUTURE PROSPECTS

Both single-cell and spatial multi-omics have a promising future. The limits of resolution, productivity, and multimodality are being explored by emerging technology. In the same cells and tissues, initiatives are being introduced to profile not simply RNA and proteins but also chromatin architecture, signaling dynamics, and epigenetic indicators.

Due to its capabilities to integrate with clinical data as well as comprehend intricate datasets in real time, AI will turn to be extremely important. Precision medicine will improve as multi-omics, digital pathology, and imaging for medical and electronic health records come together to develop comprehensive patient profiles.

Additionally, these technologies must be made more accessible. Single-cell and spatial omics will shift from specialist research instruments to regular component of medical practice with decrease in expenses and standardized procedures. This will fulfil the prospect of completely tailored healthcare by empowering clinicians to establish therapeutic decisions on thorough, patient-specific molecular mapping.

10. CONCLUSION

The emergence of single-cell and spatial multi-omics represents a turning point in clinical practice and biological studies. These technologies offer previously inaccessible insights into the biological processes of both health and illness by capturing tissue surroundings and cellular variability at previously unobtainable resolution. By providing pharmaceuticals tailored to each patient's unique molecular profile, their incorporation into precision medicine processes has the potential to completely transform diagnosis, prognosis, and therapy (Figure 2).

Although there are obstacles for instance, technical constraints, computational difficulties, and clinical application limitations that need to be addressed. Complete understanding of the prospective benefits of these techniques can be obtained in a collaborative approach among biologists, doctors, engineers, and computational scientists.

Spatial and single-cell multi-omics will not solely expand our knowledge of biology in the decades that lies ahead, but they will additionally influence medical practice. They help us get closer to precision medicine's ultimate aim of providing the right treatment to the right patient at the right time by blending state-of-the-art technology with clinical relevance.

REFERENCES

Ahmed, F. J. (2024). Personalized Medicine: The Dawn of a New Era in Healthcare. *Futuristic Biotechnology*, *01*, 01. Advance online publication. DOI: 10.54393/fbt.v4i03.138

Aihara, G., Clifton, K. P., Chen, M., Atta, L., Miller, B. F., & Fan, J. (2024). SEraster: a rasterization preprocessing framework for scalable spatial omics data analysis. *bioRxiv*. DOI: 10.1101/2024.02.01.578436

Akhtar, A., Khatun, R., Sengupta, S., & Bhattacharya, M. (2025). Role of Multi-Omics in Disease Biology. *Applied Sciences Research Periodicals., 3*(03), 02–21. https://doi.org/DOI: 10.63002/asrp.303.925

Argelaguet, R., Arnol, D., Bredikhin, D., Deloro, Y., Velten, B., Velten, B., Marioni, J. C., Marioni, J. C., Marioni, J. C., Stegle, O., & Stegle, O. (2020). MOFA+: A statistical framework for comprehensive integration of multi-modal single-cell data. *Genome Biology*, *21*(1), 1–17. DOI: 10.1186/s13059-020-02015-1

Asp, M., Bergenstråhle, J., & Lundeberg, J. (2020). Spatially Resolved Transcriptomes-Next Generation Tools for Tissue Exploration. *BioEssays*, *42*(10), 1900221. DOI: 10.1002/bies.201900221

Athaya, T., Ripan, R. C., Li, X., & Hu, H. (2023). Multimodal deep learning approaches for single-cell multi-omics data integration. *Briefings in Bioinformatics*, *24*(5), bbad313. Advance online publication. DOI: 10.1093/bib/bbad313

Augustyn, D. R., Wyciślik, Ł., & Mrozek, D. (2021). Perspectives of using Cloud computing in integrative analysis of multi-omics data. *Briefings in Functional Genomics*, *20*(4), 198–206. DOI: 10.1093/bfgp/elab007

Ayaz, A., Ali, A., Mohammed, I., & Ali, N. Alauddin, Singh, N., Jeelani, G., Mujtaba, H., & Bhat, K. A. (2024). *Multiomics for understanding neurodegenerative disorders* (pp. 95–125). Elsevier BV. https://doi.org/DOI: 10.1016/B978-0-443-23971-7.00005-5

Bai, X., Lau, B. T., Grimes, S. M., Sathe, A., & Ji, H. P. (2022). Single cell multi-omic mapping of subclonal architecture and pathway phenotype in primary gastric and metastatic colon cancers. *bioRxiv*. https://doi.org/DOI: 10.1101/2022.07.03.498616

Bécavin, C., & Benecke, A. (2011). New dimensionality reduction methods for the representation of high dimensional "omics" data. *Expert Review of Molecular Diagnostics*, *11*(1), 27–34. DOI: 10.1586/erm.10.95

Bhamidipati, K., & Wei, K. (2022). Precision medicine in rheumatoid arthritis. *Best Practice & Research. Clinical Rheumatology*, *36*(1), 101742. DOI: 10.1016/j.berh.2022.101742

Black, S., Phillips, D. J., Hickey, J. W., Kennedy-Darling, J., Venkataraaman, V. G., Samusik, N., Samusik, N., Goltsev, Y., Schürch, C. M., & Nolan, G. P. (2021). CODEX multiplexed tissue imaging with DNA-conjugated antibodies. *Nature Protocols*, *16*(8), 3802–3835. DOI: 10.1038/s41596-021-00556-8

Boe, R. H., Triandafillou, C. G., Lazcano, R., Wargo, J. A., & Raj, A. S. (2024). *Spatial transcriptomics reveals influence of microenvironment on intrinsic fates in melanoma therapy resistance*. https://doi.org/DOI: 10.1101/2024.06.30.601416

Bond, D. R., Uddipto, K., Enjeti, A. K., Enjeti, A. K., & Lee, H. J. (2020). Single-cell epigenomics in cancer: Charting a course to clinical impact. *Epigenomics*, *12*(13), 1139–1151. DOI: 10.2217/epi-2020-0046

Border, S., Ferreira, R. M., Lucarelli, N., Kumar, S., Paul, A. S., Manthey, D., Barisoni, L., Strekalova, Y. A., Ray, J., Cheng, Y., Rosenberg, A. Z., Tomaszewski, J. E., Mimar, S., Hodgin, J. B., El-Achkar, T. M., Jain, S., Eadon, M. T., & Sarder, P. (2024). *FUSION: A web-based application for in-depth exploration of multi-omics data with brightfield histology*. https://doi.org/DOI: 10.1101/2024.07.09.602778

Bowers, R., Gonzalez-Pena, V., Wardhani, K., Goudeau, D., Blow, M. J., Udwary, D., Klein, D., Vill, A. C., Brito, I. L., Woyke, T., Malmstrom, R. R., & Gawad, C. (2024). scMicrobe PTA: Near Complete Genomes from Single Bacterial Cells. *bioRxiv*. https://doi.org/DOI: 10.1101/2024.01.30.577819

Bressan, D., Battistoni, G., & Hannon, G. J. (2023). The dawn of spatial omics. *Science*, *381*(6657), eabq4964. Advance online publication. DOI: 10.1126/science.abq4964

Carmo-Fonseca, M. (2021). *Precision Medicine: Enabling Healthcare Progress in the Twenty-First Century*. Springer., DOI: 10.1007/978-3-030-70761-3_2

Cavarischia-Rega, C., Sharma, K., Fitzgerald, J. C., & Maček, B. (2025). Proteomic insights into the biology of dopaminergic neurons. *Frontiers in Molecular Neuroscience*, *18*, 1642519. Advance online publication. DOI: 10.3389/fnmol.2025.1642519

Cervilla, S., Grases, D., Perez, E., Real, F. X., Musulén, E., Esteller, M., & Porta-Pardo, E. (2024). Comparison of spatial transcriptomics technologies across six cancer types. *bioRxiv*. https://doi.org/DOI: 10.1101/2024.05.21.593407

Chang, Y. I. (2025). *A Survey: Potential Dimensionality Reduction Methods*. https://doi.org//arxiv.2502.11036DOI: 10.48550

Chen, C., Ge, Y., & Lu, L. (2023). Opportunities and challenges in the application of single-cell and spatial transcriptomics in plants. *Frontiers in Plant Science*, *14*, 1185377. Advance online publication. DOI: 10.3389/fpls.2023.1185377

Chen, K. Y., Takeshima, Y., Zorzetto-Fernandes, A. L., Makino, K., Kibayashi, T., Hata, M., Ichiyama, K., Croteau, J., Taylor, K., & Sakaguchi, S. (2025). Joint single-cell measurements of surface proteins, intracellular proteins and gene expression with icCITE-seq. https://doi.org/DOI: 10.1101/2025.01.11.632564

Chen, W.-Y., Li, S., Kulkarni, A. S., Huang, L., Cao, J., Qian, K., & Wan, J. (2020). Single Cell Omics: From Assay Design to Biomedical Application. *Biotechnology Journal*, *15*(1), 1900262. DOI: 10.1002/biot.201900262

Chen, Y., (2025). Single-cell sequencing to multi-omics: Technologies and applications. *Molecular Systems Biology*, *21*(5).

Cheng, K., & McCarrey, J. R. (2023). Profiling the Epigenetic Landscape of the Spermatogonial Stem Cell-Part 1: Epigenomics Assays. (Vol. 2656, pp. 71–108). https://doi.org/DOI: 10.1007/978-1-0716-3139-3_5

Cheung, E., Gakhar, R., & Sundaram, V. (2024). Xenium in situ analysis and reproducibility study of multiple carcinomas for clinical studies. *Cancer Research*, *84*(6_Supplement), 6830. Advance online publication. DOI: 10.1158/1538-7445.AM2024-6830

Clark, S. J., Argelaguet, R., Kapourani, C.-A., Stubbs, T. M., Lee, H. J., Krueger, F., Sanguinetti, G., Kelsey, G., Marioni, J. C., Stegle, O., & Reik, W. (2017). Joint Profiling Of Chromatin Accessibility, DNA Methylation And Transcription In Single Cells. *bioRxiv*, 138685. https://doi.org/DOI: 10.1101/138685

Close, J. L., Long, B., & Zeng, H. (2021). Spatially resolved transcriptomics in neuroscience. *Nature Methods*, *18*(1), 23–25. DOI: 10.1038/s41592-020-01040-z

Computational Methods for Single-Cell Multi-Omics Integration and Alignment. (2022). https://doi.org//arxiv.2201.06725DOI: 10.48550

Conrad, T., & Altmüller, J. (2022). Single cell- and spatial 'Omics revolutionize physiology. *Acta Physiologica (Oxford, England)*, *235*(4), e13848. Advance online publication. DOI: 10.1111/apha.13848

Cui, N., Xu, X., & Zhou, F. (2024). Single-cell technologies in psoriasis. *Clinical Immunology (Orlando, Fla.)*, *110242*, 110242. Advance online publication. DOI: 10.1016/j.clim.2024.110242

Czarnewski, P., Mahfouz, A., Calogero, R. A., Palagi, P. M., Portell-Silva, L., Gonzalez-Uriarte, A., Soneson, C., Burdett, T., Szomolay, B., Videm, P., Hotz, H.-R., Papatheodorou, I., Hancock, J. M., Grüning, B., Haerty, W., Krause, R., Capella-Gutierrez, S., Leskošek, B., Alessandri, L., & Korpelainen, E. (2022). Community-driven ELIXIR activities in single-cell omics. *F1000 Research*, *11*, 869. DOI: 10.12688/f1000research.122312.1

Datlinger, P., Rendeiro, A. F., Boenke, T., Senekowitsch, M., Krausgruber, T., Barreca, D., & Bock, C. (2021). Ultra- high-throughput single- cell RNA sequencing and perturbation screening with combinatorial fluidic indexing. *Nature Methods*, *18*(6), 635–642. DOI: 10.1038/s41592-021-01153-z

Dawson, T. (2024). *Sinai SCENT TMC - Single Cell Assay for Transposase Accessible Chromatin (ATAC-seq) v1*. https://doi.org/DOI: 10.17504/protocols.io.j8n-lkoqo1v5r/v1

Derks, S., Derks, S., & Diosdado, B. (2015). Personalized cancer medicine: Next steps in the genomic era. *Cellular Oncology (Dordrecht)*, *38*(1), 1–2. DOI: 10.1007/s13402-015-0221-0

Dezem, F., Morosini, N. S., Arjumand, W., DuBose, H., & Plummer, J. (2024). Spatially Resolved Single-Cell Omics: Methods, Challenges, and Future Perspectives. *Annual Review of Biomedical Data Science*, *7*(1), 131–153. Advance online publication. DOI: 10.1146/annurev-biodatasci-102523-103640

Dhar, R., Kumar, A., & Karmakar, S. (2025). The future of immunohistochemistry: Advancing precision medicine through artificial intelligence-driven spatial proteomics and multiplexed diagnostic technologies. *Asian Journal of Medical Sciences*, *16*(8), 1–2. DOI: 10.71152/ajms.v16i8.4674

Dimitriu, M.-A., Lazar-Contes, I., Roszkowski, M., & Mansuy, I. M. (2022). Single-Cell Multiomics Techniques: From Conception to Applications. *Frontiers in Cell and Developmental Biology*, *10*, 854317. Advance online publication. DOI: 10.3389/fcell.2022.854317

Ding, L., Li, X., Zhu, H., & Luo, H. (2022). Single-Cell Sequencing in Rheumatic Diseases: New Insights from the Perspective of the Cell Type. *Aging and Disease*, *13*(6), 1633–1651. DOI: 10.14336/AD.2022.0323

Dong, M., Wang, L., Hu, N., Rao, Y., Wang, Z., & Zhang, Y. (2025). Integration of multi-omics approaches in exploring intra-tumoral heterogeneity. *Cancer Cell International*, *25*(1), 317. Advance online publication. DOI: 10.1186/s12935-025-03944-2

Duncan, K. D., Fyrestam, J., & Lanekoff, I. (2019). Advances in mass spectrometry based single-cell metabolomics. *Analyst, 144*(3), 782–793. DOI: 10.1039/C8AN01581C

Emanuel, G., & He, J. (2021). MERFISH Enables Mapping of Cellular Diversity with Spatial Context at Subcellular Resolution. *Microscopy Today*, *29*(4), 30–33. DOI: 10.1017/S1551929521000894

Emons, M., Gunz, S., Crowell, H. L., Mallona, I., Kuehl, M., Furrer, R., & Robinson, M. D. (2025). Harnessing the potential of spatial statistics for spatial omics data with pasta. *Nucleic Acids Research, 53*(17), gkaf870. Advance online publication. DOI: 10.1093/nar/gkaf870

Fan, R., & Lin, F. (2023). Spatial Delivery. [editorial]. *GEN Biotechnology*, *2*(5), 331–332. Advance online publication. DOI: 10.1089/genbio.2023.29117.editorial

Fernández-Moya, S. M., Ganesh, A. J., & Plass, M. (2024). Neural cell diversity in the light of single-cell transcriptomics. *Transcription*, 1–19. DOI: 10.1080/21541264.2023.2295044

Firdous, A., Firdous, S., Ali, A., Nisar, B., Ahmad, H., Iqbal, G., & Ahmad, S. B. (2024). *Machine learning approaches for multiomics data analysis*. Elsevier BV., DOI: 10.1016/B978-0-443-23971-7.00013-4

Fujimoto, M., & Tanaka, T. (2025). Advancing liver metabolic zonation with single-cell and spatial omics. *Endocrine Journal*, *72*(10), 1069–1078. Advance online publication. DOI: 10.1507/endocrj.EJ25-0140

Gao, R., Yang, L., Zhu, Y., Zhang, T., Gu, W., & Yang, L. (2024). Advancing precision medicine through single-cell sequencing: Insights and implications. *Clinical and Translational Discovery*, *4*(2), e293. Advance online publication. DOI: 10.1002/ctd2.293

Gao, S., Zhao, J., Guan, Y., Tang, Y., Li, Y., & Liu, L. (2022). Mass Spectrometry Imaging technology in metabolomics: A systematic review. *Biomedical Chromatography*, *37*(7), e5494. DOI: 10.1002/bmc.5494

Goltsev, Y., Samusik, N., Kennedy-Darling, J., Bhate, S. S., Hale, M., Vasquez, G., & Nolan, G. P. (2018). Deep Profiling of Mouse Splenic Architecture with CODEX Multiplexed Imaging. *Social Science Research Network*. https://doi.org/ DOI: 10.2139/SSRN.3155542

Gong, D., Arbesfeld-Qiu, J. M., Perrault, E., Bae, J. W., & Hwang, W. L. (2024). Spatial oncology: Translating contextual biology to the clinic. *Cancer Cell*, *42*(10), 1653–1675. Advance online publication. DOI: 10.1016/j.ccell.2024.09.001

Greenstreet, L., Afanassiev, A., Kijima, Y., Heitz, M., Ishiguro, S., King, S., Yachie, N., & Schiebinger, G. (2023). DNA-GPS: A theoretical framework for optics-free spatial genomics and synthesis of current methods. *Cell Systems*, *14*(10), 844–859. e4. Advance online publication. DOI: 10.1016/j.cels.2023.08.005

Haider, S., & Pal, R. (2013). Integrated Analysis of Transcriptomic and Proteomic Data. *Current Genomics*, *14*(2), 91–110. DOI: 10.2174/1389202911314020003

He, Q. (2016). Tumor heterogeneity and drug resistance of targeted antitumor agents. *Yao Xue Xue Bao*, *51*(2), 197–201. https://pubmed.ncbi.nlm.nih.gov/29856199/

He, Y., Lu, W., Zhou, X., Mu, J., & Shen, W. (2024). Unraveling Alzheimer's disease: Insights from single-cell sequencing and spatial transcriptomic. *Frontiers in Neurology*, *15*, 1515981. Advance online publication. DOI: 10.3389/fneur.2024.1515981

Hippen, A. A., Falco, M. M., Weber, L. M., Erkan, E. P., Zhang, K., Doherty, J. A., Vähärautio, A., Greene, C. S., & Hicks, S. C. (2021). miQC: An adaptive probabilistic framework for quality control of single-cell RNA-sequencing data. *PLoS Computational Biology*, *17*(8), 1009290. DOI: 10.1371/journal.pcbi.1009290

Houseini, S. T., Mahmanzar, M., Rahimian, K., Bayat, S., Gholamzad, A., Mollapour Sisakht, M., Farhadi, A., Kuehu, D. L., & Deng, Y. (2025). Spatial Transcriptomics and Single Cell-RNASeq Reveals Cellular Heterogeneity of SARS-CoV-2 in Lung Tissues and Global Mutational Patterns in COVID-19 Patients. *Journal of Medical Virology*, *97*(9), e70586. Advance online publication. DOI: 10.1002/jmv.70586

Houston, A. (2024). *10x Genomics Xenium for FFPE User Guides v1*. https://doi.org/DOI: 10.17504/protocols.io.81wgbxjonlpk/v1

Hu, J., Schroeder, A., Coleman, K., Chen, C., Auerbach, B. J., & Li, M. (2021). Statistical and machine learning methods for spatially resolved transcriptomics with histology. *Computational and Structural Biotechnology Journal*, *19*, 3829–3841. DOI: 10.1016/j.csbj.2021.06.052

Hu, W., Zhang, Y., Mei, J., & Fang, X. (2023). Spatial transcriptomics in human biomedical research and clinical application. *Current Medicine (Cham, Switzerland)*, *2*(1), 6. Advance online publication. DOI: 10.1007/s44194-023-00023-4

Huang, L., Ma, F., Chapman, A., Lu, S., & Xie, X. S. (2015). Single-Cell Whole-Genome Amplification and Sequencing: Methodology and Applications. *Annual Review of Genomics and Human Genetics*, *16*(1), 79–102. DOI: 10.1146/annurev-genom-090413-025352

Huang, S., Ouyang, L., Tang, J., Qian, K., Chen, X., Zhang, X., Ming, J., & Xi, R. (2024). Spatial transcriptomics: A new frontier in cancer research. *Clinical Cancer Bulletin*, *3*(1), 13. Advance online publication. DOI: 10.1007/s44272-024-00018-8

Iafrate, A., Onozato, M., Elliot, H., & Yapp, C. (2020). Highly multiplexed fluorescence in situ hybridization (fish) platform for gene copy number evaluation.

Ilicic, T., Ilicic, T., Kim, J. K., Kolodziejczyk, A. A., Kolodziejczyk, A. A., Bagger, F. O., Bagger, F. O., Bagger, F. O., McCarthy, D. J., McCarthy, D. J., Marioni, J. C., Marioni, J. C., Marioni, J. C., Teichmann, S. A., Teichmann, S. A., & Teichmann, S. A. (2016). Classification of low quality cells from single-cell RNA-seq data. *Genome Biology*, *17*(1), 29. DOI: 10.1186/s13059-016-0888-1

Ivanisevic, T., & Sewduth, R. N. (2023). Multi-Omics Integration for the Design of Novel Therapies and the Identification of Novel Biomarkers. *Proteomes*, *11*(4), 34. Advance online publication. DOI: 10.3390/proteomes11040034

Jager, R. (2023). Special Issue on Whole Genome Amplification. *International Journal of Molecular Sciences*, *24*(11), 9626. DOI: 10.3390/ijms24119626

Jiménez-Gracia, L., Marchese, D., Nieto, J. C., Caratu, G., Melón-Ardanaz, E., Gudiño, V., Wise, K., Ryan, N. S., Hernando-Momblona, X., Bernardes, J. P., Tran, F., Sievers, L. K., Schreiber, S., van den Berge, M., Kole, T. M. T. M., Nawijn, M. C., Rosenstiel, P., Batlle, E., Parish, I. A., & Martelotto, L. G. (2023). FixNCut: Single-cell genomics through reversible tissue fixation and dissociation. *bioRxiv*. DOI: 10.1101/2023.06.16.545221

Jung, N., & Kim, T. K. (2023). Spatial transcriptomics in neuroscience. *Experimental & Molecular Medicine*, *55*(10), 2105–2115. DOI: 10.1038/s12276-023-01093-y

Jyothsna, V., Sandhya, E., Pydala, B., Swetha, N. N., & Sree, T. S. D. (2024). *Visual data analysis and inference through dimensionality reduction techniques*. 21–68. https://doi.org/DOI: 10.1201/9781003542735-2

Kalisky, T., & Quake, S. R. (2011). Single-cell genomics. *Nature Methods*, *8*(4), 311–314. DOI: 10.1038/nmeth0411-311

Kamsalem, A. S., Ubah, O. M., Olatokun, T. P., Oluwaniyi, R. O., Philips, O. A., Ojo-omoniyi, D. S., & Oluwasakinju, O. L. (2025). Unraveling tumor microenvironments with spatial genomics: Implications for precision breast cancer therapy. *International Journal of Life Science Research Archive, 9*(1), 043–063. https://doi.org/DOI: 10.53771/ijlsra.2025.9.1.0045

Kaur, P., Singh, A., & Chana, I. (2021). Computational Techniques and Tools for Omics Data Analysis: State-of-the-Art, Challenges, and Future Directions. *Archives of Computational Methods in Engineering*, *28*(7), 1–37. DOI: 10.1007/s11831-021-09547-0

Kaur, R. P., Ludhiadch, A., & Munshi, A. (2019). *Single-Cell Genomics: Technology and Applications*. Academic Press., DOI: 10.1016/B978-0-12-814919-5.00009-9

Kennedy-Darling, J., Nolan, G. P., Goltsev, Y., & Samusik, N. (2016). *Abstract A089: Multiparametric immunofluorescence analysis of the tumor microenvironment using CODEX*. *4*. https://doi.org/DOI: 10.1158/2326-6066.IMM2016-A089

Khairnar, M., & Marda, S. (2024). Integrative Multi-Omics in Precision Medicine: From Molecular Interconnectivity to Single-Cell Resolution. *International Journal of Scientific Research & Technology*, *2*(7).

Kim, Y.-S., Choi, J., & Lee, S. H. (2023). Single-cell and spatial sequencing application in pathology. *Journal of Pathology and Translational Medicine*, *57*(1), 43–51. DOI: 10.4132/jptm.2022.12.12

Kulasinghe, A., Berrell, N., Donovan, M. L., & Nilges, B. S. (2025). *Spatial-Omics Methods and Applications*. https://doi.org/DOI: 10.1007/978-1-0716-4276-4_5

Kuta Alam, T. P. J. (2022). *Visualizing the Interactions Shaping the Imaging of the Microenvironment in Human Cancers* (pp. 67–79). *Methods in Molecular Biology (Clifton, N.J.)*. Advance online publication. DOI: 10.1007/978-1-0716-2703-7_5

LaFave, L. M., Savage, R. E., & Buenrostro, J. D. (2022). Single-Cell Epigenomics Reveals Mechanisms of Cancer Progression. *Annual Review of Cancer Biology*, *6*(1), 167–185. DOI: 10.1146/annurev-cancerbio-070620-094453

Lareau, C. A., Duarte, F. M., Chew, J. G., Kartha, V. K., Burkett, Z. D., Kohlway, A. S., Pokholok, D., Aryee, M. J., Steemers, F. J., Lebofsky, R., & Buenrostro, J. D. (2019). Droplet- based combinatorial indexing for massive- scale single- cell chromatin accessibility. *Nature Biotechnology*, *37*(8), 916–924. DOI: 10.1038/s41587-019-0147-6

Le, J., Dian, Y., Zhao, D., Guo, Z., Luo, Z., Chen, X., Zeng, F., & Deng, G. (2025). Single-cell multi-omics in cancer immunotherapy: From tumor heterogeneity to personalized precision treatment. *Molecular Cancer*, *24*(1), 221. Advance online publication. DOI: 10.1186/s12943-025-02426-3

Lee, D., Jung, Y., Lee, J. E., Lim, M., Lee, D., Na, K. J., & Choi, H. (2025). Abstract 6276: Comparative analysis and integration of image-based and high-resolution sequence-based spatial transcriptomics platforms for tumor microenvironment analysis. *Cancer Research*, *85*(8, Supplement_1), 6276. DOI: 10.1158/1538-7445.AM2025-6276

Lee, J., Yoo, M., & Choi, J. (2022). Recent advances in spatially resolved transcriptomics: Challenges and opportunities. *BMB Reports*, *55*(3), 113–124. DOI: 10.5483/BMBRep.2022.55.3.014

Lee, J. H. (2025). From single-cell maps to diagnostics: Enabling biomarker discovery in precision medicine. *Academia Molecular Biology and Genomics.*, *2*(3). Advance online publication. DOI: 10.20935/AcadMolBioGen7859

Lee, R. Y., Ng, C. W., Rajapakse, M., Ang, N., Yeong, J. P. S., & Lau, M. C. (2023). The promise and challenge of spatial omics in dissecting tumour microenvironment and the role of AI. *Frontiers in Oncology*, *13*, 1172314. Advance online publication. DOI: 10.3389/fonc.2023.1172314

Lerch, J. P. (2024). *eLife Assessment: Three-dimensional single-cell transcriptome imaging of thick tissues*. https://doi.org/DOI: 10.7554/elife.90029.3.sa0

Li, T., Gu, Z., & Zhou, G. (2025). Next-Generation Barcoding for Single-Cell Omics. *Analytical Chemistry*, *97*(31), 16708–16713. Advance online publication. DOI: 10.1021/acs.analchem.5c03587

Li, W., Chen, Z., Wu, A., Qin, F. X.-F., & Zhang, L. (2020). *Systems immunology meets epigenetics*. Academic Press., DOI: 10.1016/B978-0-12-817964-2.00010-1

Li, X., & Hou, J. (2018). A richer understanding of intratumoral heterogeneity: Single-cell genomics put it within reach. *Journal of Thoracic Disease*, *10*(3), 1178–1182. DOI: 10.21037/jtd.2018.03.17

Lin, W. N., Tay, M. Z., Lu, R., Liu, Y., Chen, C.-H., & Cheow, L. F. (2020). The Role of Single-Cell Technology in the Study and Control of Infectious Diseases. *Cells*, *9*(6), 1440. DOI: 10.3390/cells9061440

Liu, B., Wu, F.-X., & Zou, X. (2020). scASK: A novel ensemble framework for classifying cell types based on single-cell RNA-seq data. *bioRxiv*. https://doi.org/DOI: 10.1101/2020.06.07.138271

Liu, L., Chen, A., Li, Y., Mulder, J., Heyn, H., & Xu, X. (2024). Spatiotemporal omics for biology and medicine. *Cell*, *187*(17), 4488–4519. DOI: 10.1016/j.cell.2024.07.040

Liu, Q., Martínez-Jarquín, S., & Zenobi, R. (2022). Recent Advances in Single-Cell Metabolomics Based on Mass Spectrometry. *CCS Chemistry*, *5*(2), 310–324. DOI: 10.31635/ccschem.022.202202333

Liu, Z., Zhang, S., James, B. T., Galani, K., Mangan, R. J., Fass, S. B., Liang, C., Wagle, M. M., Boix, C. A., Tanigawa, Y., Yun, S., Sung, Y., Xiong, X., Sun, N., Hou, L., Wohlwend, M., Qiu, M., Han, X., Xiong, L., & Kellis, M. (2025). Single-cell multiregion epigenomic rewiring in Alzheimer's disease progression and cognitive resilience. *Cell*, *188*(18), 4980–5002.e29. Advance online publication. DOI: 10.1016/j.cell.2025.06.031

Liyanage, S. H., & Guo, J. (2024). *Single-cell spatial proteomics.* 18861. https://doi.org/DOI: 10.14670/hh-18-861

Lo, P.-K., & Zhou, Q. (2018). Emerging techniques in single-cell epigenomics and their applications to cancer research. 1(1), 1–8. https://doi.org/DOI: 10.4172/JCG.1000103

Lomakin, A., Svedlund, J., Strell, C., Gataric, M., Shmatko, A., Rukhovich, G. D., Park, J. S., Ju, Y. S., Dentro, S. C., Kleshchevnikov, V., Vaskivskyi, V., Li, T., Bayraktar, O. A., Pinder, S. L., Richardson, A. L., Santagata, S., Campbell, P. J., Russnes, H. G., Gerstung, M., & Yates, L. R. (2022). Spatial genomics maps the structure, nature and evolution of cancer clones. *Nature*, *611*(7936), 594–602. DOI: 10.1038/s41586-022-05425-2

Loscalzo, J. (2024). Multi-Omics and Single-Cell Omics: New Tools in Drug Target Discovery. *Arteriosclerosis, Thrombosis, and Vascular Biology*, *44*(4), 759–762. DOI: 10.1161/ATVBAHA.124.320686

Lu, N., Qiao, Y., Lu, Z., & Tu, J. (2023). Chimera: The spoiler in multiple displacement amplification. *Computational and Structural Biotechnology Journal*, *21*, 1688–1696. DOI: 10.1016/j.csbj.2023.02.034

Luo, T., Rosasco, M. G., Hoeman, C. M., Khare, S., Gorski, D. J., Rand, T., Klinghoffer, R. A., Guinney, J., & Ho, C.-S. (2025). Abstract 2085: Visium HD combined with deep-learning-based cell segmentation on H&E images yield accurate cell annotation at single-cell resolution. *Cancer Research*, *85*(8, Supplement_1), 2085. DOI: 10.1158/1538-7445.AM2025-2085

Ma, S.-X., & Lim, S. B. (2021). Single-Cell RNA Sequencing in Parkinson's Disease. *Biomedicines*, *9*(4), 368. DOI: 10.3390/biomedicines9040368

Ma, Y., Shi, W., Dong, Y., Sun, Y., & Jin, Q.-G. (2024). Spatial Multi-Omics in Alzheimer's Disease: A Multi-Dimensional Approach to Understanding Pathology and Progression. *Current Issues in Molecular Biology*, *46*(5), 4968–4990. DOI: 10.3390/cimb46050298

Manzoni, C., Kia, D. A., Vandrovcova, J., Hardy, J., Wood, N. W., Lewis, P. A., Lewis, P. A., & Ferrari, R. (2018). Genome, transcriptome and proteome: The rise of omics data and their integration in biomedical sciences. *Briefings in Bioinformatics*, *19*(2), 286–302. DOI: 10.1093/bib/bbw114

Marconato, L., Palla, G., Yamauchi, K. A., Virshup, I. H., Heidari, E., Treis, T., Vöhringer, H., Huber, W., Gerstung, M., Moore, J., Theis, F. J., & Stegle, O. (2023). SpatialData: an open and universal data framework for spatial omics. *bioRxiv*. https://doi.org/DOI: 10.1101/2023.05.05.539647

Marshall, C., Farrow, M. A., Djambazova, K. V., & Spraggins, J. M. (2023). Untangling Alzheimer's disease with spatial multi-omics: A brief review. *Frontiers in Aging Neuroscience*, *15*, 1150512. Advance online publication. DOI: 10.3389/fnagi.2023.1150512

Martinelli, G., & Foreman, N. (2015). Advancing precision medicine through multi-omics: An integrated approach to tumor profiling. *Science*, *349*(6253), 1246. DOI: 10.1126/science.349.6253.1246-c

Matsubayashi, J., & Takano, T. (2025). Proximity labeling uncovers the synaptic proteome under physiological and pathological conditions. *Frontiers in Cellular Neuroscience*, *19*, 1638627. Advance online publication. DOI: 10.3389/fncel.2025.1638627

Milošević, V. (2023). Different approaches to Imaging Mass Cytometry data analysis. *Bioinformatics Advances*, *3*(1), vbad046. Advance online publication. DOI: 10.1093/bioadv/vbad046

Moffitt, J. R., Hao, J., Wang, G., Chen, K. H., Babcock, H. P., & Zhuang, X. (2016). High-throughput single-cell gene-expression profiling with multiplexed error-robust fluorescence in situ hybridization. *Proceedings of the National Academy of Sciences of the United States of America*, *113*(39), 11046–11051. DOI: 10.1073/pnas.1612826113

Morse, D. B., Michalowski, A. M., Ceribelli, M., De Jonghe, J., Vias, M., Riley, D., Davies-Hill, T., Voss, T. C., Pittaluga, S., Muus, C., Liu, J., Weitz, D. A., Brenton, J. D., Buenrostro, J. D., Knowles, T. P. J., & Thomas, C. J. (2023). Positional influence on cellular transcriptional identity revealed through spatially segmented single-cell transcriptomics. Cell Systems, 14 6(6), 464-481.e7. https://doi.org/DOI: 10.1016/j.cels.2023.05.003

Myers, R. (2022). scNMT-seq v2 v1. https://doi.org/DOI: 10.17504/protocols.io.q26g7yrz3gwz/v1

Myers, R. J., Tretter, Z. M., Daffron, A. G., Fritschi, E., Santos, W. T., Foster, M., Klotz, M. G., Stafford, K. M., Kasch, C., Taylor, T. J., Tellefson, L. C., Hartman, T. P. V., Hackler, D., Specque, F., Barre, A., Nikolski, M., & Chalopin, D. (2025). SeuratIntegrate: An R package to facilitate the use of integration methods with Seurat. *Bioinformatics (Oxford, England)*, *41*(6), btaf358. Advance online publication. DOI: 10.1093/bioinformatics/btaf358

Nadaf, N., & Rayan, N. (2025). *Slide-Tags using Tissue-Tape v1*. https://doi.org/DOI: 10.17504/protocols.io.36wgqdpdkvk5/v1

Ning, Z., & Xu, G. (2022). Can spatially resolved metabolomics uncover weak points in tumors? *Expert Review of Proteomics*, *19*(4–6), 227–230. DOI: 10.1080/14789450.2023.2176754

Nunes, F., Behrens, L. M. P., Weimer, R. D., Gonçalves, G. F., Fernandes, G., & Dorn, M. (2025). Deep learning methods and applications in single-cell multimodal data integration. *Molecular Omics*, *21*(6), 545–565. Advance online publication. DOI: 10.1039/D5MO00062A

Ospina, O. E., Soupir, A., & Fridley, B. L. (2023). *A Primer on Preprocessing* (Vol. 2629). Visualization, Clustering, and Phenotyping of Barcode-Based Spatial Transcriptomics Data., DOI: 10.1007/978-1-0716-2986-4_7

. Otto, E. A. (2019). Single cell RNA sequencing (scRNA-seq). https://doi.org/DOI: 10.17504/protocols.io.7dthi6n

Park, J., Foox, J., Hether, T., Danko, D., Warren, S., Kim, Y., Reeves, J., Butler, D., Mozsary, C., Rosiene, J., Shaiber, A., Afshin, E. E., MacKay, M., Rendeiro, A. F., Bram, Y., Chandar, V., Geiger, H., Craney, A., Velu, P., & Mason, C. E. (2022). System-wide transcriptome damage and tissue identity loss in COVID-19 patients. *Cell Reports Medicine*, *3*(2), 100522. DOI: 10.1016/j.xcrm.2022.100522

Pozojevic, J., & Spielmann, M. (2023). Single-Cell Sequencing in Neurodegenerative Disorders. *Molecular Diagnosis & Therapy*, *27*(5), 553–561. DOI: 10.1007/s40291-023-00668-9

Pribluda, A., de la Cruz, C., & Jackson, E. L. (2015). Intratumoral Heterogeneity: From Diversity Comes Resistance. *Clinical Cancer Research*, *21*(13), 2916–2923. DOI: 10.1158/1078-0432.CCR-14-1213

Ptacek, J., Geyer, F. C., Mignault, A., Deeds, J., Giedt, J., Gu, J., McLaughlin, M. E., Sigal, Y., Tarolli, J. G., Aksoy, M., Zhang, Y., Hav, M., Finck, R., & Finn, J. (2020). 48 Advances in multiplexed ion beam imaging (MIBI) for immune profiling of the tumor microenvironment. *Journal for Immunotherapy of Cancer*, *8*, A30.1–A30. Advance online publication. DOI: 10.1136/jitc-2020-SITC2020.0048

Qin, S., Miao, D., Zhang, Y., & Bai, Y. (2023). Methods developments of mass spectrometry based single cell metabolomics. *Trends in Analytical Chemistry*, *117086*, 117086. Advance online publication. DOI: 10.1016/j.trac.2023.117086

Qu, H., Kao, C., & Hakonarson, H. (2023). Single-Cell RNA Sequencing Technology Landscape in 2023. *Stem Cells (Dayton, Ohio)*. Advance online publication. DOI: 10.1093/stmcls/sxad077

Rademacher, A., Huseynov, A., Bortolomeazzi, M., Wille, S. J., Schumacher, S., Sant, P., Keitel, D., Okonechnikov, K., Ghasemi, D. R., Pajtler, K. W., Mallm, J., & Rippe, K. (2025). Comparison of spatial transcriptomics technologies using tumor cryosections. *Genome Biology*, *26*(1), 176. Advance online publication. DOI: 10.1186/s13059-025-03624-4

Rahimzadeh, N., Srinivasan, S. S., Zhang, J., & Swarup, V. (2024). Gene networks and systems biology in Alzheimer's disease: Insights from multi-omics approaches. *Alzheimer's & Dementia*, *20*(5), 3587–3605. DOI: 10.1002/alz.13790

Raimundo, F., Papaxanthos, L., Vallot, C., Vert, J.-P., & Vert, J.-P. (2021). Machine learning for single cell genomics data analysis. *bioRxiv*. https://doi.org/DOI: 10.1101/2021.02.04.429763

Raufaste-Cazavieille, V., Santiago, R., & Droit, A. (2022). Multi-omics analysis: Paving the path toward achieving precision medicine in cancer treatment and immuno-oncology. *Frontiers in Molecular Biosciences*, *9*, 962743. Advance online publication. DOI: 10.3389/fmolb.2022.962743

Rawat, S. (2023). Spatial Omics Spotlights the Players in the Tumor Microenvironment. *GEN Biotechnology*, *2*(5), 342–346. DOI: 10.1089/genbio.2023.29115.sro

Rendeiro, A. F., Ravichandran, H., Bram, Y., Kim, J., Meydan, C., Park, J., Foox, J., Hether, T., Warren, S., Kim, Y., Reeves, J., Salvatore, S., Mason, C. E., Swanson, E., Borczuk, A., Elemento, O., & Schwartz, R. E. (2021). *The spatial landscape of lung pathology during COVID-19 progression - targeted spatial transcriptomics data*. https://doi.org/DOI: 10.5281/zenodo.4635285

Rocque, B., Guion, K., Singh, P., Bangerth, S., Pickard, L., Bhattacharjee, J., Eguizabal, S., Weaver, C., Chopra, S., Zhou, S., Kohli, R., Sher, L., Akbari, O., Ekser, B., & Emamaullee, J. A. (2024). Technical optimization of spatially resolved single-cell transcriptomic datasets to study clinical liver disease. *Scientific Reports*, *14*(1), 3612. Advance online publication. DOI: 10.1038/s41598-024-53993-2

Sabit, H., Arneth, B., Pawlik, T. M., Abdel-Ghany, S., Ghazy, A., Abdelazeem, R. M., Alqosaibi, A., Al-Dhuayan, I. S., Almulhim, J., Alrabiah, N. A., & Hashash, A. (2025). Leveraging Single-Cell Multi-Omics to Decode Tumor Microenvironment Diversity and Therapeutic Resistance. *Pharmaceuticals (Basel, Switzerland)*, *18*(1), 75. DOI: 10.3390/ph18010075

Saul, D., & Kosinsky, R. L. (2023). Spatial transcriptomics herald a new era of transcriptome research. *Clinical and Translational Medicine*, *13*(5), e1264. Advance online publication. DOI: 10.1002/ctm2.1264

Schlecht, A., Boneva, S., Salié, H., Killmer, S., Wolf, J., Hajdu, R. I., Hajdu, R. I., Auw-Haedrich, C., Agostini, H., Reinhard, T., Schlunck, G., Bengsch, B., & Lange, C. (2021). Imaging mass cytometry for high-dimensional tissue profiling in the eye. *BMC Ophthalmology*, *21*(1), 338. DOI: 10.1186/s12886-021-02099-8

Schulte-Schrepping, J., Ferreira, H. J., Saglam, A., Hinkley, E. R., Schultze, J. L., & Schultze, J. L. (2020). *Advances in single-cell epigenomics of the immune system*. Academic Press., DOI: 10.1016/B978-0-12-817964-2.00008-3

Shang, Y., Wang, Z., Xi, L., Wang, Y., Liu, M., Feng, Y., Wang, S., Wu, Q., Xiang, X., Chen, M., & Ding, Y. (2024). Droplet-based single-cell sequencing: Strategies and applications. *Biotechnology Advances*, *77*, 108454. DOI: 10.1016/j.biotechadv.2024.108454

Sharma, A., & DasGupta, R. (2019). Tracking tumor evolution one-cell-at-a-time. *Molecular & Cellular Oncology*, *6*(3), 1590089. DOI: 10.1080/23723556.2019.1590089

Shi, X., Wei, F., Mehrpouyan, M., Chen, Y., D'Cruz, L. M., Widmann, S., & Tyznik, A. J. (2023). Flow cytometry analysis of protein expression using antibody-derived tags followed by CITE-Seq. *Cytometry. Part A*. Advance online publication. DOI: 10.1002/cyto.a.24792

Shimagami, H., Nishimura, K., Matsushita, H., Metsugi, S., Kato, Y., Kawasaki, T., Tsujimoto, K., Edahiro, R., Itotagawa, E., Naito, M., Kawada, S., Nakatsubo, D., Matsukawa, K., Namba-Hamano, T., Inoue, K., Takahashi, A., Mizui, M., Kato, S., Hikita, H., & Nishide, M. (2025). Single-cell analysis reveals immune cell abnormalities underlying the clinical heterogeneity of patients with systemic sclerosis. *Nature Communications*, *16*(1), 4949. Advance online publication. DOI: 10.1038/s41467-025-60034-7

Singh, V. (2025). *Single-Cell Proteomics: Recent Advancement and Prospects.* https://doi.org/DOI: 10.20944/preprints202505.0303.v1

Soja, E., Bhargava, S., Sivajothi, S., Flynn, W. F., & Courtois, E. T. (2025). KAPP-Sen TMC: Xenium - Cell DIVE - H&amp. *E (Norwalk, Conn.)*, *v1*, •••. DOI: 10.17504/protocols.io.n92ld6zj7g5b/v1

Stephen, P., & Sumner, L. W. (2025). Spatially Resolved Plant Metabolomics. *Metabolites*, *15*(8), 539. DOI: 10.3390/metabo15080539

Stoeckius, M., Hafemeister, C., Stephenson, W., Houck-Loomis, B., Chattopadhyay, P. K., Swerdlow, H., Satija, R., & Smibert, P. (2017). Simultaneous epitope and transcriptome measurement in single cells. *Nature Methods*, *14*(9), 865–868. DOI: 10.1038/nmeth.4380

Sun, Y., Piya, S., Ouyang, Z., Chen, Y., Gagnon, J., Cao, S., Zhang, H., Song, B., Zhu, J., Chandratre, K., Yu, H., Hu, W., Ryals, M., Casey, F., Huh, D., & Zhang, B. (2025). *SpaceSequest: A unified pipeline for spatial transcriptomics data analysis.* https://doi.org/DOI: 10.1101/2025.09.15.676389

Suo, Y., Thimme, R., & Bengsch, B. (2025). Spatial single-cell omics: new insights into liver diseases. *Gut*, gutjnl-332105. https://doi.org/DOI: 10.1136/gutjnl-2024-332105

Swanson, E., Lord, C., Reading, J., Heubeck, A. T., Genge, P. C., Thomson, Z., Da Weiss, M., Li, X., Savage, A. K., Green, R., Green, R., Torgerson, T. R., Torgerson, T. R., Bumol, T. F., Graybuck, L. T., & Skene, P. J. (2021). Simultaneous trimodal single-cell measurement of transcripts, epitopes, and chromatin accessibility using TEA-seq. *eLife*, *10*, e63632. Advance online publication. DOI: 10.7554/eLife.63632

Tedesco, M., Giannese, F., Lazarevic, D., Giansanti, V., Rosano, D., Monzani, S., Catalano, I., Grassi, E., Zanella, E. R., Botrugno, O. A., Morelli, L., Panina Bordignon, P., Caravagna, G., Bertotti, A., Martino, G., Aldrighetti, L., Pasqualato, S., Trusolino, L., Cittaro, D., & Tonon, G. (2021). Chromatin Velocity reveals epigenetic dynamics by single-cell profiling of heterochromatin and euchromatin. *Nature Biotechnology*, 1–10. DOI: 10.1038/s41587-021-01031-1

Tian, L., Su, S., Dong, X., Amann-Zalcenstein, D., Biben, C., Seidi, A., Hilton, D. J., Naik, S. H., & Ritchie, M. E. (2018). scPipe: a flexible R/Bioconductor preprocessing pipeline for single-cell RNA-sequencing data. *bioRxiv*, 175927. https://doi.org/DOI: 10.1101/175927

Vandereyken, K., Sifrim, A., Thienpont, B., & Voet, T. (2023). Methods and applications for single-cell and spatial multi-omics. *Nature Reviews. Genetics*, *24*(8), 1–22. DOI: 10.1038/s41576-023-00580-2

Waldman, S. A., & Terzic, A. (2011). Widening the path to personalized medicine. *Clinical and Translational Science*, *4*(5), 392–394. DOI: 10.1111/j.1752-8062.2011.00345.x

Walker, H. J. (2021). *Metabolite imaging by mass spectrometry: A new discovery tool* (Vol. 98). Academic Press., DOI: 10.1016/bs.abr.2020.09.022

Walling, M. A., & Shepard, J. R. E. (2011). Cellular heterogeneity and live cell arrays. *Chemical Society Reviews*, *40*(7), 4049–4076. DOI: 10.1039/c0cs00212g

Wang, G., Moffitt, J. R., & Zhuang, X. (2017). Multiplexed imaging of high density libraries of RNAs with MERFISH and expansion microscopy. *bioRxiv*, 238899. https://doi.org/DOI: 10.1101/238899

Wang, N., & He, Q. (2024). *Artificial Intelligence and Bioinformatics Applications in Precision Medicine and Future Implications*. Elsevier BV., DOI: 10.1016/B978-0-12-824010-6.00058-7

Wang, Q., Zhu, H.-Y., Deng, L., Xu, S., Xie, W., Li, M., Wang, R., Liang, T., Zhan, L., & Yu, G. (2025). Spatial Transcriptomics: Biotechnologies, Computational Tools, and Neuroscience Applications. *Small Methods*, *9*(5), 2401107. Advance online publication. DOI: 10.1002/smtd.202401107

Wei, M., Wang, H., Zeng, Y., Ye, D., Wang, G., & Shi, L. (2025). Multi-Modal Spatial Metabolomics with Mid-Infrared Microscope Guided Mass Spectrometry Imaging. *Journal of Visualized Experiments*, *221*(221). Advance online publication. DOI: 10.3791/68709

Williams, C. G., Lee, H. J., Asatsuma, T., Vento-Tormo, R., & Haque, A. (2022). An introduction to spatial transcriptomics for biomedical research. *Genome Medicine*, *14*(1), 68. Advance online publication. DOI: 10.1186/s13073-022-01075-1

Wu, X., Yang, X., Dai, Y., Zhao, Z., Zhu, J., Guo, H., & Yang, R. (2024). Single-cell sequencing to multi-omics: Technologies and applications. *Biomarker Research*, *12*(1), 110. Advance online publication. DOI: 10.1186/s40364-024-00643-4

Xu, J., Jiang, M., Koga, S., Zhang, N., & Huang, Z. (2025). *SpatialFinder: A Human-in-the-Loop Vision-Language Framework for Prioritizing High-Value Regions in Spatial Transcriptomics*. https://doi.org/DOI: 10.1101/2025.08.16.670684

Xu, T., Cai, H., Liu, J., Mao, X., Chen, Y., Chen, M., & Wang, Y. (2025). Exploring the role of unconventional T cells in rheumatoid arthritis. *Frontiers in Immunology*, *16*, 1656994. Advance online publication. DOI: 10.3389/fimmu.2025.1656994

Yan, H., Shi, J., Dai, Y., Li, X., Wu, Y., Zhang, J., Gu, Z., Zhang, C., & Leng, J. (2022). Technique integration of single-cell RNA sequencing with spatially resolved transcriptomics in the tumor microenvironment. *Cancer Cell International*, *22*(1), 155. Advance online publication. DOI: 10.1186/s12935-022-02580-4

Yeo, Y. Y., Chang, Y., Qiu, H., Yiu, S. P. T., Michel, H. A., Wu, W., Jin, X., Kure, S., Parmelee, L., Luo, S., Cramer, P., Lee, J. L., Wang, Y., Yeung, J., El Ahmar, N., Simsek, B., Mohanna, R. A., Van Orden, M., Lu, W. S., & Jiang, S. (2024). *Same-Slide Spatial Multi-Omics Integration Reveals Tumor Virus-Linked Spatial Reorganization of the Tumor Microenvironment*. https://doi.org/DOI: 10.1101/2024.12.20.629650

Yuan, G.-C., Cai, L., Elowitz, M. B., Enver, T., Fan, G., Guo, G., Irizarry, R. A., Kharchenko, P. V., Kim, J., Orkin, S. H., Orkin, S. H., Orkin, S. H., Quackenbush, J., Saadatpour, A., Schroeder, T., Shivdasani, R. A., & Tirosh, I. (2017). Challenges and emerging directions in single-cell analysis. *Genome Biology*, *18*(1), 84. DOI: 10.1186/s13059-017-1218-y

Zhang, C., Feng, H., Zhu, J., & Hong, W. (2025). Application of spatial and single-cell omics in tumor immunotherapy biomarkers. *LabMed Discovery.*, *2*(2), 100076. DOI: 10.1016/j.lmd.2025.100076

Zhang, X., Marand, A. P., Yan, H., & Schmitz, R. J. (2023). Massive-scale single-cell chromatin accessibility sequencing using combinatorial fluidic indexing. *bioRxiv*. https://doi.org/DOI: 10.1101/2023.09.17.558155

Zhang, Y. (2022). An omic and multidimensional spatial atlas from serial biopsies of an evolving metastatic breast cancer. *Cell Reports Medicine*, *3*(2), 100525. DOI: 10.1016/j.xcrm.2022.100525

Zhao, T., Chiang, Z. D., Morriss, J. W., LaFave, L. M., Murray, E. M., Del Priore, I., Meli, K., Lareau, C. A., Nadaf, N. M., Li, J., Earl, A. S., Macosko, E. Z., Jacks, T., Buenrostro, J. D., & Chen, F. (n.d.). *Spatial Genomics Enables Multi-Modal Study of Clonal Heterogeneity in Tissues*. https://doi.org/DOI: 10.1038/s41586-021-04217-4

Zheng, X. (2023). Spatial transcriptomic characterization of COVID-19 pneumonitis identifies immune circuits related to tissue injury. *JCI Insight*, *8*(2), e157837. Advance online publication. DOI: 10.1172/jci.insight.157837

Zhou, S., Wang, Y., Wakhloo, D., Dikshit, A., Wang, L.-C., & Srinivasan, M. (2025). Abstract 6471: Spatial multiomics reveals T-cell activation and exhaustion states in the tumor microenvironment. *Cancer Research*, *85*(8, Supplement_1), 6471. DOI: 10.1158/1538-7445.AM2025-6471

Zhou, X., Xu, Y., Zhu, L., Su, Z., Han, X., Zhang, Z., Huang, Y., & Liu, Q. (2020). Comparison of Multiple Displacement Amplification (MDA) and Multiple Annealing and Looping-Based Amplification Cycles (MALBAC) in Limited DNA Sequencing Based on Tube and Droplet. *Micromachines*, *11*(7), 645. DOI: 10.3390/mi11070645

Zong, C., Lu, S., Chapman, A. R., & Xie, X. S. (2012). Genome-wide detection of singlenucleotide and copy-number variations of a single human cell. Science. 338:1622–1626. https://doi.org/. 1.DOI: 10.1126/science.1229164

Chapter 2
Precision Genomics Integrates Advanced Genomic Technologies

Manish Kumar
https://orcid.org/0000-0003-2042-1243
Department of Pharmaceutics, ISF College of Pharmacy, Moga, India

Rishi Chhabra
https://orcid.org/0009-0002-4026-5292
Department of Quality Assurance, ISF College of Pharmacy, Moga, India

Shruti Srivatava
School of Pharmacy, Chandigarh University, Unnao, India

Uma Shanker Maurya
Goel Institute of Pharmacy and Sciences, Lucknow, India

Rupesh K. Gautam
https://orcid.org/0000-0001-5580-5410
Centre for Pharmacology, Amity Institute of Pharmacy, Amity University, Noida, India

ABSTRACT

Precision genomics is a new paradigm, a combination of high-throughput genomic and multi-omics technologies, which facilitates proactive, personalized prediction, prevention, diagnostics and treatment of disease. These include; incomplete variant identification with short-read sequencing, low statistical power to analyze high-dimensional omics data, high computation and storage needs, data misalignments across platforms and complex issues in privacy, ownership and consent of genomic

DOI: 10.4018/979-8-3373-4849-0.ch002

data. Future directions are focused on precision genomics combined with digital twin technology, diagnostics via nanotechnology and nano-genomics, and quantum computing on intractable genome problems, protein folding, and drug discovery. These innovations indicate the direction towards more predictive and preventive, as well as personalized healthcare eco systems.

1 INTRODUCTION TO PRECISION GENOMICS

Genomic medicine is an innovative medical field that utilizes an individual's genomic information in clinical care for prediction, prevention, early diagnosis, or personalized treatment within the framework of precision medicine, as defined by the National Human Genome Research Institute (NHGRI) (Hasanzad et al., 2022). Precision genomics refers to the integration of genomics into healthcare to provide patients with the most suitable treatment. This approach marks a shift from a traditional, reactive method of illness control to proactive disease avoidance and health maintenance (Mani et al., 2025). Through sophisticated measurement and techniques, advances in genetics bring genetic medicine and genetic data into clinical practice, improving the detection of uncommon diseases, reducing risk, and increasing treatment effectiveness (Khan et al., 2025).

In terms of comprehension of the genetic underpinnings of life, the 20th century represented a revolution. Human biology and genomics entered a new era after the double-helix structure of DNA was discovered and the fundamental principles of molecular biology were clarified (Athanasopoulou et al., 2025). Later, Sanger sequencing—which makes use of dideoxynucleotides (ddNTPs)—was introduced. Chemically altered dNTPs, or ddNTPs, have no hydroxyl groups at the 2' and 3' locations and stop extending after incorporation. Applied Biosystems (ABI) initially offered automated Sanger sequencing in 1986 after releasing the ABI Prism 310, a commercially available DNA sequencing device that used automated capillary electrophoresis and a sophisticated imaging system (Akintunde et al., 2024). The Human Genome Project (HGP) in 2003 was a significant turning point in the field of genomic medicine; nevertheless, this achievement is only one of several significant turning points in the field's development from Mendel to next-generation sequencing (NGS). The HGP accomplishment marked the start of the post-genomic era rather than its conclusion (Hasanzad et al., 2022). The US National Institutes of Health provided funding for the HGP, an ambitious joint team effort comprising 20 institutes in six different nations. It was finished in 2003 after being started in 1990. The open chromatic region of the human genome was sequenced using the Sanger sequencing method in this effort, which produced a sequence of 2.85 billion nucleotides (about 85–90% of the human genome) with few disruptions (Mani et al., 2025). A major advancement was made in the mid-2000s with the introduction of Next-Generation Sequencing (NGS technologies. Massive datasets were produced by the quick and affordable sequencing of complete genomes or transcriptomes made possible by NGS systems like Illumina and Roche 454. More sophisticated bioinformatics tools and methods for data processing, analysis, and storage were required as a result of the explosion of data brought about by NGS (Sarkar et al., 2024).

Since the DNA structure was discovered, advances in omics technology have revolutionized precision medicine by providing previously unheard-of insights into the intricate molecular systems underlying human health and illness (Marques et al., 2024). Because omics research provides high-throughput data at multiple molecular levels, such as genomics, metabolomics, and epigenomics, it has transformed our knowledge of biological systems. Combining omics layers provides a comprehensive understanding of biological functions and disease mechanisms (Cockburn et al., 2025).

2 CORE PRINCIPLES OF PRECISION GENOMICS

The basic ideas and processes behind heredity, gene expression, and genetic variants are included in the field of genetics/genomics. Deciphering the intricacies of genetics and their consequences for biological processes requires an understanding of these principles and mechanisms. Improved risk assessment, genetic counseling, and diagnostic methods were made possible by the discovery of genetics' intricacies. Additionally, it has established the groundwork for the creation of customized medicine and targeted medicines, which allow for more accurate and successful treatments (Biswas et al., 2023). In order to predict phenotypic reactions to environments and connect genetic information to phenotypic outcomes, these omics data are essential. To completely understand the complexity and interdependencies of biological systems, it is crucial to combine all layers of omics data, even if each type of omics offers a distinct viewpoint on molecular processes within cells, tissues, or organisms (Wu & Xie, 2025). While there arc various ways to personalize care, such as making predictions based on clinical traits and patient preferences, genomics-based personalized care has the greatest promise and is currently undergoing a paradigm change (Basu et al., 2016). One of the primary obstacles to creating individualized treatments is stratification, which is a significant objective of biomedical research. Subgroups can be identified through stratification based on their physiological, genetic, genomic, and clinical characteristics. Clinical outcomes should be comparable for patients with similar omics profiles and/or clinical signatures. Networks naturally represent pairwise similarities between patients (Petti & Farina, 2023). The shift from a one-size-fits-all strategy to a more customized treatment paradigm is being propelled by these technological advancements (Jamalinia & Weiskirchen, 2025).

In order to provide a thorough understanding of biological systems, multiomics research incorporates data from genomics and other omics technologies. This method is revolutionary in the biological sciences (Luo et al., 2024). When combined, omics technologies provide a thorough understanding of an organism's biological state and changes, particularly when illness is involved. Thus, another key component of precision genomics is the integration of data from these various sources (Gutierrez Reyes et al., 2024). In order to get valuable insights into cellular activities, analysis of multi-omics data in conjunction with clinical information has taken center stage. The method must carry out an integrative step in which several data sets are simultaneously processed (parallel incorporation of data sets, not sequential). The strategy or method should be easily accessible as a tool or package so that it may be applied to any data source (Subramanian et al., 2020). Our capacity to decipher intricate biological datasets, forecast the course of illness, and identify new treatment targets has been completely transformed by these technologies. Through the use of (Artificial Intelligence) AI and (Machine Learning) ML algorithms, researchers can glean valuable insights from vast amounts of data that are difficult for conventional approaches to process effectively (Vyas & D. M. S., 2024).

The Ethical, Legal, and Social Implications (ELSI) Research Program was one of the novel elements of the US Human Genome Project when it was started in 1990. Bioethics, history, law, medicine, genetics, economics, philosophy, and the behavioral and social sciences are just a few of the many disciplinary viewpoints that are included in "ELSI research". ELSI researchers have a "translational" mandate to pursue studies that help manage practical policy issues affecting human genetics, just as basic genomic scientists are urged to do "translational" research that permits the invention of medically relevant technologies (Burke et al., 2015). Initially, genomic information is information about the genetic or epigenetic composition of a specific individual. When interpreted in this way, it brings up concerns such as ownership, privacy and secrecy, right to be aware or not to be aware, and informed permission to the use of personal data (Francis, 2014). A combination of these main principles constitutes the operational environment of precision genomics, which influences the application of genomic technologies, both in the research environment and in medicine.

3 ADVANCED GENOMIC TECHNOLOGIES ENABLING PRECISION GENOMICS

3.1 NGS Platforms

NGS is a novel technique for sequencing DNA and RNA and detecting variants and mutations. This method combines the benefits of bioinformatics technology, distinct sequencing chemistries, and various sequencing matrices. NGS entails multiple significant sequencing steps. For instance, large parallel sequencing, bioinformatics analysis, variant/mutation annotation and interpretation, library preparation, and DNA fragmentation are all part of DNA NGS. Interrogating hundreds, thousands, or even millions of targets simultaneously is one of the benefits of NGS. Because of this ability, NGS has enormous promise for use in therapeutic contexts. For instance, in the treatment of cancer patients, a tumor may have more than one mutation. These many assays might require more tissue. NGS technology allows for the examination of these targets in a single test (Qin, 2019). NGS has revolutionized genomics, enabling quick genome sequencing, reducing prices, and having a number of therapeutic applications (Khan et al., 2025). By providing high-throughput and affordable sequencing options, NGS and third-generation sequencing (TGS) platforms such as those created by Oxford Nanopore Technologies (ONT) and Illumina®, respectively, revolutionized the gathering of genomic data. These developments sped up research and therapeutic applications by enabling the sequencing of whole genomes in a matter of hours (Athanasopoulou et al., 2025). The limitations of more conventional techniques like the Sanger method, karyotyping, and chromosomal arrays for rare genetic diseases have been overcome by recent developments in NGS, which include whole exome sequencing, whole genome sequencing, targeted exam sequencing, whole mtDNA sequencing, and RNA sequencing (Khan et al., 2025). The application of various platforms used in sequencing are given in (Table 1).

A) Whole Genome Sequencing (WGS)

The process of identifying and assembling an organism's entire genomic code is known as WGS. It depends on NGS technologies, which fall into two categories: third generation (like PacBio or Nanopore) and second generation (like Illumina). Without a doubt, WGS in microbiology has ushered in a new era of advanced microbial genome analysis, enabling detailed examinations of the genetic material present in microbial populations (Bianconi et al., 2023). WGS makes it possible to identify genetic differences, from single-nucleotide polymorphisms (SNPs) to more significant structural alterations like insertions, deletions, and rearrangements, by sequencing all of an organism's DNA. Numerous applications in a variety of sectors are made possible by the abundance of information acquired by WGS. Large whole-genome sequencing, which deciphers bigger genomes of >5 Mb, such as those of eukaryotes, and short whole-genome sequencing, which deciphers smaller genomes of <5 Mb, primarily of prokaryotes, are the two types of sequencing methodologies offered by WGS (Satam et al., 2024).

B) Whole Exome Sequencing (WES)

The sequence that includes all of the exons of the genes that code for proteins is known as the exome. Depending on the species, this region can make up 1% to 2% of the genome. In solution-based whole-exome sequencing (WES), biotinylated oligonucleotide probes, or baits, are employed to preferentially hybridize to specific areas of the genome after DNA samples are broken apart (Warr et al., 2015). Single-nucleotide variants (SNVs), insertions, deletions, and copy number variations (CNVs) within protein-coding genes can all be found by using WES to sequence the exome. In WES, target-specific amplification or hybrid capture methods are used to enrich exonic regions, and high-throughput sequencing is then performed. Since WES is a component of WGS, the bioinformatic method utilized for WES data analysis is the same as that used for WGS (Satam et al., 2024).

C) Targeted Sequencing (TS)

TS panels have lower data loads with respect to bioinformatics or computational requirements than WGS and WES, and they concentrate on a specific cluster of genetic areas. It is crucial to include the extra step of selective enrichment for the genomic regions of interest and compare them to the genomic background before developing and using the TS panels. This critical stage guarantees that the NGS procedure is specially engineered to swiftly and precisely sequence the genomic targets (Pei et al., 2023). Overcoming the constraints of single-gene assays, target-ed next-generation sequencing (NGS) employing multigene panels has emerged as a useful technique for thorough genomic characterization in cancer. However, the longer turnaround time (about three weeks) needed to collect findings and the outsourcing of these assays to outside laboratories may hinder the prompt clinical management of cancer patients (Das et al., 2025).

3.2 Third Generation Sequencing

The most recent developments in DNA sequencing are represented by third-generation sequencing technologies, which provide fresh strategies to get beyond the drawbacks of earlier generations (Satam et al., 2024). Sequencing in real-time and

Single-molecule sequencing (SMS) are the two essential and unique characteristics of the recently developed TGS techniques that allow nucleotide molecules (DNA or RNA) to be sequenced without requiring amplification of the template by using the Polymerase chain reaction (PCR) technique (Athanasopoulou et al., 2021).

A) Single-molecule real-time sequencing (SMRT)

The first known SMRT sequencing technique was introduced by Pacific Biosciences (PacBio) in early 2011 with the release of their PacBio RS sequencer.PacBio's Sequel System underwent incremental enhancements throughout time, leading to the creation of two more Sequel platforms: the Sequel II System and Sequel IIe System, the latter of which is the most potent platform in the Sequel family (Athanasopoulou et al., 2021). In addition to a few reads greater than 50 kbp, SMRT sequencing often produces reads with an average median length of roughly 10–20 kbp. During library preparation, double-stranded DNA molecules are circularized by ligating certain hairpin adapters to both ends, creating a structure called SMRTbell.Higher sequencing costs per base have resulted from SMRT sequencing's advantages over NGS.2019 saw the release of PacBio's Sequel II, an upgraded platform built on new chemicals and outfitted with a unique 8 million ZMW micro wells SMRT cell that could theoretically produce a throughput of 160 Gb per SMRT cell while simultaneously reducing sequencing expenses by up to eight times (Giani et al., 2020).

B) Nanopore Sequencing

Oxford Nanopore sequencing is a further technique that relies on nanopore technology. This method involves passing a single-stranded DNA molecule through a nanopore and measuring changes in electrical current to identify the DNA sequence. Oxford Nanopore sequencing offers mobility, real-time analysis, and extended read lengths (Satam et al., 2024). The process of nanopore sequencing involves submerging a biological membrane with nanopores in an ionic solution, driving the unraveled single-stranded molecules to be tested through the nanopores after applying a voltage to both sides of the phospholipid bilayer to create an ionic current through the nanopore channels. For instance, only single-stranded RNA or DNA may pass through the most widely utilized α-hemolysin nanopore channels, which have a pore size of 2.6 nm (Chen & Xu, 2023).

3.3 Functional Genomics Technologies

Based on genomic data, functional genomics examines the functional characteristics of genes and their expression to explain transcription, translation, interactions between protein complexes, etc. Functional genomics research helps us comprehend how DNA functions and interacts. Functional genomics has the ability to show how an organism's biological function is influenced by its DNA. In the future, this might be a useful tool for treating genetic illnesses in humans (Khan et al., 2019).

A) RNA sequencing (RNA-seq)

RNA-Seq makes use of high-throughput sequencing techniques to reveal information about a cell's transcriptome. RNA-Seq offers far better coverage and resolution of the dynamic nature of the transcriptome than earlier Sanger sequencing and microarray-based techniques. Researchers have been able to better understand the functional complexity of transcription thanks to recent developments in the RNA-Seq workflow, which include sample preparation, library creation, and data analysis.RNA-Seq can be used to examine various RNA populations, such as total RNA, pre-mRNA, and noncoding RNA, such as microRNA and long ncRNA, in addition to polyadenylated messenger RNA (mRNA) transcripts (Kukurba & Montgomery, 2015).

B) Assay for Transposase-Accessible Chromatin using sequencing (ATAC-seq)

The development of genomic and transcriptome resources for both model and non-model species has been fueled by the introduction and quick development of DNA sequencing technologies. A flexible and popular technique for locating putative regulatory areas, including promoters and enhancers, inside a genome is the ATAC-seq. A hyperactive Tn5 transposase, which preferentially binds to accessible nuclear DNA, is used in ATAC-seq, which starts with the isolation of cell nuclei. Usually, synthesized oligonucleotide sequences that can be used as sequencing adapters for Illumina short-read sequencing are pre-loaded into the transposase. Furthermore, there is a great deal of promise for connecting accessible chromatin areas with transcriptional activity through the integration of ATAC-seq and gene expression data (Erdoğan et al., 2025).

C) Chromatin Immunoprecipitation Sequencing (ChIP-seq)

Chromatin immunoprecipitation followed by sequencing is a method for profiling DNA-binding proteins, histone modifications, or nucleosomes over the entire genome. Compared to its array-based predecessor, ChIP-chip, ChIP-seq delivers higher resolution, less noise, and greater coverage because of the remarkable advancements in

NGS technology. ChIP-seq is now a vital technique for researching gene regulation and epigenetic mechanisms due to the declining cost of sequencing (Park, 2009).

3.4 Structural Genomics and Chromatin Architecture Tools

A) Hi-C and 3D genomics approaches

Depending on the area of interest, multiple approaches can be used to investigate the genome at different levels. Until recently, the majority of DNA mapping techniques relied on linear genome exploration. Nevertheless, the genome is a dynamic three-dimensional (3D) structure that varies according to the different situations of the cell. There are two primary ways to examine the 3D genome: ligation-based approaches, also known as chromosomal conformation capture (3C)-based methods, and ligation-free methods. Subsequently, high-throughput chromosomal conformation capture (Hi-C), chromosome conformation capture-on-chip (4 C), and chromosome conformation capture carbon copy (5 C) were created. Hi-C has made it possible to map every potential interaction between loci throughout the whole genome. The bulk of ligation-based approaches is based on Hi-C, the most widely used and well-liked 3C technique. The importance of chromatin structure and three-dimensional organization on gene expression, regulatory element function, and other characteristics that either directly or indirectly affect gene regulation has been made clear by ligation-based techniques (Tavallaee & Orouji, 2025).

B) Optimal Genome Mapping (OGM)

OGM complements Hi-C by mapping large-scale structural changes in intact DNA molecules using fluorescent tagging and nanochannel arrays. Large insertions, deletions, translocations, and intricate rearrangements that sequencing technologies can miss are particularly well-detected by OGM. Furthermore, OGM provides the possibility of a single experiment that would otherwise necessitate chromosomal microarray analysis, numerous rounds of FISH testing, and karyotyping (Levy et al., 2025).

Table 1. Comparison of sequencing technologies

Sequencing Technology	Throughput	Resolution	Accuracy	Major Applications	Ref.
First Generation (Sanger Sequencing)	Low	Single base resolution read length<1kb	Very high base pair accuracy	Targeted gene sequencing and mutation/ variant confirmation	(Heather & Chain, 2016)
NGS (Illumina Platforms)	Very high (Gb-Tb)	35-300bp	99-99.5% per base accuracy	WGS, WES, Targeted Sequencing, RNA-seq	(Goodwin et al., 2016)
PacBio (SMRT Sequencing)	Sequel II/ IIe – 500 Gb per run, 4-8 million ZMWs	10-15kbp	High error rate, but HiFi circular consensus sequences achieve >99% accuracy	Targeted sequencing, Metagenomics, structural variants, De novo genome assembly and full-length transcriptomics	(Athanasopoulou et al., 2021)
Nanopore (Oxford Nanotechnology)	2.8 Gb to 14 Tb	Single nucleotide via 1-2nm biological nanopores	99% (Q20 simplex) / 99.99% (Q30 duplex)	Metagenomics, Pulmonary Pathogen ID	(Chen & Xu, 2023)
Spatial Transcriptomics	1007 features, 2k-17k genes	100µm spots, 200µm spacing	r=0.94-0.97; 7% sensitivity	Breast cancer heterogeneity, Brain layer profiling	(Ståhl et al., 2016)

4 INTEGRATIVE OMICS IN PRECISION GENOMICS

Large-scale, thorough analysis of a variety of biomolecular data is made possible by the omics high-throughput pipeline, which is revolutionizing biological research as shown in Figure 1. These cutting-edge technologies produce vast amounts of data at several omics levels, such as metabolomics, proteomics, transcriptomics, and genomes (Vitorino, 2024). Researchers can comprehend illness heterogeneity, find new disease biomarkers, and create tailored treatment plans by utilizing multi-omics data. Multi-omics integration has proven crucial in precision medicine for determining therapy responses and individual-specific disease risks. For example, multi-omics techniques in oncology have made it possible to classify tumors according to their molecular profiles rather than conventional histological features, resulting in more potent, focused treatments (Khodadadian et al., 2020).

4.1 Genomics and Transcriptomics Integration

The German botanist Hans Winkler initially used the term "genome" to refer to a collection of haploid chromosomes in his 1920 book "Spread and cause of pathogenesis in plant and animal kingdoms." These days, any genetic material found in any kind of organism is referred to as the genome. The chromosomal and extra-chromosomal DNA material that makes up the genome can be classified as either non-coding or coding (including both coding and non-coding genes that may or may not produce proteins). The study of the entire genome is known as genomics, an interdisciplinary branch of science. In contrast to genetics, which focuses on the individual traits of genes, genomics seeks to determine how genes interact and have synergistic effects, how the environment interacts with genes, and how the non-coding sections of the genome are studied (Khodadadian et al., 2020).

A cell or tissue's entire collection of RNA transcripts is referred to as its transcriptome. There are different splice variants in the transcriptome. The study of transcriptomics examines how cells, tissues, or organisms function under a variety of biological circumstances. An organism's information is expressed by RNA (the transcriptome) and stored in DNA (the genome), according to the link between the two. The primary goal of transcriptomics is to understand how disease or environmental factors (such as medications, hormones, etc.) affect the transcripts of a cell, tissue, or living organism. The study of the transcriptome (transcriptomics) and genome (genomics) is crucial for identifying disease pathways and creating precise and potent medications. Treating illnesses that are resistant to current therapies may benefit from a deeper understanding of these topics. One of the most crucial tools for researching these levels is the advancement of nucleic acid sequencing techniques, especially next-generation sequencing technology (Khodadadian et al., 2020).

4.2 Proteomics and Genomics Integration (Proteogenomics)

Proteogenomics has developed during the past 20 years and functions at the nexus of proteomics and genomics. Any nucleotide sequence that has the capacity to encode a protein expressed in a sample, including sequences from the exome, genome, translatome, and transcriptome, is considered significant to proteogenomics. The simultaneous advancements in top-down proteomics and third-generation sequencing point to a future in proteogenomics where transcriptomes and proteomes might be examined with high sensitivity and resolution. For example, an entire sample-specific, full-length protein database might be predicted using third-generation sequencing data on a human transcriptome, providing further insight into the individual protein isoforms that might be expressed in a sample (Sheynkman et al., 2016).

4.3 Epigenomics

The study of epigenetic modifications, heritable changes in gene expression patterns that do not entail changes in the DNA sequence, is known as epigenomics. DNA methylation, histone modification, and RNA methylation (epi-transcriptome) are the most often researched types of epigenetic alterations. These chemical markers then modify nucleosome placement, chromatin remodeling, and DNA accessibility (Satam et al., 2024). Thus, through echoed complex mechanisms like the interaction between DNA methylation and histone changes, epigenetic and genomic instability are indirectly linked. For example, the frequent DNA methylation modification 5-methylcytosine (5mC) might affect DNA accessibility by forming a more compact chromatin structure that suppresses gene expression (Baumann et al., 2025).

4.4 Multi-omics Data Harmonization and Standardization

The goal of multi-omics, also known as integrated omics, pan-omics, and trans-omics, is to integrate two or more omics data sets to support data analysis, visualization, and interpretation in order to identify a biological process's mechanism. Over the past ten years, there has been considerable success in applying various individual omics studies (such as genomics, epigenomics, transcriptomics, proteomics, and metagenomics) to understand a specific issue in human disease, agriculture, plant science, microbiology, and the environment. In the age of precision medicine, multi-omics can aid in the identification of prognostic or predictive biomarkers as well as potentially innovative and repurposed therapeutic targets. Therefore, the ultimate goal of applied multi-omics is to use a solid understanding of the genotype-to-phenotype link to improve agricultural outputs, better disease prognosis, and boost the diagnostic yield for health (Krassowski et al., 2020).

Figure 1. Flowchart of omics integration process

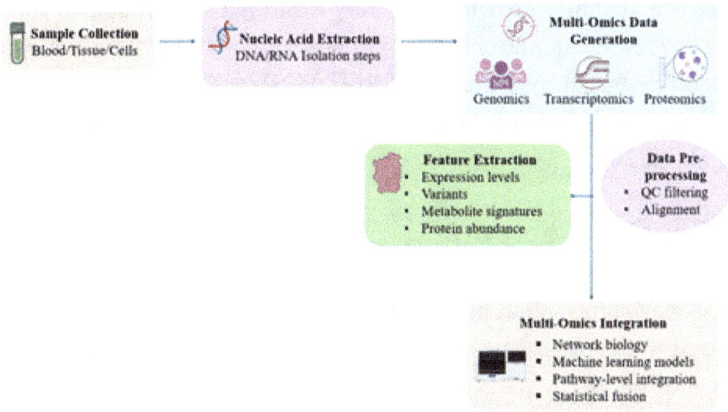

5 CLINICAL APPLICATION OF PRECISION GENOMICS

Precision genomics is applicable in the diagnosis of various diseases as well as in the treatment of rare genetic disorders and requires a certain workflow as depicted in Figure 2.

A) Precision Oncology

The development of precision oncology is largely dependent on genomic profiling. AI, gene editing, and genomics are coming together to provide more individualized, effective, and comprehensive cancer therapy. Cancer is a very diverse illness with intricate genetic and molecular differences between patients. The development of NGS and other genomic profiling methods has greatly advanced our understanding of the genetic changes and mutations that propel the growth and spread of tumors (Jamalinia & Weiskirchen, 2025). The identification of actionable changes that result in new treatments has been made possible by developments in tumor genome sequencing. Gene amplification, point mutations, and translocations are examples of genetic changes that can be linked to authorized targeted medicines for a variety of tumor types. NGS technology has reduced prices and increased throughput, leading to the development of tools to identify these genetic changes. Beyond what is provided by traditional histopathologic techniques, genomic sequencing technologies have made it possible to identify actionable targets (such as BRAF in melanoma and EGFR in lung cancer), which has facilitated treatment selection (Waarts et al., 2022). Precision oncology's transformational power is exemplified by lung cancer, especially non-small-cell lung cancer (NSCLC). The field of targeted medications is constantly growing, not only for lung cancer but also for breast and colorectal cancers. Particularly intriguing new targeted agents, such as molecular glues, proteolysis targeting chimeras (PROTACS), and PARP inhibitors, still need accompanying test applications to provide the best possible treatment customization (Ferrari et al., 2024).

B) Rare Genetic Disorders

Rare disease diagnosis has always been quite difficult. Biochemical characteristics were used to diagnose uncommon disorders for the first time in the early 20th century. In addition to the particular genetic disease's rarity, its genetic and clinical variability contribute to some of the complexity. There is a lack of infrastructure, coordination, resources, and understanding about rare genetic disorders in nations with tiny populations, where the number of people with uncommon diseases is also low. Genetic testing was transformed by NGS technologies. It has made a substantial contribution to the identification of Mendelian disease genes. Only about 40 genes were known to cause hereditary illnesses prior to 1986. More than 1000 genes, including those for Huntington's disease and cystic fibrosis, were found using the positional cloning technique, which involved identifying the gene in the genome, cloning the region of interest, and sequencing the candidate genes. In addition to finding novel gene targets for human diseases, NGS has revolutionized healthcare systems and greatly enhanced the diagnosis of uncommon genetic disorders (Vinkšel et al., 2021).

C) Infectious Disease Genomics

Clinical care, infection control, and public health response have long placed a high priority on the quick and accurate detection of infectious pathogens. The most direct clinical and public health benefits, such as precise pathogen diagnosis and an understanding of isolates within an epidemiologic framework, can be obtained through genome sequencing. One of the most valuable therapeutic uses of sequencing is the genomic prediction of antibiotic resistance from pathogen genomes, which can reduce the time to successful medication, direct antimicrobial stewardship, and identify and monitor high-risk resistance determinants (Hasan et al., 2025).

D) Pharmacogenomics

Long-term pharmacogenetic research has shown how genetic variations have a wide range of effects on drug responsiveness. The availability of high-density genomic SNP maps and the growing number of identified functional polymorphisms raise the possibility that pharmacogenetics will be able to optimize patient-specific drugs. With the increasing use of genomes and other omics technologies, the term "pharmacogenomics" has emerged to characterize this developing approach to drug development and therapy. The pharmaceutical sector may employ pharmacogenomics as a tool. It is a significant breakthrough in medical history. Its primary goals are to identify new targets for pharmaceuticals, enhance efficacy and decrease adverse drug reactions, correlate genetic material with clinical genetic data, and pharmacogenetically profile individuals to predict medication response and illness risk. One essential element of customized treatment is pharmacogenomics. The basic premise is that variability between individuals in drug response is influenced by a patient's age, gender, and/or concurrent medications, as well as environmental factors, genetics, and epigenomics. Developments in pharmacogenomics, sometimes referred to as "omics" techniques, have revolutionized our knowledge of the origins and susceptibility of illnesses and hold enormous potential for the creation of novel therapeutic strategies (Qahwaji et al., 2024).

E) Predictive and Preventive Healthcare

Implementing early illness diagnosis, prevention, and intervention requires an understanding of the psychological risk factors that lead to disease onset. Using inherited DNA variation to stratify individual genetic vulnerability is a key strategy for precision medicine. As human genetics has advanced, this strategy has been created. The gold standard for measuring genetic liability in illness risk prediction is the polygenic risk score (PRS). Single-nucleotide polymorphisms (SNPs) with genetic risks determined by genome-wide association studies (GWASs) are used in PRS, which is computed as the weighted average scores of these SNPs with genetic risks utilizing the effect sizes from GWASs as their weights (Konuma & Okada, 2021).

Figure 2. Clinical decision-making workflow

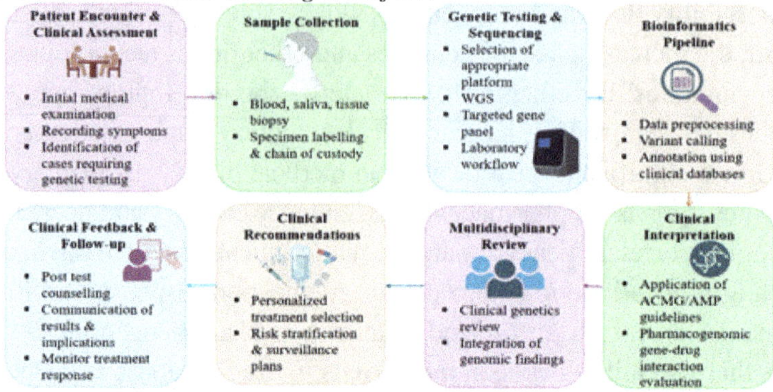

6 CHALLENGES, LIMITATIONS, AND FUTURE DIRECTIONS

6.1 Challenges and Limitations

A) Technical Challenges

The intrinsic difficulties posed by NGS's technical and methodological features may have an effect on the accuracy and dependability of genomic data in a clinical context. One technical challenge is maintaining uniform coverage throughout the genome. Although the majority of clinical NGS applications use short-read sequencing technologies, which have inherent limits, WGS strives for full coverage. Long repeat sequences, triplet repeat expansions, polyploidy, mosaicism, complex structural variants (such as those larger than 300–400 base pairs), and variants found in regions with high sequence identity (such as identical genes and pseudogenes) or high guanine or cytosine (GC) content are examples of these limitations. This indicates that even with high throughput, a totally uniform genomic readout is not possible due to inherent technical limitations, particularly for extremely complicated or uncommon changes. In order to obtain a full molecular picture, these technical limitations directly affect the thoroughness and accuracy of genomic profiling, requiring ongoing optimization and the creation of supplementary technologies (Isaic et al., 2025).

B) Sample size and Statistical power

To make trustworthy inferences, different omics data require varying sample sizes. The number of measured entities (i.e., transcripts, proteins, and metabolites) affects the false discovery rate (FDR), which determines reliability. For smaller omics data-generating platforms, like micro RNAs, a power of 0.8 at a fold change of 1.5 with FDR<0.1 may require roughly 19 samples per experimental group. In contrast, using ssize RNA with parameters pi0=0.8 disp=0.1, a set of 10,000 transcripts, each with at least 10 counts, would require at least 35 samples per group for the same effect size at the same power and FDR control level. For every omics, the power computation is not as simple. There are fewer tools specifically designed for metabolomics and proteomics research, despite the fact that many were developed for transcriptomics/genomics power analysis (Krassowski et al., 2020).

C) Infrastructure challenges

Due to the complexity and number of data involved, multimodal omics data analysis needs significant computational resources. Terabytes (TB) of space are frequently needed to store raw and processed data, including sequence readings, MS data sets, and imaging data, for omics data processing. Another crucial element is the networking architecture, where effective data transfer between storage and processing nodes is ensured by high-speed networks (e.g., 10 GbE or greater). Platforms for cloud computing, including AWS, Google Cloud, and Azure, provide scalable computing resources that can be modified based on workload demands, offering cost-effectiveness and flexibility. Lastly, high-performance CPUs and GPUs, vast memory and storage capacities, fast networking, and a stable, scalable software environment are all necessary for multimodal omics data analysis (Luo et al., 2024).

D) Data Inconsistency

When working with multi-omics data, data inconsistency is frequently another bottleneck because these data are produced using a variety of platforms and technologies (e.g., targeted versus untargeted, high resolution versus single cell, or sequencing versus mass spectrometry). Pre-processing procedures relevant to specific data sets may barely help overall, especially when democratizing them under a uniform framework still remains problematic (Krassowski et al., 2020).

E) Ethical and Privacy Issues

Research on common disorders also makes use of genomic data. In order to identify strong genetic risk factors that may serve as therapeutic targets, researchers may employ a strategy similar to that employed for rare diseases, analyzing genomic data from families with extremely severe or early-onset forms of the disease. Participants may have enormous expectations about the potential benefits of genomic data research, either for themselves or for society at large, as a result of frequent political discussions surrounding the data. This makes obtaining consent for the creation, storage, and use of such data more difficult. First of all, regardless of what their biological relatives may desire, a person who donates genomic data for research invariably also provides genomic data from their relatives. For instance, even if the twin was adamantly against taking part in research, a person with an identical twin may effectively supply their twin's whole genetic code. Second, there is a tendency for the political rhetoric surrounding genetic data in research to be overly optimistic, which could lead to exaggerated expectations about the potential outcomes of genomic research in the near future. Thirdly, the prospective and future applications of genetic data are unknown, yet the providers and their families will find lasting value in the data, and the data holders will get uncertain financial compensation. Since it is difficult to predict the effects of supplying genetic data, data providers' consent must be broad and unrestricted (Horton & Lucassen, 2023).

6.2 Future Directions

A) Digital twin technology (DT)

With its cutting-edge capabilities in medical education, workflow optimization, individualized therapy, and predictive diagnostics, DT technology is gradually revolutionizing the healthcare industry. Globally, there is a significant increase in the use of DT technology in healthcare. According to recent industry statistics, 66% of executives in the healthcare sector intend to make investments in DT technology throughout the course of the following three years. In the context of product lifecycle management, NASA first used the phrase “digital twin” in 2002.DTs have developed throughout time from static digital models to dynamic, artificial intelligence-based systems that can replicate sophisticated and flexible systems (Saratkar et al., 2025).

B) Nano-genomics

Numerous overlapping molecular technologies play a significant role in the field of medicine. Numerous illnesses, including cancer, cirrhosis, and bacterial and viral infections, have been identified and diagnosed using nanoscale fast diagnostics and biosensors. Nanoparticles can be administered continuously without being harmful due to their small size and large surface area. Nanotechnology has made it easier to understand the genomes and proteomics of new viral epidemics, which was previously an enormous undertaking (Kamat et al., 2021).

C) Quantum Genomics

Drug discovery, Protein folding, structural simulations, genome assembly, and other challenging molecular biology issues can be effectively addressed by quantum computing. By carrying out numerous calculations at once, quantum computing, which is founded on the concepts of superposition and entanglement, offers the capacity to address issues that are well beyond the capabilities of traditional computers. Quantum computing holds great potential for tackling problems that were thought to be nearly impossible. For instance, a quantum computer can perform some computations more rapidly than a classical computer by applying a particular approach called Shor's approach (Ghamsari, 2025).

7 CONCLUSION

Precision genomics is an unprecedented paradigm shift in the modern healthcare system, that is reinventing the manner in which diseases are detected, categorized and treated. The use of the latest sequencing technologies, such as next-generation sequencing, third-generation systems, and functional genomics as well as structural genomics, has provided unprecedented understanding of genetic variation, gene regulation, and the molecular basis of disease. When combined with multi-omics approaches, these approaches provide an integrated, holistic view of biological systems beyond individual single-gene studies to a large-scale genotype-phenotype mapping. Even though great progress has been made, precision genomics is facing serious setbacks. The barriers to the general implementation of precision genomic methodologies in the clinical setting remain in technical limitations in sequencing fidelity, data heterogeneity across platforms, and computational needs of multi-omics integration. Another ethical issue, like data confidentiality, informed consent, and long-term use of the genomic data, also makes the implementation difficult, and it is important to create robust governance frameworks. Moreover, equitable access among the wide range of populations is also a critical concern, and genomic datasets still do not reflect a large number of communities in the world.

Future perspectives suggest that the technologies currently in the pipeline, such as digital twins, nano-genomics, and quantum computing, hold a lot of potential in overcoming the current bottlenecks and launching precision genomics to greater heights. These breakthroughs hint on real-time disease modelling, very sensitive diagnostics and formidable computing power to solve complex biological problems. With the increasing rate of technological development, it is probable that precision genomics will be kept in the center of clinical routine practice, contributing to the higher predictive value, the earlier onset of the therapeutic intervention, and the truly personalized treatment regimens. Finally, it is estimated that a successful combination of these modalities will turn the practice of medicine into a proactive, data-centric and patient-centered field.

REFERENCES

Akintunde, O., Tucker, T., & Carabetta, V. J. (2024). The evolution of next-generation sequencing technologies. In *High-throughput gene screening: Methods and protocols* (pp. 3–29). Springer.

Ali, H. (2023). Artificial intelligence in multi-omics data integration: Advancing precision medicine, biomarker discovery, and genomic-driven disease interventions. *International Journal of Scientific Research Archive*, *8*(1), 1012–1030.

Athanasopoulou, K., Boti, M. A., Adamopoulos, P. G., Skourou, P. C., & Scorilas, A. (2021). Third-generation sequencing: The spearhead towards the radical transformation of modern genomics. *Life (Chicago, Ill.)*, *12*(1), 30. DOI: 10.3390/life12010030 PMID: 35054423

Athanasopoulou, K., Michalopoulou, V.-I., Scorilas, A., & Adamopoulos, P. G. (2025). Integrating artificial intelligence in next-generation sequencing: Advances, challenges, and future directions. *Current Issues in Molecular Biology*, *47*(6), 470. DOI: 10.3390/cimb47060470 PMID: 40699869

Basu, A., Carlson, J. J., & Veenstra, D. L. (2016). A framework for prioritizing research investments in precision medicine. *Medical Decision Making*, *36*(5), 567–580. DOI: 10.1177/0272989X15610780 PMID: 26502985

Baumann, A. A., Buribayev, Z., Wolkenhauer, O., Salybekov, A. A., & Wolfien, M. (2025). Epigenomic echoes—Decoding genomic and epigenetic instability to distinguish lung cancer types and predict relapse. *Epigenomes*, *9*(1), 5. DOI: 10.3390/epigenomes9010005 PMID: 39982247

Bianconi, I., Aschbacher, R., & Pagani, E. (2023). Current uses and future perspectives of genomic technologies in clinical microbiology. *Antibiotics (Basel, Switzerland)*, *12*(11), 1580. DOI: 10.3390/antibiotics12111580 PMID: 37998782

Biswas, A., Kumari, A., Gaikwad, D. S., & Pandey, D. K. (2023). Revolutionizing biological science: The synergy of genomics in health, bioinformatics, agriculture, and artificial intelligence. *OMICS: A Journal of Integrative Biology*, *27*(12), 550–569. DOI: 10.1089/omi.2023.0197 PMID: 38100404

Burke, W., Appelbaum, P., Dame, L., Marshall, P., Press, N., Pyeritz, R., Sharp, R., & Juengst, E. (2015). The translational potential of research on the ethical, legal, and social implications of genomics. *Genetics in Medicine*, *17*(1), 12–20. DOI: 10.1038/gim.2014.74 PMID: 24946153

Chen, J., & Xu, F. (2023). Application of nanopore sequencing in the diagnosis and treatment of pulmonary infections. *Molecular Diagnosis & Therapy*, *27*(6), 685–701. DOI: 10.1007/s40291-023-00669-8 PMID: 37563539

Cockburn, J. G., Mariappan, V., Loke, M. F., Al-Maleki, A. R., Muttiah, B., Vellasamy, K. M., & Vadivelu, J. (2025). Integrative research: Current trends and considerations for biomarker discovery and precision medicine. *The Microbe*, *7*, 100368. DOI: 10.1016/j.microb.2025.100368

Das, K., Tay, M. L. I., Yong, E. Y., & Chuah, K. L. (2025). A targeted next-generation sequencing panel for identification of clinically relevant mutation profiles in solid tumours. *Scientific Reports*, *15*(1), 20740. DOI: 10.1038/s41598-025-08039-6 PMID: 40593313

Erdoğan, D. E., (2025). ATAC-seq in emerging model organisms: Challenges and strategies. *Journal of Experimental Zoology. Part B, Molecular and Developmental Evolution*. PMID: 40452285

Ferrari, V., Mograbi, B., Gal, J., & Milano, G. (2024). Companion tests and personalized cancer therapy: Reaching a glass ceiling. *International Journal of Molecular Sciences*, *25*(18), 9991. DOI: 10.3390/ijms25189991 PMID: 39337479

Francis, L. P. (2014). Genomic knowledge sharing: A review of the ethical and legal issues. *Applied & Translational Genomics*, *3*(4), 111–115. DOI: 10.1016/j.atg.2014.09.003 PMID: 27294025

Ghamsari, M. S. (2025). Quantum computing applications in biology. *Discover Computing*, *28*(1), 276. DOI: 10.1007/s10791-025-09803-y

Giani, A. M., Gallo, G. R., Gianfranceschi, L., & Formenti, G. (2020). Long walk to genomics: History and current approaches to genome sequencing and assembly. *Computational and Structural Biotechnology Journal*, *18*, 9–19. DOI: 10.1016/j.csbj.2019.11.002 PMID: 31890139

Goodwin, S., McPherson, J. D., & McCombie, W. R. (2016). Coming of age: Ten years of next-generation sequencing technologies. *Nature Reviews. Genetics*, *17*(6), 333–351. DOI: 10.1038/nrg.2016.49 PMID: 27184599

Gutierrez Reyes, C. D., Alejo-Jacuinde, G., Perez Sanchez, B., Chavez Reyes, J., Onigbinde, S., Mogut, D., Hernández-Jasso, I., Calderón-Vallejo, D., Quintanar, J. L., & Mechref, Y. (2024). Multi-omics applications in biological systems. *Current Issues in Molecular Biology*, *46*(6), 5777–5793. DOI: 10.3390/cimb46060345 PMID: 38921016

Hasan, G. M., Mohammad, T., Shamsi, A., Sohal, S. S., & Hassan, M. I. (2025). Applications of genome sequencing in infectious diseases: From pathogen identification to precision medicine. *Pharmaceuticals (Basel, Switzerland)*, *18*(11), 1687. DOI: 10.3390/ph18111687 PMID: 41304932

Hasanzad, M., Sarhangi, N., Naghavi, A., Ghavimehr, E., Khatami, F., Ehsani Chimeh, S., Larijani, B., & Aghaei Meybodi, H. R. (2022). Genomic medicine on the frontier of precision medicine. *Journal of Diabetes and Metabolic Disorders*, *21*(1), 853–861. DOI: 10.1007/s40200-021-00880-6 PMID: 35673457

Heather, J. M., & Chain, B. (2016). The sequence of sequencers: The history of sequencing DNA. *Genomics*, *107*(1), 1–8. DOI: 10.1016/j.ygeno.2015.11.003 PMID: 26554401

Horton, R., & Lucassen, A. (2023). Ethical considerations in research with genomic data. *The New Bioethics : a Multidisciplinary Journal of Biotechnology and the Body*, *29*(1), 37–51. DOI: 10.1080/20502877.2022.2060590 PMID: 35484929

Isaic, A., Motofelea, N., Hoinoiu, T., Motofelea, A. C., Leancu, I. C., Stan, E., Gheorghe, S. R., Dutu, A. G., & Crintea, A. (2025). Next-generation sequencing: A review of its transformative impact on cancer diagnosis, treatment, and resistance management. *Diagnostics (Basel)*, *15*(19), 2425. DOI: 10.3390/diagnostics15192425 PMID: 41095644

Jamalinia, M., & Weiskirchen, R. (2025). Advances in personalized medicine: Translating genomic insights into targeted therapies for cancer treatment. *Annals of Translational Medicine*, *13*(2), 18. DOI: 10.21037/atm-25-34 PMID: 40438512

Kamat, S., Kumari, M., & Jayabaskaran, C. (2021). Nano-engineered tools in the diagnosis, therapeutics, prevention, and mitigation of SARS-CoV-2. *Journal of Controlled Release*, *338*, 813–836. DOI: 10.1016/j.jconrel.2021.08.046 PMID: 34478750

Khan, A., Barapatre, A. R., Babar, N., Doshi, J., Ghaly, M., Patel, K. G., Nawaz, S., Hasana, U., Khatri, S. P., Pathange, S., Pesaru, A. R., Puvvada, C. S., Billoo, M., & Jamil, U. (2025). Genomic medicine and personalized treatment: A narrative review. *Annals of Medicine and Surgery (London)*, *87*(3), 1406–1414. DOI: 10.1097/MS9.0000000000002965 PMID: 40213198

Khan, N., Jameel, N., & Khan, M. (2019). Functional genomics: Linking genotype with phenotype on a genome-wide scale. *International Journal of Applied Pharmaceutical Sciences and Research*, *4*(1), 4–12. DOI: 10.21477/ijapsr.4.1.2

Khodadadian, A., Darzi, S., Haghi-Daredeh, S., sadat Eshaghi, F., Babakhanzadeh, E., Mirabutalebi, S. H., & Nazari, M. (2020). Genomics and transcriptomics: Powerful technologies in precision medicine. *International Journal of General Medicine*, *13*, 627–640. DOI: 10.2147/IJGM.S249970 PMID: 32982380

Konuma, T., & Okada, Y. (2021). Statistical genetics and polygenic risk score for precision medicine. *Inflammation and Regeneration*, *41*(1), 18. DOI: 10.1186/s41232-021-00172-9 PMID: 34140035

Krassowski, M., Das, V., Sahu, S. K., & Misra, B. B. (2020). State of the field in multi-omics research: From computational needs to data mining and sharing. *Frontiers in Genetics*, *11*, 610798. DOI: 10.3389/fgene.2020.610798 PMID: 33362867

Kukurba, K. R., & Montgomery, S. B. (2015). RNA sequencing and analysis. *Cold Spring Harbor Protocols*, *2015*(11), 951–969. DOI: 10.1101/pdb.top084970 PMID: 25870306

Levy, B., Burnside, R. D., & Akkari, Y. (2025). Optical genome mapping: A new tool for cytogenomic analysis. *Genes*, *16*(8), 924. DOI: 10.3390/genes16080924 PMID: 40869972

Luo, Y., Zhao, C., & Chen, F. (2024). Multiomics research: Principles and challenges in integrated analysis. *BioDesign Research, 6*, 0059.

Mani, S., Lalani, S. R., & Pammi, M. (2025). Genomics and multiomics in the age of precision medicine. *Pediatric Research*, *97*(4), 1–12. DOI: 10.1038/s41390-025-04021-0 PMID: 40185865

Marques, L., Costa, B., Pereira, M., Silva, A., Santos, J., Saldanha, L., Silva, I., Magalhães, P., Schmidt, S., & Vale, N. (2024). Advancing precision medicine: A review of innovative in silico approaches for drug development, clinical pharmacology, and personalized healthcare. *Pharmaceutics*, *16*(3), 332. DOI: 10.3390/pharmaceutics16030332 PMID: 38543226

Park, P. J. (2009). ChIP-seq: Advantages and challenges of a maturing technology. *Nature Reviews. Genetics*, *10*(10), 669–680. DOI: 10.1038/nrg2641 PMID: 19736561

Pei, X. M., Yeung, M. H. Y., Wong, A. N. N., Tsang, H. F., Yu, A. C. S., Yim, A. K. Y., & Wong, S. C. C. (2023). Targeted sequencing approach and its clinical applications for the molecular diagnosis of human diseases. *Cells*, *12*(3), 493. DOI: 10.3390/cells12030493 PMID: 36766834

Petti, M., & Farina, L. (2023). Network medicine for patients' stratification: From single-layer to multi-omics. *WIREs Mechanisms of Disease*, *15*(6), e1623. DOI: 10.1002/wsbm.1623 PMID: 37323106

Qahwaji, R., Ashankyty, I., Sannan, N. S., Hazzazi, M. S., Basabrain, A. A., & Mobashir, M. (2024). Pharmacogenomics: A genetic approach to drug development and therapy. *Pharmaceuticals (Basel, Switzerland), 17*(7), 940. DOI: 10.3390/ph17070940 PMID: 39065790

Qin, D. (2019). Next-generation sequencing and its clinical application. *Cancer Biology & Medicine, 16*(1), 4–10. DOI: 10.20892/j.issn.2095-3941.2018.0055 PMID: 31119042

Saratkar, S. Y., Langote, M., Kumar, P., Gote, P., Weerarathna, I. N., & Mishra, G. V. (2025). Digital twin for personalized medicine development. *Frontiers in Digital Health, 7*, 1583466. DOI: 10.3389/fdgth.2025.1583466 PMID: 40851640

Sarkar, D. I. (2024). The evolution of bioinformatics: From gene sequencing to personalized medicine. *Innovation and Integrative Research Center Journal, 2*(6), 307–337.

Satam, H., (2024). Correction: Next-generation sequencing technology: Current trends and advancements. *Biology (Basel), 13*(5), 286. DOI: 10.3390/biology13050286 PMID: 38785841

Sheynkman, G. M., Shortreed, M. R., Cesnik, A. J., & Smith, L. M. (2016). Proteogenomics: Integrating next-generation sequencing and mass spectrometry to characterize human proteomic variation. *Annual Review of Analytical Chemistry (Palo Alto, Calif.), 9*(1), 521–545. DOI: 10.1146/annurev-anchem-071015-041722 PMID: 27049631

Ståhl, P. L., Salmén, F., Vickovic, S., Lundmark, A., Navarro, J. F., Magnusson, J., Giacomello, S., Asp, M., Westholm, J. O., Huss, M., Mollbrink, A., Linnarsson, S., Codeluppi, S., Borg, Å., Pontén, F., Costea, P. I., Sahlén, P., Mulder, J., Bergmann, O., & Frisén, J. (2016). Visualization and analysis of gene expression in tissue sections by spatial transcriptomics. *Science, 353*(6294), 78–82. DOI: 10.1126/science.aaf2403 PMID: 27365449

Subramanian, I., Verma, S., Kumar, S., Jere, A., & Anamika, K. (2020). Multi-omics data integration, interpretation, and its application. *Bioinformatics and Biology Insights, 14*, 1177932219899051. DOI: 10.1177/1177932219899051 PMID: 32076369

Tavallaee, G., & Orouji, E. (2025). Mapping the 3D genome architecture. *Computational and Structural Biotechnology Journal, 27*, 89–101. DOI: 10.1016/j.csbj.2024.12.018 PMID: 39816913

Vinkšel, M., Writzl, K., Maver, A., & Peterlin, B. (2021). Improving diagnostics of rare genetic diseases with NGS approaches. *Journal of Community Genetics*, *12*(2), 247–256. DOI: 10.1007/s12687-020-00500-5 PMID: 33452619

Vitorino, R. (2024). Transforming clinical research: The power of high-throughput omics integration. *Proteomes*, *12*(3), 25. DOI: 10.3390/proteomes12030025 PMID: 39311198

Vyas, R., & D. M. S. (2024). Harnessing AI and machine learning for advanced biological data analysis. *African Journal of Biomedical Research*, *27*(1S), 1694–1701.

Waarts, M. R., Stonestrom, A. J., Park, Y. C., & Levine, R. L. (2022). Targeting mutations in cancer. *The Journal of Clinical Investigation*, *132*(8), e154943. DOI: 10.1172/JCI154943 PMID: 35426374

Warr, A., Robert, C., Hume, D., Archibald, A., Deeb, N., & Watson, M. (2015). Exome sequencing: Current and future perspectives. *G3 (Bethesda, Md.)*, *5*(8), 1543–1550. DOI: 10.1534/g3.115.018564 PMID: 26139844

Wu, Y., & Xie, L. (2025). AI-driven multi-omics integration for multi-scale predictive modeling of genotype–environment–phenotype relationships. *Computational and Structural Biotechnology Journal*, *27*, 265–277. DOI: 10.1016/j.csbj.2024.12.030

Chapter 3
Spatial Multi-Omics in Drug Discovery

Kazi Asraf Ali
https://orcid.org/0000-0002-1753-7504
Maulana Abul Kalam Azad University of Technology, West Bengal, India

Dipika Mandal
Maulana Abul Kalam Azad University of Technology, West Bengal, India

Sourav Dutta
https://orcid.org/0009-0003-3632-8622
Maulana Abul Kalam Azad University of Technology, West Bengal, India

Syed Nazimul Haque
https://orcid.org/0009-0009-5899-4304
Maulana Abul Kalam Azad University of Technology, West Bengal, India

Basudev Sahana
https://orcid.org/0000-0002-6725-0683
Institute of Pharmacy, Jalpaiguri, India

ABSTRACT

Spatial multi-omics (SMO) is revolutionizing drug discovery by integrating molecular profiling with spatial tissue context to provide deep insights into cellular heterogeneity, molecular interactions, and disease microenvironments. Combining spatial transcriptomics, proteomics, metabolomics, and epigenomics, SMO maintains tissue architecture while mapping genes, proteins, metabolites, and regulatory elements at single-cell or subcellular resolution. This approach enables accurate identification of therapeutic targets, elucidation of drug mechanisms, and discovery of spatial biomarkers predictive of drug response or resistance. Widely applied across

DOI: 10.4018/979-8-3373-4849-0.ch003

oncology, neuroscience, autoimmune, and infectious diseases, SMO reveals tumor heterogeneity, immune landscapes, localized cytokine activity, and host-pathogen dynamics. Enhanced by AI, computational modeling, and network analysis, SMO supports integrated data visualization and tissue-level analysis. Its growing use in companion diagnostics, patient stratification, and translational models bridges preclinical and clinical research.

1. INTRODUCTION

The development of targeted and effective therapeutics depends on a deep understanding of how diseases impact the molecular networks in cells and tissues. Spatial multi-omics is a ground-breaking milestone in biomedical sciences that combines spatial information (such as morphological appearance) and molecular layers (including DNA, RNA, proteins, metabolites, and epigenetic changes) in native tissues. Single-omics methods capture only a fraction of the complexity of biological systems. On the other hand, multi-omics offers an integrated view of how cell works and drug mechanisms through transcriptomics, proteomics, metabolomics, epigenomics, genomics, and so on. This enables personalized treatment decisions, mechanistic explanations, and better identification of pharmacological targets (Chen et al., 2023).

The development of spatial multi-omics, which preserves the spatial layout of biomolecules within tissues while examining their molecular states, is a significant advance in this field. Spatial multi-omics maps transcripts, proteins, metabolites, and epigenetic traits within their original microenvironment. It employs imaging, sequencing, and mass spectrometry techniques. This differs from traditional bulk or single-cell omics, which often lose spatial context when samples are separated. To understand cellular differences, how cells interact with their environment, and localized therapeutic responses in complex tissues like tumors, we need this spatial detail (Liu et al., 2024a).

Researchers can explore how drugs impact molecular pathways at both cellular and subcellular levels using spatial multi-omics in drug discovery. This gives insights into off-target effects, resistance, and effectiveness. Today, tools such as mass spectrometry imaging (MSI), multiplex immunofluorescence, imaging mass cytometry (IMC), and spatial transcriptomics help identify intratumor heterogeneity, delineate tissue structure, and link genetic markers to treatment outcomes. These technologies simplify the discovery of geographical biomarkers for research on mechanisms of action, patient grouping, and predicting drug responses. The improved data interpretation enabled by combining spatial multi-omics with artificial intelligence and computational modeling (Ali, Choudhuri, Bishal, et al., 2025), high-resolution mapping of drug-target interactions and biological processes is now achievable (Xu et al., 2025). Spatial data is valuable in oncology drug development (Ali et al., 2024) for therapy design, target validation, and biomarker discovery. Spatial multi-omics integrates molecular details with location and treatment, a crucial approach for precision pharmacology (Cao et al., 2025).

1.1. Definitions and scope: Single-Cell vs Spatial Multi-Omics

Biological systems are complex assemblages of various cell types forming tissues and organs. Each cell's activity is shaped by its position, interactions with neighbors, and genetics. Conventional methods lose important spatial details, whereas spatial multi-omics preserve these connections (Z. Chen et al., 2024).

Single-cell multi-omics integrates multiple molecular layers from a single cell, giving insights into cellular variations and disease progression. It aids in identifying key cell types for drug discovery, treatment responses, and drug resistance (Xue et al., 2025). Since single-cell multi-omics can produce molecular data from single cells, performing single-cell studies will also involve sectioning the tissue sample to obtain cells. While isolating cells, the researcher loses spatial information that can be valuable for investigating how cells communicate or how tissue is structured. To address the loss of spatial information, spatial omics technologies have been developed (Nassar et al., 2021).

Spatial multi-omics extends the single-cell paradigm by retaining the spatial context of molecular information within intact tissues. It combines high-resolution imaging and sequencing technologies to map the distribution and interactions of genes, proteins, metabolites, and other molecules directly in situ. Spatial multi-omics emerged as an evolution of single-cell analysis to overcome its spatial limitations, allowing researchers to determine not only what molecular features exist but also where they are localized within a tissue architecture (Liang et al., 2024).

1.2. The rationale for spatial resolution in drug discovery

The strength of spatial multi-omics is its ability to identify molecular complexity within natural tissue architecture, providing information not possible with bulk or single-cell omics. Drugs frequently act in highly heterogeneous tissues composed of diverse cell types and microenvironments, where spatial context dictates signaling pathways, cellular behavior, and therapeutic response (Gough et al., 2017). Researchers can see how genes, proteins, and metabolites are arranged and interact within disease-relevant niches by integrating transcriptomic, proteomic, metabolomic, and epigenomic data while maintaining spatial information. Patterns of drug distribution, detection of off-target effects, identification of cell-specific drug targets, and elucidation of the mechanisms of efficacy or resistance are all facilitated by this spatial resolution. Spatial multi-omics also facilitates biomarker discovery, patient stratification, and precision drug design (Ali, Choudhuri, Chakraborty, et al., 2025) by connecting molecular information with functional outcomes through correlations between tissue histopathology and molecular signatures. Hence, the paradigm of drug development and discovery is being revolutionized by spatially resolved multi-omics, which provides a global perspective on drug action in situ (Saliba et al., 2024; Wanichthanarak et al., 2015).

1.3. Historical evolution and emerging technologies

Spatial multi-omics demonstrates the transition from traditional tissue analysis to more sophisticated molecular mapping. Older approaches, such as immunohistochemistry, could provide only limited information, whereas single-cell sequencing lost spatial information. Recent technologies, such as Slide-seq and 10x Genomics Visium, preserve the spatial architecture as they map the gene expression. Soon after, spatial proteomics platforms such as CODEX, CyTOF, and MIBI improved high-plex protein detection (T.-Y. Chen et al., 2023; Williams et al., 2022).

In contrast, imaging-based methods such as MERFISH and seq FISH enabled single-molecule resolution of RNA transcripts. Concurrent advances in spatial metabolomics, such as MALDI-MSI and DESI-MSI (Desorption Electrospray Ionization Mass Spectrometry Imaging), enabled direct visualization of metabolic and drug distribution patterns in tissue sections. Chromatin accessibility and histone modification profiling in tissue microenvironments have become possible more recently through spatial epigenomics technologies such as spatial ATAC-seq and spatial CUT &Tag. By combining these modalities-transcriptomic, proteomic, metabolomic, and epigenomic-under the general heading of spatial multi-omics, a potent framework for drug discovery has been established, providing previously unheard-of insights into cellular heterogeneity, disease mechanisms, drug-target interactions, and therapeutic response in the context of native tissue (M. J. He et al., 2022).

2. SPATIAL MULTI-OMICS TECHNOLOGIES

In spatial multi-omics, several technologies (Figure 1) are used:

Spatial transcriptomics, which maps RNA expression across tissues.
Spatial proteomics profiles protein localization and abundance.
Spatial epigenomics, which reveals chromatin accessibility and regulatory landscapes.

Spatial metabolomics/lipid-omics, which visualizes the distribution of small molecules, metabolites, and drugs.

Figure 1. Classification of spatial multi-omics technologies. This figure provides an overview of the significant classes of spatial multi-omics technologies, grouped by their methodological principles and analytical outputs. Spatial transcriptomics platforms (left panel) are categorized into laser-capture microdissection (LCM-based), in situ hybridization (ISH-based), in situ sequencing (ISS-based), and barcoding-based approaches. These include a wide range of methods such as sm-FISH, MERFISH, seqFISH, HDST, Slide-seq, Stereo-seq, Geo-seq, and Tomo seq, each enabling spatially resolved gene expression profiling with varying resolution, throughput, and molecular coverage

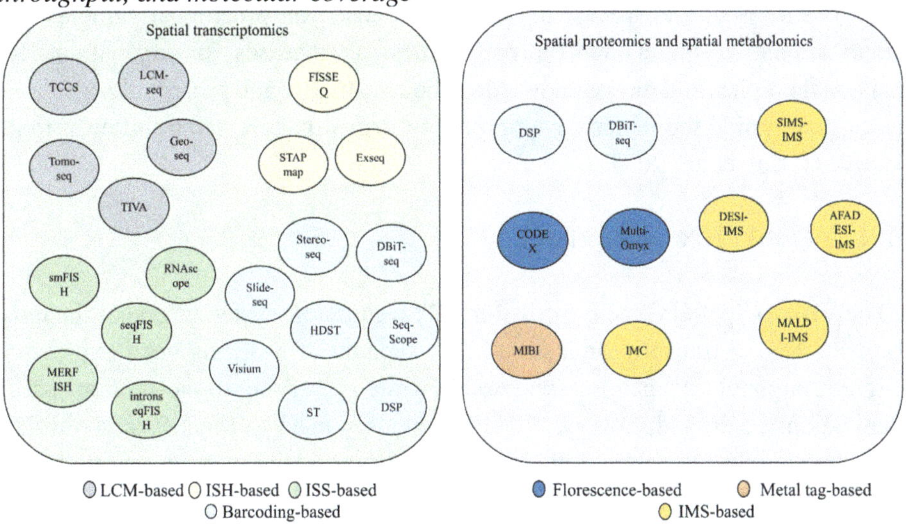

The right panel highlights spatial proteomics and spatial metabolomics technologies that rely on fluorescence-based detection (e.g., CODEX, MIBI), metal-tag-based imaging (e.g., IMC), and mass-spectrometry-based imaging modalities (e.g., MALDI-IMS, DESI-IMS, SIMS-IMS, and AFADESI-IMS). These platforms allow quantitative, spatially resolved mapping of proteins, peptides, lipids, and metabolites within tissue sections.

2.1. Spatial transcriptomics

Recently, a newly developed high-throughput sequencing method called spatial transcriptomics (ST) has become a game-changer for drug discovery, providing genome-wide gene expression profiling of cells within the context of tissues. Through in situ mapping of transcripts, ST identifies where (through what attractors), when (with what kinetics), and how much distinct cell populations interact, leaving marks that single-cell and bulk RNA-seq cannot yield. This spatially resolved information facilitates the discovery and validation of targets by localizing known disease drivers to specific microenvironments, supports biomarker development by uncovering location-specific expression patterns that predict therapeutic response, and enables characterization of drug effects in tissue to identify off-target effects, resistance niches, or immune modulation. Technologies such as Visium (10x Genomics), Slide-seq, and MERFISH are widely used on preclinical samples and clinical samples to understand the pathogenesis of diseases, stratify patients for precision therapies, and merge knowledge about spatiality and proteomic/metabolic regulation to guide the better development of drugs that are safer and even more effective (J. Cao et al., 2024a).

2.1.1. Visium (10x Genomics)

The Visium Spatial Gene Expression (10x Genomics) system is one of the most promising spatial transcriptomics methods, capable of profiling gene expression at high throughput on a genome-wide scale while preserving tissue-section spatial organization. In drug discovery, Visium is widely used to identify and validate therapeutic targets by mapping the location of disease-driving genes and pathways within intact tissue, thereby revealing cellular heterogeneity and microenvironmental niches that can influence drug response or resistance. Researchers can use Visium on preclinical models or biopsy samples to assess drug effects, observe off-target effects, or track responses (Lee et al., 2025a).

2.1.2. Slide-seq

Slide-seq is a high-resolution technique that reads RNA from tissue slices and maps it to DNA barcoded beads on a glass puck. It provides gene expression data at approximately 10 µm resolution and is valuable for drug discovery. This method helps identify drug-resistant subclones and can track treatment effects in patients, aiding in more accurate treatment. (Marshall et al., 2022).

2.1.3. Multiplexed Error-Robust Fluorescence in Situ Hybridization (MERFISH)

At subcellular resolution, hundreds of RNA species in intact tissues can be directly visualized and quantified using the MERFISH imaging-based spatial transcriptomics technique. In drug development, MERFISH provides ultra-high-resolution gene expression mapping to identify rare cell states, intracellular signaling alterations, and cell-cell interactions that are critical to disease progression and treatment effectiveness. By analyzing tissue samples collected pre- and post-treatment, researchers can monitor how drugs influence gene expression at single-cell or even subcellular resolution, reveal off-target effects, and discover spatial biomarkers with predictive value [10.1073/pnas. 1912459116].

2.2. Spatial proteomics

Spatial proteomics is a revolutionary platform in drug discovery, enabling high-dimensional, spatially resolved mapping of protein expression within tissues. While classical proteomics evaluates homogenized samples and sacrifices contextual information, spatial proteomics preserves tissue architecture and the microenvironment, enabling scientists to analyze proteins in situ at single-cell or even subcellular resolution. State-of-the-art platforms like CyTOF (imaging mass cytometry), CODEX (Co-Detection by Indexing), and MIBI (Multiplexed Ion Beam Imaging) allow for the detection of dozens of protein markers from a single tissue section in parallel, producing multiplexed images that can map cell phenotypes, functional states, and their positions. The technology aids in identifying new drug targets by uncovering disease-specific cellular niches, mechanisms of drug action and resistance, and immune infiltration and tumor heterogeneity in cancer research. By bridging molecular profiling and histopathology, spatial proteomics offers a powerful tool to accelerate target validation, optimize therapeutic strategies, and reduce the risk of failure in later stages of drug development (Alshehri et al., 2025).

2.2.1. Cytometry by Time-of-Flight (CyTOF)

CyTOF, or mass cytometry imaging, is a new technology that integrates flow cytometry with time-of-flight mass spectrometry to measure dozens of protein markers simultaneously at single-cell resolution. CyTOF can facilitate high-throughput measurement of immune and tumor microenvironments during drug discovery and uncover important characteristics of cellular heterogeneity and functional states associated with drug response. CyTOF can capture in situ protein cell-surface expression, determine functional states, and measure and describe signaling pathways in a multiplexed format, supporting the discovery of novel drug targets, the uncovering of mechanisms of action and resistance, and the assessment of drug off-target effects (Subrahmanyam & Maecker, 2017).

2.2.2. Co-Detection by Indexing (CODEX)

CODEX is an imaging technology that uses DNA-barcoded antibodies to detect over 50 protein markers in a single tissue section. It helps to understand disease processes and therapy effects by showing tissue structure, cell communities, and how immune cells interact and change during treatment. (Black et al., 2021). Additionally, by improving our understanding of how different biological pathways may mediate or contribute to treatment, we can better select patients for clinical trials, develop combination therapies, and identify predictive biomarkers.

2.2.3. Multiplexed Ion Beam Imaging (MIBI)

MIBI uses secondary-ion mass spectrometry to image tissues labeled with metal-labeled antibodies. It can also detect 30-40 protein markers with subcellular resolution simultaneously. Since MIBI imaging offers exquisite spatial resolution, it is particularly well-suited for examining complex tissue microenvironments, such as malignancies, injured or inflammatory organs, or fibrotic tissue, in drug discovery (Keren et al., 2019). Retrospective studies of archival clinical samples benefit greatly from MIBI's high spatial fidelity and multiplexing capabilities, which enable the validation of potential biomarkers and enhance drug development (Zhang et al., 2025).

2.3. Spatial metabolomics

In the setting of underlying histology, spatial metabolomics uses high-resolution mass spectrometry imaging (MSI) to directly detect the distribution of metabolites, lipids, and small molecules on tissue slices. This technique is applied in drug development to map endogenous metabolites and exogenously supplied substances (such as potential medications and their metabolites) in tissues, malignancies, or microenvironments. It provides options for mechanistic research, toxicity, and resistance by enabling the monitoring of local drug penetration, off-target accumulation, and alterations in metabolic pathways at the tissue or cellular level (Du et al., 2024a).

Among MSI techniques, MALDI-MSI (Matrix-Assisted Laser Desorption/Ionization Mass Spectrometry Imaging) offers extensive analyte coverage and very high spatial resolution (usually 5-10 μm) by desorbing analytes from a laser-matrix co-crystallized with them. In contrast, DESI-MSI is a matrix-free MS technique performed under ambient conditions that can afford fast profiling analysis of unfixed or frozen samples with modest spatial resolution (35-200 μm), which is more suitable for high-throughput screening of drug penetration and off-target accumulation as well as metabolic pathway variation in nearly natural status (Xie et al., 2024). Together, MALDI-MSI and DESI-MSI (Figure 2) provide complementary ADME/Tox (absorption, distribution, metabolism, excretion, and toxicity) and mode-of-action insights by integrating spatial transcriptomics or proteomics, offering a multi-level view of tissue biology that supercharges target validation, biomarker discovery, and safer drug design (Grzeski et al., 2025).

Figure 2. A schematic overview of the MALDI-MSI analytical platform contributes to drug discovery and development (Alolga et al., 2024).

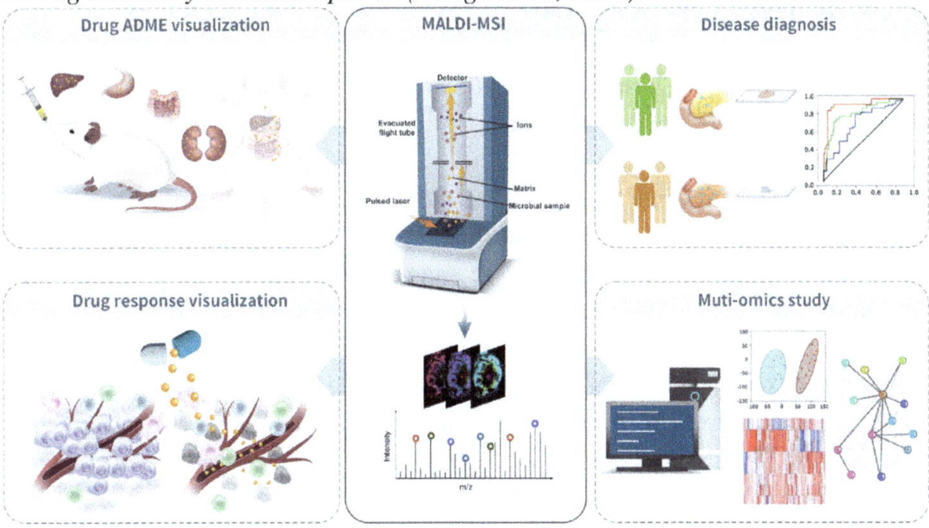

2.4. Spatial epigenomics

Spatial epigenomics is a novel discipline that combines precise tissue mapping with the profiling of epigenetic alterations, such as DNA modifications. It preserves tissue architecture, unlike other approaches that lose context when examining mixed cell populations. Spatial epigenomics reveals how epigenetic states differ across tissues, influence gene expression, and evolve with treatment, facilitating the discovery of drugs and understanding disease. It is also used to classify patients and to identify novel therapeutic targets. (Dai et al., 2024).

2.4.1. Assay for Transposase-Accessible Chromatin sequencing (Spatial ATAC-seq)

Spatial ATAC-seq combines ATAC-seq with spatial transcriptomics to map chromatin accessibility in tissue slices while preserving cell organization. This technique helps visualize active regulatory elements and aids in drug development by analyzing gene expression in relevant cell types and environments. (J. Cao et al., 2024a). It also identifies personalized therapies and classifies patients using epigenetic profiles (Cheng et al., 2025).

3. DATA INTEGRATION AND COMPUTATIONAL TOOLS

Spatial multi-omics has revolutionized drug discovery by enabling scientists to map the spatial distribution of molecular features across tissue structures at unprecedented levels of detail. (Goodwin et al., 2024). This technology combines traditional omics data (transcriptomics, proteomics, metabolomics) with spatial coordinates, providing a comprehensive understanding of how cellular and molecular interactions contribute to disease mechanisms and therapeutic responses(Liu et al., 2024b).

Multi-omics data integration in space offers a unique system for understanding drug discovery by revealing spatially resolved biomarkers, cellular states of heterogeneity, and therapeutic targets in their natural tissue context. Spatial multi-omics technologies currently include spatial transcriptomics systems (such as Visium and Xenium), spatial proteomics systems (such as Imaging Mass Cytometry), and emerging spatial metabolomics systems (Lee et al., 2025b).

3.1. Challenges in aligning multi-modal spatial data

One of the most significant challenges of spatial multi-omics is combining datasets measured on heterogeneous platforms with different resolutions, feature mixtures, and noise distributions. (R. Zhou et al., 2023a). Examples include single-cell RNA-based spatial techniques such as MERFISH or CosMx, which provide subcellular resolution; capture-based techniques such as 10x Genomics Visium, which measure many cells at once; and mass spectrometry imaging (MSI), which offers lower spatial resolution but enables measurements of metabolites and drugs at the individual(Shu et al., 2024). Different resolutions make cross-modality alignment difficult. Each modality has biases that require specific normalization, and alignment issues are worsened by batch effects and tissue deformation, requiring robust image processing techniques (Y. Zhou et al., 2025).

3.2. Statistical and machine learning models for spatial analysis

The multi-omics data associated with spatial data has motivated the design of specialized statistical and machine learning methods capable of effectively modelling spatial associations while incorporating multiple data modalities.(Sanches et al., 2024). Graph neural networks can be used for analyzing spatial omics, which is used to study intricate interactions within cells. This is improved by the PRAGA model, which uses graph-adaptive aggregation to reveal unseen relationships and minimize noise (Valous et al., 2024). Advanced deep learning models designed explicitly for spatial data have shown superior performance compared to traditional statistical methods(Heydari & Sindi, 2023). The DeepST framework uses convolutional neural networks and autoencoders to achieve high-accuracy spatial-domain detection. Likewise, STAIG combines gene expression, spatial coordinates, and histological images with graph-contrastive learning to maximize spatial domain recognition. (Heydari & Sindi, 2023; C. Xu et al., 2022). Traditional spatial statistics for omics data analysis have decreased. Spatial GE software uses methods such as Moran's I and Geary's C to analyse tumor environments. (Ospina et al., 2022).

3.3. Visualization of spatial relationships and heterogeneity

It is essential to have efficient visualization of spatial multi-omics data to comprehend complex tissue structures and cellular interactions. (Ye et al., 2023). Current visualization methods need to handle data with many dimensions while maintaining connections between locations and enabling engaging exploration. (Martinelli & Rapsomaniki, 2022). The Spatial-Live tool is an important addition to spatial omics visualization and demonstrates a layer-stacking approach to address the shortcomings of classical 2D orthographic modes. The method enables rich, interactive rendering of various data types in a common three-dimensional space, enabling better visualization with richer information.(Moon et al., 2019). ATHENA is a tool that measures and visualizes tumor heterogeneity, offering scores that reflect the complexity and diversity of tumor ecosystems (Martinelli & Rapsomaniki, 2022). Spatial omics data is stored, processed, and visualized using the Squidgy framework. It combines image analysis and omics tools to manage large datasets (Palla et al., 2022).

3.4. Spatial network biology and neighborhood analysis

Spatial network biology studies how cells interact and work together in their physical space. It uses network theory and spatial omics data to identify treatment targets and understand how drugs work (Ren et al., 2024). The Spatial Hierarchical Network framework connects small molecular structures with large biomedical networks for drug repositioning(Recanatini & Cabrelle, 2020). Advanced computational techniques are analyzing cellular neighborhoods and their functions. The SOTIP approach represents microenvironments as graphs, enabling the study of their interactions and roles in disease pathways and drug responses. (Hansson et al., 2025). Spatial omics, combined with network biology, accelerates drug discovery by analyzing protein interactions and pathways to identify therapeutic targets in specific tissues. (MacDonald & Westwick, 2006). The COSMOS algorithm integrates various omics data with graph neural networks, improving results by combining the strengths of different modalities rather than analyzing them separately(Coletti et al., 2024).

4. SPATIAL MULTI-OMICS IN TARGET DISCOVERY AND VALIDATION

Conventional omics methods, despite their power, tend to ignore the specific positions of cells within tissues by averaging signals. The location of a cell implies its functionality and state, as in the brain or a tumor. The tumor microenvironment comprises multiple interacting cells that influence disease and treatment.

The genes are studied using spatial multi-omics methods, such as in situ sequencing and highly multiplexed immunofluorescence, including CODEX and Imaging Mass Cytometry. The chapter will discuss the effects of spatial resolution on target discovery, validation, and the interpretation of disease biology.

4.1. Identifying spatially-defined druggable targets

A fundamental aim of drug discovery is to identify targets with a good therapeutic window and a robust causal link to disease. Spatially multi-omics targets that are differentially expressed and preferentially enriched in pathogenic settings help to refine this process.

4.1.1. Application

- **Niche-Specific Oncogenic Programs:**

Using spatial transcriptomics, one may identify areas within a tumour with distinct expression signatures, such as hypoxic cores, invasive fronts, or immune-excluded zones. One that is clearly expressed, especially in the hypoxic, treatment-resistant centre of a tumour, is a more attractive and mechanistically linked option than one that is uniformly expressed. Spatially resolved studies have, for example, revealed that some metabolic enzymes and surface receptors are elevated at the invading front of cancers, therefore providing fresh targets for anti-metastatic therapies (S. Z. Wu et al., 2021).

- **Stromal-Targeted Therapy:**

Diseases such as cancer and fibrosis are caused by activated stromal cells, including cancer-associated fibroblasts (CAFs). CAFs, meanwhile, are varied. Multiplexed protein imaging may separate CAF subpopulations based on their spatial position and marker expression. CAF subpopulations (e.g., inflammatory vs. myofibroblast CAFs) enable the development of treatments targeting specific pathogenic CAF subsets while sparing others (Bhate et al., 2022).

4.1.2. Validation

Target validation in preclinical models relies on spatial technologies. Spatial omics helps confirm that a candidate therapy acts on its target and adjusts the pathway as intended. Use PDX or modified mouse models.

4.2. Mapping cell-cell interactions and microenvironmental niches

Signals from close neighbours and the broader surroundings control cellular activity. Utilizing spatial multi-omics is one objective means of arranging these interactions and highlighting functional niches.

4.2.1. Application

- **Deconstructing the Tumor Microenvironment (TME)**

Scientists classify cells based on their chemical and spatial characteristics to identify recurring areas, or niches. The positive immunotherapy response is associated with a niche containing B cells and T cells, whereas the cold niche, which contains some cells, indicates resistance. This knowledge helps determine objectives to modify the tumor environment(Schürch et al., 2020).

- **Neurodegenerative and Inflammatory Diseases**

Spatial mapping can reveal whether certain neuron types are lost near active microglia or lymphocytes in diseases such as multiple sclerosis or Alzheimer's, indicating harmful interactions and potential new treatment pathways.

4.3. Spatial ligand-receptor analysis:

Ligand-receptor pairs aid cell contact. Computational methods use single-cell RNA-seq data to predict interactions, but lack spatial confirmation. Spatial omics data provide necessary boundaries for medical relevance (Jackson et al., 2020).

4.3.1. Application

Using tools such as MISTY, Cell Chat, and Giotto, one can uncover statistically significant ligand-receptor interactions between nearby cell types. For example:

> An immediate contact can result in T cell exhaustion when a PD-L1-expressing cancer cell (the ligand) is in proximity to a functional immunological checkpoint. A CD8+ T cell that expresses the receptor (Ali, Choudhuri, Roy, et al., 2025).
>
> The pathway supporting tumor cell survival and retention is revealed by the interaction between a chemokine expressed by fibroblasts (e.g., CXCL12) and its receptor (CXCR4) on nearby cancer cells.

This aids in prioritizing therapeutic strategies, such as combination medicines that block these beneficial communication axes while also targeting intrinsic cancer cell pathways. (Dries et al., 2021).

4.4. Spatial biomarkers in drug resistance and sensitivity

Direct clinical use is most effective for discovering spatial indicators (Figure 3). Often, determining how well a treatment works is the spatial organization of cells rather than just their presence or absence.

Figure 3. Spatial multi-omics in translational pipeline

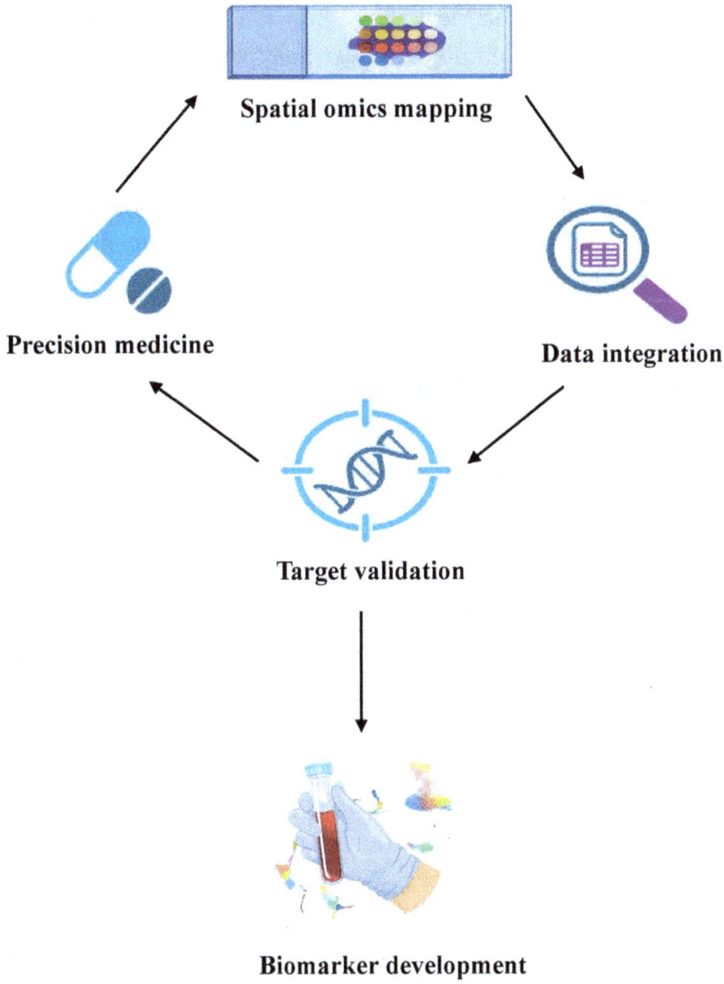

4.4.1. Application

- **Predicting Immunotherapy Response:** The location of CD8+ T cells in tumors affects how patients respond to immune checkpoint inhibitors. Those with T cells within the tumor respond better than those with them outside the tumor. Spatial biomarkers, like distance-to-nearest-neighbor, are effective (Barua et al., 2018).

- **Understanding Mechanisms of Resistance:** Spatial proteomics of biopsies taken before and after treatment shows how the tumor environment changes with drug stress. This can influence next-line treatment choices.

4.4.2. Validation in Clinical Trials

Spatial proteomics of biopsy samples shows changes in the tumor microenvironment before and after treatment. This may affect treatment choices due to therapy-resistant niches.

5. MECHANISM OF ACTION AND DISEASE MODELLING

5.1. Understanding drug response within the spatial tissue context

Spatial multi-omics technologies improve our understanding of drug actions by revealing how drugs interact with complex tissue structures, thereby filling gaps in knowledge about their effects (J. Cao et al., 2024b). The arrangement of cells in tissues affects how drugs work. Researchers can now create networks showing how different cell groups respond to treatments within their environment(Z. Wu et al., 2025). The study of varied drug responses, in which the same treatment can have quite different effects, has significantly benefited from this method. Based on the position of cells and their direct interactions with nearby cells(Liu et al., 2024b). Cell arrangement in tissues influences drug effects. Researchers use spatial transcriptomic data to study cell responses to treatments, illustrating the effects of spatial context on therapy(Du et al., 2024b). Spatial multi-omics in oncology shows how tumor environments influence drug response and cell diversity(Moon et al., 2019).

5.2. Reconstructing disease progression maps

Disease progression is complex and dynamic. Spatial multi-omics helps researchers map disease evolution across tissues to inform better treatment plans (Jain & Eadon, 2024). Spatial transcriptomics in Alzheimer's studies shows gene expression changes linked to amyloid plaques. Mouse research reveals early genetic disturbances. Additionally, profiling in amyotrophic lateral sclerosis has revealed molecular processes affecting microglial and astrocyte populations at different disease stages(Bayaraa et al., 2025). Spatial transcriptomics reveals distinct areas in tumours that affect disease progression. Research has found reduced levels of specific markers, such as ANXA1, in cells from patients with oesophageal squamous cell carcinoma. This technology also maps tumour-normal edges and immune-related structures associated with cancer progression(R. Zhou et al., 2023b). Spatial multi-omics has studied host-pathogen interactions in infections such as SARS-CoV-2, revealing spatial expression patterns and the important roles of specific cell types in lung disease(Kiessling & Kuppe, 2024).

5.3. Spatially informed in vitro and in vivo models

The use of spatial biology and advanced in vitro models has created organ-on-chip systems that mimic human body functions(Ahmed, 2022). Combined gut-liver models are important for studying drug metabolism and liver injury, which traditional methods do not effectively address (Nair & Weiskirchen, 2024). Modern models keep 3D tissue structure, allowing researchers to study drug effects in realistic environments. AI and machine learning improve predictive models for drug discovery by optimizing complex parameters(Dave et al., 2025). Disease models that are spatially aware in vitro and in vivo (or in silico) occupy the gap between intact tissue understanding and experimental systems and enhance the relevance of translationally and mechanistically faithful models.

5.4. Application in functional genomics screening

Functional genomics screening now includes spatial information, helping researchers identify genes linked to cell traits(Heynen-Genel et al., 2012). High-content screening examines multiple factors at the single-cell level. Recent advances enable CRISPR screening and spatial transcriptomics to track gene expression and changes in context, showing how genetic modifications influence cell behaviour in specific environments(Binan et al., 2025). Spatial functional genomics screening helps identify small changes in cell behaviour that traditional methods miss(Z. He et al., 2025). It reveals new treatment targets and is especially useful for identifying disease-specific targets. Spatial multi-omics in drug discovery is advancing but faces challenges in data analysis and integration(Jiang et al., 2025).

6. CASE STUDIES IN DRUG DISCOVERY

Spatial multi-omics combines molecular profiling with spatial information to study tissue organization. It reveals disease mechanisms and therapeutic targets that traditional methods may overlook.

6.1. Oncology: tumor heterogeneity and immune landscape

Tumor heterogeneity leads to treatment resistance and different patient results. Spatial multi-omics helps researchers study this complexity at the single-cell level. (Le et al., 2025).

Spatial transcriptomics and proteomics help map cancer cells and immune cells in tumors. These studies show different tumor areas with varying gene expressions. Regions rich in cytotoxic T cells may react better to immunotherapy, while others with suppressive cells may resist treatment. (Gong et al., 2024).

Researchers can find rare tumor cell subpopulations in different microenvironments, helping develop targeted therapies and personalized cancer treatments.

6.2. Neuroscience: spatial omics in brain disorders and neuroinflammation

The brain's structural and functional complexity poses significant challenges for understanding neurological disorders. Spatial multi-omics enables mapping molecular and cellular changes within specific brain regions, linking localized pathology to functional outcomes (Neumann, 2025).

Spatial transcriptomics and proteomics help study how disease proteins accumulate in brain areas affected by Alzheimer's and Parkinson's. They reveal patterns of inflammation, suggesting new targets for treatment (Li et al., 2019).

Spatial omics helps profile immune cell infiltration, cytokine levels, and neuronal vulnerability in neuroinflammatory conditions. This data can aid in identifying disease progression and drug discovery (Xia et al., 2025).

6.3. Autoimmune diseases: mapping spatial cytokine responses

Dysregulated immune responses that differ across tissues and disease stages drive autoimmune disorders such as rheumatoid arthritis, lupus, and inflammatory bowel disease. Bulk measurements can sometimes mask spatially restricted effects, as can cytokine concentrations. (Song et al., 2025).

Spatial multi-omics shows detailed cytokine distributions and immune cell interactions in tissues. In rheumatoid arthritis, it reveals varied cytokine levels associated with distinct immune cell groups. (Miyamoto et al., 2025).

6.4. Infectious diseases: tissue-level host-pathogen interaction analysis

Infectious diseases result from complex interactions between pathogens and host tissues. Spatial multi-omics helps study these interactions for better treatments. (Gioacchino et al., 2024).

Spatial transcriptomics helps map responses during viral infections, while spatial proteomics identifies areas of bacterial colonization and inflammation during bacterial infections (Kostas et al., 2025).

7. TRANSLATIONAL POTENTIAL AND INDUSTRY APPLICATIONS

The conventional model of drug development is ineffective. The failure to identify the appropriate patient population most likely to respond to a therapy is a significant factor in the approximately 90% failure rate of clinical candidates (Hwang et al., 2016). Omics at the single-cell and bulk levels offer insights but lack the context to comprehend cellular ecosystems.

Spatial multi-omics fills this crucial void. Through mapping the complex relationships between cells in situ, SMO provides answers to "where" it is and "who" it is speaking to, going beyond the question of "what" is present (Hunter et al., 2021). This chapter describes integrating this capability into industry workflows to meet regulatory requirements for comprehensive biomarker data, de-risk development, and personalized treatment (Tan et al., 2020).

7.1. Role in biomarker discovery for clinical trials

Biomarkers are the compass of modern clinical research. Spatial biomarkers are a new class of biomarkers that are naturally helpful and predictive thanks to SMO.

Application phases:

Preclinical Target Validation: SMO analysis of patient-derived xenograft (PDX) models or treated preclinical samples confirms that a drug reaches its target in situ and has the expected effect on the tumour microenvironment (TME). Before human testing, this provides mechanistic assurance (Wang et al., 2022).

Retrospective Analysis of Biobanks: SMO platforms compatible with FFPE tissue enable the analysis of large historical biobanks with related clinical outcome data (e.g., responders vs. non-responders). This is a practical and cost-effective method of identifying possible geographic biomarkers associated with response prior to initiating a prospective trial (Carstens et al., 2017).

Prospective Biomarker Identification: SMO can be used on base and on-treatment biopsies in early-phase trials to:

- **Identify Mechanisms of Action (MoA):** Confirm that the drug is working as intended and altering the route (Nirmal et al., 2022).
- **Uncover Resistance Mechanisms:** Identify geographic changes in the TME that serve as early indicators of resistance development, such as the overexpression of alternative immunological checkpoints in adjacent cells (Bassez et al., 2021).
- **Discover Predictive Signatures:** Seek out complex, multicellular markers of a response, such as the proximity of a specific subset of immune cells to tumour cells (Pagès et al., 2018).

7.2. Companion diagnostics and patient stratification

For a related treatment product to be used safely and effectively, a medical device called a companion diagnostic (CDx) is required (Jørgensen, 2021). SMO enables a new generation of CDx that transcends individual biomarkers to complex spatial signatures.

7.2.1. Defining Spatial Signatures for Stratification

SMO can identify signatures that are potent predictors of response:

Proximity Signatures: The distance between CD8+ T cells and cancer cells predicts immunotherapy response more accurately than either cell type alone (Pagès et al., 2018).

Interface Analysis: Certain receptor-ligand interactions at the tumor-stroma interface can identify active pathways that support growth or resistance (Bassez et al., 2021).

Niche Identification: The presence of specific functional niches, like tertiary lymphoid structures, can be used to classify patients with a favourable immunological environment (Helmink et al., 2020).

7.2.2. Development of SMO-Based CDx

When switching from a discovery platform (like 10x Visium or CODEX) to a clinically authorised CDx and beyond, repeatability, scalability, and cost must be given top priority (Lewis et al., 2021; Tan et al., 2020). This usually means:

Signature Translation: Utilising a multiplexed immunofluorescence (mIF) panel with five to ten markers that can be used on standard clinical immunohistochemistry (IHC) platforms to create a manageable assay from a discovery-based SMO signature (Parra et al., 2017).
Algorithm Development: Creating automated digital pathology algorithms (such as cell segmentation, phenotyping, and distance computation) to reliably and impartially quantify the complex spatial signature (Bera et al., 2019).
Clinical Validation: The test and algorithm must be locked down and validated in a large, independent clinical cohort in order to firmly establish a relationship between the spatial signature and patient outcome (Jørgensen, 2021; Pagès et al., 2018).

7.3. Regulatory considerations and clinical implementation

Even after a biomarker or companion diagnosis has been developed and clinically validated, securing regulatory approval and implementing it can be challenging (Table 1).

7.3.1. Regulatory Frameworks

FDA (USA): Regulates companion diagnostics (CDx) as medical devices; they must be cleared or approved under PMA (premarket approval) or 510(k), depending on their level of risk. The FDA's In Vitro Diagnostic (IVD) Guidance lays out requirements (Cooper & Chen, 2022; Kang et al., 2023).
EMA (Europe): Through the *In Vitro* Diagnostic Regulation (IVDR), which went into effect in May 2022 and calls for clinical evidence, performance evaluation, and notified bodies' involvement in order to categorize IVDs based on risk (Spitzenberger et al., 2022).
Other regions: There are ongoing efforts to harmonize the regulatory bodies across countries, including China, Japan, and India (such as the IMDRF).

Table 1. Summary of key regulatory requirements

Requirements	Key Elements
Analytical Performance	Sensitivity, specificity, precision, reproducibility, limit of detection/ quantification
Clinical Performance (Validation)	Retrospective and prospective studies; positive and negative predictive value; association with clinical endpoints
Risk Assessment and Classification	Risk to patient if false positive/negative; risk classification under local regulation (e.g., Class II, III IVD)
Manufacturing Quality	Good Manufacturing Practices / Quality Management Systems (ISO 13485, etc.)
Labeling & Indications for Use	The clinical context, population, and limitations
Post-market Surveillance	Monitoring performance in real-world use, reporting adverse results, and ensuring lot-to-lot consistency

7.3.2 Implementation in Clinical Settings

Laboratory accreditation: Labs that conduct tests must be certified, such as the CLIA in the USA.

Reimbursement and health economics: Payers demand evidence of cost-effectiveness, and biomarker tests must be covered.

Clinical guidelines and physician acceptance: Inclusion in treatment guidelines (like NCCN and ESMO) facilitates uptake. Clinicians must receive education.

Patient access and equity: Offering tests in a range of settings, such as low- and middle-income countries.

Quality control & proficiency testing: Preserving assay uniformity and inter-laboratory comparability.

7.4. Future perspectives:

Spatial multi-omics (SMO) is rapidly evolving from an innovative research technology into a foundational engine for drug discovery and translational pharmacology (Lewis et al., 2021). By mapping the functional geography of tissues at single-cell or subcellular resolution, SMO enables the discovery of spatially constrained disease mechanisms that were previously undetectable using traditional bulk or single-cell methods. These spatial insights provide a unique opportunity to uncover why patients differ in drug response, identify microenvironment-specific therapeutic vulnerabilities, and reveal spatial biomarkers that predict treatment outcomes (Bassez et al., 2021; Hunter et al., 2021). Pharmaceutical and biotech industries are increasingly adopting SMO to de-risk clinical development, stratify patients with unprecedented precision, and design more effective companion diagnostics. The integration of spatial transcriptomics, proteomics, and metabolomics into drug development pipelines is accelerating the discovery of new targets, clarifying mechanisms of action, and enabling the real-time evaluation of therapeutic effects in complex tissues (Hwang et al., 2016; Lewis et al., 2021).

Looking ahead, SMO is expected to become an even more powerful driver of drug discovery. Platform-agnostic standardization of spatial data, harmonized regulatory frameworks, and seamless interoperability across analytical tools will be essential to enable cross-study comparison and clinical translation. In parallel, AI-driven analysis will automate the extraction of spatial signatures, turning high-dimensional geographic biomarkers into actionable insights that can be incorporated directly into clinical decision-making (Tan et al., 2020). Ultimately, the future of SMO lies in embedding the geographic context of disease biology as a standard and indispensable dimension in drug discovery, therapeutic optimization, and patient care—transforming spatial biology from a cutting-edge innovation into a core pillar of precision medicine.

8. CONCLUSION

A paradigm shift in drug discovery, spatial multi-omics (SMO) combines molecular profiling with spatial context, offering higher-resolution views of cellular heterogeneity, microenvironmental interactions, and tissue-specific molecular landscapes than ever before. In contrast to traditional bulk or single-cell omics, SMO does not lose positional information about biomolecules, enabling accurate mapping of gene expression, protein localization, metabolite localization, and epigenetic changes in intact tissues. This spatial understanding enables niche-related drug targets, understanding of cell-cell interactions, and the identification of predictive spatial biomarkers, which ultimately improve target validation, therapeutic development, and patient stratification.

SMO has been used in oncology, neuroscience, autoimmune, and infectious disease studies to demonstrate key spatial patterns in disease development, drug action, and drug resistance. With multi-modal data and state-of-the-art computational technologies, such as machine learning and network biology, scientists can reconstruct disease cross-over maps, understand pharmacologic mechanisms in vivo, and develop individualized interventions. Moreover, SMO enhances the development of companion diagnostics, regulatory compliance, and, hence, the connection of preclinical evidence to clinical practice.

Spatial multiomics will play an important role in precision medicine as technological advancements advance. It will help convert intricate tissue architecture into viable therapies, thereby improving drug discovery and targeted therapies.

REFERENCES

Ahmed, T. (2022). Organ-on-a-chip microengineering for bio-mimicking disease models and revolutionizing drug discovery. *Biosensors & Bioelectronics: X, 11*, 100194. DOI: 10.1016/j.biosx.2022.100194

Ali, K. A., Choudhuri, S., Bishal, A., Das, T., Nandy, S., & Hossain, C. M. (2025). Approaches of AI and Advanced Computational Biology for Combating HIV Infections. In J.-T. Chen, *AI and Precision Medicine in Infectious Disease Management* (1st ed., pp. 149–166). CRC Press. DOI: 10.1201/9781003615699-8

Ali, K. A., Choudhuri, S., Chakraborty, R., Pradhan, A., Kotal, A., & Shaharyar, M. A. (2025). AI Technology-Powered Precision Medicine and Smart Healthcare. In J.-T. Chen, *AI and Precision Medicine in Infectious Disease Management* (1st ed., pp. 51–70). CRC Press. DOI: 10.1201/9781003615699-3

Ali, K. A., Choudhuri, S., Roy, S. K., Reza, K. H., Mandal, D., Chakraborty, R., & Hossain, C. M. (2025). AI-Generated Nanoparticles in Immunological Research: Opportunities and Challenges. In Chen, J.-T. (Ed.), (pp. 295–322). Advances in Computational Intelligence and Robotics. IGI Global., DOI: 10.4018/979-8-3693-9725-1.ch011

Ali, K. A., Mohin, S., Mondal, P., Goswami, S., Ghosh, S., & Choudhuri, S. (2024). Influence of artificial intelligence in modern pharmaceutical formulation and drug development. *Future Journal of Pharmaceutical Sciences, 10*(1), 53. DOI: 10.1186/s43094-024-00625-1

Alolga, R. N., Wang, S.-L., Qi, L.-W., Zang, H., & Huang, F.-Q. (2024). MALDI mass spectrometry imaging in targeted drug discovery and development: The pros, the cons, and prospects in global omics techniques. *Trends in Analytical Chemistry, 178*, 117860. DOI: 10.1016/j.trac.2024.117860

Alshehri, S., Vitorino, R., Saleh, O., Al-Harthi, S., Alahmadi, A., Alotibi, R., Da Silva Rosa, S. C., Osama, A., Magedeldin, S., Alhattab, D., Emwas, A.-H., & Jaremko, M. (2025). Advances and applications of clinical proteomics in precision medicine. *Expert Review of Proteomics, 22*(10), 1–20. DOI: 10.1080/14789450.2025.2560919 PMID: 40963397

Barua, S., Fang, P., Sharma, A., Fujimoto, J., Wistuba, I., Rao, A. U. K., & Lin, S. H. (2018). Spatial interaction of tumor cells and regulatory T cells correlates with survival in non-small cell lung cancer. *Lung Cancer (Amsterdam, Netherlands), 117*, 73–79. DOI: 10.1016/j.lungcan.2018.01.022 PMID: 29409671

Bassez, A., Vos, H., Van Dyck, L., Floris, G., Arijs, I., Desmedt, C., Boeckx, B., Vanden Bempt, M., Nevelsteen, I., Lambein, K., Punie, K., Neven, P., Garg, A. D., Wildiers, H., Qian, J., Smeets, A., & Lambrechts, D. (2021). A single-cell map of intratumoral changes during anti-PD1 treatment of patients with breast cancer. *Nature Medicine*, *27*(5), 820–832. DOI: 10.1038/s41591-021-01323-8 PMID: 33958794

Bayaraa, O., Aksu, M., DeBose-Scarlett, E., Hocke, E., Jain, V., Wang, S.-H. J., Cruz, D., & Gregory, S. (2025). Multi-region spatial transcriptomics reveals region-specific differences in response to amyloid beta (Aβ) plaque-induced changes in Alzheimer's Disease (AD). *Genomics*. Advance online publication. DOI: 10.1186/s40246-025-00875-x PMID: 40766691

Bera, K., Schalper, K. A., Rimm, D. L., Velcheti, V., & Madabhushi, A. (2019). Artificial intelligence in digital pathology—New tools for diagnosis and precision oncology. *Nature Reviews. Clinical Oncology*, *16*(11), 703–715. DOI: 10.1038/s41571-019-0252-y PMID: 31399699

Bhate, S. S., Barlow, G. L., Schürch, C. M., & Nolan, G. P. (2022). Tissue schematics map the specialization of immune tissue motifs and their appropriation by tumors. *Cell Systems*, *13*(2), 109–130.e6. DOI: 10.1016/j.cels.2021.09.012 PMID: 34653369

Binan, L., Jiang, A., Danquah, S. A., Valakh, V., Simonton, B., Bezney, J., Manguso, R. T., Yates, K. B., Nehme, R., Cleary, B., & Farhi, S. L. (2025). Simultaneous CRISPR screening and spatial transcriptomics reveal intracellular, intercellular, and functional transcriptional circuits. *Cell*, *188*(8), 2141–2158.e18. DOI: 10.1016/j.cell.2025.02.012 PMID: 40081369

Black, S., Phillips, D., Hickey, J. W., Kennedy-Darling, J., Venkataraaman, V. G., Samusik, N., Goltsev, Y., Schürch, C. M., & Nolan, G. P. (2021). CODEX multiplexed tissue imaging with DNA-conjugated antibodies. *Nature Protocols*, *16*(8), 3802–3835. DOI: 10.1038/s41596-021-00556-8 PMID: 34215862

Cao, J., Li, C., Cui, Z., Deng, S., Lei, T., Liu, W., Yang, H., & Chen, P. (2024a). Spatial Transcriptomics: A Powerful Tool in Disease Understanding and Drug Discovery. *Theranostics*, *14*(7), 2946–2968. DOI: 10.7150/thno.95908 PMID: 38773973

Cao, J., Li, C., Cui, Z., Deng, S., Lei, T., Liu, W., Yang, H., & Chen, P. (2024b). Spatial Transcriptomics: A Powerful Tool in Disease Understanding and Drug Discovery. *Theranostics*, *14*(7), 2946–2968. DOI: 10.7150/thno.95908 PMID: 38773973

Cao, S., Liu, J., & Li, Y. (2025). Harnessing multi-omics and machine learning for predicting immune checkpoint blockade responses: Advances, challenges, and future directions. *Fundamental Research (Beijing)*, *S2667325825004613*. Advance online publication. DOI: 10.1016/j.fmre.2025.08.009

Carstens, J. L., Correa De Sampaio, P., Yang, D., Barua, S., Wang, H., Rao, A., Allison, J. P., LeBleu, V. S., & Kalluri, R. (2017). Spatial computation of intratumoral T cells correlates with patient survival in pancreatic cancer. *Nature Communications*, *8*(1), 15095. DOI: 10.1038/ncomms15095 PMID: 28447602

Chen, C., Wang, J., Pan, D., Wang, X., Xu, Y., Yan, J., Wang, L., Yang, X., Yang, M., & Liu, G. (2023). Applications of multi-omics analysis in human diseases. *MedComm*, *4*(4), e315. DOI: 10.1002/mco2.315 PMID: 37533767

Chen, T.-Y., You, L., Hardillo, J. A. U., & Chien, M.-P. (2023). Spatial Transcriptomic Technologies. *Cells*, *12*(16), 2042. DOI: 10.3390/cells12162042 PMID: 37626852

Chen, Z., Du, C., Liu, S., Liu, J., Yang, Y., Dong, L., Zhao, W., Huang, W., & Lei, Y. (2024). Progress in biomaterials inspired by the extracellular matrix. *Giant (Oxford, England)*, *19*, 100323. DOI: 10.1016/j.giant.2024.100323

Cheng, X., Peng, T., Chu, T., Yang, Y., Liu, J., Gao, Q., Cao, C., & Wei, J. (2025). Application of single-cell and spatial omics in deciphering cellular hallmarks of cancer drug response and resistance. *Journal of Hematology & Oncology*, *18*(1), 70. DOI: 10.1186/s13045-025-01722-1 PMID: 40605007

Coletti, R., Carrilho, J., Martins, E. P., Gonçalves, C. S., Costa, B. M., & Lopes, M. B. (2024). A Novel Tool for Multi-Omics Network Integration and Visualization: A Study of Glioma Heterogeneity. *Bioinformatics (Oxford, England)*. Advance online publication. DOI: 10.1101/2024.03.18.585500

Cooper, L., & Chen, J. (2022). Changes in Companion Diagnostic Labelling: Implementation of FDA's April 2020 Guidance for Industry for In Vitro CDx Labeling for Specific Oncology Therapeutic Groups. *Therapeutic Innovation & Regulatory Science*, *56*(5), 689–697. DOI: 10.1007/s43441-022-00422-z PMID: 35689144

Dai, W., Qiao, X., Fang, Y., Guo, R., Bai, P., Liu, S., Li, T., Jiang, Y., Wei, S., Na, Z., Xiao, X., & Li, D. (2024). Epigenetics-targeted drugs: Current paradigms and future challenges. *Signal Transduction and Targeted Therapy*, *9*(1), 332. DOI: 10.1038/s41392-024-02039-0 PMID: 39592582

Dave, R., Pandey, K., Patel, R., Gour, N., & Bhatia, D. (2025). Leveraging 3D cell culture and AI technologies for next-generation drug discovery. *Cell Biomaterials*, *1*(3), 100050. DOI: 10.1016/j.celbio.2025.100050

Dries, R., Chen, J., Del Rossi, N., Khan, M. M., Sistig, A., & Yuan, G.-C. (2021). Advances in spatial transcriptomic data analysis. *Genome Research*, *31*(10), 1706–1718. DOI: 10.1101/gr.275224.121 PMID: 34599004

Du, P., Fan, R., Zhang, N., Wu, C., & Zhang, Y. (2024a). Advances in Integrated Multi-omics Analysis for Drug-Target Identification. *Biomolecules*, *14*(6), 692. DOI: 10.3390/biom14060692 PMID: 38927095

Du, P., Fan, R., Zhang, N., Wu, C., & Zhang, Y. (2024b). Advances in Integrated Multi-omics Analysis for Drug-Target Identification. *Biomolecules*, *14*(6), 692. DOI: 10.3390/biom14060692 PMID: 38927095

Gioacchino, E., Vandelannoote, K., Ruberto, A. A., Popovici, J., & Cantaert, T. (2024). Unraveling the intricacies of host-pathogen interaction through single-cell genomics. *Microbes and Infection*, *26*(7), 105313. DOI: 10.1016/j.micinf.2024.105313 PMID: 38369008

Gong, D., Arbesfeld-Qiu, J. M., Perrault, E., Bae, J. W., & Hwang, W. L. (2024). Spatial oncology: Translating contextual biology to the clinic. *Cancer Cell*, *42*(10), 1653–1675. DOI: 10.1016/j.ccell.2024.09.001 PMID: 39366372

Goodwin, R. J. A., Platz, S. J., Reis-Filho, J. S., & Barry, S. T. (2024). Accelerating Drug Development Using Spatial Multi-omics. *Cancer Discovery*, *14*(4), 620–624. DOI: 10.1158/2159-8290.CD-24-0101 PMID: 38571424

Gough, A., Stern, A. M., Maier, J., Lezon, T., Shun, T.-Y., Chennubhotla, C., Schurdak, M. E., Haney, S. A., & Taylor, D. L. (2017). Biologically Relevant Heterogeneity: Metrics and Practical Insights. *SLAS Discovery*, *22*(3), 213–237. DOI: 10.1177/2472555216682725 PMID: 28231035

Grzeski, M., Jensen, P. M., Hempel, B.-F., Thiele, H., Lellmann, J., Schallenberg, S., Budach, V., Keilholz, U., Tinhofer, I., & Klein, O. (2025). Integrating MALDI-MSI-Based Spatial Proteomics and Machine Learning to Predict Chemoradiotherapy Outcomes in Head and Neck Cancer. *International Journal of Molecular Sciences*, *26*(18), 9084. DOI: 10.3390/ijms26189084 PMID: 41009656

Hansson, F. G., Madsen, N. G., Hansen, L. G., Jakočiūnas, T., Lengger, B., Keasling, J. D., Jensen, M. K., Acevedo-Rocha, C. G., & Jensen, E. D. (2025). Labels as a feature: Network homophily for systematically annotating human GPCR drug-target interactions. *Nature Communications*, *16*(1), 4121. DOI: 10.1038/s41467-025-59418-6 PMID: 40316519

He, M. J., Pu, W., Wang, X., Zhang, W., Tang, D., & Dai, Y. (2022). Comparing DESI-MSI and MALDI-MSI Mediated Spatial Metabolomics and Their Applications in Cancer Studies. *Frontiers in Oncology*, *12*, 891018. DOI: 10.3389/fonc.2022.891018 PMID: 35924152

He, Z., Liu, L., Han, D., Gao, K., Dong, L., Bu, D., Huo, P., Wang, Z., Deng, W., Liu, J., Guo, J., Zhao, Y., & Wu, Y. (2025). *POINT: A web-based platform for pharmacological investigation enhanced by multi-omics networks and knowledge graphs* (Version 1). arXiv. https://doi.org/DOI: 10.48550/ARXIV.2503.07203

Helmink, B. A., Reddy, S. M., Gao, J., Zhang, S., Basar, R., Thakur, R., Yizhak, K., Sade-Feldman, M., Blando, J., Han, G., Gopalakrishnan, V., Xi, Y., Zhao, H., Amaria, R. N., Tawbi, H. A., Cogdill, A. P., Liu, W., LeBleu, V. S., Kugeratski, F. G., & Wargo, J. A. (2020). B cells and tertiary lymphoid structures promote immunotherapy response. *Nature*, *577*(7791), 549–555. DOI: 10.1038/s41586-019-1922-8 PMID: 31942075

Heydari, A. A., & Sindi, S. S. (2023). Deep learning in spatial transcriptomics: Learning from the next next-generation sequencing. *Biophysics Reviews*, *4*(1), 011306. DOI: 10.1063/5.0091135 PMID: 38505815

Heynen-Genel, S., Pache, L., Chanda, S. K., & Rosen, J. (2012). Functional genomic and high-content screening for target discovery and deconvolution. *Expert Opinion on Drug Discovery*, *7*(10), 955–968. DOI: 10.1517/17460441.2012.711311 PMID: 22860749

Hunter, M. V., Moncada, R., Weiss, J. M., Yanai, I., & White, R. M. (2021). Spatially resolved transcriptomics reveals the architecture of the tumor-microenvironment interface. *Nature Communications*, *12*(1), 6278. DOI: 10.1038/s41467-021-26614-z PMID: 34725363

Hwang, T. J., Carpenter, D., Lauffenburger, J. C., Wang, B., Franklin, J. M., & Kesselheim, A. S. (2016). Failure of Investigational Drugs in Late-Stage Clinical Development and Publication of Trial Results. *JAMA Internal Medicine*, *176*(12), 1826. DOI: 10.1001/jamainternmed.2016.6008 PMID: 27723879

Jackson, H. W., Fischer, J. R., Zanotelli, V. R. T., Ali, H. R., Mechera, R., Soysal, S. D., Moch, H., Muenst, S., Varga, Z., Weber, W. P., & Bodenmiller, B. (2020). The single-cell pathology landscape of breast cancer. *Nature*, *578*(7796), 615–620. DOI: 10.1038/s41586-019-1876-x PMID: 31959985

Jain, S., & Eadon, M. T. (2024). Spatial transcriptomics in health and disease. *Nature Reviews. Nephrology*, *20*(10), 659–671. DOI: 10.1038/s41581-024-00841-1 PMID: 38719971

Jiang, W., Ye, W., Tan, X., & Bao, Y.-J. (2025). Network-based multi-omics integrative analysis methods in drug discovery: A systematic review. *BioData Mining*, *18*(1), 27. DOI: 10.1186/s13040-025-00442-z PMID: 40155979

Jørgensen, J. T. (2021). Companion and complementary diagnostics: A key to precision medicine. In *Precision Medicine and Artificial Intelligence* (pp. 97–107). Elsevier., DOI: 10.1016/B978-0-12-820239-5.00003-6

Kang, S. L., Woo, J. H., Kim, N. H., Kwon, J. Y., & Kim, S. M. (2023). Necessity of strengthening the current clinical regulatory framework for companion diagnostics: An institutional comparison of the FDA, EMA, and MFDS. *Molecular Therapy. Methods & Clinical Development, 30*, 447–458. DOI: 10.1016/j.omtm.2023.08.008 PMID: 37663648

Keren, L., Bosse, M., Thompson, S., Risom, T., Vijayaragavan, K., McCaffrey, E., Marquez, D., Angoshtari, R., Greenwald, N. F., Fienberg, H., Wang, J., Kambham, N., Kirkwood, D., Nolan, G., Montine, T. J., Galli, S. J., West, R., Bendall, S. C., & Angelo, M. (2019). MIBI-TOF: A multiplexed imaging platform relates cellular phenotypes and tissue structure. *Science Advances, 5*(10), eaax5851. Advance online publication. DOI: 10.1126/sciadv.aax5851 PMID: 31633026

Kiessling, P., & Kuppe, C. (2024). Spatial multi-omics: Novel tools to study the complexity of cardiovascular diseases. *Genome Medicine, 16*(1), 14. DOI: 10.1186/s13073-024-01282-y PMID: 38238823

Kostas, J. C., Brainard, C. S., & Cristea, I. M. (2025). A Primer on Proteomic Characterization of Intercellular Communication in a Virus Microenvironment. *Molecular & Cellular Proteomics, 24*(3), 100913. DOI: 10.1016/j.mcpro.2025.100913 PMID: 39862905

Le, J., Dian, Y., Zhao, D., Guo, Z., Luo, Z., Chen, X., Zeng, F., & Deng, G. (2025). Single-cell multi-omics in cancer immunotherapy: From tumor heterogeneity to personalized precision treatment. *Molecular Cancer, 24*(1), 221. DOI: 10.1186/s12943-025-02426-3 PMID: 40855431

Lee, Y., Lee, M., Shin, Y., Kim, K., & Kim, T. (2025a). Spatial Omics in Clinical Research: A Comprehensive Review of Technologies and Guidelines for Applications. *International Journal of Molecular Sciences, 26*(9), 3949. DOI: 10.3390/ijms26093949 PMID: 40362187

Lee, Y., Lee, M., Shin, Y., Kim, K., & Kim, T. (2025b). Spatial Omics in Clinical Research: A Comprehensive Review of Technologies and Guidelines for Applications. *International Journal of Molecular Sciences, 26*(9), 3949. DOI: 10.3390/ijms26093949 PMID: 40362187

Lewis, S. M., Asselin-Labat, M.-L., Nguyen, Q., Berthelet, J., Tan, X., Wimmer, V. C., Merino, D., Rogers, K. L., & Naik, S. H. (2021). Spatial omics and multiplexed imaging to explore cancer biology. *Nature Methods*, *18*(9), 997–1012. DOI: 10.1038/s41592-021-01203-6 PMID: 34341583

Li, K. W., Ganz, A. B., & Smit, A. B. (2019). Proteomics of neurodegenerative diseases: Analysis of human post-mortem brain. *Journal of Neurochemistry*, *151*(4), 435–445. DOI: 10.1111/jnc.14603 PMID: 30289976

Liang, A., Kong, Y., Chen, Z., Qiu, Y., Wu, Y., Zhu, X., & Li, Z. (2024). Advancements and applications of single-cell multi-omics techniques in cancer research: Unveiling heterogeneity and paving the way for precision therapeutics. *Biochemistry and Biophysics Reports*, *37*, 101589. DOI: 10.1016/j.bbrep.2023.101589 PMID: 38074997

Liu, X., Peng, T., Xu, M., Lin, S., Hu, B., Chu, T., Liu, B., Xu, Y., Ding, W., Li, L., Cao, C., & Wu, P. (2024a). Spatial multi-omics: Deciphering the technological landscape of integration of multi-omics and its applications. *Journal of Hematology & Oncology*, *17*(1), 72. DOI: 10.1186/s13045-024-01596-9 PMID: 39182134

Liu, X., Peng, T., Xu, M., Lin, S., Hu, B., Chu, T., Liu, B., Xu, Y., Ding, W., Li, L., Cao, C., & Wu, P. (2024b). Spatial multi-omics: Deciphering the technological landscape of integration of multi-omics and its applications. *Journal of Hematology & Oncology*, *17*(1), 72. DOI: 10.1186/s13045-024-01596-9 PMID: 39182134

MacDonald, M. L., & Westwick, J. K. (2006). Exploiting Network Biology to Improve Drug Discovery. In G. A. Kenneth, H. R. Jeffrey, & T. D. Lansing, *High Content Screening* (Vol. 356, pp. 221–232). Humana Press. DOI: 10.1385/1-59745-217-3:221

Marshall, J. L., Noel, T., Wang, Q. S., Chen, H., Murray, E., Subramanian, A., Vernon, K. A., Bazua-Valenti, S., Liguori, K., Keller, K., Stickels, R. R., McBean, B., Heneghan, R. M., Weins, A., Macosko, E. Z., Chen, F., & Greka, A. (2022). High-resolution Slide-seqV2 spatial transcriptomics enables the discovery of disease-specific cell neighborhoods and pathways. *iScience*, *25*(4), 104097. DOI: 10.1016/j.isci.2022.104097 PMID: 35372810

Martinelli, A. L., & Rapsomaniki, M. A. (2022). ATHENA: Analysis of tumor heterogeneity from spatial omics measurements. *Bioinformatics (Oxford, England)*, *38*(11), 3151–3153. DOI: 10.1093/bioinformatics/btac303 PMID: 35485743

Miyamoto, A. T., Shimagami, H., Kumanogoh, A., & Nishide, M. (2025). Spatial transcriptomics in autoimmune rheumatic disease: Potential clinical applications and perspectives. *Inflammation and Regeneration*, *45*(1), 6. DOI: 10.1186/s41232-025-00369-2 PMID: 39980019

Moon, K. R., Van Dijk, D., Wang, Z., Gigante, S., Burkhardt, D. B., Chen, W. S., Yim, K., Elzen, A. V. D., Hirn, M. J., Coifman, R. R., Ivanova, N. B., Wolf, G., & Krishnaswamy, S. (2019). Visualizing structure and transitions in high-dimensional biological data. *Nature Biotechnology*, *37*(12), 1482–1492. DOI: 10.1038/s41587-019-0336-3 PMID: 31796933

Nair, D. G., & Weiskirchen, R. (2024). Advanced In Vitro Models for Preclinical Drug Safety: Recent Progress and Prospects. *Current Issues in Molecular Biology*, *47*(1), 7. DOI: 10.3390/cimb47010007 PMID: 39852122

Nassar, S. F., Raddassi, K., & Wu, T. (2021). Single-Cell Multiomics Analysis for Drug Discovery. *Metabolites*, *11*(11), 729. DOI: 10.3390/metabo11110729 PMID: 34822387

Neumann, E. K. (2025). Spatial Multiomics Toward Understanding Neurological Systems. *Journal of Mass Spectrometry : JMS*, *60*(6), e5143. DOI: 10.1002/jms.5143 PMID: 40360168

Nirmal, A. J., Maliga, Z., Vallius, T., Quattrochi, B., Chen, A. A., Jacobson, C. A., Pelletier, R. J., Yapp, C., Arias-Camison, R., Chen, Y.-A., Lian, C. G., Murphy, G. F., Santagata, S., & Sorger, P. K. (2022). The Spatial Landscape of Progression and Immunoediting in Primary Melanoma at Single-Cell Resolution. *Cancer Discovery*, *12*(6), 1518–1541. DOI: 10.1158/2159-8290.CD-21-1357 PMID: 35404441

Ospina, O. E., Wilson, C. M., Soupir, A. C., Berglund, A., Smalley, I., Tsai, K. Y., & Fridley, B. L. (2022). spatialGE: Quantification and visualization of the tumor microenvironment heterogeneity using spatial transcriptomics. *Bioinformatics (Oxford, England)*, *38*(9), 2645–2647. DOI: 10.1093/bioinformatics/btac145 PMID: 35258565

Pagès, F., Mlecnik, B., Marliot, F., Bindea, G., Ou, F.-S., Bifulco, C., Lugli, A., Zlobec, I., Rau, T. T., Berger, M. D., Nagtegaal, I. D., Vink-Börger, E., Hartmann, A., Geppert, C., Kolwelter, J., Merkel, S., Grützmann, R., Van Den Eynde, M., Jouret-Mourin, A., & Galon, J. (2018). International validation of the consensus Immunoscore for the classification of colon cancer: A prognostic and accuracy study. *Lancet*, *391*(10135), 2128–2139. DOI: 10.1016/S0140-6736(18)30789-X PMID: 29754777

Palla, G., Spitzer, H., Klein, M., Fischer, D., Schaar, A. C., Kuemmerle, L. B., Rybakov, S., Ibarra, I. L., Holmberg, O., Virshup, I., Lotfollahi, M., Richter, S., & Theis, F. J. (2022). Squidpy: A scalable framework for spatial omics analysis. *Nature Methods*, *19*(2), 171–178. DOI: 10.1038/s41592-021-01358-2 PMID: 35102346

Parra, E. R., Uraoka, N., Jiang, M., Cook, P., Gibbons, D., Forget, M.-A., Bernatchez, C., Haymaker, C., Wistuba, I. I., & Rodriguez-Canales, J. (2017). Validation of multiplex immunofluorescence panels using multispectral microscopy for immune-profiling of formalin-fixed and paraffin-embedded human tumor tissues. *Scientific Reports*, *7*(1), 13380. DOI: 10.1038/s41598-017-13942-8 PMID: 29042640

Recanatini, M., & Cabrelle, C. (2020). Drug Research Meets Network Science: Where Are We? *Journal of Medicinal Chemistry*, *63*(16), 8653–8666. DOI: 10.1021/acs.jmedchem.9b01989 PMID: 32338900

Ren, Z., Zeng, X., Lao, Y., Zheng, H., You, Z., Xiang, H., & Zou, Q. (2024). A spatial hierarchical network learning framework for drug repositioning, allowing interpretation from macro to micro scale. *Communications Biology*, *7*(1), 1413. DOI: 10.1038/s42003-024-07107-3 PMID: 39478146

Saliba, A., Du, Y., Feng, T., & Garmire, L. (2024). Multi-Omics Integration in Nephrology: Advances, Challenges, and Future Directions. *Seminars in Nephrology*, *44*(6), 151584. DOI: 10.1016/j.semnephrol.2025.151584 PMID: 40216576

Sanches, P. H. G., De Melo, N. C., Porcari, A. M., & De Carvalho, L. M. (2024). Integrating Molecular Perspectives: Strategies for Comprehensive Multi-Omics Integrative Data Analysis and Machine Learning Applications in Transcriptomics, Proteomics, and Metabolomics. *Biology (Basel)*, *13*(11), 848. DOI: 10.3390/biology13110848 PMID: 39596803

Schürch, C. M., Bhate, S. S., Barlow, G. L., Phillips, D. J., Noti, L., Zlobec, I., Chu, P., Black, S., Demeter, J., McIlwain, D. R., Kinoshita, S., Samusik, N., Goltsev, Y., & Nolan, G. P. (2020). Coordinated Cellular Neighborhoods Orchestrate Antitumoral Immunity at the Invasive Front of Colorectal Cancer. *Cell*, *182*(5), 1341–1359.e19. DOI: 10.1016/j.cell.2020.07.005 PMID: 32763154

Shu, H., Chen, J., Xu, C., Hu, J., Wang, Y., Peng, J., Jiang, Q., Shang, X., & Wang, T. (2024). Efficient integration of spatial omics data for joint domain detection, matching, and alignment with stMSA. *Bioinformatics (Oxford, England)*. Advance online publication. DOI: 10.1101/2024.07.29.605604

Song, X., Liang, H., Nan, F., Chen, W., Li, J., He, L., Cun, Y., Li, Z., Zhang, W., & Zhang, D. (2025). Autoimmune Diseases: Molecular Pathogenesis and Therapeutic Targets. *MedComm*, *6*(7), e70262. DOI: 10.1002/mco2.70262 PMID: 40529617

Spitzenberger, F., Patel, J., Gebuhr, I., Kruttwig, K., Safi, A., & Meisel, C. (2022). Laboratory-Developed Tests: Design of a Regulatory Strategy in Compliance with the International State-of-the-Art and the Regulation (EU) 2017/746 (EU IVDR [In Vitro Diagnostic Medical Device Regulation]). *Therapeutic Innovation & Regulatory Science*, *56*(1), 47–64. DOI: 10.1007/s43441-021-00323-7 PMID: 34291407

Subrahmanyam, P. B., & Maecker, H. T. (2017). CyTOF Measurement of Immunocompetence Across Major Immune Cell Types. *Current Protocols in Cytometry*, *82*(1). Advance online publication. DOI: 10.1002/cpcy.27 PMID: 28967988

Tan, W. C. C., Nerurkar, S. N., Cai, H. Y., Ng, H. H. M., Wu, D., Wee, Y. T. F., Lim, J. C. T., Yeong, J., & Lim, T. K. H. (2020). Overview of multiplex immunohistochemistry/immunofluorescence techniques in the era of cancer immunotherapy. *Cancer Communications*, *40*(4), 135–153. DOI: 10.1002/cac2.12023 PMID: 32301585

Valous, N. A., Popp, F., Zörnig, I., Jäger, D., & Charoentong, P. (2024). Graph machine learning for integrated multi-omics analysis. *British Journal of Cancer*, *131*(2), 205–211. DOI: 10.1038/s41416-024-02706-7 PMID: 38729996

Wang, Y., Wei, J., Zhang, W., Doherty, M., Zhang, Y., Xie, H., Li, W., Wang, N., Lei, G., & Zeng, C. (2022). Gut dysbiosis in rheumatic diseases: A systematic review and meta-analysis of 92 observational studies. *EBioMedicine*, *80*, 104055. DOI: 10.1016/j.ebiom.2022.104055 PMID: 35594658

Wanichthanarak, K., Fahrmann, J. F., & Grapov, D. (2015). Genomic, Proteomic, and Metabolomic Data Integration Strategies. *Biomarker Insights, 10s4*. *Biomarker Insights*, *S29511*, BMI.S29511. Advance online publication. DOI: 10.4137/BMI.S29511 PMID: 26396492

Williams, C. G., Lee, H. J., Asatsuma, T., Vento-Tormo, R., & Haque, A. (2022). An introduction to spatial transcriptomics for biomedical research. *Genome Medicine*, *14*(1), 68. DOI: 10.1186/s13073-022-01075-1 PMID: 35761361

Wu, S. Z., Al-Eryani, G., Roden, D. L., Junankar, S., Harvey, K., Andersson, A., Thennavan, A., Wang, C., Torpy, J. R., Bartonicek, N., Wang, T., Larsson, L., Kaczorowski, D., Weisenfeld, N. I., Uytingco, C. R., Chew, J. G., Bent, Z. W., Chan, C.-L., Gnanasambandapillai, V., & Swarbrick, A. (2021). A single-cell and spatially resolved atlas of human breast cancers. *Nature Genetics*, *53*(9), 1334–1347. DOI: 10.1038/s41588-021-00911-1 PMID: 34493872

Wu, Z., Boen, J., Jindal, S., Basu, S., Bieniosek, M., He, S., LaPelusa, M., Mayer, A. T., Kaseb, A. O., Zou, J., Sharma, P., & Trevino, A. E. (2025). *Spatial multi-omics and deep learning reveal fingerprints of immunotherapy response and resistance in hepatocellular carcinoma*. Cancer Biology. DOI: 10.1101/2025.06.11.656869

Xia, M., Liu, Q., Zhang, W., Ge, J., & Mei, Z. (2025). Spatiotemporal Dynamics of Central Nervous System Diseases: Advancing Translational Neuropathology via Single-Cell and Spatial Multiomics. *MedComm*, *6*(9), e70328. DOI: 10.1002/mco2.70328 PMID: 40843131

Xie, C., Wang, J., Diao, X., Wang, X., & Cai, Z. (2024). *Single-Cell Resolution DESI Mass Spectrometry Imaging through 10-Fold Sample Expansion.* DOI: 10.1101/2024.10.21.619369

Xu, C., Jin, X., Wei, S., Wang, P., Luo, M., Xu, Z., Yang, W., Cai, Y., Xiao, L., Lin, X., Liu, H., Cheng, R., Pang, F., Chen, R., Su, X., Hu, Y., Wang, G., & Jiang, Q. (2022). DeepST: Identifying spatial domains in spatial transcriptomics by deep learning. *Nucleic Acids Research*, *50*(22), e131–e131. DOI: 10.1093/nar/gkac901 PMID: 36250636

Xu, X., Su, J., Zhu, R., Li, K., Zhao, X., Fan, J., & Mao, F. (2025). From morphology to single-cell molecules: High-resolution 3D histology in biomedicine. *Molecular Cancer*, *24*(1), 63. DOI: 10.1186/s12943-025-02240-x PMID: 40033282

Xue, Q., Hu, H., Wang, R., Wu, F., & Xiong, H. (2025). The applications of single-cell multiomics in drug screening. *Pharmaceutical Science Advances*, *3*, 100090. DOI: 10.1016/j.pscia.2025.100090

Ye, Z., Lai, Z., Zheng, S., & Chen, Y. (2023). Spatial-Live: A lightweight and versatile tool for single-cell spatial-omics data visualization. *Bioinformatics (Oxford, England)*. Advance online publication. DOI: 10.1101/2023.09.24.559173 PMID: 37873441

Zhang, Y., Yang, C., Chen, X., Wu, L., Yuan, Z., Zhang, F., & Qian, B. (2025). Cancer therapy resistance from a spatial-omics perspective. *Clinical and Translational Medicine*, *15*(7), e70396. DOI: 10.1002/ctm2.70396 PMID: 40677104

Zhou, R., Yang, G., Zhang, Y., & Wang, Y. (2023a). Spatial transcriptomics in development and disease. *Molecular Biomedicine*, *4*(1), 32. DOI: 10.1186/s43556-023-00144-0 PMID: 37806992

Zhou, R., Yang, G., Zhang, Y., & Wang, Y. (2023b). Spatial transcriptomics in development and disease. *Molecular Biomedicine*, *4*(1), 32. DOI: 10.1186/s43556-023-00144-0 PMID: 37806992

Zhou, Y., Xiao, X., Dong, L., Tang, C., Xiao, G., & Xu, L. (2025). Cooperative integration of spatially resolved multi-omics data with COSMOS. *Nature Communications*, *16*(1), 27. DOI: 10.1038/s41467-024-55204-y PMID: 39747840

Chapter 4
Advancing Personalized Medicine With AI-Driven Microbiome and Nutrigenomics

Ashok Kumar Sah
https://orcid.org/0000-0002-7762-4351
A' Sharqiyah University, Oman

Abhishek Maurya
Galgotias University, Greater Noida, India

Prem Shankar Mishra
https://orcid.org/0000-0003-3797-1575
Vidya University, India

Rakhi Mishra
Noida Institute of Engineering and Technology (Pharmacy Institute), Greater Noida, India

Ranjay Kumar Choudhary
https://orcid.org/0000-0003-2801-4079
A' Sharqiyah University, Oman

ABSTRACT

The integration of artificial intelligence (AI), microbiome research, and nutrigenomics is advancing personalised medicine by enabling precise, data-driven healthcare. AI supports the analysis of complex biological datasets, revealing important links among genes, metabolism, diet, and gut microbial communities. Machine learning

DOI: 10.4018/979-8-3373-4849-0.ch004

models detect subtle patterns that traditional methods may overlook, improving understanding of host–microbe–diet interactions and enhancing diagnostic accuracy and personalised treatment strategies. This chapter highlights recent progress in AI-enabled metagenomic and nutrigenomic integration that supports highly individualised nutritional and therapeutic interventions. Key challenges including algorithmic bias, data standardisation, ethics, and patient privacy are discussed alongside potential regulatory and collaborative solutions. Future directions emphasise portable biosensors, cloud platforms, and digital health tools capable of delivering scalable, real-time personalised care and guiding predictive, preventive healthcare.

1. INTRODUCTION

1.1 Background and Significance

Nutrition denotes to the biochemical and physiological progresses over which an organism obtains nutrition from food. It helps the delivered of digestible nutrients to organisms, thereby providing them with energy and chemical structures needed for their survival, growth, and reproductive processes (Carpenter, 2003). In the nineteenth century, A main modification have done in field of scientific study related to food and nutrition ensued with the start of the chemical revolution. The discovery of several micronutrients in the 1910s marked the beginning of modern nutrition science (Mozaffarian et al., 2018). Since related of the 20th century, the medical and healthcare industries had experienced tremendous change, commonly due to technological breakthroughs, advances in clinical and scientific research, and improved ability to treat disease (Rampton et al., 2022).[1]

Moreover, research revealed that the relationship between nutrition and illnesses varies depending on an individual's ethnicity and genetic makeup. People who have a C → T substitution in the methylene-tetrahydrofolate reductase (MTHFR) gene, for example, may need more folate than people who have the wild-type allele (Zhao et al., 2023). A study on the prevalence of hepatocellular carcinoma in the Sudanese community provides another fascinating illustration of the intricate relationships between heredity, nutrition, and illness. According to the study, Sudanese individuals with the glutathione S-transferase M1 null genotype had a higher chance of developing the disease when they consumed peanut butter tainted with aflatoxins than those without this genotype. Furthermore, the subject is expanding rapidly due to the combination of genomics and personalized nutrition. Personalized nutrition uses individual genetic information to prescribe dietary choices this, could significantly effects of disease prevention and health improvement.

Clinical trials and observational research have always been the mainstays of the nutrition sector, but AI has replaced both. These technologies, which include data analysis, deep learning, and machine learning have the capability to find patterns in big datasets, reveal intricate linkages, and produce useful information (Jimenez-Carvelo & Cuadros-Rodríguez, 2021; Taye, 2023). AI has a widespread range of possible applications in nutrition, from illness prevention models to dietary personalisation. The field of dietary treatments is changing as a result of technical breakthroughs made possible by the integration of AI applications in nutrition (Theodore Armand et al., 2024). In this data-driven age, artificial intelligence (AI) tools have the possible to completely renovate our understanding, tracking, and optimisation of nutritional outcomes. AI in nutrition involves the use of conceptual algorithms, machine learning, and deep learning approaches to the analysis, interpretation, and decision-making of various datasets pertaining to eating patterns, nutritional data, and other health aspects (Oliveira Chaves et al., 2023; Armand et al., 2024).

To study how nutrients are affect human genetic variants, genetic and nutritional data can be analysed using random forests and machine learning algorithms (Kosaraju et al., 2022; Szabo de Edelenyi et al., 2008).

The technique of customising dietary recommendations and interventions to take into account each person's unique nutritional needs, genetic makeup, health status, lifestyle, and personal preferences is known as personalised nutrition (Bush et al., 2020). In general, by tailoring dietary recommendations according to unique traits, personalised nutrition has the prospective to completely make over our approach to diet and nutrition. As our understanding of genetics, metabolism, and nutrition advances, it is expected that the importance of customised nutrition will grow in order to attain optimal health and well-being. In this book chapter, discussed the relationship of Microbiome and Nutrigenomics with AI personalized medicine and their future prospective and innovations.

1.2 Scope and Objectives

Nutrigenomic and Nutrigenetic deliver the strategies to study how components of the diet correlate with genes, and their products, to modify phenotype and, conversely, how genes and their products metabolize these constituents into nutrients, and bioactive compounds, with a mutual crucial goal to improve health through the personalization of diet. At the same time provide powerful approaches to undo the complex relationship between nutritional molecules, genetic polymorphisms, and the biological system as a whole (Mutch et al., 2005). So according to these disciplines, through the study of genetic polymorphism and mapping of genes it is probable to identify foods that can have positive or negative effects on our health, helping us to protect the organism. The development of genomic technologies and sciences is closely associated to the scope of genomic information use in medical practice. There is little clinical evidence to justify the use of medical genomics in complex disorders, which explains the relatively narrow area of therapeutic utility and the slow early acceptance. The early emphasis was on variations that had a high or near-certain genotype–phenotype correlation probability (high penetrance). The outline of genomic medicine, catalyzed by the Human Genome Project, has provided an unmatched influx of genetic data (Green et al., 2020) AI plays a pivotal role in processing and interpreting this huge amount of genomic information. Machine learning algorithms excel in identifying genetic variations, annotating genomic data, and predicting the functional implications of these variations (Shendure & Ji, 2008). Genomic sequencing procedures have been transformed by AI-powered technologies. These machine learning-powered technologies make it possible to call variants, annotate, and interpret genomic data with greater accuracy. By using AI in genomic sequencing, disease-related genetic variations can be better identified, and the genomic basis of diverse illnesses can be better understood (Zou et al., 2019). Pharmacogenomics, a subset of personalized medicine, focuses on how an individual's genetic makeup effects their response to drugs. AI applications in pharmacogenomics aim to tailor medications constructed on genetic variations, optimizing treatment efficacy and minimizing adverse effects (Zou et al., 2019). AI algorithms, particularly machine learning models, are employed in predicting drug efficacy based on individual genetic profiles. These models analyse diverse data, including genetic information, clinical parameters, and treatment outcomes, to generate predictive insights. Advanced algorithms are used in AI applications for early disease diagnosis to examine a variability of data sources, such as genomic, clinical and imaging data. Machine learning algorithms are able to recognise minute patterns that point to illnesses including cancer, heart problems, and neurological disorders. Using AI in diagnostic procedures improves accuracy and makes early intervention possible (Esteva et al., 2019).

AI aids in the creation of treatment response predicting models, which direct the personalisation of treatment regimens. AI prototypes can forecast a patient's predictable response to individual therapeutic procedures by inspecting personal patient data, such as clinical history, genetic information, and prior treatment results. There are some specific points related to scope and objectives –

1.2.1 Applications of Artificial Intelligence (AI) in Microbiome Analysis

Artificial Intelligence (AI), Machine learning (ML) and deep learning (DL) has completely transformed for microbiome analysis. They are making it possible to analyze intricate microbial data These methods make it easier to identify microbial shapes and how they relate to different states of health and illness. For example, using high-throughput sequencing data, DL models such as Deep Micro have been effective in predicting illness connections and identifying microbial communities (Rouskas et al., 2025).

1.2.2 Nutrigenomics and Personalized Nutrition

In order to create individualised dietary recommendations, nutrigenomics aims to comprehend how individual genetic variants impact responses to nutrients. AI advances this area by effective gene-diet interaction analysis. According to studies, AI-powered customised nutrition plans can improve people's gut microbiota and general health (Rouskas et al., 2025).

1.2.3 Addition of Multi-Omics Data

Comprehensive perceptions are interested in personal health profiles that can be obtained by combining data from the microbiome, metabolomics, and genetics. Through the management and analysis of the large and intricate datasets, AI plays a crucial part in this integration. Recent developments have shown AI's promise in integrating multi-modal omics data, which could result in more precise illness forecasts and individualised treatment plans.

1.2.4 Clinical Applications and Therapeutic Strategies

Significant clinical uses of AI-driven studies of nutrigenomic and microbiome data include the creation of individualised treatment regimens, illness prevention, and prediction. To optimise treatment approaches, AI models have been used, for instance, to forecast how each person will react to probiotic therapies (D'Urso & Broccolo, 2024).

1.2.5 Ethical, Privacy, and Accessibility Considerations

AI's incorporation into personalised medicine presents significant privacy and ethical issues. Data protection, informed consent, and fair access to AI-powered medical solutions are critical issues. Resolving these problems is critical to assuring the ethical and comprehensive use of AI in healthcare (Farhud & Zokaei, 2021).

1.3 The Convergence of AI, Microbiome Research, and Nutrigenomics

The convergence of nutrigenomics, microbiome research, and artificial intelligence (AI) is revolutionising personalised medicine by allowing for accurate, individualised health interventions based on a person's dietary requirements, gut microbiota, and genetic profile. Researchers may now better comprehend microbial interactions and forecast disease risks by exhausting AI technologies, mostly machine learning and deep learning, to interpret dense microbiome information. For example- for instance, demonstrated the prospective of AI in clinical microbiome research by introducing the Deep Micro framework, which uses deep representation learning to predict diseases based on microbiome data. In addition to helping with disease detection, AI can also help with appropriate therapy of certain conditions. AI methods are frequently used to evaluate a patient's response to a particular medication and choose the best development of action based on the patient's clinical and genetic characteristics (Pushkaran & Arabi, 2024).

All of these might be viewed as a means of reaching therapy optimisation, which is regarded as some of the significant advancements in the field of personalised medicine that can regulate to better patient outcomes and a higher value of life. The relationship between nutrition genomics and the microbiota is another significant one. The collective genetic material of all the microorganisms that reside in and on the human body is known as the microbiome. Despite its complexity, the human digestive system provides numerous examples that how the gut microbiota has been connected to overall health. The relationship between genes and bioactive substances from various food sources is made clear by the study of nutrigenomics. Additionally, using nutritional systems biology to identify and find biomarkers—the "stress signs" that predispose to diseases associated to food (Freudenheim & Gower, 2017).

2. FOUNDATIONS OF PERSONALIZED MEDICINE

2.1 Definition and Principles

Personalized medicine are related to method treatment that also known as precision medicine, represents a standard modification in healthcare. Precision medicine is becoming more incorporated into clinical practice, and it has become a major focus of biomedical research. In order to improve disease prevention, therapeutic efficacy, and diagnostic accuracy, medical care is customised to each patient's unique molecular, phenotypic, and contextual traits. The framework includes information from environmental exposures, behavioural factors, social determinants of health, clinical phenotypes, transcriptomes, proteomes, metabolomes, epigenomes, and microbiomic analyses (Topol, 2019). Pharmaceutical and medical device firms seem to be driving the growing trend of personalised remedy in American modern medicine. Personalised medicine is well-defined by the U.S. Food and Drug Administration (FDA) as giving "the appropriate patient the right drug at the right dose at the right time." More generally, "personalised medicine" can be defined as the process of altering medical care to each patient's unique needs, preferences, and appearances at every step of care - prevention,, treatment, diagnosis and follow-up. In keeping with its crucial objective, the FDA is collaborating with researchers, pharmaceutical and medical device and biologics producers, medical professionals, and others to better comprehend and adjust to the promise of personalised medicine (Gronowicz, 2016). The two vital principles of personalized medicine are as follows:-

2.1.1 Individualization

An emerging concept in "contemporary American medicine" is personalised medicine, which supports therapies tailored to an individual's genetic composition while taking into reason the separate response related to genome. This individualised approach recognises that although the majority of genetic differences have little effect on well-being, when paired with environmental circumstances and lifestyle choices, they can have a significant impact on an individual's health.

2.1.2 Precision

Personalised medicine can find certain DNA variants associated with diseases by using cutting-edge technologies like genome sequencing. This allows for more accurate diagnosis and treatment that marks these particular mutations.

These principles joined greatly reduce the trial and error inefficiencies, allowing healthcare professionals to well understanding related disease and address it.

By detecting genetic predispositions and risk factors, personalised medicine also supports preventive healthcare by enabling focused preventative measures. This preventative strategy not only reduces the prevalence of disease but also enhances longevity and general well-being. The ability to thoroughly analyse a person's genetic code through genomic sequencing is a key element of personalised treatment. Through the identification of genetic mutations and variations related with medication metabolism, illness risk, and responsiveness to treatment, medical professionals can tailor interventions created on the genetic profile of each patient.

Figure 1. Aspects of personalized nutrition

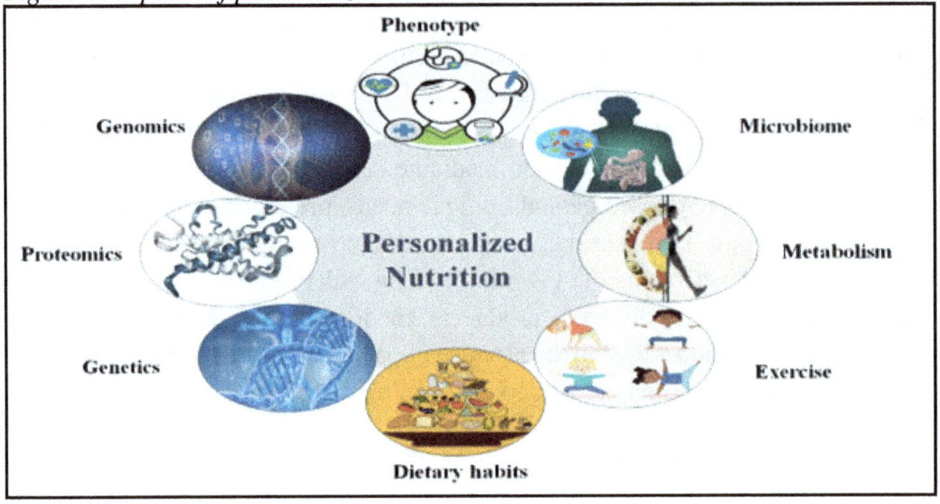

2.2 Historical Development

Several convergent technical sectors have influenced the evolution of precision medicine. Developments in human genetics, computational biology, and molecular biology have laid the groundwork for personalised health care strategies. The historical and technical foundations of precision medicine are grounded in the maturation of genomic science, the evolution of multi-modal health data platforms, and the expanding role of computational analytics. Continued progress will depend on sustained efforts to ensure data representativeness, methodological rigor, and cross-sector collaboration in building interoperable and ethically governed precision health systems.

2.2.1 Ancient Foundations

Ancient civilizations are the source of personalised medicine. Individualised treatment plans based on symptoms and constitution are mentioned in early medical literature as the Ebers Papyrus (c. 1550 BC). In Homer's Odyssey, treatments were also reportedly customised for each patient, indicating a knowledge of patient-specific strategies. Practitioners in Ayurveda and Traditional Chinese Medicine (TCM) identified and treated patients based on their particular energies, or doshas, demonstrating early holistic techniques that acknowledged each person's biological individuality. The conceptual underpinnings of contemporary personalised care were established by these antiquated methods (Visvikis-Siest et al., 2020).

2.2.2 Hippocratic Insights

The father of modern medicine, Hippocrates (460–370 BC), placed a strong emphasis on comprehending the patient as opposed to merely the illness. According to his theory, each person has a different makeup of the four humors like "blood, phlegm, yellow bile and black bile". These fluids must be in balance for their health. Despite being scientifically out of date, this idea recognised biological individuality and the necessity for customised treatments. Patient-centered care, the cornerstone of modern personalised medicine, was brought into focus by Hippocratic beliefs (Stoykova & Koeva-Dimitrova, 2019).

2.2.3 Emergence of Pharmacogenetics

The 1950s marked the beginning of pharmacogenetics, the study of how genetic similarities affect specific responses to medications. One significant outcome concerned genetic deviations in drug-metabolizing enzymes, like the cytochrome P450 family. Researchers observed that patients metabolized medications such isoniazid and debrisoquine differently, leading to varying efficacy and side effects. These results showed how inadequate and rarely harmful a "one-size-fits-all" method to medication prescription was. The scientific paradigm for treatment changed during this time from population-wide to individualised.

2.2.4 Human Genome Project (HGP)

The Human Genome Project, which was ended in 2003, was a major development in biomedical science. It recognised all 20,000–25,000 genes in the human DNA sequence and produced a thorough map of it. The HGP established the foundation for genomic medicine by revealing the genetic causes of illnesses and interindividual variation. It made it possible for researchers to connect certain genetic abnormalities to medication metabolism, treatment outcomes, and disease risk. The HGP accelerated the development of genetic diagnostics and treatments, bringing personalised medicine from a theoretical concept to a clinical reality (Barrera-Saldaña, 2020).

2.2.5 Next-Generation Sequencing (NGS) Technologies

Genetic analysis was transformed by the initiation of next-generation sequencing (NGS) technologies in the late 2000s. NGS democratised genomic medicine and made real-time personalised care possible by lowering sequencing costs significantly and enabling whole exome sequencing (WES) and whole genome sequencing (WGS) for individual patients, which allowed clinicians to diagnose rare genetic disorders, predict the progression of cancer, and guide treatment decisions in real time (Popova & Carabetta, 2024).

2.2.6 Precision Medicine Initiative (PMI)

President Barack Obama established the Precision Medicine Enterprise in the United States in 2015. By integrating lifestyle, environmental, and genetic information into preventive and treatment plans, the PMI sought to revolutionise healthcare. Large-scale data gathering (like the All of Us research programme) is supported, and studies on the relationship between these factors and disease are encouraged. By encouraging fair access, public involvement, and the combination of AI and big data to improve health outcomes for a variety of groups, the PMI is an example of a policy-driven acceleration of personalised medicine.

2.3 Current Applications and Challenges

The development of personalized medicine will result in a more cohesive therapeutic strategy tailored to each patient's unique genetic makeup. Better diagnosis with earlier intervention, more effective medication development and more focused treatments could with the help of personalized medicine.

2.3.1 Diagnosis and intervention

The capacity is examine each patient individually will enable a more precise diagnosis and tailored treatment plan. Genotyping is the technique of determining a person's DNA sequence through the use of biological assays. A comprehensive description of a person's DNA sequence allows their genome to be compared to a reference genome, such as the Human Genome Project's, in order to evaluate the genetic differences that may be responsible for certain disorders. The public can now obtain Direct-to-Consumer genome sequencing thanks to several private companies like Illumina, Navigenics, and 23andMe. Pharmacogenomics, which uses a person's DNA to deliver a more customized and informed medication prescription, is one facet of this. Personalized medicine can significantly support the development of preventive care in addition to targeted therapy. For example, many women who are predisposed because to a family history of ovarian or breast cancer are already being genotyped for specific mutations in the BRCA1 and BRCA2 genes. Companion diagnostics is now a tool used to examine the safety and effectiveness of a medicine tailored to a certain patient group or subgroup. This test technology, which was created during or after a drug's release onto the market, aids in improving the therapeutic treatment options available to each patient. In an attempt to help the patient make the best treatment choice possible, these companion diagnostics have integrated the drug's pharmacogenomic information into their prescription label.

2.3.2 Drug development and usage

When creating medications that are awaiting FDA approval for general use, having a person's genetic information can be important. When determining whether to include or exclude a patient from the final stages of a clinical study, having a thorough description of the patient's genetic composition can be very helpful. Finding the patients who will gain the most from a clinical study would improve patient safety from unfavorable outcomes brought on by the product during testing and enable smaller, quicker trials, which will reduce total expenses.

Additionally, even if a medication may only be needed by a small portion of the population, the FDA can approve treatments that are considered useless for the general population by using individual genomes to qualify the effectiveness and need for that particular prescription or therapy (Hamburg & Collins, 2010). The targeting tactics of nanocarriers vary depending on the ailment. To achieve recognition and binding, for instance, if the disease is cancer, a common method is to find the biomarker expressed on the surface of cancer cells and load its corresponding targeting vector onto a nanocarrier. The nanocarriers' size scale will also be designed to achieve the enhanced permeability and retention effect (EPR) in tumor targeting.

To accomplish organ-targeting medication delivery and prevent non-specific absorption, the surface of the nanocarriers can be coated with a particular ligand that binds to the receptors inside the organ, such as the kidney, if the disease is localized there.

Even though this medication delivery method based on nanoparticles has a lot of promise, the field has not advanced much, and the nanocarriers are still being studied and adjusted to clinical standards.

2.3.3 Theranostics

In nuclear medicine, theranostics is a customized method that uses comparable molecules for both imaging (diagnostic) and treatment. “Diagnostics” and “therapeutics” are combined to form the phrase. Its most popular uses are for electron emitters in radiotherapy or for attaching radionuclides (either positron or gamma emitters) to molecules for SPECT or PET imaging. Examples Treating thyroid cancer patients with radioactive iodine (Langbein et al., 2019). Radium-223 is used to diagnostic bone metastases and Lutetium-177 DOTATATE is used to diagnostic neuroendocrine tumors, Lutetium-177 PSMA is used to treat prostate cancer, and radio-labelled anti-CD20 antibodies (such as Bexxar) are used to treat lymphoma.

2.3.4 Respiratory proteomics

Humans are impacted by respiratory disorders all over the world. Lung cancer and long-lasting lung conditions like asthma, chronic obstructive pulmonary disease, and idiopathic pulmonary fibrosis are between the conditions that cause the most morbidity and mortality. Because they are so diverse, these illnesses need to be diagnosed as soon as possible. However, the earliest symptoms are vague, and the medical diagnosis is sometimes made too late. Using cutting-edge technology, customized medicine has been popular in recent years with the aim of tailoring remedies to each patient's unique medical requirements (Priyadharshini & Teran, 2016). Proteomics aids in early diagnosis since proteins regulate the body's biological processes, including those related to health and illness. Proteomics examines a variety of biological materials in respiratory disease cases, such as serum, sputum, blood cells, bronchoalveolar lavage fluids (BAL), nasal lavage fluids (NLF), and others (Priyadharshini & Teran, 2020). Mass spectrometry and sophisticated analytical methods are used to identify and measure the whole protein expression from these biological materials. In recent years, respiratory proteomics has advanced significantly in the field of customized medicine to enhance healthcare. For instance, the proteomics-based method significantly improved the identification of several lung cancer biomarkers that can be utilized to customize individualized therapy for each patient (Lazzari et al., 2012).

2.3.5 Cancer genomics

The genetic diversity of cancer types that have similarities in traditional pathology has been extensively studied in recent decades. The model of tumor heterogeneity or the genetic variation have within a single tumor, has also gained more attention. These findings suggest the possibility that medications that have not been effective when used on a wide population of patients could nevertheless work for a percentage of individuals with certain genetic profiles, among other possibilities. Personalized medicine applied to cancer genomics is called personalized oncogenomics. In order to better understanding related disease pathophysiology and advance therapeutic development, high-throughput sequencing techniques are engaged to describe genes linked to cancer. Given its implications for pharmacological therapy, oncogenomics is one of the most promising areas of genomics. Some instances of this are as follows:-

- As that way give the treatment of chronic myeloid leukemia (CML), where used the BCR-ABL fusion gene—the result comes as reciprocal translocation between chromosomes 9 and 22—is present in more than 95% of cases and causes hyperactivated abl-driven protein signaling, tyrosine kinase inhibitors like imatinib (marketed as Gleevec) have been developed. These drugs are an excellent illustration of "rational drug design" based on understanding of disease pathophysiology since they precisely inhibit the Ableson tyrosine kinase (ABL) protein.

- A high mutation burden directs an immune-therapy response, and some mutation patterns have also been linked to prior exposure to cytotoxic cancer medications (Pleasance et al., 2020).

Challenges

There are several difficulties when personalized treatment becomes increasingly popular. To adapt to the changes that personalized medicine will bring to healthcare, the existing frameworks for regulatory oversight, patient privacy, payment policies, intellectual property rights, data biases, and confidentiality will need to be redesigned. One UK survey, for example, found that 63% of respondents in the country felt uncomfortable with the usage of their personal info for the purpose of applying AI in the medical profession (Vayena et al., 2018). Future advancements in medicine will necessitate the multidisciplinary collaboration of specialists from several disciplines of study, including medicine, clinical oncology, biology, and artificial intelligence, in demand to accelerate the acceptance of personalized medicine.

3. THE HUMAN MICROBIOME: A NEW FRONTIER

3.1 Overview of Human Microbiome

The populations of microorganisms (as well as viruses and fungi) that make up the human microbiome are more complex than the human genome. Associated to the approximately 23,000 genes found in the human genome, large-scale metagenomic initiatives (community and environmental genomics), like the Human Microbiome Project and the European Metagenomics of the Human Intestinal Tract, have identified 3.3 million distinct protein-encoding genes. The positive effects of a healthy gut microbiota on health, even at the genetic level, have been detailed in these research (Qin et al., 2010). The human microbiome is an ecosystem of about 40 trillion microorganisms, including bacteria, viruses, yeast, and fungi, that live inside our bodies and affect a variability of functions, such as immune function, metabolism, and hematopoiesis. Although the microbiome remains relatively stable throughout adulthood, it has been verified to change due to a variety of factors, including infection, antibiotic exposure, lifestyle choices (such as exercise), surgical procedures, diet, and different disease states, including cancer. According to research, some bacteria may help the host at specific levels, while others may be linked to clinical diseases. We can use the atherosclerotic metabolic syndrome as an example. Atherosclerotic plaques from people with high levels of Veillonella and Streptococcus in their digestive systems contained DNA from these oral microbiota. There are several ecosystems within human organisms that are home to a huge amount of microbial species known as microbiota. These microbiota plays a crucial biological role in maintaining health and execution multiple essential tasks. The regular activity and metabolic cellular turnover required for our organism's growing and appropriate operation depend equally on the ongoing changes in the many habitats.

Figure 2. The human microbiome shows an essential role in control of vital homeostatic mechanisms in the body. These include improved metabolism, resistance to infection and inflammation, prevention against autoimmunity as well as an influence on the gut–brain axis. SCFA, short-chain fatty acid

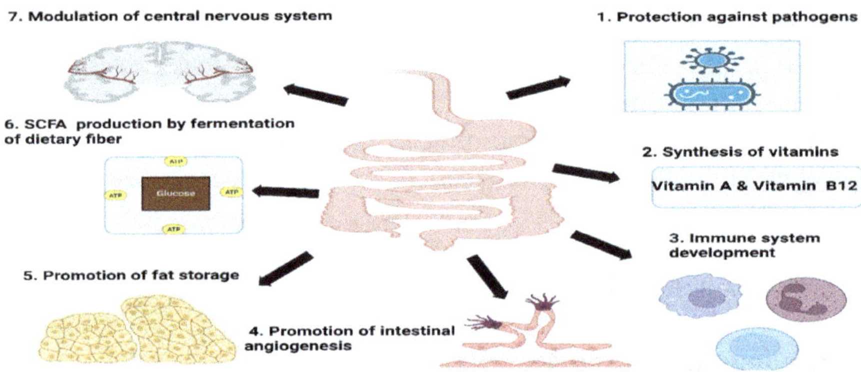

3.2 Role in Health and Disease

The gut microbiota is often mentioned to as the "forgotten organ" (O'Hara & Shanahan, 2006) because to mounting evidence that it shows a significant role in supportive health. The host and microbiota have a mutualistic, symbiotic connection in which both parties gain from one another. Although mutualism and symbiosis are related concepts, mutualism is defined as "an connection between species that is advantageous to both of them," while symbiosis is defined as "the living together of two creatures in intimate association," (Ghosh, 2013). The host offers the microbiota a non-toxic and nutrient-rich environment, and the microbiota develops things like immunity, digestion, and brain development.

3.2.1 Gastric microbiome

The idea that the stomach is sterile was disproved when it was found that H. pylori was suited to live in the acidic gastric environment. The diversity of gastric microbiota is high in H. pylori-negative individuals; the majority of the major gastric phylotypes (Streptococcus, Actinomyces, Prevotella, and Gemella) are also prevalent in these individuals' oropharynx, suggesting that either a large number of constituents are swallowed from more proximal sites or that close relatives of the oral microbiota colonise more distally. On the other side, H. pylori often used for over 90% of sequence delivers from the stomach microbiota individualy who are positive for the bacteria, thus lowering the microbiota's overall diversity. Its presence raises the risk of gastric adenocarcinoma, stomach mucosa associated lymphoid tissue (MALT) cancers, and peptic ulcer disease (El-Serag & Sonnenberg, 1998).

3.2.2 The colonic microbiota and colorectal cancer

Short-chain fatty acids (SCFAs) and other the metabolites formed by the intestinal microbiota It have played a role in the development of colorectal cancers (Plottel & Blaser, 2011). Through Wnt (Wingless-related integration site) signalling (Lazarova et al., 2004). SCFAs have especially butyrate that can effect differentiation of cell cycle arrest and their death. Because Enterococcus faecalis induces aneuploidy and tetraploidy, microbes may also be genotoxic to colonic epithelial cells. Additionally, the colonic microbiota may contribute to colorectal cancer by inducing host reactions, such as triggering heightened immune responses, possibly through Th17 cells.

3.2.3 The gut microbiota and rheumatoid arthritis

By activating host immunological responses, dysregulation of host responses brought on by DYSBIOSIS in the gut lumen may have an impact on distant anatomical regions. Rheumatoid arthritis, another long-term idiopathic inflammatory disease, may have this mechanism. When SFBs are present in the gut microbiome of mice, Th17 cells (Ivanov et al., 2009) locally proliferate, move to peripheral immunological compartments, and activate B cells to produce antibody-producing plasma cells. Similar to rheumatoid arthritis, antibody production results in immune-mediated joint damage (Scher & Abramson, 2011).

3.2.4 Acute Gastroenteritis and Infectious Colitis

Diarrheal illnesses brought on by bacteria, viruses, and protozoal pathogens are all included in acute gastroenteritis. The causative agent of acute gastroenteritis may be difficult to diagnose using standard microbiology techniques. Children under three are more likely to be infected by viral agents, while children over three are more likely to be infected by bacterial pathogens Only 47% of stool samples that were subjected to comprehensive diagnostic testing in one investigation produced a particular etiologic agent (Klein et al., 2006). Acute gastroenteritis pathobiology involves intestinal villi effacement, increased intestinal permeability brought on by pathogen-gut epithelial interactions, and the synthesis of toxins that promote immune cell infiltration and intestinal barrier disruption.

3.3 Technologies for Microbiome Analysis (e.g., Metagenomics, 16S rRNA)

The microbiome can be studied using a variety of different technologies. We will go over the basic techniques first in this overview, followed by the main technology.

3.3.1 Marker Gene Analysis

The most popular sequencing techniques include internal transcribed spacer (ITS) region that used for sequencing for 16S ribosomal RNA gene sequencing for bacteria and fungus, among other, less popular target genes. These genes have despite being highly conserved, have diverged over time and offer a individual barcode that can be attributed to particular taxonomies and used to determine the rate of each member in the microbial community. Finding and establishing a distinct sequence is one difficulty that comes up while analyzing these marker genes.

There are hypervariable areas in both the 16S and ITS genes that can alter quickly and even vary within a single cell. Furthermore, various taxa may have identical gene sequences (such as Escherichia and Shigella), which makes it more difficult to distinguish between distinct taxa and define a unique sequence (Větrovský & Baldrian, 2013). The Illumina MiSequence allows for a sequencing length that is most commonly used for current 16S and ITS sequencing. The 2×300 sequencing length is commonly used to cover as many variable regions as possible or achieve the maximum accuracy.This sequencing technique generates a complete amplicon for every taxon in the microbial group by sequencing from both the forward primer and the reverse primer using region-specific primers, such as V1–V3 or V4.

3.3.2 Shotgun Metagenomics

Marker gene sequencing techniques are the type of techniques that used to analysis of a small units of microbial genomes. It have provided amazing insights into the function of the microbiome in health and illness. Shotgun metagenomics techniques include techniques that use untargeted sequencing techniques to identify all microbial genome found in the sample (Quince et al., 2017; Sharpton, 2014). It is possible to analyze DNA unitof viruses, bacteria, fungi, and other microorganisms by capturing the entire repertoire of genetic information from a microbiome sample; however, this process is reliant on reference genomes and scientific expertise (Huffnagle & Noverr, 2013; Stern et al., 2019).Finding gene coding sequences and mining the functional potential of the microbial community (Qin et al., 2010), or determining in this taxa are extant and their comparative abundance (Xia et al., 2011) can be accomplished with this information in a manner similar to that of amplicon sequence-based techniques. The optimal assembly techniques for shotgun metagenomic data are still up for debate. Metagenomic shotgun assemblies can be made using reference genomes, de novo, or a combination of the two. Short read data can be combined using overlapping reads to create contigs (Overlap, Layout, Consensus assembly), much like whole genome sequencing, although this approach is impracticable due to its high computing requirements.

3.3.3 Metatranscriptomics, Metabolomics, and Metaproteomics

Like shotgun metagenomics, metatranscriptomic techniques use comparable analytical concepts, but they focus on capturing the RNA that is transcribed from microbial cells, enabling evaluations of the expression activities of these organisms (Bashiardes et al., 2016). Shotgun metagenomics and metatranscriptomic methods primarily rely upon Illumina sequencing methodologies, most often the HiSeq or NovaSeq families of instruments due to the high throughput and low cost per base. However, there has also been a movement toward PacBio and Oxford Nanopore sequencing technologies to leverage longer read lengths that aid in gene calling and genetic mapping to reference genomes. A standard workflow involves the isolation of the total RNA from the microbiome sample, RNA enrichment, fragmentation, cDNA synthesis, and preparation of transcriptome libraries for sequencing. RNA sequence readings are usually mapped to various genomes and pathways (KEGG) (Kanehisa & Goto, 2000). In order to determine the taxonomy of organisms that are transcriptionally active and the role of the genes that are expressed in these organisms. Metatranscriptomic data from microbiome samples has been aligned and assembled using bioinformatics tools such as SOAP denovo. To ascertain which pathways are up- and down-regulated or during various health and illness states, comparisons are done between various groups or variables. The crucial goal of metabolomics investigations is to describe the metabolites that the microbiota generates a methods in which these products affect the microbiota and host metabolism (Lamichhane et al., 2018). Minor compounds like antibiotics, antibiotic biproducts, and intermediates of bacterial and/or host metabolism are commonly quantified using these techniques. To identify known compounds, mass spectrometry is normally used in metabolomics. Although metaproteomics also creates the use of mass spectrometry, its primary goal is to detect and measure the proteins that are present in a microbiome (Lai et al., 2019; Blakeley-Ruiz et al., 2019). The tools of metabolomics and metaproteomics for microbiome research are both evolving quickly.

Table 1.

Technology	Technology Platform	Common Software	Pros	Cons
Gene Marker Analysis	Next Generation Sequencing	QIIME, Mothur, VEGAN, phyloseq, DADA2	Cost-effective, analytical pipelines widely accepted	Lacks clear functional information, potential errors in differentiation of taxa
Shotgun Metagenomics	Next Generation Sequencing	IDBA-UD, SPAdes, MEGAHIT, MetaPhlAn2, MG-RAST, HUMAnN2	Captures all microbial genomes present within a sample	Expensive, computationally demanding, no consensus on analytical pipelines
Metabolomics	LC/GC-MS	–	Allows for profiling of the metabolites microbiota produce, semi-quantitative	Origin of metabolite unknown
Metaproteomics	LC/GC-MS	–	Allows for identification and quantification of the proteins in a sample	Origin of protein unknown
Metatranscriptomics	Next Generation Sequencing	SOAP denovo	Allows assessments of gene expression	Protein expression may depend on translation and post-translational modifications

4. NUTRIGENOMICS AND ITS ROLE IN HEALTH

4.1 Gene-Diet Interactions

Genetic variations influenced food metabolism and specific responses to dietary intake, which is crucial for improving health and preventing disease. A number of elements of human diets, most remarkably the Modern Western Diet (MWD), have the potential to result in detrimental gene–diet interactions when paired with the influence of various genotypes on the metabolism of specific nutrients. The expression of genes linked to metabolism is influenced by these interactions, which in turn affects the amounts or functions of the enzymes that produce or catabolize that nutrient. In the end, this can change clinical phenotypes, such as human disease, and molecular phenotypes, such as the amounts of bioactive nutrition products and their metabolites. A more thorough comprehension is required because to the intricate relationships between genes and nutrients, which are impacted by a number of hereditary and environmental reasons. From fundamental research to practical uses, the fields of nutrigenetics and nutrigenomics investigate the connection between genes and nutrition. Scientists have started to interpret the code of good health by comprehending how genes affect the body's reaction to nutrition or how nourishment changes the body's reaction to genetic coding variances. Determining which particular meals fit a person's DNA to produce the best biological response might be aided by personal profiling of genetic and nutritional responses. Individuals' metabolic regulation is directly compressed by their genomic makeup. Nutritional genomics sheds light on how genetic variants can help people and societies modify their diets (Tekpınar & Erdem, 2019; Farhud & Yeganeh, 2010).

We are discuss few example:-

Genes and cytochrome P450 enzymes

P450 genes encode a class of enzymes known as cytochrome P450 (CYP) enzymes, which are membrane-bound proteins mostly present in the liver's endoplasmic reticulum (Elsamanoudy et al., 2016). As monoxygenases, CYP enzymes influence oxidation by transferring one oxygen atom concluded a series of processes.

For example, the CYP1A2 gene can be induced by indole-3-carbinol found in cruciferous vegetables, by heterocyclic amines in prepared meats, and by PAHs found in grilled meat. CYP1A2 is also inhibited by the compound naringenin, which is found in grapefruit (Malcomson & Mathers, 2023).

Glutathione S-transferases (GSTs)

A significant family of cytosolic enzymes known as the GSTs catalyses the conversion of reduced glutathione to many electrophilic chemicals that are produced by "cytochrome P450 enzymes" . Since the ability of electrophiles have to attach with DNA and create adducts and possibly DNA mutations, GSTs are essential for shielding cells from the mutagenic and cytotoxic significances of these reactive substances. Therefore, it has been proposed that GST induction reduces overall cancer susceptibility and that, on the other hand, defective detoxification by GST increases disease susceptibility.

Example, GSTA makes up 80% of the total hepatic GST protein, GSTM makes up 10% to 20%, and GSTP makes up less than 5%. GST and GSTA isozymes make up just 7–8% of the lung, while GSTP is the most common type (85%).

4.2 Nutritional Modulation of Gene Expression

By directly interacting with regulatory regions in the genome, certain dietary components can affect gene expression and alter a gene's transcription rate (Figure 1). Examples include zinc, short- and long-chain fatty acids, metabolites including retinoic acid, a derivative of vitamin A, and calcitriol, a derivative of vitamin D, as well as steroids.

However, by physiological manipulation of secondary mediators, including metabolites generated by gut microbiota, modifications in hormone signaling, mechanical stimulation, and post-translational modification of proteins, dietary fiber components may indirectly affect gene expression. For example:- butyric acid (Velázquez et al., 1996). By using molecular biologic techniques, the study of food control and gene expression has advanced to the forefront of nutrition research. It is challenging to determine how dietary fiber affects gene expression since these effects are most likely indirect and involve the relations of many different variables (e.g. enzymes, hormones, and cellular secondary messengers).

Nonetheless, significant data has been gathered from research associating the impact of diets at high fiber to low fiber on particular gene products. As like the study of gene expression and dietary regulation has been by attaching to proteins that interact with metal response elements in the gene promoter, dietary zinc is known to control the expression of the metallothionein gene. This connection offers one of the most thoroughly studied instances of a nutrient's ability to regulate gene expression. Here, it is examined as an illustration of a gene–nutrient relationship that is reasonably well understood. Zinc's impact on other genes' expression, however, has not been thoroughly studied (Cousins, 1999).

Figure 3. Gene expression

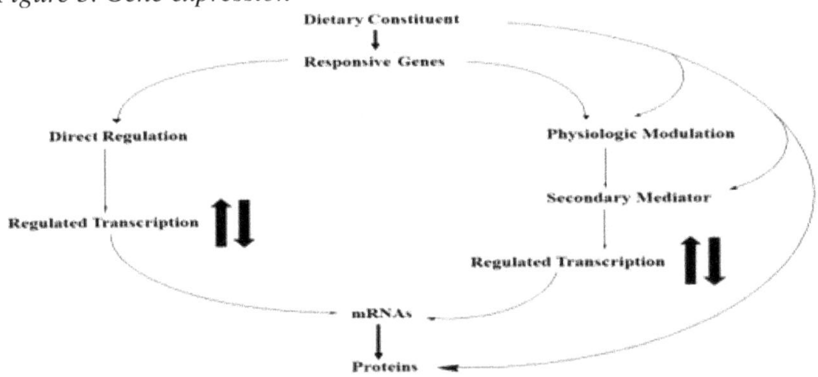

4.3 Personalized Nutrition Strategies

Recent research, on the connection between nutrition and genome that produced positive results. There is enough scientific proof to provide individualized dietary guidance based on genotypic data. "Personalized nutrition aims to improve or adjust official guidelines for each person where there is evidence to support it, rather than replacing them. This practice is as old as the guidelines themselves, thus it is not a recent development: Lactose-intolerant persons are advised to limit or avoid fresh dairy products, while overweight people are encouraged to consume less calories than the recommended amount. Now, we have the chance to examine the data from gene-diet interaction studies. Nutrigenetics is a component of a broader discussion about personal genetics that started in the UK in 2001 with the establishment of Sciona Ltd. and continued after 23andMe, deCode, and Navigenics were founded in 2007 (Görman et al., 2013).

4.3.1 Nutrigenetics based diet planning

Nutrigenetics is the study for genetic that how to show differences and specific responses to nutrients, forming the basis of personalized diet planning. Unlike general dietary guidelines, nutrigenetics-based diets are tailored to a person's genetic makeup, optimizing health outcomes and disease prevention through individualized nutrition (Ordovas & Corella, 2004). Single nucleotide polymorphisms (SNPs) is a type of genes influence metabolic rate, nutrient absorption, and disease susceptibility. For example, individuals with polymorphisms in the FTO (Fat mass and obesity associated) gene are more related to obesity due to increased appetite and reduced satiety response. Personalized interventions in such individuals may include high-fiber, low-energy-density diets to promote fullness and reduce caloric intake (Loos & Bouchard, 2008). MTHFR is another extensively researched gene that influences the metabolism of folate. Because they may have greater homocysteine levels, people with the 677C>T mutation need to consume more folate to avoid cardiovascular problems (Bailey & Gregory III, 1999). Based on genetic inclination, nutrigenetics-based planning guarantees that these nutrient gaps are proactively addressed.

Applications of nutrigenetics have been improved by recent developments in artificial intelligence (AI). Large-scale genomic data can be analysed by AI systems to find nutrient-gene correlations and given the best diets. Then, depend on the user's genetic profile, mobile apps and online platforms can deliver real-time nutritional recommendations (Kussmann & Fay, 2008).

4.3.2 Gene diet interaction model

The gene-diet interaction model looks at how a genetic feature affect the human and how they react physiologically and metabolically toward particular foods. Gene-diet interaction demonstrates how genetic variations might alter nutrient metabolism, illness risk, and dietary requirements, in contrast to conventional nutrition, which presumes consistent dietary effects (Kaput, 2008). One well-known instance is the TCF7L2 gene, where specific polymorphisms are highly associated with an elevated threat of type 2 diabetes. Research indicates that high-fiber diets, which enhance insulin sensitivity and glycemic control, may be more advantageous for people with the risk allele. The APOA2 gene is another well-established interaction, whereby certain alleles affect lipid profiles and weight gain in response to consumption of saturated fat. Compared to generic recommendations, customised diets based on such genetic insights can help lower cardiovascular risk more successfully (Lai et al., 2019). By facilitating the combination of extensive genetic, nutritional, and health data, artificial intelligence (AI) and machine learning (ML) have improved the analysis of gene-diet interactions. Based on a person's genotype, these techniques can create predictive models for nutritional effects and spot trends in complicated datasets (Afman & Müller, 2006). For instance, using a mix of lifestyle factors, gut bacteria, and gene variations, AI can forecast blood glucose responses.

4.3.3 DNA based weight management program

Genetic insights are used by DNA-based weight control programmes to develop individualised plans for reaching and sustaining a healthy body weight. Compared to generalised weight-loss regimens, these programmes allow for more targeted interventions by examining genetic variants that affect appetite, fat metabolism, energy expenditure, and response to diet or exercise (Loos & Bouchard, 2008).

The 'FTO' (Fat Mass and Obesity Associated) gene is one of the best researched genes in this area. Increased appetite and calorie consumption are associated with variations in FTO, which raises the risk of obesity. Research indicates that people with the risk allele gain more from low-glycemic, high-protein diets because they help control hunger hormones and increase feelings of fullness (de Toro- Martín et al., 2017).

Additionally, genes like ADRB2 (beta-2 adrenergic receptor), which influences lipolysis and exercise response, are evaluated using DNA-based programmes. Resistance exercise may not be as helpful in helping people with certain variations shed weight as endurance-based training. Furthermore, UCP1, a gene that affects thermogenesis, establishes metabolic efficiency and can help with food planning and calorie requirements.

By combining genetic information with lifestyle indicators like stress, sleep, and food records, AI technologies improve these programmes by offering dynamic, genotype-specific advice. This degree of personalisation is provided by commercial applications available on platforms such as DNAfit and Nutrigenomix.

5. AI AND MACHINE LEARNING IN BIOMEDICAL RESEARCH

5.1 Fundamentals of AI and ML

The ultimate goal of artificial intelligence (AI) was copied and understanding like the human brain in order toward identify future issues and real-world problems using the most intelligent possible approach. Researchers and scientists around the world were extremely excited about the developments in inventions that resulted from the innate drive to create newer and more advanced technology. In the past, artificial intelligence has drawn a lot of users since it helps with scientific research, enterprises, and medical sciences, among other things. When used carefully, these systems produced unexpectedly precise findings and prevented (Forestier et al., 2017).

AI in health care and drugs may be a new analysis approach that mixes advanced realistic and computational approaches with the insights of professional doctors to deliver new innovations for rising health care in the form of tools. AI is the study of concepts, that modifies computers to try and to do the items that create individuals intelligent (Salman et al., 2018). Machine learning (ML) is a field of artificial intelligence (AI) that deals with the implementation of computational algorithms that increase performance upon experience; in other words, a ML system learns from data (Salman et al., 2018; Jordan & Mitchell, 2015). In many situations, multidimensional data can be considered using unsupervised learning, and then supervised learning is used to the resulting dataset to obtain more insight. Coordinating transcriptomes sequenced from individual cells to identify transcriptionally comparable cell clusters is a common use case for unsupervised learning (Plemel et al., 2020).

5.2 Applications in Genomics and Microbiomics

Artificial Intelligence and Machine learning algorithms in microbial genomics-

In 1959, "Arthur Samuel" presented machine learning (ML), and artificial intelligence that examines huge amounts of various data using algorithms to forecast actual occurrences . Instead than following precisely defined instructions like traditional software programs do, it learns from a variety of datasets to find patterns and generate predictions (Bi et al., 2019; Jordan & Mitchell, 2015). Using genomic data to classify pathogenic organisms has revolutionized microbiological research. It enhances the analysis of microbial group conformation and optimizes the strategy of genetic circuits by forecasting optimal route layouts and regulatory components, allowing for the quick prototyping and analysis of newly created biological systems (Blanco-Míguez et al., 2023). It's used to investigate the functional potential of microorganisms and methodically clarify interactions between microflora and hosts.

5.2.1 Adding AI with multi-omics for functional annotation

There are Multi-omics data, such as transcriptomics, metabolomics, proteomics and genomes, are increasingly being included via AI and ML models to develop functional annotations and found a very deep empathetic of microbial ecosystems (Farhud & Zokaei, 2021). In order to extract biologically significant insights from the massive and complicated datasets generated by multi-omics methodologies, advanced computational techniques are needed. The functional characterisation of genes and their regulatory networks is enhanced by AI-driven techniques that make it easier to correlate metabolic pathways and gene expression levels (Ferrocino et al., 2023).

5.2.2 Machine learning for antimicrobial resistance

Microbes' variety and flexibility make them crucial markers of ecosystem shifts. They change rapidly according to physical and chemical circumstances. Microbial communities have a remarkable ability to accurately report changes in their environment. Machine learning approaches have the potential to uncover hidden evidence in microbial data, improve ecosystem comprehension, and make it easier to identify crucial environmental changes. By studying environmental microbiomes, microbial ecology seeks to comprehend species interactions, such as symbiosis, competition, predation, and facilitation, as well as their distribution and maintenance processes. It is crucial to comprehend how the microbial network is structured and stable in various environments.

5.3 Data Integration and Predictive Modeling

To maintain a competitive edge and derive actionable insights, it is imperative to effectively integrate and manage this enormous volume of heterogeneous data. Modern data settings are becoming more complicated, and traditional data management techniques are becoming less effective. This has prompted the development of sophisticated data platforms that usage state-of-the-art technologies to expedite the data management and integration process.

The processing, analysis, and use of data have been completely transformed by artificial intelligence (AI) and machine learning (ML), which are now crucial parts of these contemporary data platforms. Algorithms based on AI and ML provide advanced techniques for intelligent data governance, data quality assurance, and task automation in data integration. The efficiency and efficacy of data management systems are improved by AI and ML through the automation of repetitive operations, pattern recognition, and predictive analytics.

Figure 4. AI in biomedical research

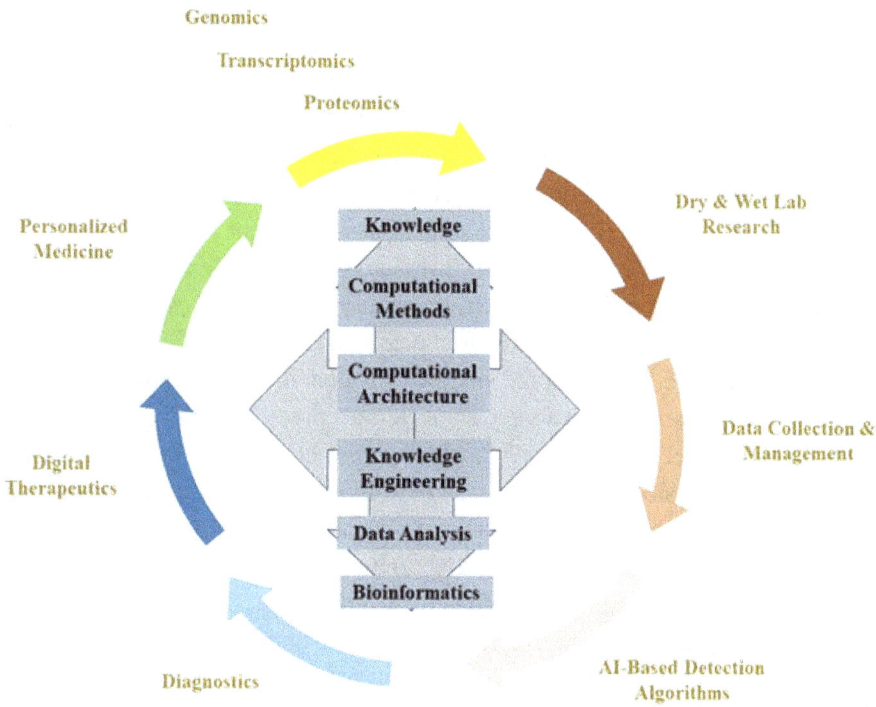

6. AI-DRIVEN MICROBIOME ANALYSIS

6.1 Pattern Recognition and Microbial Profiling

Rapid pattern detection and in-depth microbial profiling from complex metagenomic data are made possible by artificial intelligence (AI), which has greatly improved microbiome analysis. The statistical approaches that were used extensively in traditional microbiome analysis were unable to effectively handle the high dimensionality and heterogeneity of microbial datasets (Lloyd-Price et al., 2016). Researchers are able to find microbial signatures and community structures that are associated with health or illness situations by using artificial intelligence (AI), namely machine learning (ML) and deep learning (DL) algorithms (Topçuoğlu, 2022). For example, even in cases when conventional alignment-based methods are ineffective, convolutional neural networks (CNNs) have been used to consider microbiome data in order to detect patterns of microbial diversity and abundance. Microbial population patterns linked to food practices, medicine use, and regional variances can be found using unproven learning systems such as principal component analysis (PCA) and clustering. These techniques use latent patterns in 16S rRNA or shotgun sequencing data to identify previously unidentified taxa in addition to identifying recognized microorganisms (Franzosa et al., 2015). AI-powered microbial profiling enables precise genus or species-level classification of microbiome composition. Labelled microbiome samples have been used to train support vector machines (SVMs) and casual forest classifiers to predict host characteristics like inflammatory bowel disease or obesity (Duvallet et al., 2017). Additionally, AI models make it possible to include metadata like age, food, and drug use, which enhances the contextual comprehension of microbiome dynamics.

By distinguishing between harmful and benign microbe patterns, this method aids personalised medicine by enabling the customisation of nutritional treatments or probiotics. Therefore, using microbiome data for precision health applications is based on AI-driven pattern detection.

6.2 Disease Risk Prediction

AI-driven microbiome analysis, which finds microbial signatures linked to certain diseases, is revolutionary in predicting illness risk. Inflammatory bowel disease (IBD), diabetes, and colorectal cancer can all be predicted by patterns found in highly -dimensional microbial information that machine learning prototypes can analyse (Pasolli et al., 2016). To identify people into risk categories with a higher degree of accuracy, these models learn from labelled microbiome datasets. For instance, utilizing characteristics like microbial diversity, Bacteroides abundance, and the presence of pro-inflammatory species, random forest and gradient boosting classifiers have been trained using gut microbiome profiles to expect the threat of type 2 diabetes (Qin et al., 2010). In certain studies, these models show the predictive power of microbiological data, outperforming conventional indicators such as BMI or glucose levels (Wirbel et al., 2019).

Unsupervised pattern detection in microbiome data is made possible by deep learning systems, specially autoencoders and convolutional neural networks (CNNs), which uncover hidden microbial characteristics linked to illness. By identifying subclinical microbial imbalances, or dysbiosis, before symptoms appear, these techniques offer a proactive approach to health treatment (Loomba et al., 2017). In addition, AI models can use multi-omic data to enhance disease risk assessment, including host genomes, metabolomics, and metagenomics .

6.3 Real-Time Monitoring and Feedback Systems

Artificial intelligence (AI) has transformed real-time monitoring and feedback systems in microbiome investigation, enabling ongoing health surveillance and tailored therapies. These days, data generated by wearable biosensors, smart toilets, and home-based microbiome testing kits is processed by AI algorithms to provide useful insights about disease risks, food, and gut health.

To identify dysbiosis patterns and notify users of upcoming gastrointestinal problems like flare-ups of inflammatoryi bowel disease (IBD) or occurrences of irritable bowel syndrome (IBS), for instance, AI can process sequential faecal microbiota data obtained via in-home sensors (Schupack et al., 2022). Deep learning methods that can monitor microbial changes and correlate them with dietary or lifestyle modifications make these predictions possible (Duballet et al., 2017). Furthermore, healthcare providers can remotely monitor patients by integrating cloud-based AI dashboards with real-time microbiome data, which improves the management of chronic diseases. Patients receiving immunotherapy, for example, can have their microbiome diversity tracked for changes that could impact treatment results. Monitoring public health is also aided by real-time feedback systems. Predictive public health responses are made possible by AI platforms that can examine microbiome trends across populations to monitor disease outbreaks or patterns of antibiotic resistance (Franzosa et al., 2015).

7. ARTIFICIAL INTELLIGENCE IN NUTRIGENOMICS

7.1 Personalized Diet Planning

The combination of "artificial intelligence" and "nutrigenomics"— the learning of how genes interact with foods to affect health—is accelerating the evolution of personalised diet planning. Metabolic reactions to foods like lipids, carbs, and caffeine, including APOA2, CYP1A2, and FTO. For example, high-protein diets can help people with specific FTO genotypes maintain their weight because they respond better to them. These days, AI systems are being created to combine this genetic information with personal health indicators to suggest individualised food regimens. These artificial intelligence (AI) systems can process multidimensional data, such as phenotypic, microbiome, metabolomic, and genetic information, to create meals that enhance daily wellness and prevent chronic diseases like diabetes and obesity. Additionally, AI improves real-time dietary feedback via wearable technology, smartphone apps, and ongoing monitoring, making dietary changes more responsive and dynamic. In over 75% of those research, meal suggestions were personalised using machine learning (ML) approaches based on lifestyle trends, physiological data, and eating behaviour. The book emphasised that AI technologies shown excellent potential for practical use, achieving high accuracy—average nutrient estimation errors were below 15% and food recognition accuracy went up to 99.85% .

These tools also do more than just track calories. Certain models use machineilearning algorithms, such as collaborative approaches, decisioni trees, and neural networks, to forecast glycemic reactions or nutrient deficits. This makes it possible to provide dietary advice that is more accurate, timely, and physiologically suited to each person's requirement.

7.2 AI in Nutrient-Gene Mapping

The concept of "nutrient-gene mapping," which describes how genes and nutrients interact, is being redefined by artificial intelligence (AI). Precision nutrition, which customises nutritional interventions based on each person's unique genetic profile, depends on this process. It was accustomed examine only nucleotide polymorphisms (SNPs) and forecast their effects on metabolic reactions. The FTO gene, for example, is linked to the metabolism of fat and the risk of obesity, whereas CYP1A2 influences the metabolism of caffeine. Through the use of supervised learning algorithms, AI is able to categorise people according to SNP patterns and suggest individualised dietary intakes, such as the ideal amounts of protein or coffee. The capacity of AI in this area to handle large and intricate datasets that contain lifestyle factors, biomarkers, and genetic variations is one of its main advantages.

These models increase the accuracy of predictions and enable more comprehensive and scalable study into gene-diet connections. Patterns that would be very hard to find manually are detected using techniques like neural networks and decision trees (Ramos-Lopez, 2024). Genes & Nutrition explores how medical professionals—dietitians in particular—view the practical application of AI-powered nutrigenomics. Although there is hope for its future, the study highlights real-world issues, primarily the absence of standardised procedures, scant clinical validation, and inadequate practitioner training (Mozaffarian et al., 2018). Dietitians also pointed out that for AI solutions to be successfully used in real-world situations, they need to be understandable and easy to use. Additionally, emphasis was placed on ethical concerns such as algorithmic prejudice and data privacy. In order to nearby the gap among invention and clinical application, the research advocates for collaborative frameworks involving clinicians, AI developers, and policymakers (Karantonis, 2025).

7.3 Optimization Algorithms for Health Outcomes

Artificial intelligence (AI) systems can create adaptive, data-driven, and individually tailored dietary plans that are aimed at enhancing health outcomes thanks to optimisation algorithms in nutrigenomics. These algorithms create dynamic nutrition programs that change over time by analysing complicated biological data, such as metabolic indicators, microbiota profiles, and genetic information. One method makes procedure of machine learningi(ML) and deep learning (DL) models, such as neural networks, casual forests, and maintenance vector machines, which forecast how each person would react to particular foods based on phenotypic and genetic information. These models have been used to predict cholesterol levels, postprandial glucose responses, and even inflammation markers, enabling clinicians to suggest diets that are most likely to be beneficial to each patient's health. Real-time optimisation through smartphone applications that use computer vision (e.g., convolutional neural networks) to evaluate food photos, calculate nutritional value, and provide substitutes is another development. In order to improve future recommendations, these systems frequently make use of feedback loops, in which user input—such as blood glucose or activity data—is continuously given back into the model.

8. INTEGRATIVE APPROACHES: MERGING AI, MICROBIOME, AND NUTRIGENOMICS

8.1 Multi-Omics Data Integration

The humanimicrobiome is an involved ecosystem made up of bacteria, viruses (including bacteriophages), fungus, archaea, and occasionally a range of micro-eukaryotes. These organisms function differently in different ecological niches within the body and change depending on the environment and population. Many of the startling findings into microbiome–host interactions that have already been discovered by multi-omic investigations have come from dietary circumstances. For instance, by increasing the number of bacteria that make propionate rather than butyrate, the combination of metabolomics (SCFA, cytokines, bile acids, hormones),I metagenomics, and amino-acid labelling has shown how fermentable fibers improve satiety (Deehan et al., 2022). The study of microbiome data also has problems, like over-reliance on databases and sensitivity to analysis procedures. During the bioinformatic analysis process, while employing several differential analysis software, varying settings, or conforming to various reference materials, researchers could get various outcomes. Additionally, integrative multi-omics has made it easier to comprehend the intricate interactions between diseases and provides a clear path for implementing treatment plans. In nutrigenomics, multi-omic data integration is essential for determining how nutrition interacts with genetic, epigenetic, transcriptomic, proteomic, metabolomic, and microbiome components to affect personal health outcomes. Researchers can comprehend gene-nutrient interactions and forecast the influence of specific dietary components on the metabolic pathways and illness hazard by integrating these layers (de Toro-Martín et al., 2017). In multi-omics research, artificial intelligence (AI) has become a game-changing tool that provides answers to numerous problems relating to data integration, analysis, and interpretation. Multi-omics applications benefit greatly from the higher performance of AI-powered algorithms, especially deep learning models, when processing high-dimensional, diverse, and sparse datasets. AI helps researchers find new biomarkers, forecast disease trajectories, and find treatment targets that would be challenging to find using conventional statistical methods by learning complicated patterns across several omics layers. A Deep learning, which uses neural networks to identify significant characteristics from high-dimensional data, is one popular AI technique in multi-omics analysis. One kind of unofficial deep learning model that has been especially successful at compacting multi-omics data while maintaining physiologically significant information is autoencoders (Debbadi & Boateng, 2025).

8.2 Development of Personalized Health Platforms

Personalized health platforms use nutrition models and sophisticated algorithms to create personalized meal plans constructed on a patient's age, gender, body mass index (BMI), metabolic rate, disease kind, and severity. According to research, Higher fiber consumption from whole grains, cereals, and vegetables may help stabilize blood sugar levels and decrease the absorption of carbohydrates in patients with insulin resistance. Real-time food intake tracking is possible with the platform thanks to sensors and mobile devices. Through the use of a mobile app, patients input food photographs, and the system uses image recognition and a food database to automatically identify different types of food, determine calorie consumption, and analyze nutritional content (Giuntini et al., 2022). Personalized health platforms can identify acute complications and anticipate disease risks by utilizing AI-driven algorithms and big data analytics. These models evaluate the risk of acute events like diabetic ketoacidosis or hypertensive crises by taking into account lifestyle, genetics, medical history, and real-time health data (McCoy et al., 2021). Although the features of these AI-based platforms vary greatly, they all aim to use AI to increase the quality and cost-effectiveness of healthcare. AI programs created to improve or customize current medication regimens. These evaluate a great deal of information on a single patient, such as their lifestyle, medical background, and genome, and modify treatment to fit that patient. The distinctive feature of these platforms is that they take into account a lot more variables than just the medication; the objective is to better tailor treatment to the patient while maximizing efficacy and minimizing negative effects.

8.3 Case Studies and Clinical Applications

8.3.1. Glycemic management and personalised nutrition

Using machine learning models to predict people's postprandial glycemic responses, the PREDICT 1 study is one of the most well-known instances of AI integration with microbiome and nutrigenomics. This was accomplished by examining the genomic profiles, lifestyle choices, and gut microbiome makeup of the subjects. According to the study, those who ate the same meals experienced notably varied rises in blood sugar levels, underscoring the need for tailored dietary treatments in the action of metabolic diseases such as - type 2 diabetes (Zeevi et al., 2015). This was a significant step in converting multi-omic data powered by AI into useful, applicable dietary recommendations.

8.3.2. Early Intervention and Disease Risk Prediction

By using a deep learning model that combines host genomes, microbiome data, and nutritional habits to forecast the risk of colorectal cancer demonstrated a unique clinical application. The model outperformed conventional risk assessment techniques by achieving an accuracy of above 90%. This method is an example of how predictive healthcare might be made possible by AI-facilitated multi-omic integration, which may lead to proactive disease prevention measures and early screening.

8.3.3. TheiImpact of Microbiota on Drug Reaction

The gut microbiota effects the safety and effectiveness of drugs in precision pharmacotherapy, which is another potent application. Digoxin and irinotecan, for example, have different metabolisms depending on the gut microbial mix. The ability of AI-based pharmaco-genomic models that incorporate microbiome factors to forecast drug response, minimise adverse effects, and enhance treatment efficacy was emphasized. This illustrates how pharmaco-genetics has crucially expanded into personalised medicine influenced by the microbiota (ElRakaiby et al., 2014).

8.3.4. Predicting Youth Obesity Risk

To counteract the rising obesity rates, particularly among younger populations identified adolescents at risk for obesity and metabolic syndrome by combining AI algorithms with SNP and gut microbiome data. AI can support personalised preventative care initiatives from an early age, as demonstrated by the creation of targeted food and lifestyle treatments that are geared at prevention rather than therapy thanks to this multi-omic approach (Sugimoto et al., 2020).

8.3.5. Enhancement of Cardiovascular Health

According to their microbiome profiles and genetic predispositions examined how people react differently to dietary fibres, particularly in terms of their impact on inflammation and cholesterol. They were able to forecast cardiovascular outcomes and create customised dietary therapies for heart health by using AI models. This example shows how combining nutrigenomic and microbiome data can improve nutritional therapy for preventing cardiovascular disease.

9. ETHICAL, LEGAL, AND SOCIAL IMPLICATIONS (ELSI)

This section begins with the statement that conventional Ethical, Legal, and Social Implications (ELSI) issues emphasized the hazards of using data for secondary research outside the bounds of conversant consent, particularly in relation to advanced Big Data technology. In example, data privacy risks (such as those resulting from illegal usage or data breaches characterize an out-of-consent scope of issues where possible patient harm is frequently difficult to measure. A new set of ELSI concerns relating to the performance, efficacy, risk, interpretability, governance, and acceptance of AI/ML has surfaced as the technology becomes increasingly likely to impact the health of human patients. A set of dangers may arise that fall within the parameters of consent and intended usage, or models may perform poorly or have large error margins, or they may favorably benefit or hurt a group of people. Some of these more recent hazards are, at least in theory, directly quantifiable and manageable, as we shall see in this section even be unknown in certain circumstances.

9.1 Data-Privacy and Consent

Data privacy and conversant consent become crucial issues as AI becomes more and more integrated into personalised medicine, particularly in microbiome and nutrigenomics research. These domains entail gathering private genetic, microbiological, and lifestyle information from people—often over extended periods of time and at high resolution. Such information may be misused or exploited by third parties, like employers or insurance, if it is managed improperly (Shabani & Borry, 2018). There must be more than one instance of informed consent. People can continuously control their data-sharing preferences with the use of dynamic consent models, which are being promoted. ELSI talks have placed a strong emphasis on privacy. It has connections to other issues. The above foreshadows some of such junctions. One of the most important legal issues is privacy. The necessity for regulation and the challenge of permitting data gathering and use for good while prohibiting bad ones are acknowledged in privacy discussions. The legal and medical literatures largely concur that this is a troublesome issue, even though there may be disagreements on exactly how to achieve it. In order to ensure user-centric methods in biomedical AI applications, institutions must submit with rules such as the "General Data Protection Regulation" (GDPR) and "India's Digital Personal Data Protection Act", 2023, while ranking respect for autonomy.

9.2 Algorithmic Bias and Transparency

When using AI in microbiome and nutrigenomics research, algorithmic bias is a significant obstacle. AI prototypes can be simply as objective as the data that under represented communities may receive suggestions from the AI that are harmful or inaccurate if the training datasets are biased towards particular age groups, races, or socioeconomic backgrounds (Obermeyer et al., 2019). Individuals and the categories they are assigned by precision medicine algorithms based on data that purports to reflect them are both affected by fairness and bias considerations. Reducing prejudice and inadvertent bias in treatment decisions can be achieved by basing care on data. Nowadays, it is hard, if not impossible, to understand the data or the algorithms that are produced by artificial intelligence.

When algorithms are proprietary, important facets of their creation, testing, and verification may not be made public due to intellectual property rights. Clinicians are more hesitant to employ an algorithm if they lack faith in it and don't fully comprehend the rationale behind its recommendations (Kaplan, 2020).

9.3 Accessibility and Health Equity

Ensuring equitable access for all communities is crucial since AI shows a major role in personalised health, mainly in the areas of microbiome and nutrigenomics. Although these advancements hold promise for improved nutrition and illness prevention, they mostly benefit affluent, metropolitan, and technologically savvy populations at the moment. Access to sophisticated genetic testing or microbiome analysis methods is frequently limited for marginalised, low-income, and rural populations (Khoury et al., 2015). Many lack the literacy necessary to use AI health apps, cellphones, and dependable internet. Furthermore, because the majority of AI systems are trained on Western data, their suggestions for diet or microbiota are irrelevant or incorrect for diets that are culturally varied, particularly in places like India or Africa (Vayena et al., 2018).

Health disparity affects populations that have worse health outcomes and a higher burden of disease, including preventable diseases, injuries, and violence. It have many primary care consultants, health-care advocates, and policy officials want to improve health equity by addressing social determinants of health and access to care. Common examples of characteristics that influence health even-handedness include race, location, age, poverty, mental health, gender, and disability. By itself, access to healthcare predicts health effects and is impacted by many of the similar factors (Pratt et al., 2023).

10. CHALLENGES AND LIMITATIONS

10.1 Data Standardization and Quality

Lack of standardisation among datasets is a major obstacle when using AI to microbiome and nutrigenomic data. Inaccurate cross-study comparisons might result from batch effects and biases introduced by various sequencing platforms, sample handling techniques, and annotation tools. Furthermore, poor genome coverage and inconsistent dietary reporting are common problems in nutrigenomic investigations, which might result in heterogeneous findings.

Microbial communities are extremely dynamic and impacted through a change of influences, including age, location and nutrition (Lloyd-Price et al., 2016). It makes data from microbiome research particularly sensitive. Pre-processing, normalisation, and functional annotation processes must be standardised immediately to guarantee scalability and reproducibility.

10.2 Clinical Validation and Regulatory Hurdles

Regulatory and validation issues hinder the clinical application of AI-driven microbiome and nutrigenomics tools, despite encouraging outcomes in research settings. The majority of studies are still in the pilot or observational phase, and there aren't many randomised controlled trials (RCTs) to verify their effectiveness. Microbiome-profile-based personalised food therapies, for instance, have demonstrated advantages; yet, few of these interventions have received FDA approval or national guidelines' endorsement (Guan et al., 2025).

It also makes regulatory approval more difficult due to the lack of agreement on clinical diagnostics based on nutrigenomic or microbiome data. Additionally, AI algorithms need to clear, understandable, and obedient with laws like HIPAA (Health insurance protability and accountabilityact) and the General Data Protection Regulation (GDPR), particularly when handling genomic data and medical information. For researchers and entrepreneurs looking to commercialise these technologies, the regulatory environment is a barrier because it is still lagging behind technological innovation.

10.3 Interdisciplinary Collaboration

The fields of bioinformatics, computational biology, clinical nutrition, microbiology, and healthcare must work closely together to successfully integrate AI, nutrigenomics, and microbiome science. However, innovation and execution are hampered by disciplinary silos, disparate terminologies, and a lack of mutual understanding (Jacobsen et al., 2020).

While data scientists might not have a thorough understanding of clinical settings, clinicians could not fully comprehend omics data or AI models. AI-generated insights can therefore be underutilised or mistrusted. Collaborative research ecosystems and interdisciplinary education and training programmes are crucial to overcoming this.

11. FUTURE PERSPECTIVES AND INNOVATIONS

11.1 Emerging Technologies

Personalised medicine could undergo a revolution thanks to emerging technologies like digital twins and synthetic biology. Using AI, genetics, microbiome data, and medical records, a digital matching is a effective patient that mimics the development of a disease and the effectiveness of treatment (Bruynseels et al., 2018). This makes it possible to use customised diet simulations in nutrigenomics to assess long-term impacts on gut microbiota and metabolic health.

Digital health innovations are expressively reforming our healthcare systems by integrating a variety of advanced technologies that raise the delivery of more accurate, personalized, and active care. At the forefront of these innovations, digital sensors and IoT devices play a crucial role by constantly capturing data from patients and their environments. This data is transmitted to the cloud, where sophisticated big data tools and analytics extract actionable insights. Such a near-real-time flow of information is essential for creating a active digital imitation of a patient, effectively reflecting various stages of disease progression and providing clinical with a deeper understanding of a patient's health journey.

New treatment options are made possible by synthetic biology, which allows gut bacteria to be engineered to either create advantageous metabolites or decrease detrimental ones. Engineered probiotics can introduce gene-editing tools straight into the gut, enhance nutritional absorption, or regulate inflammatory responses Synthetic biology has accelerated the pharmaceutical industry's progress by increasing the capacity to manufacture treatments. Pharmaceutical goods were mostly restricted to tiny molecules prior to the development of recombinant DNA technologies. Previously, insulin was extracted from animals; however, in 1982, the first commercial synthetic human insulin was manufactured using E. coli that had been modified. More biologics, including protein and RNA-based medicinal solutions, have now been developed and manufactured thanks to recombinant DNA technologies (Johnson-Laird, 1983).

11.2 Predictive and Preventive Healthcare Models

The practice of medicine has a lot of potential for the future thanks to predictive, preventative, and personalized medicine (PPPM). This strategies key elements are well-planned population screening procedures that make use of cutting-edge diagnostic biomarkers of illness states, focused prevention of simple human diseases, efficient cure planning, and personalized medicine, all of which contribute to a significant improvement in quality of life. Since it challenges social and ethical concerns about healthcare affordability and access, this strategy also has the potential benefit of providing care at lower prices to the general public. A large percentage of the population may be able to live to and beyond 100 years of age while maintaining good physical and mental health and being active contributors to society if there is widespread distribution and regular clinical application of cutting-edge technological approaches. The screening of healthy people for probable diseases later in life is one of the main topics covered. With interdisciplinary competence and timely preventative interventions, the general population's identified risk groups can have a reasonable chance for targeted diagnosis. The PPPM perspective takes into account the novel generation of molecular and nanotechnology established diagnostic techniques significant economic and social effects, possible application to therapeutic procedures, and the validity of the suggested blood tests (Golubnitschaja, 2012).

11.3 Global Impact and Public Health Potential

A revolutionary age in global public health is being helped in by the conjunction of artificial intelligence (AI), nutrigenomics and microbiome science, which has the probable to improve health outcomes for a wide series of populations. By facilitating a better comprehension of the ways in which genetic and microbiological factors impact dietary requirements and illness risks, these technologies pave the way for preventive, individualised, and fair healthcare solutions. Globally, AI-powered nutrigenomic models are able to examine genetic markers in various populations to identify the best dietary habits for preventing non-communicable diseases containing obesity, diabetes and heart disease. This has a particularly significant effect in areas where undernutrition and an increase in chronic diseases linked to diet result in the "double burden" of malnutrition (World Health Organinzation, 2021). microbiome-based strategies can also aid in the management of worldwide problems such as gastrointestinal illnesses and antibiotic resistance by encouraging customised dietary or probiotic therapies. By using these insights, policymakers can create national nutrition programmes that are effective and promote the Sustainable Development Goals for health equity set forth by the United Nation (Kumar et al., 2020).

In the end, the trinity of AI, microbiome, and nutrition promotes individualised treatment and democratises access to cutting-edge medical care, providing scalable solutions for global population wellness and disease prevention.

12. CONCLUSION

12.1 Key Takeaways

The fields of microbiome science, nutrigenomics, and artificial intelligence (AI) are radically changing personalised medicine. This trinity enables a thorough comprehension of the ways in which gut microbiota and individual genetic variants impact food responses, illness susceptibility, and health outcomes. AI is essential because it makes it possible to analyse large, intricate multi-omic datasets, find predictive biomarkers, and create customised treatment and diet plans.

Genetic and microbiological profiling-based personalised diet has already demonstrated promise in lowering inflammation, improving metabolic indicators, and controlling chronic illnesses like diabetes, obesity, and cardiovascular disease. Another level of personalisation is added by real-time health monitoring using wearable sensors and AI algorithms, which provide dynamic feedback and recommendations for changing behaviour.

12.2 Recommendations for Research and Policy

Strong interdisciplinary cooperation and policy support are crucial for this field to realise its full potential. In order to prevent bias in AI-driven models, future studies should prioritise ethnic and geographic diversity in genomic and microbiome datasets.

It is needed to found criteria for permission, privacy, and data interoperability. In order to create legal frameworks that guarantee patient autonomy and data protection in AI-health applications, policymakers must collaborate with engineers and ethicists. Furthermore, before being incorporated into accepted medical practices, the clinical value of microbiome and nutrigenomics findings needs to be confirmed by population-scale and longitudinal research.

Governments and healthcare systems should make investments in mobile health platforms and digital infrastructure, especially in areas with limited resources. These tools have the potential to democratise individualised treatment and close access gaps. Programmes for education and training are also required to give medical personnel the abilities to decipher AI-generated insights and use them in clinical practice.

12.3 Vision for the Future of Personalized Medicine

Personalised medicine is moving from a theoretical promise to a real-world application. In the future, real-time, adaptive healthcare platforms that are available through wearables or smartphones will incorporate AI, microbiome, and nutrigenomic findings. Personal health dashboards for each person will be updated continually by biological feedback, enabling them to make lifestyle decisions with previously unheard-of precision. AI will also help identify illness risks years before symptoms appear, supporting predictive and preventative healthcare approaches. The ultimate goal is to generate a healthcare system that is pre-emptive, inclusive, and sustainable. a healthcare system that is both globally equal and personalised, allowing everyone to take advantage of precision medical advancements regardless of background, region, or income.

RFERENCES

Armand, T. P. T., Deji-Oloruntoba, O., Bhattacharjee, S., Nfor, K. A., & Kim, H.-C. (2024). Optimizing longevity: Integrating smart nutrition and digital technologies for personalized anti-aging healthcare. In *Proceedings of the 2024 International Conference on Artificial Intelligence in Information and Communication* (pp. 243–248). IEEE. DOI: 10.1109/ICAIIC60209.2024.10463262

Bashiardes, S., Shapiro, H., Rozin, S., Shibolet, O., & Elinav, E. (2016). Non-alcoholic fatty liver and the gut microbiota. *Molecular Metabolism*, *5*(9), 782–794. DOI: 10.1016/j.molmet.2016.06.003

Blakeley-Ruiz, J. A., Erickson, A. R., Cantarel, B. L., Xiong, W., Adams, R., Jansson, J. K., Fraser, C. M., & Hettich, R. L. (2019). Metaproteomics reveals persistent and phylum-redundant metabolic functional stability in adult human gut microbiomes of Crohn's remission patients despite temporal variations in microbial taxa, genomes, and proteomes. *Microbiome*, *7*(1), 18. DOI: 10.1186/s40168-019-0631-8

Blanco-Míguez, A., Beghini, F., Cumbo, F., McIver, L. J., Thompson, K. N., Zolfo, M., Manghi, P., Dubois, L., Huang, K. D., Thomas, A. M., Nickols, W. A., Piccinno, G., Piperni, E., Punčochář, M., Valles-Colomer, M., Tett, A., Giordano, F., Davies, R., Wolf, J., & Segata, N. (2023). Extending and improving metagenomic taxonomic profiling with uncharacterized species using MetaPhlAn 4. *Nature Biotechnology*, *41*(11), 1633–1644. DOI: 10.1038/s41587-023-01688-w

Bruynseels, K., Santoni de Sio, F., & Van den Hoven, J. (2018). Digital twins in health care: Ethical implications of an emerging engineering paradigm. *Frontiers in Genetics*, *9*, 31. DOI: 10.3389/fgene.2018.00031

Bush, C. L., Blumberg, J. B., El-Sohemy, A., Minich, D. M., Ordovás, J. M., Reed, D. G., & Behm, V. A. Y. (2020). Toward the definition of personalized nutrition. *Journal of the American College of Nutrition*, *39*(1), 5–15. DOI: 10.1080/07315724.2019.1685332

Carpenter, K. J. (2003). A short history of nutritional science: Part 1 (1785–1885). *The Journal of Nutrition*, *133*(3), 638–645. DOI: 10.1093/jn/133.3.638

Cousins, P. D. (1999). Supply base rationalisation: myth or reality?. *European journal of purchasing & supply management, 5*(3-4), 143-155.

D'Urso, F., & Broccolo, F. (2024). Applications of artificial intelligence in microbiome analysis and probiotic interventions—An overview and perspective based on the current state of the art. *Applied Sciences (Basel, Switzerland)*, *14*(19), 8627. DOI: 10.3390/app14198627 de Toro-Martín, J., Arsenault, B. J., Després, J. P., & Vohl, M. C. (2017). Precision nutrition: A review of personalized nutritional approaches for the prevention and management of metabolic syndrome. *Nutrients*, *9*(8), 913. DOI: 10.3390/nu9080913

Debbadi, R. K., & Boateng, O. (2025). Optimizing end-to-end business processes by integrating machine learning models with UiPath for predictive analytics and decision automation. *Int J Sci Res Arch*, *14*(2), 778–796. DOI: 10.30574/ijsra.2025.14.2.0448

Deehan, E. C., Zhang, Z., Riva, A., Armet, A. M., Perez-Muñoz, M. E., Nguyen, N. K., Krysa, J. A., Seethaler, B., Zhao, Y.-Y., Cole, J., Li, F., Hausmann, B., Spittler, A., Nazare, J.-A., Delzenne, N. M., Curtis, J. M., Wismer, W. V., Proctor, S. D., Bakal, J. A., & Walter, J. (2022). Elucidating the role of the gut microbiota in the physiological effects of dietary fiber. *Microbiome*, *10*(1), 77. DOI: 10.1186/s40168-022-01248-5

Duvallet, C., Gibbons, S. M., Gurry, T., Irizarry, R. A., & Alm, E. J. (2017). Meta-analysis of gut microbiome studies identifies disease-specific and shared responses. *Nature Communications*, *8*(1), 1784. DOI: 10.1038/s41467-017-01973-8

El-Serag, H. B., & Sonnenberg, A. (1998). Opposing time trends of peptic ulcer and reflux disease. *Gut*, *43*(3), 327–333. DOI: 10.1136/gut.43.3.327

ElRakaiby, M., Dutilh, B. E., Rizkallah, M. R., Boleij, A., Cole, J. N., & Aziz, R. K. (2014). Pharmacomicrobiomics: The impact of human microbiome variations on systems pharmacology and personalized therapeutics. *OMICS: A Journal of Integrative Biology*, *18*(7), 402–414. DOI: 10.1089/omi.2014.0018

Elsamanoudy, A. Z., Neamat-Allah, M. A. M., Mohammad, F. A. H., Hassanien, M., & Nada, H. A. (2016). The role of nutrition related genes and nutrigenetics in understanding the pathogenesis of cancer. *Journal of Microscopy and Ultrastructure*, *4*(3), 115–122. DOI: 10.1016/j.jmau.2016.02.002

Esteva, A., Robicquet, A., Ramsundar, B., Kuleshov, V., DePristo, M., Chou, K., & Dean, J. (2019). A guide to deep learning in healthcare. *Nature Medicine*, *25*(1), 24–29. DOI: 10.1038/s41591-018-0316-z

Farhud, D. D., & Yeganeh, M. Z. (2010). Nutrigenomics and nutrigenetics. *Iranian Journal of Public Health*, *39*(4), 1.

Farhud, D. D., & Zokaei, S. (2021). Ethical issues of artificial intelligence in medicine and healthcare. *Iranian Journal of Public Health*, *50*(11), i. DOI: 10.18502/ijph.v50i11.7600

Ferrocino, I., Rantsiou, K., McClure, R., Kostic, T., De Souza, R. S. C., Lange, L., FitzGerald, J., Kriaa, A., Cotter, P., Maguin, E., Schelkle, B., Schloter, M., Berg, G., Sessitsch, A., & Cocolin, L. (2023). The need for an integrated multi-OMICs approach in microbiome science in the food system. *Comprehensive Reviews in Food Science and Food Safety*, *22*(2), 1082–1103. DOI: 10.1111/1541-4337.13103

Forestier, H., Grenet, M., Borel, A., & Celiberti, V. (2017). Les productions lithiques de l'Archipel indonésien. *Journal of Lithic Studies*, *4*(2), 231–303. DOI: 10.2218/jls.v4i2.2544

Franzosa, E. A., Hsu, T., Sirota-Madi, A., Shafquat, A., Abu-Ali, G., Morgan, X. C., & Huttenhower, C. (2015). Sequencing and beyond: Integrating molecular'omics' for microbial community profiling. *Nature Reviews. Microbiology*, *13*(6), 360–372. DOI: 10.1038/nrmicro3451

Freudenheim, J. L., & Gower, E. (2017). Interaction of genetic factors with nutrition in cancer. In *Nutrition in the Prevention and Treatment of Disease* (pp. 733–747). Academic Press. DOI: 10.1016/B978-0-12-802928-2.00033-3

Giuntini, E. B., Sardá, F. A. H., & de Menezes, E. W. (2022). The effects of soluble dietary fibers on glycemic response: An overview and futures perspectives. *Foods*, *11*(23), 3934. DOI: 10.3390/foods11233934

Green, E. D., Gunter, C., Biesecker, L. G., Di Francesco, V., Easter, C. L., Feingold, E. A., & Manolio, T. A. (2020). Strategic vision for improving human health at the forefront of genomics. *Nature*, *586*(7831), 683–692. DOI: 10.1038/s41586-020-2817-4

Hamburg, M. A., & Collins, F. S. (2010). The path to personalized medicine. *The New England Journal of Medicine*, *363*(4), 301–304. DOI: 10.1056/NEJMp1006304

Jimenez-Carvelo, A. M., & Cuadros-Rodríguez, L. (2021). Data mining/machine learning methods in foodomics. *Current Opinion in Food Science*, *37*, 76–82. DOI: 10.1016/j.cofs.2020.09.008

Johnson-Laird, P. N. (1983). *Mental models: Towards a cognitive science of language, inference, and consciousness* (No. 6). Harvard University Press.

Jordan, M. I., & Mitchell, T. M. (2015). Machine learning: Trends, perspectives, and prospects. *Science*, *349*(6245), 255–260. DOI: 10.1126/science.aaa8415

Kanehisa, M., & Goto, S. (2000). KEGG: Kyoto encyclopedia of genes and genomes. *Nucleic Acids Research*, *28*(1), 27–30. DOI: 10.1093/nar/28.1.27

Kaput, J. (2008). Nutrigenomics research for personalized nutrition and medicine. *Current Opinion in Biotechnology, 19*(2), 110–120. DOI: 10.1016/j.copbio.2008.02.005

Khoury, M. J., Iademarco, M. F., & Riley, W. T. (2015). Precision public health for the era of precision medicine. *American Journal of Preventive Medicine, 50*(3), 398–401. DOI: 10.1016/j.amepre.2015.08.031

Kosaraju, C., Chiruguri, S., Mohammad, N., Jetty, B. T., & Gali, G. (2022). A model for analysis of diseases based on nutrition deficiency using random forest. In *Proceedings of the 7th International Conference on Communication and Electronics Systems* (pp. 1527–1531). IEEE. DOI: 10.1109/ICCES54183.2022.9835805

Lai, Y., Xue, J., Liu, C. W., Gao, B., Chi, L., Tu, P., Lu, K., & Ru, H. (2019). Serum metabolomics identifies altered bioenergetics, signaling cascades in parallel with exposome markers in Crohn's disease. *Molecules (Basel, Switzerland), 24*(3), 449. DOI: 10.3390/molecules24030449

Lamichhane, J. R., Osdaghi, E., Behlau, F., Köhl, J., Jones, J. B., & Aubertot, J. N. (2018). Thirteen decades of antimicrobial copper compounds applied in agriculture. A review. *Agronomy for Sustainable Development, 38*(3), 28. DOI: 10.1007/s13593-018-0503-9

Langbein, T., Weber, W. A., & Eiber, M. (2019). Future of theranostics: An outlook on precision oncology in nuclear medicine. *Journal of Nuclear Medicine, 60*(Supplement 2), 13S–19S. DOI: 10.2967/jnumed.118.220566

Lazarova, M., Labaj, J., Kovacikova, Z., & Slamenova, D. (2004). Diet containing fungal (1 3)—D-glucan derivative exhibits protective effects against DNA lesions induced in freshly isolated rat cells. *Neoplasma, 51*, 431–435.

Lloyd-Price, J., Abu-Ali, G., & Huttenhower, C. (2016). The healthy human microbiome. *Genome Medicine, 8*(1), 51. DOI: 10.1186/s13073-016-0307-y

Loomba, R., Seguritan, V., Li, W., Long, T., Klitgord, N., Bhatt, A., Dulai, P. S., Caussy, C., Bettencourt, R., Highlander, S. K., Jones, M. B., Sirlin, C. B., Schnabl, B., Brinkac, L., Schork, N., Chen, C.-H., Brenner, D. A., Biggs, W., Yooseph, S., & Nelson, K. E. (2017). Gut microbiome-based metagenomic signature for non-invasive detection of advanced fibrosis in human nonalcoholic fatty liver disease. *Cell Metabolism, 25*(5), 1054–1062. DOI: 10.1016/j.cmet.2017.04.001

McCoy, D. C., Cuartas, J., Behrman, J., Cappa, C., Heymann, J., López Bóo, F., Lu, C., Raikes, A., Richter, L., Stein, A., & Fink, G. (2021). Global estimates of the implications of COVID-19-related preprimary school closures for children's instructional access, development, learning, and economic wellbeing. *Child Development, 92*(5), e883–e899. DOI: 10.1111/cdev.13658

Mozaffarian, D., Rosenberg, I., & Uauy, R. (2018). History of modern nutrition science—Implications for current research, dietary guidelines, and food policy. *BMJ (Clinical Research Ed.), 361*, k2392. DOI: 10.1136/bmj.k2392

Mutch, D. M., Wahli, W., & Williamson, G. (2005). Nutrigenomics and nutrigenetics: The emerging faces of nutrition. *The FASEB Journal, 19*(12), 1602–1616. DOI: 10.1096/fj.05-3911rev

O'Hara, A. M., & Shanahan, F. (2006). The gut flora as a forgotten organ. *EMBO Reports, 7*(7), 688–693. DOI: 10.1038/sj.embor.7400731

Obermeyer, Z., Powers, B., Vogeli, C., & Mullainathan, S. (2019). Dissecting racial bias in an algorithm used to manage the health of populations. *Science, 366*(6464), 447–453. DOI: 10.1126/science.aax2342

Oliveira Chaves, L., Gomes Domingos, A. L., Louzada Fernandes, D., Ribeiro Cerqueira, F., Siqueira-Batista, R., & Bressan, J. (2023). Applicability of machine learning techniques in food intake assessment: A systematic review. *Critical Reviews in Food Science and Nutrition, 63*(7), 902–919. DOI: 10.1080/10408398.2021.1956425

Ordovas, J. M., & Corella, D. (2004). Nutritional genomics. *Annual Review of Genomics and Human Genetics, 5*(1), 71–118. DOI: 10.1146/annurev.genom.5.061903.180008

Pasolli, E., Truong, D. T., Malik, F., Waldron, L., & Segata, N. (2016). Machine learning meta-analysis of large metagenomic datasets: Tools and biological insights. *PLoS Computational Biology, 12*(7), e1004977. DOI: 10.1371/journal.pcbi.1004977

Pleasance, E., Titmuss, E., Williamson, L., Kwan, H., Culibrk, L., Zhao, E. Y., Dixon, K., Fan, K., Bowlby, R., Jones, M. R., Shen, Y., Grewal, J. K., Ashkani, J., Wee, K., Grisdale, C. J., Thibodeau, M. L., Bozoky, Z., Pearson, H., Majounie, E., & Marra, M. A. (2020). Pan-cancer analysis of advanced patient tumors reveals interactions between therapy and genomic landscapes. *Nature Cancer, 1*(4), 452–468. DOI: 10.1038/s43018-020-0050-6

Plemel, J. R., Stratton, J. A., Michaels, N. J., Rawji, K. S., Zhang, E., Sinha, S., Baaklini, C. S., Dong, Y., Ho, M., Thorburn, K., Friedman, T. N., Jawad, S., Silva, C., Caprariello, A. V., Hoghooghi, V., Yue, J., Jaffer, A., Lee, K., Kerr, B. J., & Yong, V. W. (2020). Microglia response following acute demyelination is heterogeneous and limits infiltrating macrophage dispersion. *Science Advances*, *6*(3), eaay6324. DOI: 10.1126/sciadv.aay6324

Plottel, C. S., & Blaser, M. J. (2011). Microbiome and malignancy. *Cell Host & Microbe*, *10*(4), 324–335. DOI: 10.1016/j.chom.2011.10.003

Popova, L., & Carabetta, V. J. (2024). The use of next-generation sequencing in personalized medicine. *High Throughput Gene Screening: Methods and Protocols*, 287-315.

Priyadharshini, V. S., & Teran, L. M. (2016). Personalized medicine in respiratory disease: Role of proteomics. *Advances in Protein Chemistry and Structural Biology*, *102*, 115–146. DOI: 10.1016/bs.apcsb.2015.11.008

Pushkaran, A. C., & Arabi, A. A. (2024). From understanding diseases to drug design: Can artificial intelligence bridge the gap? *Artificial Intelligence Review*, *57*(4), 86. DOI: 10.1007/s10462-024-10714-5

Qin, J., Li, R., Raes, J., Arumugam, M., Burgdorf, K. S., Manichanh, C., ... & Wang, J. (2010). A human gut microbial gene catalogue established by metagenomic sequencing. *nature, 464*(7285), 59-65.

Rampton, V., Böhmer, M., & Winkler, A. (2022). Medical technologies past and present: How history helps to understand the digital era. *The Journal of Medical Humanities*, *43*(2), 343–364. DOI: 10.1007/s10912-021-09699-x

Rouskas, K., Bocher, O., Simistiras, A., Emmanouil, C., Mantas, P., Skoulakis, A., ... & Dimas, A. S. (2025). Periodic dietary restriction of animal products induces metabolic reprogramming in humans with effects on cardiometabolic health. *npj Metabolic Health and Disease, 3*(1), 14.

Salman, R. A. S., Frantzias, J., Lee, R. J., Lyden, P. D., Battey, T. W., Ayres, A. M., & Lanfermann, H. (2018). Absolute risk and predictors of the growth of acute spontaneous intracerebral haemorrhage: A systematic review and meta-analysis of individual patient data. *Lancet Neurology*, *17*(10), 885–894. DOI: 10.1016/S1474-4422(18)30253-9

Schupack, D. A., Mars, R. A., Voelker, D. H., Abeykoon, J. P., & Kashyap, P. C. (2022). The promise of the gut microbiome as part of individualized treatment strategies. *Nature Reviews. Gastroenterology & Hepatology, 19*(1), 7–25. DOI: 10.1038/s41575-021-00499-1

Shabani, M., & Borry, P. (2018). Rules for processing genetic data for research purposes in view of the new EU General Data Protection Regulation. *European Journal of Human Genetics : EJHG, 26*(2), 149–156. DOI: 10.1038/s41431-017-0045-7

Shendure, J., & Ji, H. (2008). Next-generation DNA sequencing. *Nature Biotechnology, 26*(10), 1135–1145. DOI: 10.1038/nbt1486

Stoykova, Z., & Koeva–Dimitrova, L. (2019). History and development of personalized medicine. *Health Economics and Management, 19*(1).

Sugimoto, K., Moriyasu, F., Oshiro, H., Takeuchi, H., Abe, M., Yoshimasu, Y., Kasai, Y., Sakamaki, K., Hara, T., & Itoi, T. (2020). The role of multiparametric US of the liver for the evaluation of nonalcoholic steatohepatitis. *Radiology, 296*(3), 532–540. DOI: 10.1148/radiol.2020192665

Szabo de Edelenyi, F., Goumidi, L., Bertrais, S., Phillips, C., MacManus, R., Roche, H., & Lairon, D. (2008). Prediction of metabolic syndrome status based on dietary and genetic parameters using random forest. *Genes & Nutrition, 3*(3-4), 173–176. DOI: 10.1007/s12263-008-0097-y

Taye, M. M. (2023). Understanding machine learning with deep learning: Architectures, workflow, applications and future directions. *Computers, 12*(5), 91. DOI: 10.3390/computers12050091

Tekpınar, L., & Erdem, R. (2019). Kişiselleştirilmiş Tip Ve Genom Araştirmalarinin Sağlik Çiktilari Bağlaminda Değerlendirilmesi. *Hacettepe Sağlık İdaresi Dergisi, 22*(4), 843–862.

Theodore Armand, T. P., Kim, H. C., & Kim, J. I. (2024). Digital anti-aging healthcare: Applications of digital technologies in diet management. *Journal of Personalized Medicine, 14*(3), 254. DOI: 10.3390/jpm14030254

Topçuoğlu, M. A. (2022). Stroke epidemiology and near future projection in Turkey: Analysis of Turkey data from the Global Burden of Disease Study. *Turkish Journal of Nephrology, 28*(4), 200–211.

Topol, E. (2019). *Deep medicine: How artificial intelligence can make healthcare human again*. Hachette UK.

Vayena, E., Blasimme, A., & Cohen, I. G. (2018). Machine learning in medicine: Addressing ethical challenges. *PLoS Medicine*, *15*(11), e1002689. DOI: 10.1371/journal.pmed.1002689

Velázquez, O. C., Seto, R. W., & Rombeau, J. L. (1996). The scientific rationale and clinical application of short-chain fatty acids and medium-chain triacylglycerols. *The Proceedings of the Nutrition Society*, *55*(1B), 49–78. DOI: 10.1079/PNS19960011

Větrovský, T., & Baldrian, P. (2013). The variability of the 16S rRNA gene in bacterial genomes and its consequences for bacterial community analyses. *PLoS One*, *8*(2), e57923. DOI: 10.1371/journal.pone.0057923

Visvikis-Siest, S., Theodoridou, D., Kontoe, M. S., Kumar, S., & Marschler, M. (2020). Milestones in personalized medicine. *Frontiers in Genetics*, *11*, 569175. DOI: 10.3389/fgene.2020.569175

Wirbel, J., Pyl, P. T., Kartal, E., Zych, K., Kashani, A., Milanese, A., Fleck, J. S., Voigt, A. Y., Palleja, A., Ponnudurai, R., Sunagawa, S., Coelho, L. P., Schrotz-King, P., Vogtmann, E., Habermann, N., Niméus, E., Thomas, A. M., Manghi, P., Gandini, S., & Zeller, G. (2019). Meta-analysis of fecal metagenomes reveals global microbial signatures that are specific for colorectal cancer. *Nature Medicine*, *25*(4), 679–689. DOI: 10.1038/s41591-019-0406-6

Zeevi, D., Korem, T., Zmora, N., Israeli, D., Rothschild, D., Weinberger, A., & Segal, E. (2015). Personalized nutrition by prediction of glycemic responses. *Cell*, *163*(5), 1079–1094. DOI: 10.1016/j.cell.2015.11.001

Zhao, L., Li, T., Dang, M., Li, Y., Fan, H., Hao, Q., & Zhang, G. (2023). Association of methylenetetrahydrofolate reductase (MTHFR) rs1801133 (677C>T) gene polymorphism with ischemic stroke risk in different populations: An updated meta-analysis. *Frontiers in Genetics*, *13*, 1021423. DOI: 10.3389/fgene.2022.1021423

Zou, Z., Jin, Y., Nevalainen, P., Huan, Y., Heikkonen, J., & Westerlund, T. (2019, March). Edge and fog computing enabled AI for IoT-an overview. In *2019 IEEE International Conference on Artificial Intelligence Circuits and Systems (AICAS)* (pp. 51-56). IEEE. DOI: 10.1109/AICAS.2019.8771621

ENDNOTES

1 Rampton, V., Böhmer, M., & Winkler, A. (2022). Medical technologies past and present: how history helps to understand the digital era. *Journal of Medical Humanities*, *43*(2), 343-364.

Chapter 5
Spatial Multi-Omics in Neurodegenerative Diseases

Arthi Gunasekaran
https://orcid.org/0009-0002-7355-6387
Bharathiar University, India

Abidharini Jothi Dheivasikamani
Bharathiar University, India

Arthi Boro
https://orcid.org/0000-0001-7921-3620
Bharathiar University, India

Loganathan Murugesan
Bharathiar University, India

Antony Prabhu Jeyabal Philomenathan
Bharathiar University, India

Gowri Subramaniam
Dr. N.G.P. Arts and Science College, Coimbatore, India

Alwin Rober Asirvatham
Prince Sultan Military Medical City, Ministry of Defense, Riyadh, Saudi Arabia

Vijaya Anand Arumugam
https://orcid.org/0000-0001-7485-1586
Bharathiar University, India

ABSTRACT

Neurodegenerative diseases represent a significant global health burden, characterized by the progressive deterioration of neuronal dysfunction and complex molecular pathologies. Traditional omics approaches have provided valuable insights into the genetic, transcriptomic, and proteomic alterations involved; however, they often lack spatial resolution, obscuring the tissue-specific and cellular-contextual dynamics of disease progression. Spatial multi-omics technologies integrate spatial transcriptomics, proteomics, epigenomics, and metabolomics to enable high-resolution mapping of molecular features within intact tissues. This chapter explores recent advance-

DOI: 10.4018/979-8-3373-4849-0.ch005

ments in spatial multi-omics technologies and also the role of spatial multi-omics involving metabolomics, epigenomics, and their application to neurodegenerative disease models, as well as the challenges and future directions in translating these insights into clinical impact.

INTRODUCTION

Spatial multi-omics can be defined as the integration of at least two spatially resolved molecular profiling technologies such as spatial transcriptomics, spatial proteomics, spatial metabolomics, and spatial epigenomics with the aim of understanding the organization of biological systems within intact tissue architecture. This is powerful in the context of NDs, in which the distribution of molecular pathology across brain regions, cortical layers, and local microenvironments is highly variable. Neurodegenerative disorders like AD, PD, ALS, and FTD are highly region-specific and cell type-specific vulnerabilities, and spatially resolved molecular methods are crucial for their disease progressions. Traditional single-cell omics suffer from a lack of spatial relationships, while bulk tissue sequencing averages out critical heterogeneity; spatial multi-omics surpasses these by linking molecular states to specific anatomic locations in the brain.

1. SPATIAL TRANSCRIPTOMICS

Spatial Transcriptomics (ST) captures mRNA expression profiles from tissue sections in situ, without dissociating the tissue, thereby preserving cellular arrangement and anatomy. The technique originally utilized barcoded capture arrays to spatially calibrate gene expression profiles, as described by Ståhl et al. 2016. More modern, high-resolution ST platforms such as Slide seq, Slide seq2, and DBiT seq have enabled near single cell or subcellular resolution mapping of transcriptomic activity (Rodriques et al., 2019; Liu et al., 2020). Spatial transcriptomics has been transformative for ND research:

- ST revealed the specific microglial activation zones around amyloid β plaques in AD brain samples, including the upregulation of inflammatory and lipid metabolism pathways.
- Layer-specific transcriptional changes have been mapped in the hippocampus and cortex that capture the early vulnerability in memory-related circuits.
- ST in PD samples has highlighted dopaminergic neuron-specific stress signatures and glial inflammatory niches (Zhang et al., 2023).

2. SPATIAL PROTEOMICS

Spatial proteomics maps protein abundance, distribution, and interactions in intact brain sections. Approaches include

- Imaging Mass Spectrometry (IMS): allows label free detection of hundreds of proteins or peptides across tissue (Giesen et al., 2014).
- Multiplexed Immunofluorescence (e.g., CODEX, MIBI, IMC): utilizes antibody panels for the visualization of dozens of proteins at subcellular resolution. Proteins represent the major executors of cellular function and the main components of pathological aggregates in neurodegenerative disorders (Giesen et al., 2014; Neumann, 2025).
- Spatial proteomics in AD cortex detected layer-specific tau and synaptic protein depletion, showing how tauopathy spreads through cortical microcircuits (Hu et al., 2023).
- Mass spectrometry imaging identified lipid binding proteins that are enriched near amyloid plaques, supporting plaque-associated metabolic disruptions (Liu et al., 2024).
- Spatial proteomics in PD identified α synuclein accumulation that matched mitochondrial dysfunction in vulnerable neurons, according to Marshall et al. (2023).

Because protein aggregation is at the center of NDs, spatial proteomics provides direct insight into pathological hotspots.

3. SPATIAL METABOLOMICS AND LIPIDOMICS

Spatial metabolomics relies on MSI to map small molecules, metabolites, and lipids within brain tissue. Unlike transcriptomics and proteomics, metabolites reflect cellular physiology energy metabolism, oxidative stress, neurotransmitter activity, and lipid homeostasis. Metabolic dysregulation is a hallmark of most NDs.

- Spatial metabolomics has further revealed altered lipid metabolism in close proximity to amyloid plaques, including ceramides and sphingolipids involved in promoting neuroinflammation (Liu et al., 2024).
- In PD, impaired dopamine metabolism and mitochondrial byproducts were spatially mapped to degenerating nigral neurons (Zhang et al., 2023).
- Altered energy metabolites, including lactate, glutamate, TCA intermediates, have been spatially correlated with axonal degeneration and microglial acti-

vation (Marshall et al., 2023). It reveals the biochemical microenvironments that drive neuronal stress and glial responses.

4. SPATIAL EPIGENOMICS

Spatial epigenomics couples ATAC, DNA methylation, and histone modifications with spatial coordinates. These approaches, although still in development, can be employed to determine epigenetic states across brain regions without the need for tissue dissociation.

Examples include:

- CUT&Tag-based spatial mapping of chromatin marks
- In situ sequencing of histone modifications (Marx, 2021). Epigenetic regulation is critical in both ND susceptibility and brain aging.
- Spatial profiling indicates that neurons in proximity to amyloid plaques have chromatin compaction and transcriptional repression, while synaptic genes are altered (Chen et al., 2022).
- Microglial epigenetic states in AD are distinguished between plaque-adjacent and plaque-distant regions, thus indicating epigenomic remodeling tied to local inflammation (Mathys et al., 2019).

Spatial epigenomics provides an additional layer of regulatory context that is complementary to transcript and protein patterns.

5. INTEGRATED SPATIAL MULTI OMICS

The full power of spatial multi-omics arises from integrating transcriptomics, proteomics, metabolomics, and epigenomics into unified spatial maps.

Examples of integrated applications include:

- Transcript protein integration to map gene-protein networks around plaques (Chen et al., 2022).
- Protein metabolite overlays showing how lipid metabolism influences plaque-associated immune responses (Liu et al., 2024).
- Region specific multi-omic signatures of early AD vulnerability by Zhang et al. (2023)
- Integration pipelines like Giotto enable spatial multi-layer computation (Dries et al., 2021).

Integrated multi-omics reveals the following:

- Disease specific cellular neighborhoods, for example, microglia astrocyte endothelial clusters in AD.
- Propagating gradients of pathology along neural circuits.
- Molecular pathways that correlate with clinical severity and regional degeneration.
- Novel therapeutic targets within spatially vulnerable niches.

6. SPATIAL TRANSCRIPTOMICS (ST, MERFISH, SEQFISH)

Spatial transcriptomics (ST) is a powerful set of technologies that enable the study of gene expression while maintaining the spatial context of cells within intact tissues. Unlike traditional RNA sequencing (RNA-seq) or single-cell RNA sequencing (scRNA-seq), which require tissue dissociation and thereby lose spatial information, ST techniques preserve the tissue architecture, allowing researchers to map where specific transcripts are expressed (Ståhl *et al.*, 2016). This spatial preservation is crucial for understanding complex biological processes such as tissue development, tumor heterogeneity, and cellular communication. Sequencing-based ST methods use spatially barcoded oligonucleotides printed on glass slides to capture RNA from defined spots, each corresponding to a specific location within a tissue section. These methods provide transcriptome-wide coverage but generally have limited resolution, as each spot may represent multiple cells. Imaging-based methods such as sequential fluorescence in situ hybridization (seqFISH) and multiplexed error-robust fluorescence in situ hybridization (MERFISH) overcome these limitations by directly imaging RNA molecules at subcellular resolution. In seqFISH, transcripts are labeled through multiple rounds of hybridization and imaging, generating a unique barcode pattern for each gene that can be decoded to identify thousands of mRNAs in a single tissue sample (Shah *et al.*, 2016). This approach provides single-molecule sensitivity and enables researchers to map gene expression across individual cells within their native tissue environments. SeqFISH+ (an improved version) has been used to profile up to 10,000 genes per cell, offering deep insights into the molecular organization of tissues (Eng *et al.*, 2019).

MERFISH, developed by Xia and Zhuang, uses a combinatorial binary barcoding scheme that incorporates error-correction features to minimize false positives and negatives during imaging (Xia *et al.*, 2019). Each gene is assigned a unique binary code, and through multiple rounds of hybridization and imaging, thousands of genes can be simultaneously detected with high accuracy. This technique enables comprehensive mapping of spatial transcriptomes at single-cell resolution, revealing not only which genes are expressed but also how they are spatially arranged within cells and tissues. MERFISH has been applied to study brain organization, RNA localization dynamics, and tissue microarchitecture with remarkable precision. Collectively, ST, MERFISH, and seqFISH provide complementary strengths in spatial resolution, throughput, and transcriptome coverage. Sequencing-based ST offers broad transcriptome profiling across large tissue areas, whereas imaging-based approaches such as MERFISH and seqFISH deliver higher resolution and single-cell precision. Together, these techniques have revolutionized spatial genomics, enabled the construction of detailed cellular maps, and advanced our understanding of tissue biology, disease progression, and cell–cell communication. As computational methods and experimental technologies continue to evolve, ST is expected to play a pivotal role in multi-omic integration and precision medicine.

6.1 Imaging mass spectrometry and spatial proteomics

Imaging mass spectrometry (IMS) and spatial proteomics are transformative technologies that allow researchers to visualize and quantify proteins, peptides, lipids, and metabolites directly within tissue sections while preserving spatial context. Unlike traditional proteomic approaches such as liquid chromatography mass spectrometry (LC–MS), which require homogenization and thereby lose spatial information, IMS enables *in situ* molecular analysis by measuring the mass-to-charge ratios of ionized molecules across defined tissue coordinates (Caprioli *et al.*, 1997). By generating molecular maps, IMS can reveal biochemical heterogeneity within tissues, making it an invaluable tool for studies in pathology, neuroscience, and cancer biology. The most commonly used IMS techniques include matrix-assisted laser desorption/ionization (MALDI) and secondary ion mass spectrometry (SIMS), each offering different spatial resolutions and molecular sensitivities.

MALDI imaging mass spectrometry (MALDI-IMS) combines histology and mass spectrometry to produce detailed maps of biomolecular distributions without the need for labels or antibodies. It uses a laser to desorb and ionize analytes from matrix-coated tissue sections, followed by time-of-flight (TOF) or orbitrap-based mass analysis. MALDI-IMS can detect proteins, lipids, and small metabolites across large tissue areas, providing molecular signatures that correlate with histopathological features (Chaurand *et al.*, 2006). This capability has proven particularly useful for identifying disease biomarkers, classifying tumor subtypes, and tracking drug localization in pharmacological studies. Recent advancements have improved both the spatial resolution (down to a few micrometers) and sensitivity of MALDI-IMS, enabling high-definition molecular imaging of single cells and subcellular compartments.

Spatial proteomics, on the other hand, extends beyond molecular imaging to map the entire proteome within its spatial and cellular context. This field combines mass spectrometry-based proteomics with spatially resolved sampling or labeling techniques to quantify protein localization and abundance in tissues or cells (Thul *et al.*, 2017). Methods such as laser capture microdissection (LCM) coupled with LC–MS allow region-specific protein profiling, while emerging technologies like imaging mass cytometry (IMC) and multiplexed ion beam imaging (MIBI) use metal-tagged antibodies and mass spectrometry to image dozens of proteins simultaneously with subcellular resolution (Giesen *et al.*, 2014). These methods have been instrumental in mapping cellular microenvironments, signaling networks, and protein localization patterns in both healthy and diseased tissues. Together, IMS and spatial proteomics provide a multidimensional view of the molecular landscape within tissues. While IMS offers label-free detection of a broad range of biomolecules, spatial proteomics provides targeted and quantitative insights into protein localization and interaction networks. The integration of these techniques with other spatial omics approaches, such as spatial transcriptomics and metabolomics, is enabling the creation of comprehensive molecular atlases of tissues and organs. These advances hold immense promise for biomarker discovery, drug development, and precision medicine, offering unprecedented insight into how spatial organization influences cellular function and disease progression.

6.2 Spatial metabolomics and lipidomics

Spatial metabolomics and spatial lipidomics are emerging branches of spatial omics that aim to map the spatial distribution of small molecules and lipids within biological tissues. Unlike bulk metabolomic and lipidomic profiling, which provide average molecular information across homogenized samples, spatial metabolomics preserves the native biochemical landscape, revealing how metabolic pathways and lipid species vary across distinct tissue regions (Rappez *et al.*, 2021). This spatially resolved information is essential for understanding tissue microenvironments, metabolic heterogeneity in cancer, and organ physiology. Techniques such as matrix-assisted laser desorption/ionization imaging mass spectrometry (MALDI-IMS), desorption electrospray ionization (DESI), and secondary ion mass spectrometry (SIMS) have become the primary tools for spatially resolved detection of metabolites and lipids directly from tissue surfaces.

MALDI imaging mass spectrometry (MALDI-IMS) has been extensively used in spatial metabolomics and lipidomics to visualize molecular distributions in tissues without labeling or prior knowledge of analytes. By applying a matrix that assists in ionization, MALDI-IMS enables the mapping of hundreds of metabolites and lipids across histological sections with spatial resolutions down to 10–20 μm (Caprioli *et al.*, 1997). This technology has revealed how metabolites such as neurotransmitters, fatty acids, and phospholipids vary across brain regions, tumors, and developing organs. DESI imaging, by contrast, allows ambient ionization of samples, eliminating the need for a matrix and minimizing sample preparation. Both MALDI and DESI have been successfully applied to study metabolic gradients in tumors and metabolic zonation in the liver, highlighting spatially distinct biochemical processes linked to physiology and pathology (Prideaux & Stoeckli, 2012).

Spatial lipidomics focuses on the distribution and dynamics of lipid species, key regulators of membrane structure, energy metabolism, and signaling. Recent advances in high-resolution mass spectrometry have enabled precise identification and localization of thousands of lipid molecules within complex tissues (Niehaus *et al.*, 2019). Lipid imaging has revealed, for instance, phospholipid reorganization in neurodegenerative diseases, altered sphingolipid metabolism in cancer, and region-specific lipid composition in brain and heart tissues. Combining lipidomics with histology or fluorescence microscopy allows for the correlation of lipid distributions with cell types and histopathological features. These insights are reshaping our understanding of how lipid metabolism supports tissue function and disease progression at the cellular level. Together, spatial metabolomics and lipidomics provide powerful tools for exploring the chemical architecture of biological systems. Integration with other spatial omics approaches, such as spatial transcriptomics, proteomics, and imaging mass cytometry, offers a holistic view of how molecular pathways are organized and regulated within tissues. These multimodal approaches are particularly promising for biomarker discovery, drug response studies, and personalized medicine, where spatial biochemical heterogeneity often determines therapeutic outcomes. As imaging mass spectrometry and computational analysis continue to improve, spatial metabolomics and lipidomics are expected to play pivotal roles in systems biology and clinical diagnostics.

6.3 Epigenomics mapping in spatial context

Epigenomic mapping in a spatial context seeks to understand how epigenetic features such as DNA methylation, histone modifications, and chromatin accessibility are distributed within the structural organization of tissues. Traditional bulk and single-cell sequencing techniques lose spatial information because they require tissue dissociation, thereby obscuring how cellular microenvironments influence gene regulation. Spatial epigenomics preserves tissue architecture while profiling epigenetic marks, providing a powerful way to study gene regulation across different anatomical regions. This approach enables the visualization of how chromatin states correspond to tissue-specific functions and developmental processes (Deng *et al.*, 2022). Technological innovations have advanced spatial epigenomics, with techniques such as spatial-ATAC-seq enabling genome-wide mapping of chromatin accessibility while maintaining spatial coordinates. This method applies in-situ Tn5 transposition and barcoding to tissue sections, revealing distinct chromatin accessibility patterns that correspond to anatomical and functional domains. For example, spatial-ATAC-seq profiling in mouse and human brain tissues has identified epigenetically distinct regions linked to neural activity and development (Deng *et al.*, 2022). Other emerging methods have expanded this concept to allow simultaneous measurement of chromatin accessibility and gene expression within the same tissue section, bridging regulatory mechanisms and transcriptional outcomes (Ma *et al.*, 2023).

Beyond chromatin accessibility, spatial epigenomic technologies now target other epigenetic layers such as histone modifications and DNA methylation. The Spatial Chromatin Accessibility Sequencing (SCA-seq) technique integrates chromatin accessibility mapping with three-dimensional genome conformation, uncovering that active chromatin regions are spatially organized in distinct nuclear domains (Xie *et al.*, 2024). Additionally, co-profiling approaches that combine epigenomic and transcriptomic data in developing tissues have revealed region-specific enhancer activity and epigenetic priming events that occur before transcriptional activation (Ma *et al.*, 2023). These advances demonstrate how spatial organization contributes to cell fate determination and tissue function. Despite rapid progress, spatial epigenomic mapping faces technical and analytical limitations. Current methods often balance trade-offs between spatial resolution, molecular coverage, and tissue compatibility. Tissue preparation processes such as fixation and barcoding can introduce bias, while computational tools for integrating multimodal data remain under development. New approaches like spatial FFPE-ATAC-seq now enable chromatin accessibility profiling in archived formalin-fixed, paraffin-embedded samples, expanding the potential for clinical research applications (Guo *et al.*, 2023). As methodologies mature, spatial epigenomics promises to transform our understanding of how epigenetic regulation shapes development, disease progression, and tissue homeostasis (Xie *et al.*, 2024).

7. COMPUTATIONAL TOOLS AND DATA INTEGRATION

7.1 Image registration and spatial alignment

Image registration and spatial alignment are fundamental processes in biomedical imaging that ensure multiple images of the same specimen or subject are accurately overlaid in a common coordinate system. These techniques are essential for comparing images acquired at different times, from different modalities, or under varying conditions. In medical imaging, for instance, aligning magnetic resonance imaging

(MRI), computed tomography (CT), or positron emission tomography (PET) scans enables integrated analysis of anatomical and functional information (Maintz & Viergever, 1998). The goal of image registration is to determine the optimal spatial transformation, rigid, affine, or non-linear, that minimizes differences between images while preserving anatomical correspondence (Sotiras *et al.*, 2013). Modern registration algorithms often rely on feature-based and intensity-based approaches. Feature-based methods extract landmarks such as edges, contours, or key points and use them for geometric alignment, while intensity-based techniques optimize similarity metrics (e.g., mutual information or cross-correlation) between pixel intensities. Mutual information, in particular, has become a widely used criterion for multimodal registration because it measures statistical dependence between images rather than absolute intensity values (Pluim *et al.*, 2003). This approach allows the integration of heterogeneous data sources, such as combining MRI with PET or histological imaging, to capture both structural and molecular information from tissues (Maes *et al.*, 1997).

In recent years, deep learning has revolutionized spatial alignment through data-driven registration models that learn complex transformations directly from annotated datasets. Convolutional neural networks (CNNs) and transformer-based models have been developed to predict deformation fields that warp moving images to fixed references without explicit feature extraction (Haskins *et al.*, 2020). These methods significantly reduce computation time and can adapt to variations in image quality and modality. For example, deep-learning-based frameworks like VoxelMorph achieve high accuracy in aligning MRI brain scans while maintaining computational efficiency suitable for large-scale studies (Balakrishnan *et al.*, 2019). Spatial alignment is also crucial in emerging spatial omics and multimodal tissue imaging platforms, where data from multiple molecular layers must be precisely co-registered to the same spatial frame. Accurate alignment allows researchers to integrate gene expression, epigenetic, and protein data within the same tissue architecture, revealing spatially resolved molecular interactions (Seidel *et al.*, 2024). Despite these advances, challenges remain in registering images with non-linear distortions, intensity inhomogeneity, and tissue deformation during sample preparation. Future developments combining deep learning, probabilistic modeling, and multi-scale optimization promise further to enhance registration robustness, accuracy, and interpretability (Haskins *et al.*, 2020).

7.2 Multi-layer data fusion and visualization

Multi-layer data fusion and visualization refer to the process of integrating heterogeneous datasets derived from different experimental modalities to create a comprehensive representation of complex biological systems. In the context of biomedical research, these layers often include genomics, transcriptomics, proteomics, metabolomics, and spatial imaging data. Integrating such diverse data sources provides a holistic view of cellular processes, tissue organization, and disease mechanisms that cannot be captured by any single modality (Wang *et al.*, 2019). The goal of multi-layer data fusion is to uncover relationships across scales from molecular interactions to tissue-level organization, while maintaining the integrity and spatial resolution of each dataset (Lee & Yoon, 2017). Data fusion approaches can generally be categorized into three levels: early, intermediate, and late integration. Early integration merges raw or pre-processed data into a single feature space, while intermediate integration combines latent representations learned from each dataset. Late integration, on the other hand, synthesizes results from independently analysed layers to draw combined biological insights (Rappoport & Shamir, 2018). Each strategy has trade-offs in terms of interpretability, noise propagation, and computational complexity. For example, early fusion is sensitive to differences in scale and dimensionality, whereas late fusion may overlook cross-layer dependencies that could reveal key biological mechanisms. Visualization plays a critical role in interpreting multi-layer fused data, enabling researchers to explore relationships and trends across molecular and spatial dimensions. Advanced visualization frameworks, such as multi-omics dashboards and dimensionality reduction methods like t-SNE or UMAP, facilitate intuitive exploration of integrated datasets (Chamber *et al.*, 2025). In spatial biology, visualization tools combine gene expression, protein distribution, and histological images into unified maps that reveal spatial correlations and micro-environmental structures. For example, integrating single-cell transcriptomics with spatial imaging data helps identify tissue-specific cell states and signalling interactions that drive disease phenotypes (Kaur *et al.*, 2025). Despite advances, multi-layer data fusion and visualization face several challenges, including data heterogeneity, missing values, and computational scalability. Different data modalities often vary in resolution, dynamic range, and noise characteristics, making direct integration non-trivial. Emerging machine learning and deep learning frameworks now offer scalable solutions for cross-modal alignment and feature extraction, such as graph neural networks and variational autoencoders that can model complex, non-linear relationships across datasets. As these methods mature, integrating visualization with analytical pipelines will become increasingly important for hypothesis generation and for translating multi-layer biological insights into clinical applications (Kaur *et al.*, 2025).

7.3 AI and machine learning approaches

Artificial intelligence (AI) and machine learning (ML) have become transformative tools across multiple scientific and technological disciplines, particularly in data-rich fields like biomedical research, healthcare, and engineering. AI refers to systems capable of performing tasks that typically require human intelligence, such as pattern recognition, reasoning, and decision-making, while ML focuses on algorithms that learn from data to improve their performance over time (Jordan & Mitchell, 2015). By analysing complex and high-dimensional datasets, ML methods can detect subtle patterns, predict outcomes, and generate models that guide data-driven decision-making. These approaches have been instrumental in automating data analysis pipelines, improving diagnostic accuracy, and accelerating discovery in genomics, imaging, and drug development (LeCun *et al.*, 2015). Machine learning algorithms can generally be categorized into supervised, unsupervised, and reinforcement learning approaches. Supervised learning models, such as support vector machines (SVMs), random forests, and neural networks, are trained using labelled datasets to perform classification or regression tasks. Unsupervised learning techniques like k-means clustering, principal component analysis (PCA), and auto-encoders identify hidden structures or patterns in unlabelled data (Murphy, 2022). Reinforcement learning, on the other hand, enables agents to learn optimal strategies through trial-and-error interactions with their environments. These methods have been applied successfully in diverse domains from medical image segmentation to predictive modelling of patient outcomes, demonstrating their adaptability and scalability (Goodfellow *et al.*, 2016).

Recent advances in deep learning have further expanded the capabilities of AI systems. Deep neural networks, especially convolutional neural networks (CNNs) and transformer-based architectures, can process large-scale data with remarkable accuracy. In healthcare, CNNs have achieved expert-level performance in radiology and histopathology image classification, while recurrent neural networks (RNNs) and attention mechanisms have improved temporal and sequential data analysis, such as electronic health record prediction (Esteva *et al.*, 2019). The integration of deep learning with multimodal data, including genomics, imaging, and clinical information, allows for holistic modelling of biological systems and personalized medicine applications (Topol, 2019). Despite these successes, AI and ML approaches face challenges related to interpretability, bias, and data privacy. Many deep learning models operate as "black boxes," providing high accuracy but limited transparency in how predictions are made. Efforts such as explainable AI (XAI) aim to enhance model interpretability and trustworthiness in clinical settings (Samek *et al.*, 2017). Furthermore, model generalization remains limited when training data lack diversity or contain biases that reflect existing inequalities. To overcome these issues, researchers are developing hybrid models that combine statistical reasoning with ML flexibility and applying federated learning to preserve privacy while leveraging distributed datasets (Topol, 2019). As AI continues to evolve, its responsible and ethical application will be crucial for achieving equitable and impactful outcomes across disciplines.

8. APPLICATIONS IN NEURODEGENERATIVE DISEASE RESEARCH:

Research on neurodegenerative diseases involves studying disorders that mainly cause the gradual death or degeneration of neurons, the essential units of the nervous system, including the brain and spinal cord. These diseases often develop slowly, with symptoms that gradually improve over time. Research in this field has significantly enhanced our understanding, diagnostic abilities, and treatment options for conditions such as Huntington's, Parkinson's, Alzheimer's, and Amyotrophic Lateral Sclerosis (Muddapu, V. R. *et al.,* 2020).

8.1 Precision Medicine Approaches

Precision medicine classifies people based on their risk, expected response to treatment, and illness progression using "omics' data like genomes, epigenomics, and lifestyle factors. Unlike the traditional "one-drug-fits-all" method, computational techniques and integrated databases support decision-making and tailor therapies, enabling earlier and more personalized treatments (Strafella, C. *et al.,* 2018). Neuroimaging methods that provide measurable biomarkers for disease detection and progression include diffusion tensor imaging, PET scans, EEG, MEG, and structural MRI. Deep learning models especially offer high-accuracy diagnosis, subtype classification, and early detection of changes such as region-specific atrophy and white matter loss. Emphasizing multi-modal integration and validation across different groups, these approaches also employ explainable AI frameworks to address standards and interpretation challenges (Zhang, Y. *et al.,* 2025).

8.2 Molecular and Cellular Therapies

In conditions like Alzheimer's and Parkinson's, stem cell-based therapies using neural stem cells (NSCs) and pluripotent stem cells (PSCs) have the potential to repair damaged neurons, reduce pathological changes, and promote neurogenesis. Recent studies show they can enhance synaptic connections, decrease amyloid buildup, regulate neuroinflammation, and repair neurons (Kim, I.K. *et al.,* 2021). Exosomes derived from stem cells, especially mesenchymal and neural types, are at the forefront of experimental treatments for neurodegenerative diseases. Their regenerative potential is being demonstrated by ongoing clinical trials that assess their impact on neuronal survival, neuroinflammation, and functional recovery. Research into their safety and effectiveness is still in progress (Tanaka, M. *et al.,* 2025). Current research focuses on key biological pathways like protein aggregation, mitochondrial dysfunction, and synaptic failure, along with efforts to develop gene, immunological, and neurotrophic factor therapies. These new approaches promise more comprehensive disease modification and patient benefits (Gadhave, D. G. *et al.,* 2024).

9. CLINICAL AND TRANSLATIONAL PERSPECTIVES

In neurodegenerative disease research, clinical and translational viewpoints concentrate on connecting fundamental scientific findings to useful clinical applications with the ultimate goal of enhancing patient outcomes, diagnosis, and therapy.

9.1 Clinical Perspectives

Neurodegenerative disorders are clinically defined by a gradual loss of neurons, which results in symptoms like behavioural abnormalities, cognitive decline, and motor dysfunction. Finding biomarkers for precise and early diagnosis, comprehending a variety of phenotypes, and creating tailored treatments are the main goals of clinical research. Electrophysiological instruments (EEG), imaging methods (MRI, PET), and fluid biomarkers (plasma, cerebrospinal fluid) are crucial clinical tools for tracking the course of a disease and the effectiveness of treatment. In order to better customise patient-specific therapies, multidimensional and trans-diagnostic techniques are employed to capture the variety in clinical presentations and link symptoms to underlying biological causes (Davenport, F. *et al.,* 2023). In order to expedite the discovery of therapies, translational research combines findings from molecular, genetic, and animal models with clinical observations. To better understand disease mechanisms and test treatments, new approaches include the use of stem cells, organoids, large animal models, and sophisticated genetic tools. Personalised medicine approaches are being improved by the growing use of artificial intelligence and machine learning techniques to categorise disease subtypes and forecast progression. In order to convert findings into workable therapies, human tissue investigations must be integrated with model systems. Additionally, this study focuses on removing obstacles in the identification of biomarkers, the creation of novel clinical technologies, and the validation of medication candidates (Palanisamy, C. P. *et al.,* 2023). Through stem cell therapies, gene editing, immunomodulation, enhanced diagnostics, and AI-driven predictive tools, translational research in neurodegenerative diseases seeks to translate basic understanding into practical clinical interventions. Despite their potential, these methods need more clinical trial validation to overcome current obstacles and enhance patient outcomes (Antonioni, A. & Di Lorenzo, F., 2025).

10. CHALLENGES

10.1 Technical limitation and reproducibility

Some of the technical limitations in spatial multiomics are the preparation of the tissue samples with out damaging RNA, proteins and metabolite content in the tissues. In case of transcriptomics the study can be carried out in single cells in small areas with broader coverage avoiding the small details. In proteomics and metabolomics, the study target is lower in comparison to transcriptomics as in preparation of RNA from the cells may cause damage in the proteins and metabolites in the cells. Acquiring of high-resolution data of multiomic imaging is expensive and time-consuming and may vary in the sensitivity in obtaining the data.

10.2 Ethical consideration and data sharing

Ethical consideration and data sharing is very critical in neurodegenerative disease study using spatial multiomics. Molecular information analysis of the brain tissue should be explicitly covered in informed consent which may reveal many important information. Ethically it may not be approved for omic profiling of the tissue samples which may require re-consent and may also rise some familial implications. The sensitive genomic data obtained by spatial omics profiling require controlled access and also should promote fair acknowledgment and scientific return to contributing institutions, donor and communities.

10.3 Integration with spatial single-cell and multimodal imaging

The single multiomic analysis give the profiling of a single cell but fails to provide the spatial context in a particular tissue and the spatial omic analysis and imaging assays provide the spatial context of the target tissues but have low molecular coverage, limited throughput and integration issues. One of the main challenges is to produce unified dataset from the multiple data sets to view the architecture, cell type and interactions in the tissue.

11. CONCLUSION

Spatial multiomics in neurobiology is more descriptive as it may help in mapping the different cells like neurons, glia, astrocytes, microglia, oligodendrocytes in the brain. It may also help in understanding different diseases in brains by integrated analysis of genome, transcriptome, proteome and metabolome analysis. IT will also give the deep insight of the functions like gene expressions, regulation of gene expression, and others and also provide the deep insight of the interaction between cells in the brain. Integration of spatial multiomics may have a significant promise in future diagnostics by mapping of the architecture of the tissues in situ in the target region. It can measure multiple layers in omics including genome, transcriptome, proteome and metabolome for deeper access for diagnostics.

REFERENCES

Balakrishnan, G., Zhao, A., Sabuncu, M. R., Guttag, J., & Dalca, A. V. (2019). Voxelmorph: A learning framework for deformable medical image registration. *IEEE Transactions on Medical Imaging*, *38*(8), 1788–1800. DOI: 10.1109/TMI.2019.2897538 PMID: 30716034

Caprioli, R. M., Farmer, T. B., & Gile, J. (1997). Molecular imaging of biological samples: Localization of peptides and proteins using MALDI-TOF MS. *Analytical Chemistry*, *69*(23), 4751–4760. DOI: 10.1021/ac970888i PMID: 9406525

Caprioli, R. M., Farmer, T. B., & Gile, J. (1997). Molecular imaging of biological samples: Localization of peptides and proteins using MALDI-TOF MS. *Analytical Chemistry*, *69*(23), 4751–4760. DOI: 10.1021/ac970888i PMID: 9406525

Chambers, T. L., & Murach, K. A. (2025). A history of omics discoveries reveals the correlates and mechanisms of loading-induced hypertrophy in adult skeletal muscle. 2024 CaMPS young investigator award invited review. *American Journal of Physiology. Cell Physiology*, *328*(5), C1535–C1557. DOI: 10.1152/ajpcell.00968.2024 PMID: 40172105

Chaurand, P., Schwartz, S. A., Reyzer, M. L., & Caprioli, R. M. (2005). Imaging mass spectrometry: Principles and potentials. *Toxicologic Pathology*, *33*(1), 92–101. DOI: 10.1080/01926230590881862 PMID: 15805060

Chen, W.-T., Lu, A., & Strohl, J. J.. (2022). Spatial transcriptomics and proteomics reveal cell-type-specific regulation in Alzheimer's disease. *Nature Neuroscience*, *25*(8), 1059–1071. DOI: 10.1038/s41593-022-01077-8 PMID: 35798980

Deng, Y., Bartosovic, M., Ma, S., Zhang, D., Kukanja, P., Xiao, Y., Su, G., Liu, Y., Qin, X., Rosoklija, G. B., Dwork, A. J., Mann, J. J., Xu, M. L., Halene, S., Craft, J. E., Leong, K. W., Boldrini, M., Castelo-Branco, G., & Fan, R. (2022). Spatial profiling of chromatin accessibility in mouse and human tissues. *Nature*, *609*(7926), 375–383. DOI: 10.1038/s41586-022-05094-1 PMID: 35978191

Dries, R., Zhu, Q., & Eng, C.-H. L.. (2021). Giotto: A toolbox for integrative analysis and visualization of spatial expression data. *Genome Biology*, *22*(1), 78. DOI: 10.1186/s13059-021-02286-2 PMID: 33685491

Eng, C. H. L., Lawson, M., Zhu, Q., Dries, R., Koulena, N., Takei, Y., & Cai, L. (2019). Transcriptome-scale super-resolved imaging in tissues by RNA seqFISH+. *Nature*, *568*(7751), 235–239. DOI: 10.1038/s41586-019-1049-y PMID: 30911168

Esteva, A., Robicquet, A., Ramsundar, B., Kuleshov, V., DePristo, M., Chou, K., Cui, C., Corrado, G., Thrun, S., & Dean, J. (2019). A guide to deep learning in healthcare. *Nature Medicine, 25*(1), 24–29. DOI: 10.1038/s41591-018-0316-z PMID: 30617335

Giesen, C., Wang, H. A., Schapiro, D., Zivanovic, N., Jacobs, A., Hattendorf, B., Schüffler, P. J., Grolimund, D., Buhmann, J. M., Brandt, S., Varga, Z., Wild, P. J., Günther, D., & Bodenmiller, B. (2014). Highly multiplexed imaging of tumor tissues with subcellular resolution by mass cytometry. *Nature Methods, 11*(4), 417–422. DOI: 10.1038/nmeth.2869 PMID: 24584193

Giesen, C., Wang, H. A. O., Schapiro, D., Zivanovic, N., Jacobs, A., Hattendorf, B., Schüffler, P. J., Grolimund, D., Buhmann, J. M., Brandt, S., Varga, Z., Wild, P. J., Günther, D., & Bodenmiller, B. (2014). Highly multiplexed imaging of tumor tissues with subcellular resolution by mass cytometry. *Nature Methods, 11*(4), 417–422. DOI: 10.1038/nmeth.2869 PMID: 24584193

Goodfellow, I. (2016). Deep learning.

Guo, P., Mao, L., Chen, Y., & Zhang, W. (2023). Spatial FFPE-ATAC-seq enables epigenomic profiling of archived tissue sections. *Nature Communications, 14*(1), 1529. PMID: 36934097

Haskins, G., Kruger, U., & Yan, P. (2020). Deep learning in medical image registration: A survey. *Machine Vision and Applications, 31*(1), 8. DOI: 10.1007/s00138-020-01060-x

Hu, B., Park, J., & Lee, C. (2023). Spatial proteomics uncovers layer-specific vulnerability in Alzheimer's disease cortex. *Frontiers in Aging Neuroscience, 15*, 1150512. DOI: 10.3389/fnagi.2023.1150512

Jordan, M. I., & Mitchell, T. M. (2015). Machine learning: Trends, perspectives, and prospects. *Science, 349*(6245), 255–260. DOI: 10.1126/science.aaa8415 PMID: 26185243

Kaur, E., Lopos, L. C., & Bilichak, A. (2025). Mitigating Heat Stress Response in CRISPR/Cas-Mediated Edited Crops by Altering the Expression Pattern of Non-coding DNA. *Genome and Epigenome Editing for Stress-Tolerant Crops*, 1-31.

LeCun, Y., Bengio, Y., & Hinton, G. (2015). Deep learning. *nature, 521*(7553), 436-444.

Lee, J. S., & Yoon, S. (2017). Integrative analysis of multi-omics data for cancer research: Methods and applications. *BMC Bioinformatics, 18*(Suppl 2), 83.

Liu, X., (2024). Spatial metabolomics reveals energy-metabolic dysregulation in neurodegeneration. *Translational Neurodegeneration*, *13*(7), e00397. DOI: 10.1186/s40035-024-00397-x PMID: 38254235

Liu, Y., Li, L., & Zhang, X. (2020). Slide-seq2: Scalable spatially resolved transcriptomics at near single-cell resolution. *Nature Methods*, *17*(7), 755–758. DOI: 10.1038/s41592-020-0909-8

Ma, S., Zhang, B., LaFave, L. M., & Fan, R. (2023). Spatial co-profiling of epigenome and transcriptome reveals developmental regulatory programs. *Nature Biotechnology*, *41*(2), 250–261.

Maes, F., Collignon, A., Vandermeulen, D., Marchal, G., & Suetens, P. (2002). Multimodality image registration by maximization of mutual information. *IEEE Transactions on Medical Imaging*, *16*(2), 187–198. DOI: 10.1109/42.563664 PMID: 9101328

Maintz, J. A., & Viergever, M. A. (1998). A survey of medical image registration. *Medical Image Analysis*, *2*(1), 1–36. DOI: 10.1016/S1361-8415(01)80026-8 PMID: 10638851

Marshall, J., Farrow, J., Djambazova, K., & Spraggins, J. (2023). Untangling Alzheimer's disease with spatial multi-omics: A brief review. *Frontiers in Aging Neuroscience*, *15*, 1150512. DOI: 10.3389/fnagi.2023.1150512 PMID: 37533766

Marx, V. (2021). Method of the Year: Spatially resolved transcriptomics. *Nature Methods*, *18*(1), 9–14. DOI: 10.1038/s41592-020-01033-y PMID: 33408395

Mathys, H., Davila-Velderrain, J., Peng, Z., Gao, F., Mohammadi, S., Young, J. Z., Menon, M., He, L., Abdurrob, F., Jiang, X., Martorell, A. J., Ransohoff, R. M., Hafler, B. P., Bennett, D. A., Kellis, M., & Tsai, L.-H. (2019). Single-cell transcriptomic analysis of Alzheimer's disease. *Nature*, *570*(7761), 332–337. DOI: 10.1038/s41586-019-1195-2 PMID: 31042697

Murphy, K. P. (2022). *Probabilistic machine learning: an introduction.* MIT press.

Neumann, E. K. (2025). Spatial multi-omics toward understanding neurological systems. *Journal of Mass Spectrometry : JMS*, *60*(6), e5143. DOI: 10.1002/jms.5143 PMID: 40360168

Niehaus, M., Soltwisch, J., Belov, M. E., & Dreisewerd, K. (2019). Transmission-mode MALDI-2 mass spectrometry imaging of cells and tissues at subcellular resolution. *Nature Methods*, *16*(9), 925–931. DOI: 10.1038/s41592-019-0536-2 PMID: 31451764

Pluim, J. P., Maintz, J. A., & Viergever, M. A. (2003). Mutual-information-based registration of medical images: A survey. *IEEE Transactions on Medical Imaging*, *22*(8), 986–1004. DOI: 10.1109/TMI.2003.815867 PMID: 12906253

Prideaux, B., & Stoeckli, M. (2012). Mass spectrometry imaging for drug distribution studies. *Journal of Proteomics*, *75*(16), 4999–5013. DOI: 10.1016/j.jprot.2012.07.028 PMID: 22842290

Rappez, L., Stadler, M., Triana, S., Gathungu, R. M., Ovchinnikova, K., Phapale, P., Heikenwalder, M., & Alexandrov, T. (2021). SpaceM reveals metabolic states of single cells. *Nature Methods*, *18*(7), 799–805. DOI: 10.1038/s41592-021-01198-0 PMID: 34226721

Rappoport, N., & Shamir, R. (2018). Multi-omic and multi-view clustering algorithms: Review and cancer benchmark. *Nucleic Acids Research*, *46*(20), 10546–10562. DOI: 10.1093/nar/gky889 PMID: 30295871

Rodriques, S. G., Stickels, R. R., Goeva, A., Martin, C. A., Murray, E., Vanderburg, C. R., Welch, J., Chen, L. M., Chen, F., & Macosko, E. Z. (2019). Slide-seq: A scalable technology for measuring genome-wide expression at high spatial resolution. *Science*, *363*(6434), 1463–1467. DOI: 10.1126/science.aaw1219 PMID: 30923225

Samek, W., Wiegand, T., & Müller, K. R. (2017). Explainable artificial intelligence: Understanding, visualizing and interpreting deep learning models. *arXiv preprint arXiv:1708.08296*.

Seidel, K., Piskol, R., Thao Nguyen, T. T., Shelton, A., Cottonham, C., de la Cruz, C. C., . . . Siebel, C. (2024). Notch signaling maintains a progenitor-like subclass of hepatocellular carcinoma. bioRxiv, 2024-12.

Shah, S., Lubeck, E., Zhou, W., & Cai, L. (2016). In situ transcription profiling of single cells reveals spatial organization of cells in the mouse hippocampus. *Neuron*, *92*(2), 342–357. DOI: 10.1016/j.neuron.2016.10.001 PMID: 27764670

Sotiras, A., Davatzikos, C., & Paragios, N. (2013). Deformable medical image registration: A survey. *IEEE Transactions on Medical Imaging*, *32*(7), 1153–1190. DOI: 10.1109/TMI.2013.2265603 PMID: 23739795

Ståhl, P. L., Salmén, F., Vickovic, S., Lundmark, A., Navarro, J. F., Magnusson, J., & Frisén, J. (2016). Visualization and analysis of gene expression in tissue sections by spatial transcriptomics. *Science*, *353*(6294), 78–82. DOI: 10.1126/science.aaf2403 PMID: 27365449

Ståhl, P. L., Salmén, F., Vickovic, S., Lundmark, A., Navarro, J. F., Magnusson, J., Giacomello, S., Asp, M., Westholm, J. O., Huss, M., Mollbrink, A., Linnarsson, S., Codeluppi, S., Borg, Å., Pontén, F., Costea, P. I., Sahlén, P., Mulder, J., Bergmann, O., & Frisén, J. (2016). Visualization and analysis of gene expression in tissue sections by spatial transcriptomics. *Science*, *353*(6294), 78–82. DOI: 10.1126/science.aaf2403 PMID: 27365449

Thul, P. J., Åkesson, L., Wiking, M., Mahdessian, D., Geladaki, A., Ait Blal, H., Alm, T., Asplund, A., Björk, L., Breckels, L. M., Bäckström, A., Danielsson, F., Fagerberg, L., Fall, J., Gatto, L., Gnann, C., Hober, S., Hjelmare, M., Johansson, F., & Lundberg, E. (2017). A subcellular map of the human proteome. *Science*, *356*(6340), eaal3321. DOI: 10.1126/science.aal3321 PMID: 28495876

Topol, E. J. (2019). High-performance medicine: The convergence of human and artificial intelligence. *Nature Medicine*, *25*(1), 44–56. DOI: 10.1038/s41591-018-0300-7 PMID: 30617339

Wang, B., Mezlini, A. M., Demir, F., Fiume, M., Tu, Z., Brudno, M., Haibe-Kains, B., & Goldenberg, A. (2014). Similarity network fusion for aggregating data types on a genomic scale. *Nature Methods*, *11*(3), 333–337. DOI: 10.1038/nmeth.2810 PMID: 24464287

Xia, C., Fan, J., Emanuel, G., Hao, J., & Zhuang, X. (2019). Spatial transcriptome profiling by MERFISH reveals subcellular RNA compartmentalization and cell cycle-dependent gene expression. *Proceedings of the National Academy of Sciences of the United States of America*, *116*(39), 19490–19499. DOI: 10.1073/pnas.1912459116 PMID: 31501331

Xie, Y., Ruan, F., Li, Y., & Chen, Y. (2024). Spatial chromatin accessibility sequencing resolves high-order spatial interactions of epigenomic markers. *eLife*, *13*, e87868. DOI: 10.7554/eLife.87868 PMID: 38236718

Yang, P., Jin, L., Liao, J., Jin, K., Shao, X., Li, C., Qian, J., Cheng, J., Yu, D., Guo, R., Xu, X., Lu, X., & Fan, X. (2023). Revealing spatial multimodal heterogeneity in tissues with SpaTrio. *Cell Genomics*, *3*(12), 100446. DOI: 10.1016/j.xgen.2023.100446 PMID: 38116121

Zhang, Y., (2023). Spatial transcriptomic landscape of Alzheimer's and Parkinson's disease brains. *Nature Neuroscience*, *26*(12), 1895–1909. DOI: 10.1038/s41593-023-01347-8 PMID: 37814027

Chapter 6
Spatial Multi-Omics in Infectious Diseases

Ashok Kumar Sah
https://orcid.org/0000-0002-7762-4351
A' Sharqiyah University, Oman

Prem Shankar Mishra
https://orcid.org/0000-0003-3797-1575
Vidya University, Meerut, India

Shubhangi Mishra
Galgotias University, Greater Noida, India

Ranjay Kumar Choudhary
https://orcid.org/0000-0003-2801-4079
A' Sharqiyah University, Oman

Rahul Kumar
https://orcid.org/0009-0001-0237-3201
Galgotias University, Greater Noida, India

SK Abdul Rahaman
Galgotias University, Greater Noida, India

ABSTRACT

Fungal infections are an increasing global health concern, especially in individuals with weakened immune systems. Their complexity stems from intricate interactions between fungal pathogens and host cells, making diagnosis and treatment difficult. Although advances have been made, key aspects of host–pathogen interactions, such as invasion mechanisms, immune evasion, and variable host responses, remain unclear. Systems biology provides a holistic way to study these processes by integrating large-scale data on genes, proteins, metabolites, and immune pathways. Multiomics approaches, including genomics, transcriptomics, proteomics, metabolomics, and immunomics, offer deeper insight into the molecular events during infection. Integrating these datasets reveals pathways involved in infection progression, host immunity, and pathogen clearance. This chapter highlights how multiomics-driven systems biology enhances understanding of fungal pathogenicity, immune dysregulation, and host defense, supporting the development of improved therapeutic strategies.

DOI: 10.4018/979-8-3373-4849-0.ch006

1.INTRODUCTION

The pathogenesis of infectious diseases is characterized by very complicated and dynamic relations of the invading pathogens and the tissues of the host. These processes occur in spatially structured micro environments that control immune activation, pathogen spread, tissue injury and treatment response. Historical methods of molecular profiling, including bulk genomics and proteomics, give an average signal in heterogenous cell populations and hence conceals important spatial and cellular information that can be used to answer biology of infection (Penaranda & Hung, 2019). The advent of spatial multi-omics has changed this topography by allowing the simultaneous profiling of multiple molecular layers transcriptomes, proteomes, epigenomes, and metabolites and maintaining their localized locations in intact tissues. This combination of spatial resolution and multi-dimensional molecular data provides previously unavailable understanding of the interaction between hosts and pathogens, immunology microenvironment, tissue-specific disease pathogenesis, and spatial responses to drugs. Spatial multi-omics has emerged as the cornerstone of precision diagnostics, host directed therapy and vaccine development in infection biology (Liu et al., 2024).

1.1 Concept and Evolution of Spatial Multi-Omics

Spatial multi-omics is a concept of simultaneous combination of two (or more) spatially resolved layers of omics in intact tissues of the body. In contrast to traditional omics methods which use homogenized samples, spatial multi-omics is capable of maintaining the architectural context of molecular signals that can be used to map gene expression, protein localization, chromatin states, and metabolic activity in highly specific cellular locales. Spatial multi-omics has developed over time in response to three broad technology changes. The initial step was the bulk omics technologies (microarrays and next-generation sequencing) that allowed performing molecular profiling on a large scale but did not provide any spatial resolution. The second step came in with single-cell omics, which solved cellular heterogeneity at the cost of spatial context because of the dissociation of tissues. The third and the present phase combines both dimensions with the use of spatially resolved molecular profiling as one can directly visualize the molecular activity in intact tissues. The major advancements of this progression are laser capture microdissection, fluorescence in situ hybridization (FISH), spatial transcriptomics technology, including Slide-seq and 10x Visium, and high-dimensional spatial proteomic technologies, such as imaging mass cytometry and multiplexed ion beam imaging. Collectively, these developments have made spatial multi-omics an instrumental part of infectious disease studies (Park, 2023).

Single-cell omics and spatial integration in the year 2012, researchers from Australia and China presented the initial paper examining how telomerase influences the migration of stem cells and skin fibroblasts in vitro and in vivo. First article exploring the connection between telomerase and the migration of stem cells and skin fibroblasts in in vitro and in vivo models was published in 2012 by Australian and Chinese scientists. The single-cell omics technologies have transformed the knowledge base of cellular heterogeneity in infected tissues mainly single-cell RNA sequencing (scRNA-seq). These procedures permit the transcriptional profiling of thousands to millions of single cells, revealing uncommon population of immune cells, cells infected by a pathogen and transitional cellular states which would be concealed in bulk analyses. As illustrated in **Figure 1**, integrated single-cell and spatial multi-omics enable the characterization of cellular heterogeneity, molecular interactions, and microenvironmental organization relevant to infectious disease (Wu et al., 2022).

Figure 1. Overview of single-cell and spatial multi-omics approaches for understanding cellular heterogeneity and microenvironment in infectious disease

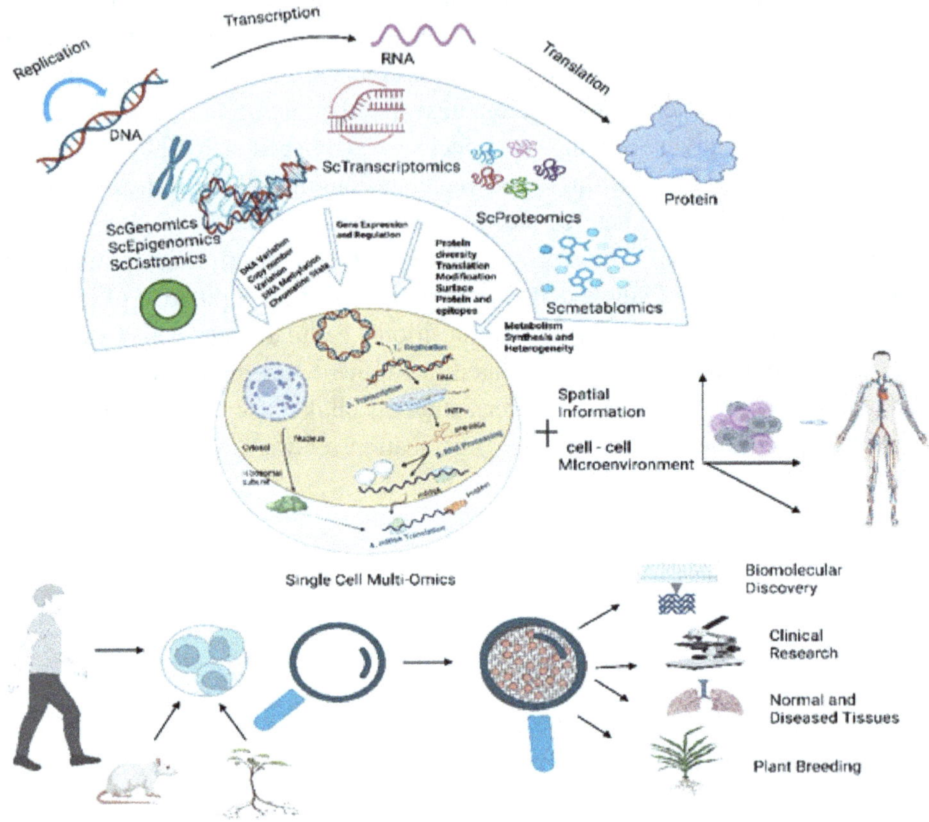

Although powerful, traditional single-cell approaches will require dissociation of the tissue and this disturbs cell-cell spatial relationships. The shortcoming resulted in spatial integration strategies that computationally or experimentally coupled single-cell data with spatial transcriptomic maps. With this kind of integration, cell types as identified by scRNA-seq can be precisely repositioned into their native tissue environment. This integrative technique in infectious diseases has made it possible to map the infiltration of immune cells in the area of the infection focus, detect a localized gradient of cytokines, and visualize a immune niche specific to the pathogen. As an example, single-cell analyses of the local immune environment in tuberculosis granuloma and region-specific antiviral immunity in COVID-19 lung tissue have been presented as spatially integrated single-cell analyses. In this way, the combination of single-cell omics and spatial technologies will fill the gap between the molecules and the tissue structure.

1.2 Advances in scRNA Sequencing

Technological Single-cell RNA sequencing has experienced a very fast development in the last ten years and made a great improvement in terms of sensitivity, throughput, and applicability to the clinic. The initial scRNA-seq technologies had poor capture efficacy and high levels of technical noise. This was, however, made possible by the introduction of droplet-based microfluidic systems, including Drop-seq and 10x Genomics Chromium, in that high-throughput profiling of tens of thousands of cells per experiment was possible. AI-based integration of multi-omics data enables biomarker discovery and personalized therapeutic decisions (**Figure 2**).

Figure 2. AI-driven integration of multi-omics and clinical data for biomarker discovery, patient stratification, and personalized treatment

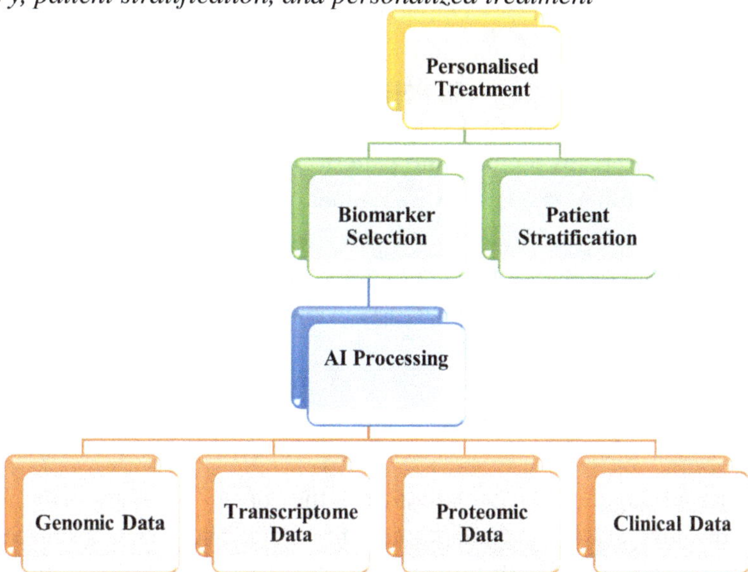

Such advances as multi-modal single-cell sequencing that simultaneously measures transcriptomic, epigenomic, and proteomic data on the same cell have been introduced in recent years. Methods like CITE-seq and REAP-seq can be used to use RNA sequencing with antibody-based protein detection so that gene expression and surface protein markers are directly correlated. Simultaneously, the technology of long-read sequencing and long full-length transcript profiling has enhanced the ability to resolve host and pathogen transcripts at the isoform level (Chukwunweike et al., 2024). ScRNA-seq has facilitated the detection of host cells that are infected by a pathogen, revealed the nature of immune exhaustion, identified interferon-stimulated gene networks and mapped tissue-specific immune signatures in the context of infectious disease. With the addition of spatial omics, these innovations can now be used to localize discrete cellular states in infection microenvironments and significantly enhance systems-level insights into disease progression.

2. ARTIFICIAL INTELLIGENCE AND MULTI-OMNICS DATA INTEGRATION

This has been facilitated by the rapid growth of omics technologies that have created massive, high-dimensional data in genomics, transcriptomics, proteomics, epigenomics and metabolomics. Even though these datasets have never been as informative on the biological level, their size and complexity present significant analysis challenges. As a new technology, artificial intelligence (AI) has demonstrated itself as a wave of change in solving these problems, allowing automated recognition of patterns, integrating multiple omics layers of data, and predicting biological systems. Regarding infectious diseases, multi-omics integration, which is promoted by AI, helps identify the pattern of host-pathogen interactions, find disease-specific biomarkers, stratify individual patient immune reactions, and predict treatment outcomes. AI allows systems-level study of infection biology with a combination of computational intelligence and spatial and single-cell omics. Magnetically, it empowers researchers to develop precision diagnostics and novel potent medicines (Wen et al., 2023).

2.1 Role of AI in Omics Research

Artificial intelligence can be exploited in the core of deriving meaningful biological knowledge out of the challenging omics data. The classical statistical tools have problems related to high dimensionality, noise, and nonlinear dynamics of the biological process. AI surpasses these shortcomings by acquiring hidden patterns directly on massive data sets without being instructed to do so (Debbadi & Boateng, 2025a). The applications of AI include feature selection, dimensionality reduction, classification, and clustering of biological samples in omics research studies. AI algorithms have been applied in several infectious disease fields to distinguish between infected and healthy states, detect disease-specific transcriptional vaccinations, and detect minor immune perturbations that relate to the severity of the disease. The use of AI in biomarker discovery has been useful as well, as the technology allows the prioritization of candidate genes, proteins, and metabolites that have great diagnostic or prognostic potential (Alves et al., 2021). Besides, AI can speed up drug discovery pipelines because it can predict drug-target interactions, applications of host-directed therapeutic targets, and pattern antimicrobial resistance models. Through the addition of spatial omics, AI is now able to interpret the molecular organization of tissues at the spatial scale, identify spatial immune niches, and predict the local response to treatment.

2.2 Deep Learning and Machine Learning Approaches

Omics analysis requires computational components based on machine learning (ML) and deep learning (DL) as the pillars of modern AI. Outcome prediction, disease classification and biomarker validation are usually conducted by supervised learning algorithms like support vectors, random forests, gradient boosting models and so on. These techniques are based on labeled data to acquire disease-specific molecular signatures. The unsupervised approaches to learning, such as k-means clustering, hierarchical clustering, or autoencoders, are frequently used in the exploratory analyses, which enabled the identification of new cellular subtypes, immune states, and patterns of host-pathogen interactions. These methods are especially useful with single-cell and spatial multi-omics, where the biological heterogeneity is extreme (Debbadi & Boateng, 2025b). Convolutional neural networks (CNNs) and recurrent neural networks (RNNs) are deep learning models that provide powerful tools in the analysis of spatial transcriptomics images, as well as histopathology slides and time-series dynamics of infections. CNNs are most effective at obtaining spatial information in tissue images, whereas RNNs are the most suitable in the modeling of disease progression and temporal immune responses. Most recently, the idea of transformer-based architectures and generative AI models has shown great possibilities in multi-omics data fusion and predictive modeling. These artificial intelligence applications have been applied in infectious disease fields to predict the severity of viral infections, antimicrobial resistance modeling on a population level, early postnatal sepsis, and stratification of COVID-19 patients (Turanli et al., 2018).

2.3 Network Biology and Graph-Based Models

Complex networks of interacting proteins, genes and signaling pathways make up the operation of biological systems. Network biology is an approach that offers a structure to model such interactions in the form of a built-in graph with nodes modeling molecular entities and edges modeling regulatory or physical interactions. Graph-based methods are especially effective in terms of multi-omics integration, as they are accurate in depicting correlations between various layers of molecules. Network models are indicated to be useful in the case of host-pathogen interaction studies, when there is a need to identify important regulatory hubs, pathway bottlenecks, and important host dependency factors utilized by pathogens. The networks of protein-protein interactions, gene regulatory networks, and metabolic networks all prevent the true picture of the infection rewiring of the host cellular machinery.

However, recent developments in graph neural networks (GNNs) enabled even the larger analysis of biological networks. GNNs are trained on graph structures, which leads to successful prediction of disease-linked genes, prescribed-drug interactions, and infection-specific signalling pathways. Graph-based models can also be used in spatial multi-omics to reconstruct cell-cell communication networks, which disclose the interactions between immune and infected cells in a specific tissue micro-environment. The approach, consisting of network biology and graph-based AI models, presents a systems-level view of how infectious diseases work by balancing molecular interactions with those of a spatial tissue and clinical phenotype (Ali, 2023).

3. PHENOTYPIC SCREENING AND DRUG DISCOVERY IN INFECTIOUS DISEASES

The discovery of drugs to treat infectious diseases is distinctly complicated by a fast rate of evolution in the pathogens, the development of antimicrobial resistance, and the complexities of interactions between the host and the pathogen. The classical methods have been dependent on target-based methods, but the methods are sometimes unable to provide the entire biological intricacy of an infection. Phenotypic screening, in contrast, assesses compound action in biologically upright structures devoid of other expectations about molecular targets, allowing the hots hotting of first-in-class and host-directed therapeutics. The application of multi-omics and artificial intelligence, integral to the field and analysis of phenotypic screening, has reappeared as a potent source of discovery. Spatial multi-omics further complements the methodology by providing visualization of drug effects in infected tissue micro-spectrums, which can provide fundamental information about drug residuation, immune-modulation, and the clearance of pathogens (Moffat et al., 2017).

3.1 Target-Based versus Phenotypic Screening

Target-based screening is concerned with the discovery of compounds that interact with a predetermined molecular target, e.g., a bacterial enzyme or a viral protease. This approach has the advantage of great specificity, clear-cut mechanisms of action, and drug design. It has been very successful in the development of drugs like HIV protease inhibitors and neuraminidase inhibitors of influenza. Its drawbacks, however, are that it cannot be translated to in vivo efficacy of the in vitro target inhibition, off-target toxicity, and there is a risk of the development of resistance mutations (Swinney & Anthony, 2011). Phenotypic screening evaluated by contrast measures, the efficacy of compounds in whole cells or organoids or models of disease physiology, for example, in an organ, animal, or human disease. The technique has the benefit of capturing complicated biological actions, such as host immune modulation, metabolic interference, and multi-target activity. It is noteworthy that it allows the identification of compounds with new modes of action that would otherwise not be developed in target-specific assays (Horvath et al., 2016). Phenotypic screening finds special application in infectious disease research in finding host-directed therapy, anti-virulence, and compounds useful against drug-tolerant or dormant populations of pathogens. When used with spatial multi-omics, phenotypic screening allows mapping of the effects of candidate drugs on the molecular state of hosts and pathogens in intact tissues in high-resolution (Vincent et al., 2015).

3.2 Discovery of Broad-Spectrum Anti-Infective Drugs

The growing popularity of multidrug-resistant pathogens has led to an augmented necessity for broad-spectrum anti-infective agents that can attack a variety of bacteria, viruses, or parasites. The screening platforms that allow the identification of compounds that have cross-pathogen activity are based on screening on phenotypic platforms that deploy a wide range of pathogen panels (Lee et al., 2017). The agents with broad-spectrums have a number of benefits, such as they can be deployed quickly in the case of an outbreak, have had fewer diagnostic delays, and can be used to treat co-infections. Nevertheless, they present significant threats like unbalanced microbiota of the host, and even toxicity, which requires thorough optimization using multi-omics profiling (Jacobs et al., 2011). Spatial multi-omics has the value of developing broad-spectrum drugs through the recognition of tissue-specific drug reactions, spatial force of pathogen elimination, regions of immune stimulation, and toxic niches. Together with AI projection systems, these platforms can be used to focus on candidate compounds that have the optimal efficacy-toxicity and high translational potential (Watson et al., 2019).

3.3 Successful Case Examples

Some phenotypic screening methods have resulted in the development of a number of anti-infective drugs that are clinically approved. Indicatively, antimalarial agents like artemisinin were discovered using whole-organism screening and subsequently depicted to be stored in a variety of parasite pathways (Grubaugh et al., 2017).

Likewise, bedaquiline, an anti-tuberculosis ATP synthase inhibitor in multidrug-resistant tuberculosis, was identified using phenotypic screening of mycobacterial growth inhibition. Phenotypic screening applications in antiviral studies were useful in speeding up hunting-down repurposed medications in the COVID-19 pandemic, such as remdesivir and host-targeting medications that regulated viral penetration and multiplication. The use of spatial transcriptomics and proteomics further suggests that these drugs changed the patterns of immune activation and viral distribution in the lung tissues (Li et al., 2022). Host-directed therapeutics against autophagy, metabolic reprogramming, and immune modulation are emerging in antibacterial development by screening platforms based on phenotypic screening and multi-omics screening. This creates a stark contrast of the effectiveness of phenotypic screening together with spatial multi-omics and AI to hasten the development of anti-infective drugs (Díaz-Menéndez et al., 2020).

4. EMERGING TECHNOLOGIES FOR SURVEILLANCE AND DISEASE MONITORING

Monitoring of the disease in real time and effective surveillance are crucial for early detection, control, and prevention of outbreaks of infectious diseases. Clinical reporting and laboratory confirmation, which have been used to formulate traditional surveillance systems, are usually characterised by delays, under-reporting, and lack of scalability. The recent genomic, digital health technologies, biosensing platforms, as well as artificial intelligence progress have reinvented criminology disease surveillance data mining into a real-time enterprise (Seers et al., 2023). The combination of genomic surveillance, epidemiological modeling, mobile health tools, and biosensors that operate rapidly allows continuous monitoring of the evolution of pathogens, their transmission patterns, and the disease patterns at the level of the whole population. These technologies can be used together with spatial multi-omics to create an effective system of exact surveillance and preparedness to epidemics (De Souza et al., 2021).

4.1 Genomic Surveillance

Genomic surveillance is the massive sequencing and analysis of the genomes of pathogens to track their evolution, transmission, and variation. Recent developments of next-generation sequencing (NGS) technology have facilitated easy, relatively inexpensive sequencing of viral, bacterial, and parasitic pathogens directly off clinical samples. By this method, variants of the emerging variants, mutations associated with anti-microbial resistance, and transmission chain can be detected in real-time (Keutzer et al., 2020). Genomic surveillance was essential during global outbreaks in recent history, including the COVID-19 infection, to identify variants of concern, trace the distribution of those variants based on the geographic area, and guide health management efforts in the population (Sharpe & Kamara, 2018). In addition to tracking outbreaks, genomic surveillance can be used when selecting vaccines, evaluating immune escape mutations, and optimizing treatment plans. Combined with the spatial data, it also allows mapping of pathogen lineages around tissues and populations (Pascucci et al., 2021).

4.2 Epidemiological Tracking

The epidemiological tracking can be defined as a systematic gathering and assessment of data connected with occurrence rates and spread of disease, in addition to risk factors of the population. Digital data streams of electronic health records, environmental surveillance, travel data, and social behavior indicators, among others, are becoming more prominent in modern epidemiology (Patel et al., 2022). Progress in computing models, such as AI-based predictive analytics and real-time dashboards, can be used to forecast the trajectory of an outbreak and health care requirements (Keilberg & Ottemann, 2016). The spatial epidemiological modeling can also be used to visualize disease hotspots, transmitters, and vulnerability regions. These are now inevitable tools in the control of specific interventions, lockdown measures, vaccination, and the decision-making of public health policies (Seo et al., 2020).

4.3 Mobile Health Applications

The mobile health (mHealth) system has become one of the tools to monitor large-scale disease and get patients involved and collect health data. The capabilities supported through smartphone applications include symptom reporting, exposure reporting, telemedicine, vaccination tracking, and treatment adherence monitoring (Liao et al., 2024).

In cases of the pandemic response, the mHealth platform provided real-time data collection of millions of users, which proved useful in terms of detecting outbreaks early and analyzing population behavior at the population level (Hadfield et al., 2018). The tools also enable community-based surveillance, especially in low-resource environments, which is limited to access to the traditional healthcare infrastructure. A combination of mHealth information with genomic and spatial surveillance also reinforces the decision-making of the population and management of disease specific to them (A. Kulkarni et al., 2019).

4.4 Biosensors of Rapid Detector

Biosensors give quick, delicate, and mobile systems of detecting pathogens, biologically harmful substances, as well as host immune mechanisms in the vicinity of the patient (Zhou et al., 2023). The latest types of biosensors combine detection on nanotechnology, microfluidics, electrochemical detection, optical sensing, and CRISPR-enhanced diagnostics to fulfill a high degree of specificity and ultra-low detection limits (Jean Beltran et al., 2017). Quick diagnostic biosensors can help detect the initial cases, contain the outbreak, and monitor the treatment progress in real-time. They are mostly applicable in decentralized operations, including the airports, local clinics, and field surveillance team (Russell et al., 2023). Biosensors can also be connected to digital reporting systems and spatial analytics when it comes to real-time disease mapping and precision surveillance networks (Singh et al., 2023).

5. SPATIAL MAPPING OF HOST-PATHOGEN INTERACTIONS

The interactions between a host and a pathogen are inherently spatial in characteristics, and they take place in highly organized tissue settings in which immune tissues, stromal elements, and pathogens are in dynamic interactions. Conventional bulk and single-cell omics methods are important to the phenotype of the infection, but do not maintain the spatial geometry by which the infection result is dictated (Spurbeck, 2024). Spatial mapping technologies are able to surmount this drawback by making it possible to map the host and pathogen molecular signatures in intact tissues with high precision.

Through the incorporation of spatial transcriptomics, spatial proteomics, and interaction-based profiling, scientists now have the opportunity to visualise pathogen niches, patterns of immune cell recruitment, signalling surfaces, and tissue-specific immune reactions (S. Kulkarni et al., 2022). This spatial view is necessary to comprehend the pathogen progression of infection, immune evasion mechanisms, pathogen persistence, and therapeutic response at the tissue level (Marshall & Warren, 1984).

5.1 Spatial Transcriptomics in infection

Spatial transcriptomics is a direct technique that allows measurement of the expression of a gene and imparts maintenance of tissue architecture. In comparison with scRNA-seq, where the tissue has to be dissociated, spatial transcriptomics records transcriptional activity in a histological context, which makes it possible to identify infection-specific cellular neighborhoods and immune microenvironment (Moss, 2017). Spatial transcriptomics has been applied in infectious disease studies to map viral RNA focalization, interferon signaling regions, as well as patterns of immune cell infiltration in infected tissues. As an example, different spatial distributions of inflammatory cytokines and antiviral-stimulated genes after respiratory viral infections have been reported in different airway and alveolar volumes (Venerito et al., 2017). On the same note, in tuberculosis, spatial transcriptomics has shown that there are heterogeneous regions of immune activation in granulomas, which include areas of immune exhaustion and persistence of the pathogen (Blaser & Atherton, 2004).

Through the combination of spatial transcriptomics and single-cell sequencing, errors in localising the location of transcriptionally defined immune and infected cell populations can be corrected and placed in their original setting in the tissue. This offers a potent framework of considering localized defense arrangements against the host and pathogen survival strategies (Rivas-Ortiz et al., 2017).

5.2. Proteomic Landscapes of Host Response

Whereas transcriptomics gives details on gene expression, proteomics records the functional cell molecular output. Applied spatial proteomics technologies, including imaging mass cytometry and multiplexed antibody-based imaging, can identify dozens of proteins at the resolutions of single cells on large tissue regions (Björkholm et al., 2001). Spatial proteomics can be applied to infection to comprehensively map the immune checkpoints, cytokine expression, antigen presentation molecules and activation markers in infected tissues. Such techniques have uncovered spatial confined immunogenic spaces, selected activation of macrophage and T cell constituents and accessory immunogenic domains in protracted infections (N. Gupta et al., 2019). As an example, local inflammatory niches rich in cytotoxic T cells and macrophages and immunosuppressive stromal sites have been identified by the spatial proteomic study of infected lung tissue. These pieces of knowledge are important in the regulation of immune dysregulation, tissue destruction and inhibition of efficient pathogen clearance (Altschuler & Wu, 2010).

5.3 Host-Pathogen Proteins and Protein Interactions

The interactions between host and pathogen proteins are directly important in causing most of the significant processes of infection, such as viral entry, immune evasion, intracellular replication, and spread of the pathogen. It is necessary to map these host-pathogen protein-protein interactions (PPIs) to determine important molecular weaknesses (R. K. Gupta & Kuznicki, 2020). Recent improvements in affinity purification mass spectrometry, proximity labeling methods, and spatially resolved interaction assays have now allowed the detection of PPIs in cellular and tissue settings. These methods have revealed wide, intricate interaction relationships between viral proteins and host signaling pathways, and among them, interferon, autophagy, and cellular metabolism (D. Wang & Bodovitz, 2010). As these interaction networks are mapped over space, the proximity of locations of important molecular struggles in tissues is revealed, such as in epithelial barriers or in immune cell synapses, or in pathogenic replication niches. The high-resolution space PPI maps are excellent targets of host-directed approaches and future anti-infective therapeutics (McClellan & King, 2010). Atherosclerosis involves complex interactions among lipid metabolism, immune cell heterogeneity, vascular cell modulation, and genetic alterations (**Figure 3**).

Figure 3. Multi-omics insights into atherosclerosis: Cellular heterogeneity, metabolic alterations, lipid species, genetic factors, and immune-mediated drivers

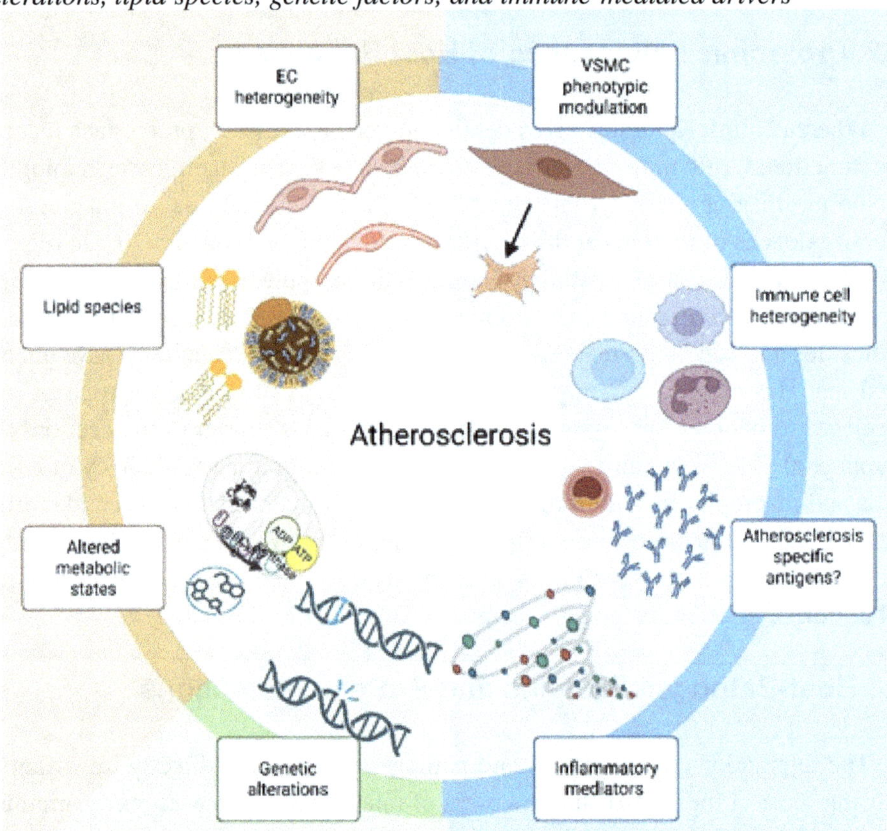

6. EPIGENOMIC AND REGULATORY MECHANISMS IN INFECTION

Host-pathogen interactions are no longer under the influence of genetic sequences but are strongly affected by epigenomic and regulatory mechanisms that can regulate the accessibility of genes and the activity of transcription. Major roles in host immune response and pathogen survival strategies are played by epigenetic modifications, including DNA methylation, histone modification, and chromatin remodeling (Mason et al., 2013). Most of the pathogens have developed advanced processes of manipulating host epigenetic machinery, hence, facilitating immune evasion, latency, and chronic infection (Huang et al., 2020). Subsequently, the epigenomics technology and spatial epigenomic technology have emerged, which allow mapping the regulatory state in whole, intact tissues. Such methods indicate the compositions of epigenetic programs in cell types, tissue compartments, and during various stages of infection, which brings novel understanding of disease pathogenesis, immune memory, and resistance to therapy. Epigenetics has an important function in the interactions between the epithelial cells of the gut and the bacteria, as discussed in this section (A. Gupta et al., 2020).

6.1 Role of Epigenetics in Host-Pathogen Interactions

The control of the activation and repression of immune-related genes is regulated by epigenetics when a person is infected (Vabret et al., 2020). The patterns of DNA methylation regulate the expression of cytokines and pattern recognition receptors, and antiviral defence genes, whereas histone modifications mediate access to immune locus chromatin. Pathogens use these regulatory strata either to prevent host defense or reprogram host cells (Aoshi et al., 2011). Indicatively, mycobacterium tuberculosis-induced bacterial pathogens trigger chromatin remodeling of hosts to suppress antigen presentation and interferon signaling. Viral pathogens such as the herpesviruses and HIV cause epigenetically controlled latent states by altering the chromatin structure of the sites where the virus integrates (Stambas et al., 2020).

Likewise, parasites can control the host epigenetic programs to support intracellular survival. Another role of epigenomic alterations in immune training and tolerance is the development of innate immune cells to memory-like characteristics of long-term immune responses after infection. Trained immunity. This attribute mediates training of memory with stable metabolic and inflammatory gene locus histone changes, facilitated by epigenetic regulation to host defense plasticity (Kanneganti, 2020).

6.2 Epigenomics-Based Diagnostic Advances

The epigenomic profiling has become an effective means of diagnostic infection and disease stratification. Agnosine methylation patterns, histone modification patterns, and chromatin accessibility profiles are potentially sensitive biomarkers of infection status, disease severity, and response to treatment (Abbott & Crotty, 2020). Single-cell and spatial epigenomics, including single-cell ATAC-seq and space-chromatin, make it possible to map the localization of regulatory states during cases of infection in infected tissues. They can be used to identify epigenetically distinct groups of immune cells at locations of infection, which disclose spatially confined regulatory platforms that drive inflammation, immune suppression, or repair (Dlamini et al., 2020). Epigenomic biomarkers have also found to be promising in the comparison of latent versus active infections, therapeutic responsiveness prediction, and immune reprogramming after vaccination or immunotherapy (Biondi-Zoccai et al., 2025). In combination with transcriptomics, proteomics, and spatial imaging, epigenomics contributes to the systems-level grasp over the biology of infectious diseases with a fundamental regulation dimension (Sahl & Moerner, 2013).

7. IMMUNE HETEROGENEITY AND DISEASE-SPECIFIC CASE STUDIES

The infectious diseases have a great heterogeneity of immunity, which is applied by the variations in host genetic composition, tissue microenvironment, pathogen range, and disease progression stage. In general, immune responses are not homogenous in different tissues but develop spatial-structured immune niches, which include heterogeneous pools of innate and adaptive immune cells with various functional conditions (Swain et al., 2020). Single-cell profiling, together with spatial multi-omics, also demonstrated that immune activation, immune suppression, immune exhaustion, and the process of immune memory can frequently coexist at the same location in products of infection (Melms et al., 2021). This heterogeneity is therefore important in understanding the different outcomes of disease, the different reactions to treatment, as well as the development of immune evasion mechanisms (Theodosiou & Read, 2023). The spatial resolution of immune profiling has now formed part and parcel in the dissection of tissue-specific immune response in chronic, acute, and emergent infectious diseases. Fundamentally, T lymphocytes have the ability to detect, eliminate, and possibly neutralize infections within tissues and organs (as depicted by the human body) (Khare & Pandey, 2022).

7.1 Cellular Heterogeneity in Immune Defense

There is a highly varied repertoire of cells composing the immune system that comprises macrophages, dendritic cells, natural killer cells, T lymphocytes, B lymphocytes, and stromal immune-supporting cells. These populations can be activated, differentiate, migrate, and polarize their functions very rapidly during the process of infection (Pavlou et al., 2024). Single-cell and spatial multi-omics have shown that not only within the identical immune cell type, but various transcriptional and epigenetic states are co-existent. As an example, pro-inflammatory, anti-inflammatory, tissue-repair, and pathogen-tolerant phenotypes can be observed in the same infected tissue by macrophages. T cells exhibit patterns of spatial activity gradients of activation, cytotoxicity, and exhaustion relative to pathogen reservoirs. The spatial proteomics also reveals that the process of cytokine secretion, immune checkpoints expression, and presentation of antigens are strongly localized (Mihalić et al., 2024). This heterogeneity of the immune system is critical when an infection is successfully cleared, when it establishes a chronic form of the infection, or when it causes tissue damage mediated by the immune system (Al-Talib et al., 2024). It is thus highly essential to map this diversity at cellular and spatial levels to determine immune correlates of protection and therapy. Helicobacter pylori is a primarily subcutaneous pathogen that can cause acute infection, chronic infection, and acute infections in humans (Corey, 2009) (Jonjić, 2024).

7.2 Helicobacter pylori Host-Pathogen Dynamics

Helicobacter pylori is a largely subcutaneous pathogen capable of causing acute infection, chronic infection, and acute infections in humans (Wayne & Hayes, 1996). A classic case of immune heterogeneity in the interaction between the host and the pathogen is the Helicobacter pylori infection of the gastric epithelium. Although there is a great immune reaction, H. pylori continues to stay for decades in most individuals, which causes gastritis, peptic ulcer disease, and gastric cancer (Lipworth et al., 2016). Transcriptomic and proteomic analyses of the space of infected gastric samples have demonstrated unique immune microdomains, such as the areas of active T-cell, macrophage, and epithelial remodelling. These investigations indicate that H. pylori selectively alters host signaling mechanisms, including NF- kB, interferon signaling, and apoptosis to develop localized areas of immune tolerance (Ge et al., 2025). Additionally, it has been shown in spatial analyses that the colonization of bacteria is not homogeneous and is concentrated as clusters in the glands of the stomach, and the local suppression of immune activity and metabolic adaptation are conducive to survival (Ha et al., 2022). These results illustrate the spatial immune heterogeneity in the survival of bacteria in the host amidst persistence of host immunity (Lac et al., 2024).

7.3 Case Studies: COVID-19, Tuberculosis, and Malaria Case Studies

a. COVID-19

A series of experiments of spatial multi-omics analysis of SARS-CoV-2-infected lung tissues has revealed dramatic immune heterogeneity among different organs. Devastated alveoli contain a heavy inflammation of macrophages and cytotoxic T cells, a robust interferon signaling, as well as cytokine production (McClymont et al., 2024). Conversely, immunosuppression, dysfunctional endothelium, and fibrotic remodeling are commonly exhibited in the surrounding areas (Ward et al., 2021). These geometric pattern of immunity reveals the concomitant occurrence of viral clearance, cytokine storm, and tissue damage in the same lung (Graham, 2020).

b. Tuberculosis

Immune heterogeneity is best seen in granulomas in the case of tuberculosis. Both spatial transcriptomics and proteomics have shown that the various areas of the same granuloma have different immune programs- some with highly activated bactericidal capabilities and others with immunosuppressed necrotic cores with frozen Mycobacterium tuberculosis. This is one of the key causes of failure to treat and persistence of the disease through these spatially separated immune states (Lurie et al., 2020).

c. Malaria

Spatial immune profiling in the liver and the spleen of malaria subjects on infection has shown heterogeneous activation of Kupffer, T cells, and B cells and spatially confined inflammatory and metabolic responses. These immunological niches control the clearance of parasites, formation of immune memory, and the severity of the disease (Rahate et al., 2024).

8. COMPUTATIONAL FRAMEWORKS FOR MULTI-OMICS ANALYSIS

The basis of multi-omics data becomes extremely challenging to compute and analyze as biological datasets contain a large number of observations with high levels of heterogeneity and dimensionality. The complexity of spatial multi-omics gets even more complicated with the addition of the spatial layer of information. Computational paradigms are thus needed in the preprocessing of data, normalization, data integration, visualization, and interpretation of the data by a biologist (Adossa et al., 2021).

Contemporary multi-omics analysis is based on using a mix of analytics modeling, machine learning, network-based inference, and high-performance computing. These methods allow the extraction of biologically meaningful patterns of complex data sets and allow predictive modelling of host-pathogen interactions, immune responses, and treatment results (Kaur et al., 2021). As shown in **Figure 4**, generative AI is a subset of deep learning, which in turn falls under machine learning and the broader field of artificial intelligence.

Figure 4. Hierarchical relationship between artificial intelligence, machine learning, deep learning, and generative AI

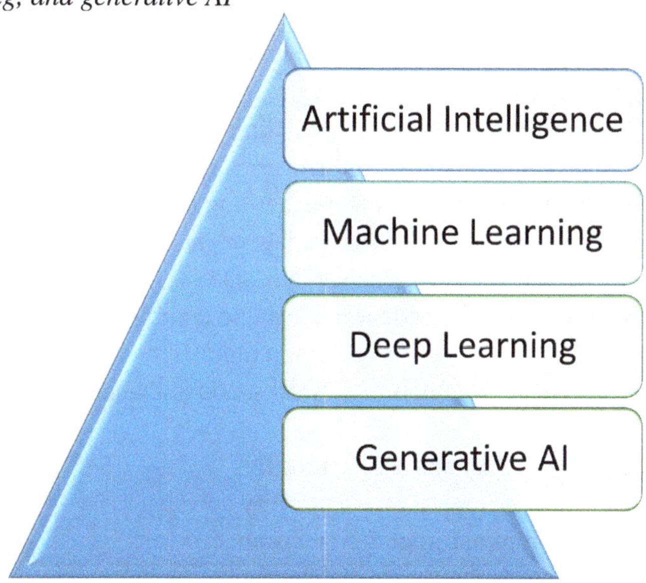

8.1 Statistical Methods of Omics Data

Omics data analysis is based on statistical techniques. The main steps are data normalization, correction of the batch effects, feature selection, and differential expression analysis. Use of linear models, generalized linear models, and Bayesian inference are the most common methods used in transcriptomics to establish the presence of differentially expressed genes between the infected and the control samples (Misra et al., 2019).

In the case of proteomics and metabolomics, statistical constructs consider missing data, technical noise, and non-linear distributions of signal. The most popular techniques of multivariate statistics, including principal component analysis (PCA), partial least squares regression (PLS), and canonical correlation analysis (CCA), are commonly used in dimensionality reduction and cross-omics association analyses (Davenport, 2018). In spatial multi-omics, spatially informed statistical models use physical proximity and tissue structure in inference, which makes it possible to identify spatially varying genes, proteins, and metabolites. These methods disclose localized areas of immune response and a gradient of metabolic activity and pathogen-associated molecular signature (Du et al., 2022).

8.2 Multi-Omics Integration with the use of machine learning

Machine learning is one of the focal points of the combination of various layers of omics into cohesive predictive models. Imagining that only each of the omics modalities can capture conflicting biological dimensions, integration methods have the goal to assemble complementary information and reduce noise and redundancy (Adegoke et al., 2024). Common machine learning methods are early integration, which fuses features across multiple omics layers; intermediate integration, which learns latent representations across each of the datasets and then fuses them; and late integration, which learns independent models and combines them on the decision level (Farnham, 2009). Some of the common algorithms include random forests, support vector machines, deep neural networks, and autoencoders.

a. Alleviate illness severity and clinical results.
b. Win host genetic and immune susceptibility correlates.
c. Identify the pattern of antimicrobial resistance.
d. Categorize patients to treat them differently.

These models can also determine tissue-specific responses and local efficacy of therapeutic responses when spatial information is added.

8.3 Predictive Modeling for Epidemic Surveillance

Epidemic surveillance is carried out through predictive modeling to identify when an outbreak could occur and the potential causes of the outbreak. In addition to the manner of molecules, computational models are of vital importance in epidemic prediction and population-level surveillance. Genomic data, epidemiological data, environmental data, and mobility are combined to predict the spread of the disease and estimate the effectiveness of the intervention using mathematical and AI-based models (Radke et al., 2022).

14.2 The study of epigenomics

The classical epidemiological frameworks, i.e., SIR, SEIR, and agent-based models, have been refined in terms of machine learning and deep learning methods to achieve better predictive power and versatility. These hybrid models provide the opportunity to forecast outbreaks with real-time schemes, early warnings, and the allocation of healthcare resources (Want et al., 2005). During the spatial multi-omics age, molecular signatures of virulence, immune escape, and drug resistance are being incorporated into population-level predictive models in increasing amounts. This is a robust new science of accuracy in public health and pandemic preparedness with a multi-scale system of a combination of molecular networks to the patterns of transmission at the global level.

9. TRANSLATIONAL APPLICATIONS: DIAGNOSTICS, VACCINES, AND PRECISION MEDICINE

The final aim of spatial multi-omics and data integration using artificial intelligence is to make molecular findings turn into clinical implementation. These technologies are transforming the way diagnostics are created, the way vaccines are tested, and the way therapies are designed by involving the spatial, cellular, and molecular complexity of an infectious disease (Fiehn & Kind, 2007). Translational spatial multi-omics delivers the bench-bedside gap by facilitating accurate stratification of disease, real-time feedback on therapeutic facilitation, and design of next-generation interventions rationally. As illustrated in **Figure 5**, integrating heterogeneous data with modeling technologies, data harmonization, and AI-driven computational methods can enhance vaccine–pathogen compatibility prediction and improve vaccine development pipelines.

Figure 5. Integrated data, modeling, and AI approaches for enhancing vaccine–pathogen compatibility and accelerating vaccine development

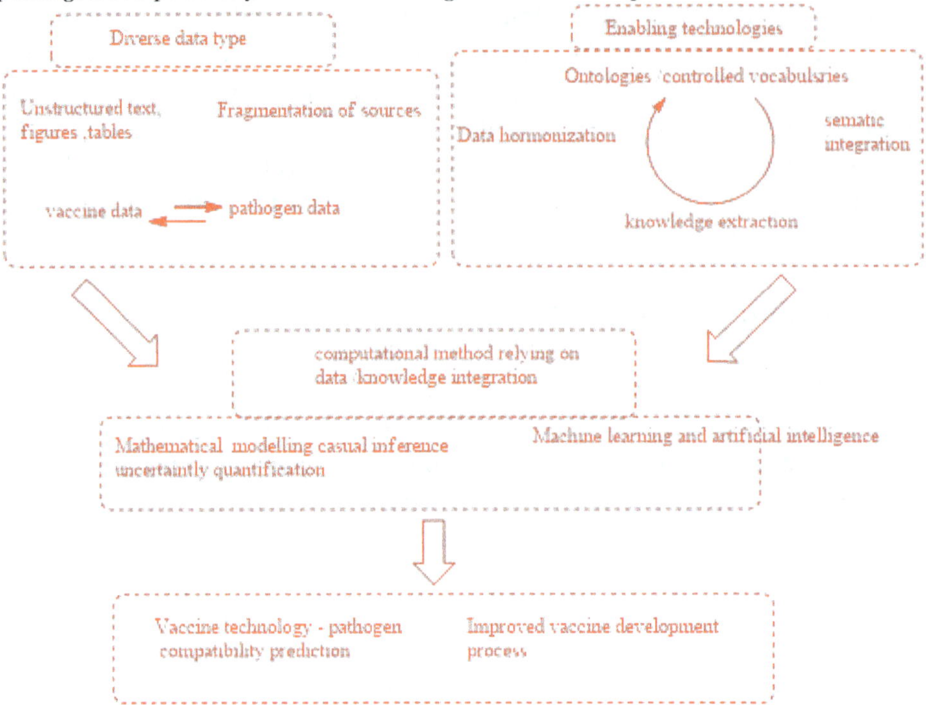

Spatially resolved molecular profiling is now becoming a considerable boost in the area of infectious disease research in precision diagnostics, targeted vaccines, and targeted treatment development, eventually leading to a better patient outcome and preparedness for a disease outbreak

9.1 Precision Medicine by Measuring Spatial Biomarkers

Biomarkers are of key importance in the diagnosis, prognosis, as well as therapeutic monitoring of infectious diseases. Conventional biomarkers frequently utilize bulk processes in a blood or tissue biopsy, which are incapable of reaching spatial and cellular heterogeneity. Spatial multi- omics makes it possible to identify spatially confined biomarkers, which indicate local immune responses, pathogen niches, and tissue-disease processes (Sielaff et al., 2017). Spatial transcriptomics, as an example, could be used to probe region-specific signatures of interferon stimulation genes during viral infections, whereas spatial proteomics could be used to unveil localized expression patterns of immune checkpoints and immune stimulants (Kafle & Ojha, 2024). These spatial markers would be of great use, especially when differentiating between active and latent infections, in the treatment prognosis, as well as tracking treatment progress.

a. With spatial biomarkers, it is possible in precision medicine to:
b. Immune architecture stratification of patients at a tissue level.
c. Timely detection of treatment effectiveness or ineffectiveness.
d. Discovery of focal niches of drug resistance.
e. Immunization reprogramming in therapy.

Combining these biomarkers and clinical as well as imaging evidence, clinicians can use more specific and adaptive care approaches.

9.2 Interpretation of Spatial Omics in Vaccine Development

The effectiveness of the vaccines is determined by both systemic and tissue-based immune responses and locally available antigen presentation. Spatial multi-omics fabricates potent technologies to sample the impacts of vaccines at the tissue level on the architecture of immunity (Vetter et al., 2018). New vaccine platform technologies include cellular vaccines, subunit vaccines, whole-pathogen adapted vaccines, recombinant vector vaccines, nucleic acid-based vaccines, engineered vaccines, and a variety of adjuvant-driven or synthetically derived vaccine combinations that combine the advantages of multiple platform technologies (**Figure 6**)

a. With the use of spatial transcriptomics and proteomics, it becomes possible to map:
b. Apoptotic sites of APCs.
c. Lymphoid Tissue Germinal center structure.
d. T-cell niches in the tissues.
e. Spatial or separation coordination of innate and adaptive immunity.

Figure 6. Overview of common vaccine technologies, including nucleic acid, vector-based, whole-pathogen, subunit, cellular, and engineered vaccine platforms

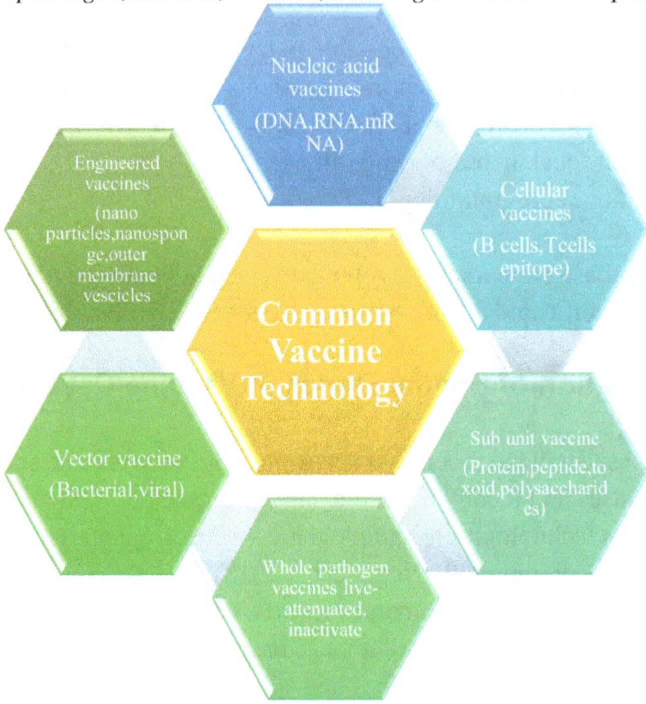

The insights are critical in optimizing the vaccine formulations, adjuvants and the delivery routes. As an illustration, spatial profiling has demonstrated the way intranasal and intramuscular vaccines produce unique lung immune landscapes in respiratory infections. Spatial omics similarly enables faster assessment of vaccine-induced immune protection in various organs and distinct groups of patients in the case of emerging infectious diseases. Moreover, an AI-based combination of vaccine-induced spatial immune signatures aids in the intelligent design of the next-generation vaccines with increased durability, breadth, and tissue-directed protection (Stern, 2020).

9.3 Clinical Application of Multi-Omics Technologies

Development of automation, low costs, and regulatory standards is driving the swift pace of clinical translation of spatial multi-omics. Coupled with regular histopathology, spatial transcriptomic and proteomic platforms are being updated for any clinical pathology lab, and thus allow direct biological profiling of biopsy samples. Examples of important fields of clinical translation are:

a. Accurate diagnosis of complex infections like tuberculosis, sepsis, and viral pneumonia.
b. Spatial immune and pathogen profiling, treatment response monitoring.
c. Optimization of host-directed therapy using tissue-specific immune conditions.
d. Mapping of antimicrobial resistance in the tissues that are infected.

Simultaneously, clinical decision support systems that will run on AI are under development to combine spatial multi-omics data with electronic health records, imaging studies, and laboratory outcomes. These systems are helping clinicians to choose the best forms of therapy, forecast complications, and develop individualized patterns of therapy. With the establishment of regulatory regimes and the development of standard workflows, spatial multi-omics will become an essential technology of precision infectious disease medicine

10. ETHICAL CHALLENGES, SURVEILLANCE LIMITATIONS, AND FUTURE PERSPECTIVES

The fast development of spatial multi-omics, artificial intelligence, and real-time surveillance technologies has presented unprecedented opportunities to learn and manage infectious diseases. But these technological innovations also have significant ethical, regulatory, social, and business issues (Agrebi & Larbi, 2020). This is necessary to address these concerns because responsible innovation, equitable access, and long-term sustainability of spatial multi-omics-based public health systems are needed (Wei et al., 2023).

10.1 Limitations of Conventional Surveillance Systems

The traditional methods of infectious disease surveillance are mostly based on clinical case reporting, lab confirmation, and post facto epidemiological investigation (Obande & Banga Singh, 2020). Though these systems have played a historic role in supporting public health, they have a number of augmented limitations:

a. Defeated reporting that necessitates early detection of an outbreak.
b. Underreporting and gaps in data, particularly in low-resource set-ups.
c. Low molecular resolution, which did not allow for following the evolution of pathogens in real time.
d. Weak connection with host biological and tissue expression information.

The systems used traditionally cannot capture the spatial heterogeneity of infection at the tissue and cellular level, which are vital in the context of comprehending the severity of the disease, immune escape, and failure in the treatment (Giuffrida & Spoto, 2017). Even though digital and genomic surveillance have enhanced responsiveness to outbreaks, integration with spatial multi-omics is still a challenge (De Falco et al., 2025).

10.2 Ethical and Data Privacy Considerations

Several ethical considerations arise in the context of the use of spatial multi-omics and AI in infectious disease studies due to the ownership, privacy, informed consent, and fair use of new technologies (Böhmer et al., 2009). Spatial omics data have very sensitive molecular and spatial data that may, potentially, be connected to an individual's identity, disease predisposition, and genetic predisposition. The major ethical issues to take into account are:

a. Safety of patient privacy of high-resolution high-performance molecular datasets.
b. Clinical predictive models on AI-driven predictive models: Ethical use.
c. Spatial and genomic data secondary use informed consent.
d. Prejudice by AI algorithms against vulnerable populations.

Secondly, the dual-use issues should be considered, because detailed molecular and spatial pathogen data may be used for malicious intentions (Fakruddin et al., 2012). The use of these potent technologies should be secured by well-established systems of governance, global coordination of regulations, and transparent ethical controls to prevent adverse, irresponsible transfers of these technologies (Lillis et al., 2016).

10.3 Future Perspectives of Spatial Multi-Omics towards Infectious Diseases

The future of spatial multi-omics in infectious disease investigations is ready to change in a more significant way. Emerging trends include:

a. Spatial profiling with 3D and 4D visualization through which the death of infections can be monitored over time.
b. Pathologically modeling inorganic infection models, mechanistic studies, integration with organoids and tissue-engineered infection models
c. Quick and real-time spatial diagnostic via fast molecular imaging and biosensor systems.

d. Artificial intelligence-based virtual twins to model patient-specific infections.
e. Multi-scale modeling, the connection between molecular networks and population-scale dynamics of transmission.

Spatial multi-omics should become an inseparable part of clinical diagnostics, medical outbreak management, and vaccine development pipeline as the costs are reduced and the computational tools reach the maturity stage (Y. Wang et al., 2015). Such innovations will not only help save patients but also enhance global resilience to pandemics and resilience of the global population to health (Polnaszek et al., 2016).

11. CONCLUSION

The spatial multi-omics have become a ground-breaking tool of fertile research into infectious diseases that has fundamentally altered the way host-pathogen interactions are studied, understood, and converted into clinical practice. This method provides a systems-level perspective of the biology of infection in intact tissues in an unprecedented, robust, and scalable manner with the integration of spatially resolved transcriptomics, proteomics, epigenomics, and metabolomics with artificial intelligence and computational modeling. In this chapter, we have noted the ways in which spatial multi-omics addresses the major gaps that prevail with the conventional bulk and single-cell technologies. It allows mapping immune microenvironments, pathogen niches, regulatory landscapes, and metabolic gradients with precision, creating an overwhelming pattern of an infection. The use of AI and machine learning can further augment the strength of such datasets because it will facilitate predictive modeling, the discovery of biomarkers, the identification of drug targets, and real-time surveillance.

We have shown how phenotypic screening and spatial omics can be used together in their quest to discover and rapid testing on anti-infective drugs, how new surveillance technologies can be used to enhance outbreak diagnosis, and epigenomic and immune heterogeneity determine disease progress and treatment outcome. Spatial multi-omics translationally can be considered most effectively in their precision diagnostics, vaccine-making, and personalized treatment planning, wherein spatial biomarkers and tissue-specific immune signatures currently lead the clinical decision-making procedure. Simultaneously in this chapter, it is also clear to highlight the critical role of managing ethical, regulatory, and surveillance constraints of large-scale and high-resolution molecular research and spatial data. The responsible development of spatial multi-omics should be ensured through increasing the privacy of data, eliminating all manifestations of algorithmic bias, permitting equal access to sophisticated diagnostics, and reinforcing frameworks of global data sharing.

In the future, spatial biology, artificial intelligence, digital health, and systems immunology will converge to shape the next-generation type of investigative writing on infectious disease and the approaches to be suggested to the general public on disease prevention and control. The growing nature of these technologies means that spatial multi-omics will enable the establishment of a foundation of precision infectious disease medicine in the future, providing potent technologies to detect infectious diseases at an early stage, targeting intervention, pandemic readiness, and global health security.

REFERENCES

Abbott, R. K., & Crotty, S. (2020). Factors in B cell competition and immunodominance. *Immunological Reviews, 296*(1), 120–131. DOI: 10.1111/imr.12861

Adegoke, B. O., Odugbose, T., & Adeyemi, C. (2024). Data analytics for predicting disease outbreaks: A review of models and tools. *International Journal of Life Science Research Updates [Online]*, *2*(2), 1–9. DOI: 10.53430/ijlsru.2024.2.2.0023

Adossa, N., Khan, S., Rytkönen, K. T., & Elo, L. L. (2021). Computational strategies for single-cell multi-omics integration. *Computational and Structural Biotechnology Journal, 19*, 2588–2596. DOI: 10.1016/j.csbj.2021.04.060

Agrebi, S., & Larbi, A. (2020). Use of artificial intelligence in infectious diseases. In *Artificial intelligence in precision health* (pp. 415–438). Elsevier. DOI: 10.1016/B978-0-12-817133-2.00018-5

Al-Talib, M., Dimonte, S., & Humphreys, I. R. (2024). Mucosal T-cell responses to chronic viral infections: Implications for vaccine design. *Cellular & Molecular Immunology*, *21*(9), 982–998. DOI: 10.1038/s41423-024-01140-2

Ali, H. (2023). Artificial intelligence in multi-omics data integration: Advancing precision medicine, biomarker discovery and genomic-driven disease interventions. *Int J Sci Res Arch*, *8*(1), 1012–1030. DOI: 10.30574/ijsra.2023.8.1.0189

Altschuler, S. J., & Wu, L. F. (2010). Cellular heterogeneity: Do differences make a difference? *Cell, 141*(4), 559–563. DOI: 10.1016/j.cell.2010.04.033

Alves, M. A., Lamichhane, S., Dickens, A., McGlinchey, A., Ribeiro, H. C., Sen, P., Wei, F., Hyötyläinen, T., & Orešič, M. (2021). Systems biology approaches to study lipidomes in health and disease. *Biochimica et Biophysica Acta (BBA)-. Biochimica et Biophysica Acta. Molecular and Cell Biology of Lipids*, *1866*(2), 158857. DOI: 10.1016/j.bbalip.2020.158857

Aoshi, T., Koyama, S., Kobiyama, K., Akira, S., & Ishii, K. J. (2011). Innate and adaptive immune responses to viral infection and vaccination. *Current Opinion in Virology*, *1*(4), 226–232. DOI: 10.1016/j.coviro.2011.07.002

Biondi-Zoccai, G., D'Ascenzo, F., Giordano, S., Mirzoyev, U., Erol, Ç., Cenciarelli, S., Leone, P., & Versaci, F. (2025). Artificial Intelligence in Cardiology: General Perspectives and Focus on Interventional Cardiology. *The Anatolian Journal of Cardiology*, *29*(4), 152–163. DOI: 10.14744/AnatolJCardiol.2025.5237

Björkholm, B., Sjölund, M., Falk, P. G., Berg, O. G., Engstrand, L., & Andersson, D. I. (2001). Mutation frequency and biological cost of antibiotic resistance in Helicobacter pylori. *Proceedings of the National Academy of Sciences of the United States of America, 98*(25), 14607–14612. DOI: 10.1073/pnas.241517298

Blaser, M. J., & Atherton, J. C. (2004). Helicobacter pylori persistence: Biology and disease. *The Journal of Clinical Investigation, 113*(3), 321–333. DOI: 10.1172/JCI20925

Böhmer, A., Schildgen, V., Lüsebrink, J., Ziegler, S., Tillmann, R. L., Kleines, M., & Schildgen, O. (2009). Novel application for isothermal nucleic acid sequence-based amplification (NASBA). *Journal of Virological Methods, 158*(1–2), 199–201. DOI: 10.1016/j.jviromet.2009.02.010

Chukwunweike, J. N., Praise, A., & Bashirat, B. A. (2024). *Harnessing Machine Learning for Cybersecurity: How Convolutional Neural Networks are Revolutionizing Threat Detection and Data Privacy.*

Davenport, T. H. (2018). From analytics to artificial intelligence. *Journal of Business Analytics, 1*(2), 73–80. DOI: 10.1080/2573234X.2018.1543535

De Falco, M., Colella, S., Abdelkarim, J. A. Y., Antonacci, A., Scognamiglio, V., & De Felice, M. (2025). Helicase-dependent isothermal amplification turns biosensing into modern diagnostic technologies. *Microchemical Journal, 216*, 114772. DOI: 10.1016/j.microc.2025.114772

De Souza, M. L. M., Lopes, G. A., Branco, A. C., Fairley, J. K., & Fraga, L. A. D. O. (2021). Leprosy screening based on artificial intelligence: Development of a cross-platform app. *JMIR mHealth and uHealth, 9*(4), e23718. DOI: 10.2196/23718

Debbadi, R. K., & Boateng, O. (2025a). Developing intelligent automation workflows in Microsoft Power Automate by embedding deep learning algorithms for real-time process adaptation. *Int J Sci Res Arch, 14*(2), 802–820. DOI: 10.30574/ijsra.2025.14.2.0449

Debbadi, R. K., & Boateng, O. (2025b). Optimizing end-to-end business processes by integrating machine learning models with UiPath for predictive analytics and decision automation. *Int J Sci Res Arch, 14*(2), 778–796. DOI: 10.30574/ijsra.2025.14.2.0448

Díaz-Menéndez, M., Esteban, E. T., Ujiie, M., Calleri, G., Rothe, C., Malvy, D., Nicastri, E., Bissinger, A. L., Grandadam, M., Alpern, J. D., Gobbi, F., Schlagenhauf, P., Duvignaud, A., Javelle, E., Nakamoto, T., Antinori, S., & Hamer, D. H. (2020). Travel-associated chikungunya acquired in Myanmar in 2019. *Euro Surveillance : Bulletin Europeen Sur Les Maladies Transmissibles = European. Eurosurveillance, 25*(1). Advance online publication. DOI: 10.2807/1560-7917.ES.2020.25.1.1900721

Dlamini, Z., Francies, F. Z., Hull, R., & Marima, R. (2020). Artificial intelligence (AI) and big data in cancer and precision oncology. *Computational and Structural Biotechnology Journal, 18*, 2300–2311. https://doi.org/https://doi.org/10.1016/j.csbj.2020.08.019. DOI: 10.1016/j.csbj.2020.08.019

Du, B., Mao, K., Bateni, S. M., Meng, F., Wang, X.-M., Guo, Z., Jun, C., & Du, G. (2022). A novel fully coupled physical–statistical–deep learning method for retrieving near-surface air temperature from multisource data. *Remote Sensing (Basel), 14*(22), 5812. DOI: 10.3390/rs14225812

Fakruddin, M. D., Mazumdar, R. M., Chowdhury, A., & Mannan, K. S. Bin. (2012). Nucleic acid sequence based amplification (NASBA)-prospects and applications. *International Journal of Life Science and Pharma Research, 2*(1), 106–121.

Farnham, P. J. (2009). Insights from genomic profiling of transcription factors. *Nature Reviews. Genetics, 10*(9), 605–616. DOI: 10.1038/nrg2636

Fiehn, O., & Kind, T. (2007). Metabolite profiling in blood plasma. In *Metabolomics: Methods and protocols* (pp. 3–17). Springer. DOI: 10.1007/978-1-59745-244-1_1

Ge, S., Sun, S., Xu, H., Cheng, Q., & Ren, Z. (2025). Deep learning in single-cell and spatial transcriptomics data analysis: Advances and challenges from a data science perspective. *Briefings in Bioinformatics, 26*(2), bbaf136. Advance online publication. DOI: 10.1093/bib/bbaf136

Giuffrida, M. C., & Spoto, G. (2017). Integration of isothermal amplification methods in microfluidic devices: Recent advances. *Biosensors & Bioelectronics, 90*, 174–186. DOI: 10.1016/j.bios.2016.11.045

Graham, B. S. (2020). Rapid COVID-19 vaccine development. *Science, 368*(6494), 945–946. DOI: 10.1126/science.abb8923

Grubaugh, N. D., Ladner, J. T., Kraemer, M. U. G., Dudas, G., Tan, A. L., Gangavarapu, K., Wiley, M. R., White, S., Thézé, J., Magnani, D. M., Prieto, K., Reyes, D., Bingham, A. M., Paul, L. M., Robles-Sikisaka, R., Oliveira, G., Pronty, D., Barcellona, C. M., Metsky, H. C., & Andersen, K. G. (2017). Genomic epidemiology reveals multiple introductions of Zika virus into the United States. *Nature, 546*(7658), 401–405. DOI: 10.1038/nature22400

Gupta, A., Madhavan, M. V., Sehgal, K., Nair, N., Mahajan, S., Sehrawat, T. S., Bikdeli, B., Ahluwalia, N., Ausiello, J. C., Wan, E. Y., Freedberg, D. E., Kirtane, A. J., Parikh, S. A., Maurer, M. S., Nordvig, A. S., Accili, D., Bathon, J. M., Mohan, S., Bauer, K. A., & Landry, D. W. (2020). Extrapulmonary manifestations of COVID-19. *Nature Medicine, 26*(7), 1017–1032. DOI: 10.1038/s41591-020-0968-3

Gupta, N., Maurya, S., Verma, H., & Verma, V. K. (2019). Unraveling the factors and mechanism involved in persistence: Host-pathogen interactions in Helicobacter pylori. *Journal of Cellular Biochemistry, 120*(11), 18572–18587. DOI: 10.1002/jcb.29201

Gupta, R. K., & Kuznicki, J. (2020). Biological and medical importance of cellular heterogeneity deciphered by single-cell RNA sequencing. *Cells, 9*(8), 1751. DOI: 10.3390/cells9081751

Ha, T., Kaiser, C., Myong, S., Wu, B., & Xiao, J. (2022). Next generation single-molecule techniques: Imaging, labeling, and manipulation <em>in vitro</em> and <em>in cellulo</em>. *Molecular Cell, 82*(2), 304–314. DOI: 10.1016/j.molcel.2021.12.019

Hadfield, J., Megill, C., Bell, S. M., Huddleston, J., Potter, B., Callender, C., Sagulenko, P., Bedford, T., & Neher, R. A. (2018). Nextstrain: Real-time tracking of pathogen evolution. *Bioinformatics (Oxford, England), 34*(23), 4121–4123. DOI: 10.1093/bioinformatics/bty407

Horvath, P., Aulner, N., Bickle, M., Davies, A. M., Del Nery, E., Ebner, D., Montoya, M. C., Östling, P., Pietiäinen, V., Price, L. S., Shorte, S. L., Turcatti, G., von Schantz, C., & Carragher, N. O. (2016). Screening out irrelevant cell-based models of disease. *Nature Reviews. Drug Discovery, 15*(11), 751–769. DOI: 10.1038/nrd.2016.175

Huang, C., Wang, Y., Li, X., Ren, L., Zhao, J., Hu, Y., Zhang, L., Fan, G., Xu, J., Gu, X., Cheng, Z., Yu, T., Xia, J., Wei, Y., Wu, W., Xie, X., Yin, W., Li, H., Liu, M., & Cao, B. (2020). Clinical features of patients infected with 2019 novel coronavirus in Wuhan, China. *Lancet, 395*(10223), 497–506. DOI: 10.1016/S0140-6736(20)30183-5

Jacobs, R. T., Nare, B., Wring, S. A., Orr, M. D., Chen, D., Sligar, J. M., Jenks, M. X., Noe, R. A., Bowling, T. S., Mercer, L. T., Rewerts, C., Gaukel, E., Owens, J., Parham, R., Randolph, R., Beaudet, B., Bacchi, C. J., Yarlett, N., Plattner, J. J., & Don, R. (2011). SCYX-7158, an orally-active benzoxaborole for the treatment of stage 2 human African trypanosomiasis. *PLoS Neglected Tropical Diseases*, *5*(6), e1151. DOI: 10.1371/journal.pntd.0001151

Jean Beltran, P. M., Federspiel, J. D., Sheng, X., & Cristea, I. M. (2017). Proteomics and integrative omic approaches for understanding host–pathogen interactions and infectious diseases. *Molecular Systems Biology*, *13*(3), 922. DOI: 10.15252/msb.20167062

Jonjić, S. (2024). Tissue-specific antiviral immunity. *Cellular & Molecular Immunology*, *21*(9), 941–942. DOI: 10.1038/s41423-024-01200-7

Kafle, A., & Ojha, S. C. (2024). Advancing vaccine development against Opisthorchis viverrini: A synergistic integration of omics technologies and advanced computational tools. *Frontiers in Pharmacology*, *15*, 1410453. DOI: 10.3389/fphar.2024.1410453

Kanneganti, T.-D. (2020). Intracellular innate immune receptors: Life inside the cell. *Immunological Reviews*, *297*(1), 5–12. DOI: 10.1111/imr.12912

Kaur, P., Singh, A., & Chana, I. (2021). Computational Techniques and Tools for Omics Data Analysis: State-of-the-Art, Challenges, and Future Directions. *Archives of Computational Methods in Engineering*, *28*(7), 4595–4631. DOI: 10.1007/s11831-021-09547-0

Keilberg, D., & Ottemann, K. M. (2016). How H elicobacter pylori senses, targets and interacts with the gastric epithelium. *Environmental Microbiology*, *18*(3), 791–806. DOI: 10.1111/1462-2920.13222

Keutzer, L., Wicha, S. G., & Simonsson, U. S. H. (2020). Mobile health apps for improvement of tuberculosis treatment: Descriptive review. *JMIR mHealth and uHealth*, *8*(4), e17246. DOI: 10.2196/17246

Khare, K., & Pandey, R. (2022). Cellular heterogeneity in disease severity and clinical outcome: Granular understanding of immune response is key. *Frontiers in Immunology, Volume 13-2022*. DOI: 10.3389/fimmu.2022.973070

Kulkarni, A., Anderson, A. G., Merullo, D. P., & Konopka, G. (2019). Beyond bulk: A review of single cell transcriptomics methodologies and applications. *Current Opinion in Biotechnology*, *58*, 129–136. https://doi.org/https://doi.org/10.1016/j.copbio.2019.03.001. DOI: 10.1016/j.copbio.2019.03.001

Kulkarni, S., Arumugam, T., Chuturgoon, A., An, P., & Ramsuran, V. (2022). Epigenetics of infectious diseases. [). Frontiers Media SA.]. *Frontiers in Immunology, 13*, 1054151. DOI: 10.3389/fimmu.2022.1054151

Lac, L., Leung, C. K., & Hu, P. (2024). Computational frameworks integrating deep learning and statistical models in mining multimodal omics data. *Journal of Biomedical Informatics, 152*, 104629. DOI: 10.1016/j.jbi.2024.104629

Lee, J. A., Carragher, N. O., & Berg, E. L. (2017). Empirical drug discovery: A view from the proteome. *Drug Discovery Today. Technologies, 23*, 1–5. DOI: 10.1016/j.ddtec.2017.05.003

Li, M., Guo, M., Peng, Y., Jiang, Q., Xia, L., Zhong, S., Qiu, Y., Su, X., Zhang, S., Yang, C., Mijiti, P., Mao, Q., Takiff, H., Li, F., Chen, C., & Gao, Q. (2022). High proportion of tuberculosis transmission among social contacts in rural China: A 12-year prospective population-based genomic epidemiological study. *Emerging Microbes & Infections, 11*(1), 2102–2111. DOI: 10.1080/22221751.2022.2112912

Liao, H., Lyon, C. J., Ying, B., & Hu, T. (2024). Climate change, its impact on emerging infectious diseases and new technologies to combat the challenge. *Emerging Microbes & Infections, 13*(1), 2356143. DOI: 10.1080/22221751.2024.2356143

Lillis, L., Siverson, J., Lee, A., Cantera, J., Parker, M., Piepenburg, O., Lehman, D. A., & Boyle, D. S. (2016). Factors influencing recombinase polymerase amplification (RPA) assay outcomes at point of care. *Molecular and Cellular Probes, 30*(2), 74–78. DOI: 10.1016/j.mcp.2016.01.009

Lipworth, S., Hammond, R. J. H., Baron, V. O., Hu, Y., Coates, A., & Gillespie, S. H. (2016). Defining dormancy in mycobacterial disease. *Tuberculosis (Edinburgh, Scotland), 99*, 131–142. https://doi.org/https://doi.org/10.1016/j.tube.2016.05.006. DOI: 10.1016/j.tube.2016.05.006

Liu, X., Peng, T., Xu, M., Lin, S., Hu, B., Chu, T., Liu, B., Xu, Y., Ding, W., Li, L., Cao, C., & Wu, P. (2024). Spatial multi-omics: Deciphering technological landscape of integration of multi-omics and its applications. *Journal of Hematology & Oncology, 17*(1), 72. DOI: 10.1186/s13045-024-01596-9

Lurie, N., Saville, M., Hatchett, R., & Halton, J. (2020). Developing Covid-19 vaccines at pandemic speed. *The New England Journal of Medicine, 382*(21), 1969–1973. DOI: 10.1056/NEJMp2005630

Marshall, B., & Warren, J. R. (1984). UNIDENTIFIED CURVED BACILLI IN THE STOMACH OF PATIENTS WITH GASTRITIS AND PEPTIC ULCERATION. *Lancet, 323*(8390), 1311–1315. https://doi.org/https://doi.org/10.1016/S0140-6736(84)91816-6. DOI: 10.1016/S0140-6736(84)91816-6

Mason, C. E., Porter, S. G., & Smith, T. M. (2013). Characterizing multi-omic data in systems biology. *Systems Analysis of Human Multigene Disorders*, 15–38.

McClellan, J., & King, M.-C. (2010). Genetic heterogeneity in human disease. *Cell, 141*(2), 210–217. DOI: 10.1016/j.cell.2010.03.032

McClymont, H., Lambert, S. B., Barr, I., Vardoulakis, S., Bambrick, H., & Hu, W. (2024). Internet-based surveillance systems and infectious diseases prediction: An updated review of the last 10 years and lessons from the COVID-19 pandemic. *Journal of Epidemiology and Global Health, 14*(3), 645–657. DOI: 10.1007/s44197-024-00272-y

Melms, J. C., Biermann, J., Huang, H., Wang, Y., Nair, A., Tagore, S., Katsyv, I., Rendeiro, A. F., Amin, A. D., Schapiro, D., Frangieh, C. J., Luoma, A. M., Filliol, A., Fang, Y., Ravichandran, H., Clausi, M. G., Alba, G. A., Rogava, M., Chen, S. W., & Izar, B. (2021). A molecular single-cell lung atlas of lethal COVID-19. *Nature, 595*(7865), 114–119. DOI: 10.1038/s41586-021-03569-1

Mihalić, A., Železnjak, J., Lisnić, B., Jonjić, S., Juranić Lisnić, V., & Brizić, I. (2024). Immune surveillance of cytomegalovirus in tissues. *Cellular & Molecular Immunology, 21*(9), 959–981. DOI: 10.1038/s41423-024-01186-2

Misra, B. B., Langefeld, C., Olivier, M., & Cox, L. A. (2019). Integrated omics: Tools, advances and future approaches. *Journal of Molecular Endocrinology, 62*(1), R21–R45. DOI: 10.1530/JME-18-0055

Moffat, J. G., Vincent, F., Lee, J. A., Eder, J., & Prunotto, M. (2017). Opportunities and challenges in phenotypic drug discovery: An industry perspective. *Nature Reviews. Drug Discovery, 16*(8), 531–543. DOI: 10.1038/nrd.2017.111

Moss, S. F. (2017). The Clinical Evidence Linking Helicobacter pylori to Gastric Cancer. *Cellular and Molecular Gastroenterology and Hepatology, 3*(2), 183–191. https://doi.org/https://doi.org/10.1016/j.jcmgh.2016.12.001. DOI: 10.1016/j.jcmgh.2016.12.001

Obande, G. A., & Banga Singh, K. K. (2020). Current and future perspectives on isothermal nucleic acid amplification technologies for diagnosing infections. *Infection and Drug Resistance, 13*, 455–483. DOI: 10.2147/IDR.S217571

Park, J. (2023). *Multi-omics modeling for the recovery of disrupted biological systems*. Weill Medical College of Cornell University.

Pascucci, M., Royer, G., Adamek, J., Al Asmar, M., Aristizabal, D., Blanche, L., Bezzarga, A., Boniface-Chang, G., Brunner, A., Curel, C., Dulac-Arnold, G., Fakhri, R. M., Malou, N., Nordon, C., Runge, V., Samson, F., Sebastian, E., Soukieh, D., Vert, J.-P., & Madoui, M.-A. (2021). AI-based mobile application to fight antibiotic resistance. *Nature Communications*, *12*(1), 1173. DOI: 10.1038/s41467-021-21187-3

Patel, S. K., Surve, J., Parmar, J., Ahmed, K., Bui, F. M., & Al-Zahrani, F. A. (2022). Recent advances in biosensors for detection of COVID-19 and other viruses. *IEEE Reviews in Biomedical Engineering*, *16*, 22–37. DOI: 10.1109/RBME.2022.3212038

Pavlou, A., Mulenge, F., Gern, O. L., Busker, L. M., Greimel, E., Waltl, I., & Kalinke, U. (2024). Orchestration of antiviral responses within the infected central nervous system. *Cellular & Molecular Immunology*, *21*(9), 943–958. DOI: 10.1038/s41423-024-01181-7

Penaranda, C., & Hung, D. T. (2019). Single-cell RNA sequencing to understand host–pathogen interactions. *ACS Infectious Diseases*, *5*(3), 336–344. DOI: 10.1021/acsinfecdis.8b00369

Polnaszek, B., Gilmore-Bykovskyi, A., Hovanes, M., Roiland, R., Ferguson, P., Brown, R., & Kind, A. J. H. (2016). Overcoming the challenges of unstructured data in multisite, electronic medical record-based abstraction. *Medical Care*, *54*(10), e65–e72. DOI: 10.1097/MLR.0000000000000108

Radke, J., Ishaque, N., Koll, R., Gu, Z., Schumann, E., Sieverling, L., Uhrig, S., Hübschmann, D., Toprak, U. H., López, C., Hostench, X. P., Borgoni, S., Juraeva, D., Pritsch, F., Paramasivam, N., Balasubramanian, G. P., Schlesner, M., Sahay, S., Weniger, M., & Heppner, F. L. (2022). The genomic and transcriptional landscape of primary central nervous system lymphoma. *Nature Communications*, *13*(1), 2558. DOI: 10.1038/s41467-022-30050-y

Rahate, K. P., Sengar, M. S., & Patel, S. (2024). Artificial Intelligence in Medical Imaging by Machine Learning and Deep Learning. In *Approaches to Human-Centered AI in Healthcare* (pp. 121–159). IGI Global. DOI: 10.4018/979-8-3693-2238-3.ch006

Rivas-Ortiz, C. I., Lopez-Vidal, Y., Arredondo-Hernandez, L. J. R., & Castillo-Rojas, G. (2017). Genetic alterations in gastric cancer associated with Helicobacter pylori infection. *Frontiers in Medicine*, *4*, 47. DOI: 10.3389/fmed.2017.00047

Russell, S. K., Harrison, J. K., Olson, B. S., Lee, H. J., O'Brien, V. P., Xing, X., Livny, J., Yu, L., Roberson, E. D. O., Bomjan, R., Fan, C., Sha, M., Estfanous, S., Amer, A. O., Colonna, M., Stappenbeck, T. S., Wang, T., Hannan, T. J., & Hultgren, S. J. (2023). Uropathogenic Escherichia coli infection-induced epithelial trained immunity impacts urinary tract disease outcome. *Nature Microbiology*, *8*(5), 875–888. DOI: 10.1038/s41564-023-01346-6

Sahl, S. J., & Moerner, W. E. (2013). Super-resolution fluorescence imaging with single molecules. *Current Opinion in Structural Biology*, *23*(5), 778–787. https://doi.org/https://doi.org/10.1016/j.sbi.2013.07.010. DOI: 10.1016/j.sbi.2013.07.010

Seers, T., Rothe, C., Hamer, D. H., Denny, S., Spindler, R., Schwartz, E., & Johnston, V. (2023). Zika virus infection in European travellers returning from Thailand in 2022: A GeoSentinel case series. *Tropical Medicine & International Health*, *28*(7), 576–579. DOI: 10.1111/tmi.13900

Seo, G., Lee, G., Kim, M. J., Baek, S.-H., Choi, M., Ku, K. B., Lee, C.-S., Jun, S., Park, D., Kim, H. G., Kim, S.-J., Lee, J.-O., Kim, B. T., Park, E. C., & Kim, S. (2020). Rapid Detection of COVID-19 Causative Virus (SARS-CoV-2) in Human Nasopharyngeal Swab Specimens Using Field-Effect Transistor-Based Biosensor. *ACS Nano*, *14*(4), 5135–5142. DOI: 10.1021/acsnano.0c02823

Sharpe, J. D., & Kamara, M. T. (2018). A systematic evaluation of mobile apps to improve the uptake of and adherence to HIV pre-exposure prophylaxis. *Sexual Health*, *15*(6), 587–594. DOI: 10.1071/SH18120

Sielaff, M., Kuharev, J., Bohn, T., Hahlbrock, J., Bopp, T., Tenzer, S., & Distler, U. (2017). Evaluation of FASP, SP3, and iST protocols for proteomic sample preparation in the low microgram range. *Journal of Proteome Research*, *16*(11), 4060–4072. DOI: 10.1021/acs.jproteome.7b00433

Singh, P. R., Dadireddy, V., Udupa, S., Kalladi, S. M., Shee, S., Khosla, S., Rajmani, R. S., Singh, A., Ramakumar, S., & Nagaraja, V. (2023). The Mycobacterium tuberculosis methyltransferase Rv2067c manipulates host epigenetic programming to promote its own survival. *Nature Communications*, *14*(1), 8497. DOI: 10.1038/s41467-023-43940-6

Spurbeck, R. R. (2024). Altered epigenetic landscape as infectious disease diagnostics. *Epigenomics*, *16*(19–20), 1269–1272. DOI: 10.1080/17501911.2024.2415282

Stambas, J., Lu, C., & Tripp, R. A. (2020). Innate and adaptive immune responses in respiratory virus infection: Implications for the clinic. *Expert Review of Respiratory Medicine*, *14*(11), 1141–1147. DOI: 10.1080/17476348.2020.1807945

Stern, P. L. (2020). Key steps in vaccine development. *Annals of Allergy, Asthma & Immunology, 125*(1), 17–27. DOI: 10.1016/j.anai.2020.01.025

Swain, S. S., Sharma, D., Hussain, T., & Pati, S. (2020). Molecular mechanisms of underlying genetic factors and associated mutations for drug resistance in Mycobacterium tuberculosis. *Emerging Microbes & Infections, 9*(1), 1651–1663. DOI: 10.1080/22221751.2020.1785334

Swinney, D. C., & Anthony, J. (2011). How were new medicines discovered? *Nature Reviews. Drug Discovery, 10*(7), 507–519. DOI: 10.1038/nrd3480

Theodosiou, A. A., & Read, R. C. (2023). Artificial intelligence, machine learning and deep learning: Potential resources for the infection clinician. *The Journal of Infection, 87*(4), 287–294. https://doi.org/https://doi.org/10.1016/j.jinf.2023.07.006. DOI: 10.1016/j.jinf.2023.07.006

Turanli, B., Karagoz, K., Gulfidan, G., Sinha, R., Mardinoglu, A., & Arga, K. Y. (2018). A network-based cancer drug discovery: From integrated multi-omics approaches to precision medicine. *Current Pharmaceutical Design, 24*(32), 3778–3790. DOI: 10.2174/1381612824666181106095959

Vabret, N., Britton, G. J., Gruber, C., Hegde, S., Kim, J., Kuksin, M., Levantovsky, R., Malle, L., Moreira, A., Park, M. D., Pia, L., Risson, E., Saffern, M., Salomé, B., Esai Selvan, M., Spindler, M. P., Tan, J., van der Heide, V., Gregory, J. K., & Laserson, U. (2020). Immunology of COVID-19: Current state of the science. *Immunity, 52*(6), 910–941. DOI: 10.1016/j.immuni.2020.05.002

Venerito, M., Vasapolli, R., Rokkas, T., Delchier, J., & Malfertheiner, P. (2017). Helicobacter pylori, gastric cancer and other gastrointestinal malignancies. *Helicobacter, 22*(S1), e12413. DOI: 10.1111/hel.12413

Vetter, V., Denizer, G., Friedland, L. R., Krishnan, J., & Shapiro, M. (2018). Understanding modern-day vaccines: What you need to know. *Annals of Medicine, 50*(2), 110–120. DOI: 10.1080/07853890.2017.1407035

Vincent, F., Loria, P., Pregel, M., Stanton, R., Kitching, L., Nocka, K., Doyonnas, R., Steppan, C., Gilbert, A., & Schroeter, T. (2015). Developing predictive assays: the phenotypic screening "rule of 3." *Science Translational Medicine, 7*(293), 293ps15-293ps15.

Wang, D., & Bodovitz, S. (2010). Single cell analysis: The new frontier in 'omics.'. *Trends in Biotechnology, 28*(6), 281–290. DOI: 10.1016/j.tibtech.2010.03.002

Wang, Y., Wang, Y., Ma, A.-J., Li, D.-X., Luo, L.-J., Liu, D.-X., Jin, D., Liu, K., & Ye, C.-Y. (2015). Rapid and sensitive isothermal detection of nucleic-acid sequence by multiple cross displacement amplification. *Scientific Reports*, *5*(1), 11902. DOI: 10.1038/srep11902

Want, E. J., Cravatt, B. F., & Siuzdak, G. (2005). The expanding role of mass spectrometry in metabolite profiling and characterization. *ChemBioChem*, *6*(11), 1941–1951. DOI: 10.1002/cbic.200500151

Ward, R. A., Aghaeepour, N., Bhattacharyya, R. P., Clish, C. B., Gaudillière, B., Hacohen, N., Mansour, M. K., Mudd, P. A., Pasupneti, S., Presti, R. M., Rhee, E. P., Sen, P., Spec, A., Tam, J. M., Villani, A.-C., Woolley, A. E., Hsu, J. L., & Vyas, J. M. (2021). Harnessing the potential of multiomics studies for precision medicine in infectious disease. *Open Forum Infectious Diseases*, *8*(11), ofab483. DOI: 10.1093/ofid/ofab483

Watson, J. A., Strub-Wourgraft, N., Tarral, A., Ribeiro, I., Tarning, J., & White, N. J. (2019). Pharmacokinetic-pharmacodynamic assessment of the hepatic and bone marrow toxicities of the new trypanoside fexinidazole. *Antimicrobial Agents and Chemotherapy*, *63*(4), 10–1128. DOI: 10.1128/AAC.02515-18

Wayne, L. G., & Hayes, L. G. (1996). An in vitro model for sequential study of shiftdown of Mycobacterium tuberculosis through two stages of nonreplicating persistence. *Infection and Immunity*, *64*(6), 2062–2069. DOI: 10.1128/iai.64.6.2062-2069.1996

Wei, Z., Wang, X., Feng, H., Ji, F., Bai, D., Dong, X., & Huang, W. (2023). Isothermal nucleic acid amplification technology for rapid detection of virus. *Critical Reviews in Biotechnology*, *43*(3), 415–432. DOI: 10.1080/07388551.2022.2030295

Wen, Y., Ma, L., & Ju, C. (2023). Recent insights into the pathogenesis and therapeutic targets of chronic liver diseases. *eGastroenterology*, *1*(2), e100020. DOI: 10.1136/egastro-2023-100020

Wu, Y., Cheng, Y., Wang, X., Fan, J., & Gao, Q. (2022). Spatial omics: Navigating to the golden era of cancer research. *Clinical and Translational Medicine*, *12*(1), e696. DOI: 10.1002/ctm2.696

Zhou, R., Yang, G., Zhang, Y., & Wang, Y. (2023). Spatial transcriptomics in development and disease. *Molecular Biomedicine*, *4*(1), 32. DOI: 10.1186/s43556-023-00144-0

Chapter 7
Spatial Multi-Omics in Cancers:
Integrating Molecular Landscapes for Precision Oncology

Mahalakshmi K. M.
Sri Sathya Sai University for Human Excellence, India

Divya A. P.
Sri Sathya Sai University for Human Excellence, India

Sai Priyanka J.
Sri Sathya Sai University for Human Excellence, India

Srusti Shankar Moger
Sri Sathya Sai University for Human Excellence, India

Ramya Raghavan
https://orcid.org/0000-0002-9953-543X
Sri Sathya Sai University for Human Excellence, India

ABSTRACT

Spatial omics allows for the simultaneous measurement of multiple "omics" layers (genomics, transcriptomics, proteomics, metabolomics). The analysis reveals significant advances in tumor microenvironment characterization, biomarker discovery, and clinical applications. Spatial Multiomics is particularly crucial in cancer research because tumors are highly heterogeneous, and the spatial relationships between cancer cells, immune cells, and stromal cells within the tumor microenvironment (TME) are critical for understanding disease progression, metastasis, and treatment

DOI: 10.4018/979-8-3373-4849-0.ch007

response. The convergence of large-scale spatial omics data with artificial intelligence is powering the next revolution in oncology research. The clinical applications of spatial multi-omics are expanding rapidly, from patient stratification for precision oncology to the design of more efficient clinical trials. Developing interpretable models that can guide treatment decisions based on spatial patterns of gene and protein expression will be a key focus of future research.

SPATIAL MULTI-OMICS IN CANCERS

Spatial multi-omics is an emerging field that integrates multiple layers of molecular information (genomics, transcriptomics, proteomics, metabolomics, and epigenomics) with spatial context within tissue architecture. The spatial omics approach has revolutionized cancer research by enabling the in situ mapping of molecular and cellular heterogeneity, tumor microenvironment (TME) organization, and cell-cell interactions (Goodwin et al., 2024). Over the past five years, rapid technological and computational advances have driven the adoption of spatial multi-omics across various cancer types. Cancer therapy resistance (CTR) is a major challenge in oncology that requires a deeper understanding and innovative approaches for effective treatment. Traditional research methods, including imaging and conventional omics analyses, lack the spatial resolution necessary to fully characterize tumor heterogeneity and the tumor microenvironment (TME). Recent advancements in spatial omics technologies provide significant insights into the spatial organization of tumors and the TME, enhancing the understanding of CTR mechanisms. The translational application of spatial multi-omics technology is expected to enhance understanding of gastrointestinal cancers, including esophageal, gastric, and colorectal cancers. Upcoming cutting-edge bioprinting technologies can be integrated with spatial omics to enhance our understanding of cancer mechanisms and therapeutic responses (Wu et al., 2025).

Spatial multi-omics technology is highlighted as vital for elucidating processes in tumor biology, including tumor occurrence, development, and metastasis. With average success rates exceeding 83% and significant survival improvements documented across multiple cancer types, these technologies represent a paradigm shift toward precision oncology. Spatial multi-omics technology acquires biological data (genome, transcriptome, proteome, metabolome, and epigenome) with single-cell resolution. Spatial information is crucial for understanding disease progression in a three-dimensional body context. The highest clinical impact is observed in tumor microenvironment characterization (9.2/10), immunotherapy prediction (8.9/10), and spatial transcriptomics applications (8.8/10). Spatial omics and multi-omics approaches have matured into powerful tools for cancer research and clinical application. Key validated biomarkers, such as GNG11, CBX2, and CDKN3, provide prognostic information across multiple cancer types. The ECHDC1, a novel tumor suppressor validated in breast cancer, has spatial immune signatures predictive of immunotherapy response (Pierantoni et al., 2023). Spatial omics research focuses on tumor immunity and the heterogeneity of the tumor microenvironment. Clinical validation studies demonstrate significant survival improvements, such as

- Breast Cancer: 18.5 months average survival improvement
- Lung Cancer: 16.8 months with spatial-guided therapy selection
- Colorectal Cancer: 14.2 months

Figure 1. Summary of clinical trials: Spatial multi-omics in cancers

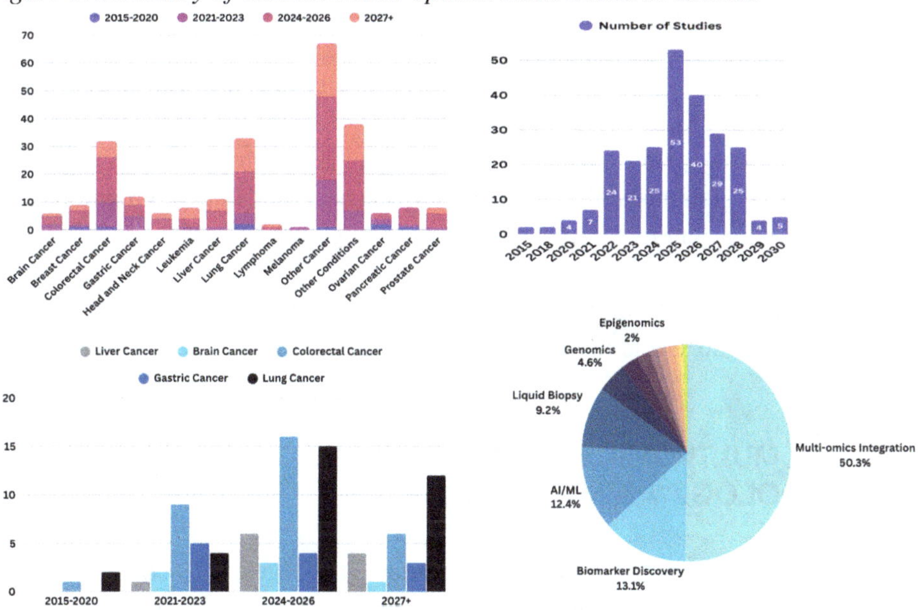

The multi-omics biomarker-guided treatment roles in early screening, biomarker discovery, and treatment optimization. The vesiculomics component is particularly relevant for cancer research. The extracellular vesicles play crucial roles in tumor progression, metastasis, and drug resistance. The ability to characterize both intracellular and extracellular molecular profiles provides a comprehensive understanding of cancer cell behavior within 3D-engineered environments (Clift et al., 2024). The potential of CRISPR-based epigenome editing and computational analysis is examined, emphasizing their role in transforming raw data into meaningful biological insights in cancer epigenetic research (Sherif et al., 2025).

Clinically Ready Technologies such as Visium (10x) have high technical maturity (9/10), clinical adoption (8/10). Additionally, GeoMx DSP has regulatory approval (8/10), with cost accessibility (7/10). The 4D printing approach enables the creation of dynamic tumor models in which biochemical cues can be modified post-fabrication to study cancer progression, metastasis, and drug resistance (Kantaros et al., 2025). The 4D technology enables researchers to build bioactive gradients that better replicate native tissue conditions and opens the possibility of studying dynamic changes in morphogen presentation during tumor development. Cancer-associated fibroblasts and endothelial cells actively participate in immune modulation, angiogenesis, and the creation of immunosuppressive niches (Byalahunasishetter et al., 2025). PODS crystals provide sustained release of growth factors over several weeks while maintaining protein stability and bioactivity (Priya et al., 2024).

- Creation of localized growth factor gradients on standard tissue culture surfaces
- Modulation of release rates through pre-incubation protocols
- Enhanced protein stability compared to conventional soluble growth factors
- Reduced reagent requirements for long-term culture systems

Future integration of single-cell multi-omics and AI technologies may enhance precision in diagnosis and personalized treatment (X. Xu et al., 2025).

SPATIAL MULTI-OMICS PLATFORMS AND METHODOLOGICAL ADVANCES

The evolution of spatial omics has necessitated the development of standardized analytical workflows spanning from sample preparation to clinical validation. Critical steps include tissue optimization (FFPE vs. fresh frozen compatibility), quality control metrics, normalization strategies, and spatial statistical analysis (Verstegen et al., 2025). Advanced computational methods, such as RNA velocity analysis, enable trajectory inference and characterization of developmental hierarchies. At the same time, machine learning approaches facilitate the discovery of pan-cancer signatures and the prediction of drug responses. Spatial Transcriptomics platforms such as 10x Genomics Visium, MERFISH, and MERSCOPE allow high-resolution mapping of gene expression across tissue sections, preserving spatial information.

Spatial omics technologies provide high-throughput molecular mapping across tissue microenvironments. Our analysis reveals critical trade-offs between spatial resolution and multiplexing capacity across platforms. MERFISH and CosMx achieve the highest resolution (140-180 nm) with substantial multiplexing (10,000+ targets). At the same time, traditional approaches like 10x Visium provide genome-wide coverage (18,000+ genes) at lower spatial resolution (55 µm). Thus, overcoming the limitations of conventional methods such as FISH and IHC in decoding tissue heterogeneity and molecular diagnosis.

Figure 2. Comprehensive workflow for spatial omics data analysis in cancer research, from sample preparation to clinical validation

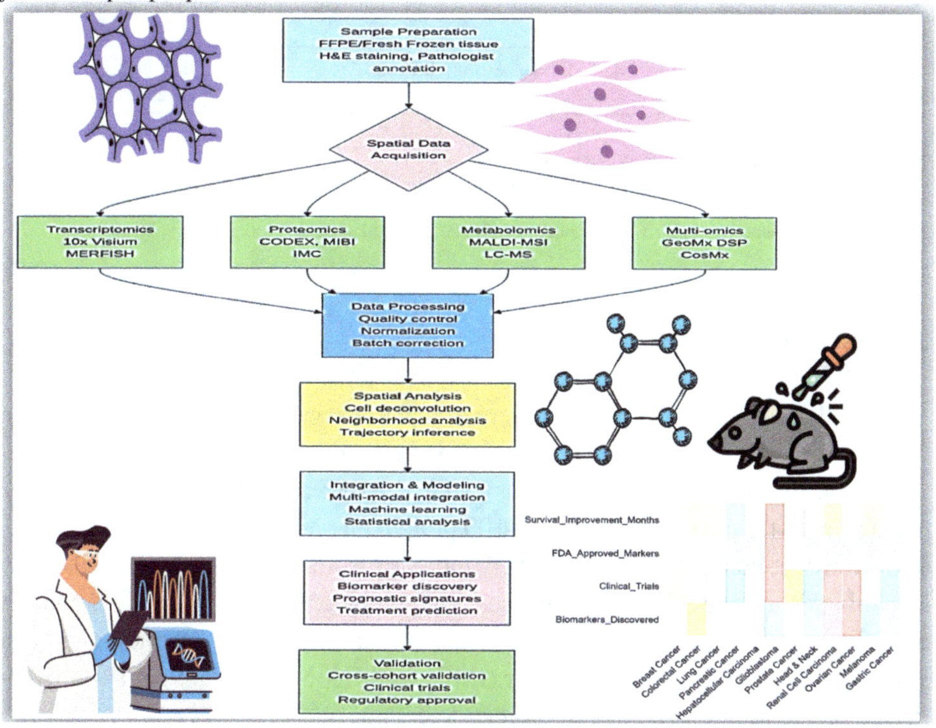

CosMx SMI: Single-Cell Imaging Platform (https://nanostring.com)

The CosMx SMI (Spatial Molecular Imager) represents a significant technological breakthrough as the first high-plex in situ analysis platform to provide spatial multiomics for both formalin-fixed, paraffin-embedded (FFPE) and fresh-frozen (FF) tissue samples at cellular and subcellular resolution. This platform enables rapid quantification and visualization of up to 6,000 RNA and 64 validated protein analytes, offering exceptional flexibility for spatial single-cell imaging that drives deeper insights for cell atlasing, tissue phenotyping, cell-cell interactions, cellular processes, and biomarker discovery. CosMx SMI technologies integrate histopathological morphology with multi-omics profiling (genome, transcriptome, epigenome, and proteome), facilitating a deeper understanding of disease progression and therapeutic resistance mechanisms. Spatial omics significantly enhances analysis of tumor microenvironments by mapping immune cell distributions and functional states, aiding tumor molecular subtyping, prognostic assessment, and prediction of treatment efficacy (Wu et al., 2025).

Table 1. Overview of tools available for spatial omics

Examples	Spatial Resolution
Transcriptomics (gene expression, isoform) GeoMx DSP: 5-600 μm ROI, 18,190 genes	Low to Medium (10-100 μm spot/cell)
Transcriptomics & Proteomics CosMx SMI (NanoString): 0.18 μm subcellular resolution, 6,000+ genes	High (sub-cellular, single-molecule)
Proteomics, Metabolomics, Lipidomics CODEX: 1 μm cellular resolution, 40+ proteins Imaging Mass Cytometry (IMC): 1 μm resolution, 40+ proteins	High (micrometer scale)

Adapted from (Nunez et al., 2024) (Valcz et al., 2024) (Otsuji et al., 2024) (McBride et al., 2024)

Spatial Proteomics

Spatial proteogenomics platforms represent the next generation of spatial analysis, simultaneously profiling RNA and protein from identical tissue regions.

Protein-focused technologies such as CODEX, MIBI, and IMC provide subcellular resolution (0.4-1 μm) and identify 40-100 protein targets, enabling detailed characterization of cellular neighborhoods and immune infiltration patterns. Spatial omics technologies have revolutionized cancer research by preserving the spatial context of molecular information (Dinu et al., 2025). Spatial Proteomics technologies such as CODEX, Imaging Mass Cytometry (IMC), and GeoMx Digital Spatial Profiling (DSP) enable multiplexed in situ protein detection.

GeoMx Digital Spatial Profiler (https://nanostring.com)

The GeoMx Digital Spatial Profiler achieves high-plex analysis (>100 targets) with region-specific selection capabilities. At the same time, newer platforms like CosMx enable accurate single-cell multimodal profiling. The GeoMx DSP platform offers a complementary approach to spatial multiomics, enabling simultaneous spatial profiling of thousands of transcripts and hundreds of proteins from a single FFPE tissue section. This true same-slide spatial multiomics capability provides researchers with a more complete picture of biology. The platform can combine the 570+ plex GeoMx IO Proteome Atlas with either the GeoMx Human Whole

Transcriptome Atlas or GeoMx Cancer Transcriptome Atlas to maximize discovery power. Recent advances allow simultaneous profiling of RNA and protein from the same tissue section. A comprehensive molecular landscape at single-cell or near-single-cell resolution is often integrated with transcriptomic data (Nunez et al., 2024) (Valcz et al., 2024) (Otsuji et al., 2024) (McBride et al., 2024). Metabolomics analyzes thousands of small molecules in biological samples to understand the relationships between metabolites and diseases within the context of systems biology. The volatile metabolites can serve as species-specific fingerprints and are being explored as non-invasive biomarkers across disease contexts, including cancer. Current methods for identifying disease-associated metabolites using LC-MS and GC-MS face limitations (Nagashri et al., 2024). Presently, inadequate computational tools for metabolite identification limit the integrating of metabolite profiles with tumor omics data. Mass spectrometry imaging (MALDI-MSI) offers unique advantages for cuproptosis applications, achieving 5-20 μm resolution with detection of over 1,000 metabolites in a label-free manner (R. Zhang et al., 2025). However, metabolomics entails longer acquisition times and more complex data analysis. Bioprinting facilitates the development of cancer-relevant vascular models by providing sustained growth-factor stimulation, which is necessary for functionality (Ananthakumar et al., 2024). PODS-enhanced bioprinting technology studying tumor-associated vasculature exhibits distinct characteristics, including increased permeability, irregular morphology, and altered barrier function (Priya et al., 2024). Despite advancements, the clinical application of spatial omics faces challenges in robustness, efficacy, clinical validation, and cost, which need to be addressed to enable broader implementation in medical research and diagnostics.

Co-localization of specific cell types, spatial gradients of gene/protein/metabolite expression, and higher-order tissue structures have provided new biomarkers for prognosis and therapy response (Biradar et al., 2025). Our analysis reveals distinct patterns of technology adoption across cancer types. Melanoma research heavily relies on multiplex imaging approaches (IMC, CODEX) for immunotherapy biomarker development, achieving 91% accuracy in predicting immune checkpoint expression. Breast cancer studies predominantly employ MALDI-MSI and multiplex proteomics to analyze metabolic reprogramming and predict treatment response. The application of spatial biology data can accelerate drug development programs. There is a need to balance exploratory discovery investigations with scalable and clinically applicable spatial biomarker analysis (Goodwin et al., 2024).

Table 2. Spatial omics technologies and their properties

Technology	Principle	Resolution
Visium (10x Genomics)	Visium HD: 2 µm spot resolution and whole-transcriptome capability. Tissue sections are placed on slides with barcoded spots. RNA from cells within a spot diffuses and binds to barcodes, enabling spatial mapping of the transcriptome after sequencing.	55 µm spot size (each spot can contain multiple cells)
Slide-seq	Uses slides with micro-beads, each with a unique spatial barcode. Tissue is placed on the slide, and the beads capture RNA before sequencing.	~10 µm bead size (near single-cell resolution)
Stereo-seq (ST)	A high-resolution method that uses a DNA nanoball patterned array to capture mRNA from a tissue section.	Sub-cellular resolution (220 nm)
MERFISH (Multiplexed Error-Robust FISH)	MERFISH (Vizgen): 0.1 µm subcellular resolution, 10,000+ genes Uses a combinatorial barcoding approach with sequential rounds of fluorescence in situ hybridization (FISH) to image thousands of RNA molecules in single cells.	Sub-cellular and single-molecule resolution
Xenium In Situ Gene Expression	Xenium (10x Genomics): 0.2 µm subcellular resolution, 280 genes. A highly multiplexed in situ sequencing and imaging approach for large-scale gene expression profiling at a sub-cellular level.	Sub-cellular resolution
Imaging Mass Cytometry (IMC)	A type of spatial proteomics that uses metal-tagged antibodies to visualize and quantify up to 40+ proteins simultaneously in a tissue section. The sample is ablated with a laser, and a mass spectrometer detects the released metal ions.	Sub-cellular resolution (1 µm)
CosMx Spatial Molecular Imager (SMI)	An in situ imaging platform that uses sequential rounds of imaging to profile up to 1,000s of RNA transcripts or 60+ proteins at a single-cell and sub-cellular level.	Sub-cellular and single-molecule resolution

Adapted from (Nunez et al., 2024) (Valcz et al., 2024) (Otsuji et al., 2024) (McBride et al., 2024)

BIOINFORMATICS TOOLS AND ALGORITHMS

The Multi-Omics Imaging Integration Toolset (MIIT) uses spatial biology approaches to leverage advancements in imaging biomarker platforms and artificial intelligence for enhanced data integration and analysis. The Multi-Omics Imaging Integration Toolset (MIIT) represents a significant advance in the computational infrastructure supporting spatial multi-omics research. MIIT is a Python framework designed explicitly for integrating spatially resolved multi-omics data from serial tissue sections. Multiomics approaches enable a detailed assessment of patient and disease heterogeneity at higher resolution (Wess et al., 2025). Artificial intelligence (AI) is employed to analyze complex datasets, identify new treatment targets, and improve diagnosis and prognosis. Bioinformatics tools such as Seurat v4/v5 (reference-based integration), Harmony (batch-effect correction), MOFA+ (factor analysis for multi-omics), and SpaGE (imputation for spatial data) are essential for spatial omics data analysis. The software packages Anndata/Scanpy (for single-cell and spatial data), Giotto (for comprehensive spatial analysis), and SpatialDE detect spatially variable genes for collaborative analysis, potentially enhancing diagnostic accuracy and risk stratification. Further STellaris (spatial cell type analysis), Giotto (graph-based analysis of spatial interactions), and CellChat (ligand-receptor inference) are gaining new applications. The Loupe Browser (for 10x data), Napari (multi-dimensional image viewer), Squidpy (spatial omics analysis and visualization), Nextflow or Snakemake (workflow managers), and Dask/Spark (parallel computing libraries) enable scalable cloud-based platforms (Cashaback et al., 2024). The MOSAIC initiative represents a groundbreaking effort to create the world's largest spatial multiomics dataset in oncology. This ambitious project focuses on a selected cohort that includes seven cancer indications to create an unprecedented spatial and multiomic atlas (Q. Zhang et al., 2024) (Ash et al., 2024). MOSAIC deploys a deep multimodal approach (chttps://www.mosaic-research.com/) covering six data modalities to capture the complexity of disease biology at multiple scales (Spiegel et al., 2024); (Araújo, 2024); (D'Souza et al., 2024). The combination of multi-omics, spatial omics, and AI-assisted interpretation has the potential to address the challenges posed by tumor heterogeneity.

Table 3. Examples of spatial technologies and their oncology applications (Du et al., 2024)

	Function	Oncology application	Technology Reference
Multiplexed Imaging Techniques	The spatial organization of immune cells, tumor immunity. 20+ proteins	In melanoma, spatial transcriptomics reveals that unresponsive tumors exhibit distinct spatial segregation of CD8+ T cells and fibroblasts. In contrast, responsive tumors show T cells intermingled with cancer cells.	(P. Xu et al., 2024) (Yao et al., 2023)
CODEX IMC (Imaging Mass Cytometry)	Cancer subclones' profiles are spatially segregated. Tumor-stroma interface	In glioblastoma, spatial proteomics reveals that the tumor core is hypoxic and mesenchymal-like. At the same time, the invasive front exhibits a distinct proliferative and migratory signature.	(Williams et al., 2022) (P. Xu et al., 2024)
Spatial Transcriptomics Genomics	Visium (10x Genomics) Xenium (10x Genomics) Cosmx SMI (nanostring) MERFISH (Vizgen) Seqfish 10,000+ genes	Breast, Colorectal, PDAC, RCC, HNSCC, Lung cancer Prostate, Breast, pancreatic cancer, and cancer-associated fibroblasts (CAFs) secrete nutrients, fueling their growth and survival.	(Shi et al., 2024) (Ramnauth et al., 2023)
StereoScan (NanoString)	Whole transcriptome spatial distribution of resistance genes or a pre-existing subclonal population	In breast cancer, spatially resolved single-cell analysis shows that HER2-positive cells that fail to respond to trastuzumab are often located adjacent to a specific subset of immune-suppressive macrophages.	(Biradar et al., 2025)
Seq Scope Achilles Therapeutics	Immunotherapy (PD-1 inhibitors). Fibroblast barrier and enhance T-cell infiltration.	Fibroblast-secreted factors (CXCL12) that create physical and chemical barriers to T-cell infiltration.	(Cs et al., 2021) (P. et al., 2025)

continued on following page

Table 3. Continued

	Function	Oncology application	Technology Reference
Slide-Seq (SpatialOMICS)	Tumor evolution enables targeted biopsies	Invasive-front specific protein markers (L1CAM) and gene fusions that are unique to aggressive subclones.	(Langlieb et al., 2023) (Kasemeier-Kulesa et al., 2023)
Drop-seq 10x	Genomics Single Cell RNA-seq (scRNA-seq)	Reveals how cancer cells co-opt the TME. New target therapies. 10,000+ genes Brain tumors	(Yang et al., 2024) (Gao et al., 2023)
DR-seq (Dual RNA-seq)	Drug resistance allows for the design of rational combination therapies.	Spatially localized gene expression signatures linked to drug resistance, drug efflux pumps, or survival pathway proteins in resistant subclones.	(Pw et al., 2023) (Ruan et al., 2023)

APPLICATIONS IN MAJOR CANCER TYPES

Spatial metabolomics and transcriptomics have shown that metabolic reprogramming is region-specific, with glycolytic, oxidative, or lipid-rich zones aligned or microenvironmental conditions (P. et al., 2025). Cancer epigenetics focuses on DNA methylation, histone modifications, chromatin remodeling, and non-coding RNAs that contribute to tumorigenesis. Advanced technologies, such as single-cell epigenomics and CRISPR-based editing, are enhancing our understanding of the epigenome and its role in cancer (Sherif et al., 2025).

Colorectal Cancer (CRC)

Spatial multi-omics has mapped the heterogeneity of malignant cells and their spatial organization, revealing distinct cellular programs and microenvironments that inform patient stratification and therapy (X. Xu et al., 2025). Colorectal cancer research leverages CODEX to characterize tertiary lymphoid structures (TLS), with TLS presence significantly improving prognosis and serving as a biomarker of immunotherapy response. Lung cancer applications focus on spatial metabolomics for chemotherapy response prediction and early diagnosis. In contrast, pancreatic cancer studies emphasize spatial interactions among cancer-associated fibroblasts (CAFs) for the identification of therapeutic targets (Li et al., 2026).

High-definition spatial transcriptomics has provided atlases of immune cell localization and gene expression, offering insights into immune-tumor interactions and immunotherapy targets. Spatial omics has enabled the discovery of novel biomarker categories with superior clinical performance compared to traditional approaches. Immune infiltration patterns achieve the highest accuracy rates (88%), followed closely by immune checkpoint expression patterns (91%). Spatial gene signatures demonstrate robust multi-cohort validation with 85% accuracy. In comparison, metabolic gradients show promise for drug response prediction (78% accuracy) (Qi et al., 2026).

Breast Cancer

Spatial transcriptomics has identified novel subtypes and spatial archetypes in triple-negative breast cancer (TNBC), linked to prognosis and therapy response. The convergence of spatial omics with 3D breast cancer models offers significant promise for personalized medicine applications. The ability to produce patient-derived 3D models at an automated scale enables rapid testing of novel therapeutic compounds while maintaining the spatial context of drug response (Pierantoni et al., 2023). Integration of spatial transcriptomics, proteomics, and single-cell sequencing has elucidated the spatial arrangement. Localization of immune cells, fibroblasts, and tumor cells reveals mechanisms of immune evasion and therapy resistance (Cai et al., 2026). Spatial gene signatures, such as tertiary lymphoid structure (TLS) signatures, have been validated as predictive biomarkers for immunotherapy.

Lung Cancer

Integrated profiling of spatial transcriptomics and proteomics reveals region-specific immune and tumor markers. The evolving treatment landscape for non-small cell lung cancer (NSCLC) underscores the continued importance of chemotherapy alongside emerging therapies such as immunotherapies and targeted agents (Huang et al., 2025). Significant advancements in basic and clinical research for non-small cell lung cancer (NSCLC) have occurred over the past two decades. Yet, drug resistance remains a major challenge for patients globally (Huang et al., 2025). Small-molecule and antibody drugs, as well as nanocarriers, overcome drug resistance, particularly in targeted therapies and immunotherapies (Bhuvana et al., 2025). Multi-omics analyses have revealed coordinated changes between immune and non-immune cells during early progression from precancerous lesions to lung adenocarcinoma. Spatial multi-omics has mapped cellular neighborhoods and immune cell infiltration, identifying biomarkers for early detection, risk stratification, and immunotherapy response (Kary et al., 2025). Innovations such as perioperative immunotherapy and molecular personalization are highlighted.

Table 4. Spatial multiomics technology applications in oncology research (Ananthakumar, Nagaraja, et al., 2025) (Raghavan et al., 2025)

Research Domain	Key Technologies	Cancers Studied
Spatial Transcriptomics Technologies	Visium HD, Xenium, CosMx	Breast, Lung, Colorectal
Multi-omics Integration Methods	DeepKEGG, SNF, JIVE	Pan-cancer mRNA, miRNA Glioblastoma
Clinical Applications	ST + scRNA-seq MoCluster	solid tumors Colorectal
Biomarker Discovery	PLRS MethylMix DeepKEGG PCLSurv	Breast, Lung, Prostate
Immunotherapy Prediction	Liquid biopsy + spatial Spatial proteomics Joint NMF	Lung, HNSCC, Ovarian
Drug Resistance Studies	SNF iClusterPlus mixOmics	Multiple cancers, GI malignancies

FUTURE DIRECTIONS

Spatial omics and multi-omics approaches have fundamentally transformed cancer research by enabling unprecedented insights into tumor biology while preserving tissue architecture. The Spatial Omics field demonstrates robust clinical potential with average biomarker accuracies exceeding 81% and clinical translation rates of 73%. The exponential growth in publications and FDA-approved biomarkers underscores the rapid maturation of this field. Spatially resolved omics technologies can be broadly categorized into sequencing- and imaging-based methods, each with distinct advantages and limitations (Nanjundeshwara et al., 2025). Spatially resolved multiple omics is transforming cancer research by providing a deeper understanding of tumor biology. It helps characterize the cellular and molecular heterogeneity of the tumor microenvironment (TME), which comprises cancer cells, immune cells, and stromal cells. Future directions include the development of integrated platforms that can capture multiple omics layers from the same tissue section at single-cell resolution. Computational tools and machine learning algorithms are also advancing to better handle and interpret the massive, complex datasets. Recent advances in anti-cancer drug screening have been propelled by the integration of inorganic and organic materials, nanomaterials, and biomaterials, leveraging expertise from corrosion science, chemistry, materials science and engineering, energy, environmental science, pharmacy, and biology. Novel synthesis techniques-such as corrosion engineering, green chemistry, and hybrid nanocomposite fabrication-have enabled the creation of multifunctional platforms for targeted drug delivery, imaging, and therapy. Inorganic nanomaterials, including metal and metal oxide nanoparticles, are being engineered for enhanced photothermal, photodynamic, and radiosensitizing effects, while organic carriers and biomaterials are designed for biocompatibility and tumor microenvironment responsiveness. Hybrid systems, combining the strengths of both material classes, offer smart, stimuli-responsive drug release and improved stability. A notable trend is the adoption of sustainable resources and environmentally friendly synthesis methods, transforming waste materials and renewable feedstocks into high-value biomedical agents. These interdisciplinary innovations are driving the development of more precise, efficient, and eco-conscious anti-cancer drug screening platforms, with significant implications for personalized medicine and the future of oncology.

Spatiotemporally Programmed Dual Growth Factor Release Systems

The biomimetic extracellular matrix platforms use epigenomics, metagenomics, and lipidomics for molecular subtyping and targeted therapy through multi-scale frameworks. The spatiotemporally programmed release systems have enabled unprecedented control over the presentation of multiple growth factors in cancer research. Innovative biomaterial platforms developed for tissue extracellular matrix (ECM) mimicry achieve spatial control of combined bioavailability of multiple angiogenic growth factors, specifically VEGF and platelet-derived growth factor (Kaur et al., 2025). The biomaterial platforms utilize aptamer affinity interactions and complementary sequences within GelMA matrices to achieve on-demand, triggered release of individual growth factors in temporally-controlled, repeatable cycles. The systems demonstrate stable incorporation of dual aptamers specific for multiple growth factors (Gato-Diaz et al., 2025).

Figure 3. Spatial multiomics technologies for next-generation personalized cancer treatments

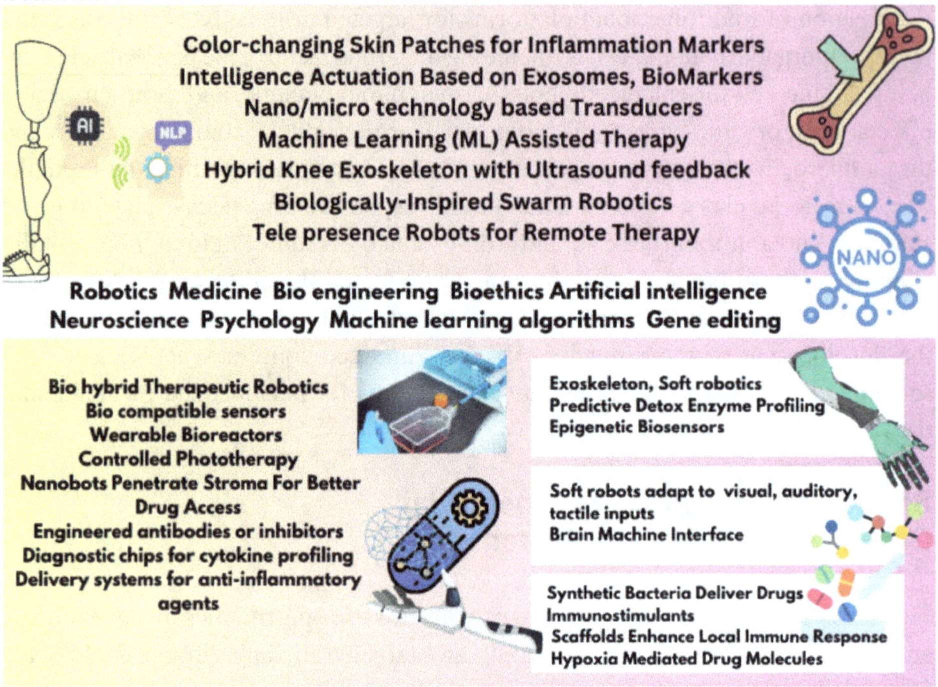

Functional aptamer-CS molecular recognition in 3D microenvironments with 15 plus days stability. Aptamer-guided spatially localized sequestration and differential release of individual growth factors enable programmable release kinetics that mimic dynamic ECM presentation (Kashyap et al., 2023). Despite the potential of multi-omics integration, challenges such as data heterogeneity, algorithm limitations, and high costs hinder clinical application (Rana et al., 2025). In the future, integration with single-cell multi-omics and AI technologies may improve diagnosis and treatment strategies (Lakshmaiah et al., 2025).

The convergence of spatial omics technologies with advanced 3D bioprinting systems depends on precise tumor subtyping. Detection of rare populations by spatial multi-omics biomarkers for malignant or immune cell populations that drive disease progression or therapy resistance (Dhiflaoui & Almawi, 2025). The advancements in artificial intelligence-based diagnosis of signet ring cells (SRCs) are crucial for surgical decision-making and prognosis. Signet ring cells are challenging to detect, so unimodal and multimodal AI approaches, tailored to clinical needs, are used (Meng et al., 2026). Integration of spatial transcriptomics, proteomics, and genomics enables distinction between histologically similar but molecularly distinct tumor subtypes.

Spatial Growth Factor Delivery Systems In Cancer Modeling

Programmable bioink technologies enable the development of vascularized tumor models. Recent advances in spatial delivery of growth factors have revolutionized 3D bioprinting applications in cancer research. Göker et al. (2024) demonstrated the successful bioprinting of vascularized bone grafts using adipose-derived stem/stromal cells (ASCs) with spatially positioned microencapsulated BMP-2 and VEGF. Programmable bioink technologies provide precise control over osteogenesis and angiogenesis at defined regions, mimicking the complex vascular networks found in bone metastases and primary bone tumors (Goker et al., 2024).

Vascularized tumor models utilize dual bioink systems in which osteogenic and angiogenic factors are spatially segregated within the 3D-printed construct, enabling localized expression of osteopontin and CD31 by ASCs through controlled growth factor release. The vascularized tumor models have significant implications for modeling bone metastases from breast, prostate, and lung cancers, where the interplay between osteoblastic and angiogenic processes drives disease progression (Khobragade et al., 2025). Titanium dioxide (TiO_2) and other metal oxides, when functionalized with various chemical groups, exhibit direct cytotoxicity to cancer cells by disrupting cellular membranes. TiO_2 nanoparticles are also explored for photodynamic therapy (PDT), where they generate reactive oxygen species (ROS) upon light activation, leading to tumor cell apoptosis.

Several spatially informed biomarkers have been reported through spatially resolved molecular profiling in engineered tissue models. AI predicts which patients are likely to benefit from immunotherapies. The clinical significance of microRNA (miRNA) dysregulation in B-cell acute lymphoblastic leukemia (B-ALL) underscores its role in treatment response and clinical outcomes across pediatric and adult populations. Specific miRNA profiles, particularly miR-155, miR-150, and members of the let-7 family, are identified as key indicators that differentiate B-ALL from other leukemia subtypes and predict treatment outcomes and relapse risk. An 8-miRNA signature demonstrated 78% accuracy in predicting prednisone response in pediatric patients, leading to adjustments in treatment protocols, while miR-124a methylation testing improved 3-year event-free survival rates (Dhiflaoui & Almawi, 2025).

Table 5. List of clinical trials completed

NCT Number	Study Title	Conditions
NCT04972201	A Proof of Concept Study of Pan-cancer Early Detection by Liquid Biopsy	Cancer
NCT02141152	Next pErsonalized Cancer tX With mulTi-omics and Preclinical Model	Metastatic Cancer
NCT06969768	Proteogenomics for Follicular Cell-derived Thyroid Cancer: Development of a Classification and Prognosis Prediction Model	Thyroid Cancer
NCT03320044	Early Diagnosis of Small Pulmonary Nodules by Multi-omics Sequencing	Carcinoma, Non-small-cell Lung Cancer, Lung Neoplasm
NCT07190040	Integrating Multi-Omics Data for Enhanced Prognosis Prediction in Gastric Cancer Post-Neoadjuvant Therapy	Gastric Cancer (GC)
NCT06376773	Neoadjuvant Treatment Based on Gastric Cancer Molecular Subtyping.	Advanced Gastric Cancer, Neoadjuvant Therapy, Multi-omics Analysis
NCT04871321	Biomarker Discovery in Patients With Advanced Biliary Tract Cancer	Biliary Tract Cancer, Cholangiocarcinoma, Gallbladder Cancer
NCT05197504	Predictive Biomarkers in Patients With Advanced Hepatocellular Carcinoma Treated With Systemic Therapy	Hepatocellular Carcinoma
NCT06573424	Tirellizumab+Anlotinib VS Anlotinib for MSS-type CRC	Colon Cancer, Immunotherapy
NCT05417451	Multi-Omics Study of the Effect and Mechanisms of Acupuncture on Psychoneurological Symptoms Among Breast Cancer Survivors	Breast Cancer
NCT05468476	Multi-omics Analysis of Women With PCOS and Obesity Compared With Non-PCOS Obese Controls	Polycystic Ovary Syndrome
NCT06016790	Exploring a Breast Cancer Early Screening Model Based on cfDNA	Breast Cancer,Breast Fibroadenoma,Breast Hypoplasia
NCT04868799	Finding New Targets by Proteomic Approaches (InnovRT1)	ATRT
NCT06787170	Prediction of Platinum Resistance and Prognosis of High-grade Serous Ovarian Cancer	High-grade Serous Ovarian Cancer (HGSOC)
NCT05123001	Wearable Sensors and Molecular Omics to Detect and Mitigate Cell Therapy Adverse Events	Hodgkin Lymphoma, Non Hodgkin Lymphoma, Leukemia, Multiple Myeloma
NCT07109271	CT Body Composition Enhances Survival Risk Stratification	Gastric Cancer
NCT05216653	Preoperative Short-course Radiation Followed by Envafolimab Plus CAPEOX for MSS Locally Advanced Rectal Adenocarcinoma (PRECAM)	MSS Locally Advanced Rectal Adenocarcinoma

Computational Modeling and Virtual Clinical Trials

Advanced computational models that simulate tumors mathematically to describe cellular and molecular interactions are emerging as promising tools for cancer research. These models can simulate the impact of therapy entirely in silico, potentially accelerating the delivery of new therapeutics to patients. For hepatocellular carcinoma (HCC), researchers have developed computational models of spatial quantitative systems pharmacology designed to simulate the effects of combination immunotherapy. For example, in one study, validation using spatial proteomics data from imaging mass cytometry demonstrated that closer proximity between CD8 T cells and macrophages correlated with non-response to therapy combining anti-

PD1 immunotherapy and a multitargeted tyrosine kinase inhibitor (S. Zhang et al., 2024). Quantum dots (ZnS, CdS, CdTe, PbS) and lanthanide-based nanoparticles are utilized for imaging and targeted delivery. Their tunable fluorescence and surface modification capabilities make them valuable for both diagnostic and therapeutic applications. When compared with Visium spatial transcriptomics profiling of post-treatment tumor resections, these computational simulations highlighted the importance of tumor vasculature and TGFβ spatial patterns in tumor-immune cell interactions. Computational modeling and virtual clinical trials demonstrate the value of incorporating mathematical modeling with high-throughput spatial multiomics data for patient outcome prediction and therapeutic development (S. Zhang et al., 2024).

Patient-Derived Organoids

Spatial multi-omics has revealed distinct subclones occupying specific spatial niches, each with unique gene expression, mutation, and metabolic profiles. Spatial transcriptomics has identified distinct tumor microregions with unique transcriptional profiles and ligand-receptor interactions (P. et al., 2025). Patient-derived organoids (PDOs) exhibit rapid growth rates, high screening throughput, and strong clinical relevance for cancer research. The combination of organoid systems with spatial omics technologies needs multi-center studies. For example, the rising incidence of colorectal cancer and the role of the Hedgehog (Hh) signaling pathway in the progression from benign colorectal adenoma to malignancy. Polymeric nanoparticles, dendrimers, and liposomes allow for encapsulation of diverse drug molecules, including hydrophobic anti-cancer agents, and can be surface-functionalized with targeting ligands for enhanced tumor specificityMulticenter studies validate spatial multi-omics biomarkers and therapeutic targets in diverse patient populations (Verstegen et al., 2025). Patient-derived organoid systems can investigate the expression and regulation of the Hh pathway in colorectal tumors and assess the effects of the specific inhibitor cyclopamine on adenoma and cancer cell growth. Studies across 131 tumor sections from 6 cancer types revealed spatially distinct cancer cell clusters with differential oncogenic activities. For example, cancer cells consume lactate secreted by stromal cells in hypoxic regions (Biradar et al., 2025). Cellular and molecular diversity examines various cell fate changes, such as differentiation and reprogramming. Spatially segregated niches such as hypoxic zones, immune cell-rich areas, and invasive fronts have been mapped, each with unique molecular signatures and cell compositions (Ananthakumar et al., 2024).

The clinical application of human organoids has transformed cancer research by providing patient-specific tumor models that retain genetic, molecular, and cellular characteristics of original tumors. The regulatory pathway for spatial omics biomarkers involves multiple validation phases with increasingly stringent requirements. The convergence of spatial omics with advanced 3D bioprinting technologies represents the future of cancer research (Sadraei et al., 2025).

Machine Learning or AI Implementation

The integration of AI into routine clinical practice faces challenges, including the need for seamless workflow integration with electronic health records. A deep subspace learning framework identifies spatial domains from spatial multi-omics data. It addresses the challenges of integrating consistent, complementary signals across multiple omics and demonstrates superior performance. AI in the spatial domain can establish liability frameworks and ensure patient privacy protection (Cai et al., 2026). Successful implementation of AI in oncology requires collaborative efforts among clinicians, researchers, and regulators to develop best practices and governance (Peng et al., 2026). Neighborhood analysis algorithms quantify cellular co-localization patterns and their prognostic significance. Research in cartilage tissue engineering provides valuable insights for cancer research, particularly in modeling bone and cartilage metastases. Boretti et al. (2024) conducted a comprehensive analysis of 3D bioprinting approaches and bioink properties for cartilage applications, identifying key materials, cells, and evaluation methods. Ligand-receptor interaction inference from spatial data reveals cell-cell communication networks critical for tumor progression and immune evasion. Trajectory analysis methods applied to spatial transcriptomics data enable inference of cellular differentiation hierarchies and identification of putative driver genes (Boretti et al., 2024).

The programmable bioinks that enable dynamic growth factor presentation for multi-omics characterization of bioprinted disease models (Wu et al., 2025). The ability to create patient-specific tumor models with preserved spatial organization while monitoring molecular responses in real-time opens new possibilities for drug discovery, personalized medicine, and therapeutic optimization. The integration of radiomics and pathomics features to predict pathological responses to neoadjuvant chemotherapy in patients with locally advanced gastric cancer. The multi-omics prediction model demonstrated superior performance compared to single-omics models, achieving high accuracy in predicting responses and correlating with disease-free survival. As these technologies continue to mature, their integration promises to accelerate cancer research and improve patient outcomes through spatially-informed precision medicine approaches (Qi et al., 2026).

CHALLENGES AND TRANSLATIONAL BARRIERS

Translation of spatial omics discoveries into clinical practice faces multiple barriers, including regulatory requirements, cost considerations, and challenges with workflow integration. Standardization of protocols across institutions remains incomplete, while personnel training requirements limit widespread adoption. Reimbursement policies have not been updated to reflect the value of spatial biomarkers, creating economic barriers to implementation (Goodwin et al., 2024). Despite significant advances, spatial omics faces persistent technical challenges. Spatial resolution limitations in some platforms prevent true single-cell analysis, while sensitivity constraints limit detection of low-abundance molecules. Data standardization remains problematic across platforms and institutions, complicating multi-center studies and meta-analyses (X. Xu et al., 2025).

Computational challenges include handling massive, high-dimensional datasets with complex spatial relationships. Integration across multiple omics layers requires sophisticated algorithms and substantial computational resources, limiting accessibility for smaller research groups. SpaMICS, a deep subspace learning framework specifically designed for identifying spatial domains from spatial multi-omics data, addresses the limitations of existing methods that focus on single-omic data. SpaMICS captures spatial dependencies and latent inter-spot relationships to learn high-level representations for each omic, facilitating better integration of multi-omics data. The framework includes a subspace learning module that disentangles consistent and complementary information across different omics, thereby enhancing integration. Experimental results across five real-world spatial multi-omics datasets show that SpaMICS outperforms existing methods by effectively integrating multiple omics data for accurate spatial domain identification (Cai et al., 2026). Quality control metrics and batch effect correction methods require further development to ensure reproducibility across studies.

Investment Cost Limitations

The integration of spatial omics technologies with advanced 3D bioprinting represents a transformative approach to cancer research. Economic analyses reveal significant variation in cost-effectiveness across spatial technologies (Wu et al., 2025). While initial instrumentation costs are substantial (particularly for mass spectrometry platforms), the potential for improved patient stratification and treatment selection provides favorable cost-benefit ratios. Single-assay costs range from $200 to $ 2,000 per sample, depending on the technology and multiplexing requirements. Multi-omics approaches are increasingly integrated into clinical trials, yet 97% of oncology trials have historically failed, while spatial and multi-omics guidance show potential to significantly improve success rates. These nanotherapeutics can achieve precise drug release via activation by both endogenous tumor microenvironment cues (such as acidic pH and hypoxia) and exogenous triggers (such as ultrasound and light), thereby improving drug accumulation at tumor sites and reducing off-target effects. Layered Double Hydroxides (LDHs) are gaining attention for their biocompatibility, chemical stability, and tunable properties. They can act as drug carriers, enabling controlled release and targeted delivery of anti-cancer agents. LDHs can also be engineered with magnetic properties (by incorporating Fe_3O_4) or functionalized with metal nanoparticles (Au, Pd, Pt, Ag) to enhance photothermal and cytotoxic effects against cancer cells.

Key applications include:

- Patient Stratification: Spatial biomarkers for treatment selection
- Response Monitoring: Real-time assessment of therapy effectiveness
- Resistance Prediction: Early identification of emerging drug resistance.

Multi-responsive systems that integrate both endogenous and exogenous mechanisms for dynamic therapeutic effects. Biometrics address current challenges in the clinical translation of these technologies for theranostics (Li et al., 2026).

Next-generation spatial technologies focus on enhanced resolution, increased multiplexing capacity, and improved multimodal integration. Transmission-mode MALDI (t-MALDI-2) achieves sub-micrometer resolution for metabolomics applications, while enhanced MERFISH protocols enable detection of 10,000+ genes at single-cell resolution. Spatial proteogenomics platforms increasingly incorporate epigenetic profiling capabilities, enabling comprehensive characterization of regulatory landscapes (Byalahunasishetter et al., 2025). The improved clinical outcomes justify implementation in specialized cancer centers. Metal-Organic Frameworks (MOFs), composed of organic linkers and metal ions, offer high surface area and adjustable pore sizes for drug encapsulation and controlled release. Hybrid MOFs can be engineered for multifunctional cancer therapies, including photothermal, photodynamic, and chemodynamic treatments. Recent studies highlight MOF-based nanoenzymes that activate therapeutic effects in response to endogenous tumor biomarkers, simplifying nanomedicine design and improving targetingOngoing improvements in spatial resolution and multi-modal integration will further enhance the power of spatial multi-omics. By integrating radiomics and pathomics features from pre-treatment contrast-enhanced CT images and biopsy whole slide images (WSIs) to forecast pathological responses to neoadjuvant chemotherapy (NAC) in patients with locally advanced gastric cancer (LAGC). A total of 146 patients were retrospectively recruited, and a multi-omics prediction model was developed using the eXtreme Gradient Boosting algorithm, demonstrating superior performance in predicting pathological responses compared with single-omics models (Qi et al., 2026). The convergence of spatial omics, digital pathology, and artificial intelligence represents a transformative opportunity for precision medicine. The multi-omics model achieved high accuracy, with areas under the receiver operating characteristic curve (AUCs) of 0.935, 0.882, and 0.836 in the training, internal validation, and external validation cohorts, respectively, and was significantly correlated with 2-year disease-free survival probabilities. The multi-omics model was identified as an independent prognostic factor in LAGC patients, suggesting its potential to help clinicians predict responses to NAC and tailor personalized neoadjuvant therapies. Combining spatial omics with histopathology and radiology images using deep learning and computer vision techniques is a promising direction (Qi et al., 2026). Deep learning models trained on histopathological images combined with spatial molecular data achieve superior diagnostic and prognostic accuracy compared to either modality alone. Integration with electronic health records enables the development of comprehensive predictive models incorporating clinical, molecular, and spatial features (Peng et al., 2026).

PERSPECTIVES AND RECOMMENDATIONS

Future development should prioritize enhanced spatial-resolution multiplexing capacity, improved sensitivity for rare cell populations, and multimodal integration workflows. Silica-coated gold nanorods and mesoporous silica nanoparticles are prominent examples, enabling simultaneous drug delivery and imaging (theranostics). Spatial omics enhance the application of stimuli-responsive nanomedicine strategies in enhancing theranostics. The mechanisms of drug release in pancreatic ductal adenocarcinoma are activated by both endogenous tumor microenvironment cues and exogenous triggers (Li et al., 2026). Smart Nanoparticles are engineered to biological cues (pH, enzymes, temperature) or external stimuli (light, magnetism), enabling on-demand drug release and minimizing off-target effects. Functionalization with tumor-specific ligands (antibodies, peptides, folic acid) further enhances targeting and therapeutic efficacy. Standardization of protocols and development of reference standards will facilitate broader adoption and inter-laboratory reproducibility. Accelerated clinical translation requires the development of companion diagnostic frameworks tailored to spatial biomarkers (Meng et al., 2026). Establishing regulatory guidance documents and fostering collaboration among technology developers, clinical researchers, and regulatory agencies will be essential to successful implementation. The complexity of spatial omics necessitates significant investment in research infrastructure, including instrumentation, computational resources, and personnel training. Nanomedicine presents promising strategies to improve drug delivery and efficacy by enhancing drug loading, counteracting drug resistance, and enabling stimuli-responsive control (Ananthakumar, Nanjundeshwara, et al., 2025). Multifunctional therapies that combine drug delivery, imaging, and synergistic treatment modalities (photothermal, photodynamic, and immunotherapy). Development of standardized training programs and certification pathways will ensure quality and consistency across institutions. Open-source software development and data sharing initiatives will democratize access to advanced analytical capabilities.

CONCLUSIONS

Spatial multi-omics technologies are revolutionizing cancer research by resolving molecular heterogeneity within the tumor microenvironment (TME) through spatially resolved transcriptomic, proteomic, and metabolomic profiling. Platforms such as CosMx SMI (6,000 RNA targets, 64 proteins) and GeoMx DSP (whole transcriptome + 570+ proteins) enable high-plex, subcellular analysis of formalin-fixed tissues, bridging clinical archives with multi-modal discovery. The MOSAIC Atlas, the largest spatial multiomics initiative in oncology, integrates six data modalities across seven cancer types to map tumor-immune dynamics, identify patient subtypes, and uncover niche-specific therapeutic targets. Computational frameworks like MIIT (Multi-Omics Imaging Integration Toolset) address integration challenges by aligning serial sections via the GreedyFHist algorithm, enabling cross-modal analysis of spatial transcriptomics and mass spectrometry imaging. In hepatocellular carcinoma (HCC), spatial proteomics and transcriptomics have revealed that CD8+ T cell-macrophage spatial proximity correlates with resistance to anti-PD1/cabozantinib therapy, while TGFβ and vascular patterning dictate immune exclusion. Virtual clinical trials, combining these spatial datasets with quantitative systems pharmacology models, predict immunotherapy responses and optimize biomarker-guided trial designs. Despite advances, challenges persist in resolution-throughput tradeoffs, batch effect mitigation, and clinical translation. Emerging directions include AI-driven cellular interaction mapping, spatially informed combination therapies (antibody-drug conjugates), and global consortia to standardize spatial biomarkers. By contextualizing molecular data within architectural and evolutionary frameworks, spatial multi-omics redefines precision oncology, transforming cancer from a static entity into a dynamic ecosystem of cellular neighborhoods, Darwinian adaptations, and targetable vulnerabilities. The convergence of spatial resolution with multi-omics depth, enhanced by artificial intelligence and machine learning, positions these approaches as essential components of future cancer care. Continued technology development, standardization efforts, and regulatory advancement will be crucial for realizing the full clinical potential of these transformative methodologies. Spatial multi-omics has fundamentally transformed cancer research and clinical oncology by providing a multidimensional, spatially resolved view of tumor ecosystems. It has enabled the discovery of new cellular and molecular interactions, spatial biomarkers, and therapeutic targets, paving the way for more effective and personalized cancer treatments. As technologies mature and become more accessible, spatial multi-omics is poised to become an integral part of routine cancer diagnostics and therapy, ultimately improving patient outcomes across diverse cancer types.

REFERENCES

Ananthakumar, B., Nagaraja, B. G., Umesh, S. M., & Raghavan, R. (2025). Predictive Medicine. *Predictive Algorithms for Rehabilitation and Assistive Systems*, 291–314. DOI: 10.4018/979-8-3373-0194-5.ch011

Ananthakumar, B., Nanjundeshwara, N., Lakshmaiah, A., & Raghavan, R. (2024). Synthetic Gene Circuits as a Promising Approach in Cancer Immunotherapy. *Critical Developments in Cancer Immunotherapy*, 391–418. DOI: 10.4018/979-8-3693-3976-3.ch011

Ananthakumar, B., Nanjundeshwara, N., Lakshmaiah, A., & Raghavan, R. (2025). Nanocarriers for Delivering Nucleic Acids in Plants: Mechanisms and Applications. *Nanocarriers in Plant Science and Agriculture*, 207–234. DOI: 10.4018/979-8-3693-9869-2.ch008

Araújo, A. (2024). From artificial intelligence to semi-creative inorganic intelligence: A blockchain-based bioethical metamorphosis. *AI and Ethics*, •••, 1–6. DOI: 10.1007/s43681-024-00471-0

Ash, S. L., Orha, R., Mole, H., Dinesh-Kumar, M., Lee, S. P., Turrell, F. K., & Isacke, C. M. (2024). Targeting the activated microenvironment with endosialin (CD248)-directed CAR-T cells ablates perivascular cells to impair tumor growth and metastasis. *Journal for Immunotherapy of Cancer*, *12*(2), e008608. DOI: 10.1136/jitc-2023-008608

Bhuvana, G. N., Umesh, S. M., & Raghavan, R. (2025). Expanding the frontiers of Schistosomiasis Treatment: The promise of Nanoparticle-Based Therapies. *Journal of Vector Borne Diseases, 10*(4103/jvbd.jvbd_180_23). DOI: 10.4103/jvbd.jvbd_180_23

Biradar, S., Suresh, C., Nanjundeshwara, N., & Raghavan, R. (2025). Spatially Variable Genes. *Spatially Variable Genes in Cancer*, 1–28. DOI: 10.4018/979-8-3693-7728-4.ch001

Boretti, G., Amirfallah, A., Edmunds, K. J., Hamzehpour, H., & Sigurjónsson, Ó. E. (2024). Advancing Cartilage Tissue Engineering: A Review of 3D Bioprinting Approaches and Bioink Properties. *Tissue Engineering. Part B, Reviews*, ten.teb.2024.0168. Advance online publication. DOI: 10.1089/ten.teb.2024.0168

Byalahunasishetter, B. I., N., V., P., S., C., M., S., S., & Raghavan, R. (2025). Spatial Omics and Bioprinting. *Innovations in Precision Medicine and Genomics*, 85–110. DOI: 10.4018/979-8-3693-5787-3.ch003

Cai, G., Chen, F., Wen, K., Li, Y., & Ou-Yang, L. (2026). Identifying spatial domains from spatial multi-omics data using consistent and specific deep subspace learning. *Information Fusion*, *125*, 103428. DOI: 10.1016/j.inffus.2025.103428

Cashaback, J. G. A., Allen, J. L., Chou, A. H.-Y., Lin, D. J., Price, M. A., Secerovic, N. K., Song, S., Zhang, H., & Miller, H. L. (2024). NSF DARE—transforming modeling in neurorehabilitation: A patient-in-the-loop framework. *Journal of Neuroengineering and Rehabilitation*, *21*(1), 23. DOI: 10.1186/s12984-024-01318-9

Clift, C. L., Blaser, M. C., Gerrits, W., Turner, M. E., Sonawane, A., Pham, T., Andresen, J. L., Fenton, O. S., Grolman, J. M., Campedelli, A., Buffolo, F., Schoen, F. J., Hjortnaes, J., Muehlschlegel, J. D., Mooney, D. J., Aikawa, M., Singh, S. A., Langer, R., & Aikawa, E. (2024). Intracellular proteomics and extracellular vesiculomics as a metric of disease recapitulation in 3D-bioprinted aortic valve arrays. *Science Advances*, *10*(9), eadj9793. Advance online publication. DOI: 10.1126/sciadv.adj9793

Cs, C. (2021). Microscopic examination of spatial transcriptome using Seq-Scope. *Cell*, *184*(13), 3559–3572.e22. Advance online publication. DOI: 10.1016/j.cell.2021.05.010

D'Souza, R. F., Mathew, M., Mishra, V., & Surapaneni, K. M. (2024). Twelve tips for addressing ethical concerns in the implementation of artificial intelligence in medical education. *Medical Education Online*, *29*(1), 2330250. DOI: 10.1080/10872981.2024.2330250

Dhiflaoui, A., & Almawi, W. Y. (2025). Mechanisms and clinical implications of microRNA associations and signaling pathways in B-cell acute lymphoblastic leukemia. *Gene*, *967*, 149730. DOI: 10.1016/j.gene.2025.149730

Dinu, M., Bita, B., Parau, A. C., Ristoscu, C., & Negut, I. (2025). Bioactive Glass and Melittin Thin Films Deposited by MAPLE for Titanium Implant Functionalization. *Materials (Basel)*, *18*(10), 2410. DOI: 10.3390/ma18102410

Du, Y., Ding, X., & Ye, Y. (2024). The spatial multi-omics revolution in cancer therapy: Precision redefined. *Cell Reports Medicine*, *5*(9), 101740. DOI: 10.1016/j.xcrm.2024.101740

Gao, M.-Y., Wang, J.-Q., He, J., Gao, R., Zhang, Y., & Li, X. (2023). Single-Cell RNA-Sequencing in Astrocyte Development, Heterogeneity, and Disease. *Cellular and Molecular Neurobiology*, *43*(7), 3449–3464. DOI: 10.1007/s10571-023-01397-7

Gato-Diaz, U., Concheiro, A., Alvarez-Lorenzo, C., & Blanco-Fernandez, B. (2025). 4D printed hydrogels for precision delivery of bioactive molecules in cancer. *Expert Opinion on Drug Delivery*, *0*(0), 1–17. DOI: 10.1080/17425247.2025.2539959

Goker, M., Derici, U. S., Gokyer, S., Parmaksiz, M. G., Kaya, B., Can, A., & Yilgor, P. (2024). Spatial Growth Factor Delivery for 3D Bioprinting of Vascularized Bone with Adipose-Derived Stem/Stromal Cells as a Single Cell Source. *ACS Biomaterials Science & Engineering*, *10*(3), 1607–1619. DOI: 10.1021/acsbiomaterials.3c01222

Goodwin, R. J. A., Platz, S. J., Reis-Filho, J. S., & Barry, S. T. (2024). Accelerating Drug Development Using Spatial Multi-omics. *Cancer Discovery*, *14*(4), 620–624. DOI: 10.1158/2159-8290.CD-24-0101

Huang, Q., Li, Y., Huang, Y., Wu, J., Bao, W., Xue, C., Li, X., Dong, S., Dong, Z., & Hu, S. (2025). Advances in molecular pathology and therapy of non-small cell lung cancer. *Signal Transduction and Targeted Therapy*, *10*(1), 186. DOI: 10.1038/s41392-025-02243-6

Kantaros, A., Petrescu, F. I. T., & Ganetsos, T. (2025). From Stents to Smart Implants Employing Biomimetic Materials: The Impact of 4D Printing on Modern Healthcare. *Biomimetics (Basel, Switzerland)*, *10*(2), 2. Advance online publication. DOI: 10.3390/biomimetics10020125

Kary, A. D., Noelle, H., & Magin, C. M. (2025). Tissue-Informed Biomaterial Innovations Advance Pulmonary Regenerative Engineering. *ACS Macro Letters*, *14*(4), 434–447. DOI: 10.1021/acsmacrolett.5c00075

Kasemeier-Kulesa, J. C., Morrison, J. A., McKinney, S., Li, H., Gogol, M., Hall, K., Chen, S., Wang, Y., Perera, A., McLennan, R., & Kulesa, P. M. (2023). Cell-type profiling of the sympathetic nervous system using spatial transcriptomics and spatial mapping of mRNA. *Developmental Dynamics*, *252*(8), 1130–1142. DOI: 10.1002/dvdy.577

Kashyap, B. K., Singh, V. V., Solanki, M. K., Kumar, A., Ruokolainen, J., & Kesari, K. K. (2023). Smart Nanomaterials in Cancer Theranostics: Challenges and Opportunities. *ACS Omega*, *8*(16), 14290–14320. DOI: 10.1021/acsomega.2c07840

Kaur, J., Thakran, A., & Naqvi, S. (2025). Recent advances in cell membrane-based biomimetic delivery systems for Parkinson's disease: Perspectives and challenges. *Asian Journal of Pharmaceutical Sciences*, *20*(4), 101060. DOI: 10.1016/j.ajps.2025.101060

Khobragade, S. S., Deshmukh, M., Vyas, U., & Ingle, R. G. (2025). Innovative Approaches in Bone Tissue Engineering: Strategies for Cancer Treatment and Recovery. *International Journal of Molecular Sciences*, *26*(9), 3937. DOI: 10.3390/ijms26093937

Lakshmaiah, A., Korrapati, C., Suresh, C., & Raghavan, R. (2025). Synthetic Biology and AI. *Spatially Variable Genes in Cancer*, 265–290. DOI: 10.4018/979-8-3693-7728-4.ch010

Langlieb, J., Sachdev, N. S., Balderrama, K. S., Nadaf, N. M., Raj, M., Murray, E., Webber, J. T., Vanderburg, C., Gazestani, V., Tward, D., Mezias, C., Li, X., Flowers, K., Cable, D. M., Norton, T., Mitra, P., Chen, F., & Macosko, E. Z. (2023). The molecular cytoarchitecture of the adult mouse brain. *Nature*, *624*(7991), 333–342. DOI: 10.1038/s41586-023-06818-7

Li, F., Tang, Z., Zheng, Y., Jiang, T., Deng, L., Dai, W., Zhao, Y., Zheng, N., Liu, S., Fan, Y., Lu, S., Chen, Y., Liu, G., Zhang, Y., & Xiong, Y. (2026). Responsive nanomedicine strategies achieve pancreatic cancer precise theranostics. *Bioactive Materials*, *55*, 334–375. DOI: 10.1016/j.bioactmat.2025.08.012

McBride, D. A., Jones, R. M., Bottini, N., & Shah, N. J. (2024). The therapeutic potential of immunoengineering for systemic autoimmunity. *Nature Reviews. Rheumatology*, *20*(4), 203–215. DOI: 10.1038/s41584-024-01084-x

Meng, Z., Dong, J., Guo, L., Su, F., Liu, J., Wang, G., & Zhao, Z. (2026). Now and future of artificial intelligence-based signet ring cell diagnosis: A survey. *Expert Systems with Applications*, *296*, 129188. DOI: 10.1016/j.eswa.2025.129188

Nagashri, N., Archana, L., & Raghavan, R. (2024). Microbial Volatile Compounds (MVOCs) in Food Industries and Food Safety Applications. *Innovations, Securities, and Case Studies Across Healthcare, Business, and Technology*, 75–96. DOI: 10.4018/979-8-3693-1906-2.ch005

Nanjundeshwara, N., Lakshmaiah, A., & Raghavan, R. (2025). Aflatoxigenic Fungi. In A. Sankara Narayanan, S. Gomathi, V. Ambikapathy, & A. Panneerselvam, *Aflatoxigenic Fungi* (1st ed., pp. 225–239). CRC Press. DOI: 10.1201/9781003487043-29

Nunez, M. D., Fernandez, K., Srinivasan, R., & Vandekerckhove, J. (2024). A tutorial on fitting joint models of M/EEG and behavior to understand cognition. *Behavior Research Methods*, *56*(6), 1–31. DOI: 10.3758/s13428-023-02331-x

Otsuji, R., Fujioka, Y., Hata, N., Kuga, D., Hatae, R., Sangatsuda, Y., Nakamizo, A., Mizoguchi, M., & Yoshimoto, K. (2024). Liquid Biopsy for Glioma Using Cell-Free DNA in Cerebrospinal Fluid. *Cancers (Basel)*, *16*(5), 1009. DOI: 10.3390/cancers16051009

Peng, N., Song, S. J., Hui, V. W.-K., Lai, J. C.-T., Wong, G. L.-H., Wong, V. W.-S., & Yip, T. C.-F. (2026). Foundations of Artificial Intelligence in Hepatology: What a Clinician Needs to Know. *Journal of Clinical and Experimental Hepatology*, *16*(1), 103183. Advance online publication. DOI: 10.1016/j.jceh.2025.103183

Pierantoni, L., Reis, R. L., Silva-Correia, J., Oliveira, J. M., & Heavey, S. (2023). Spatial -omics technologies: The new enterprise in 3D breast cancer models. *Trends in Biotechnology*, *41*(12), 1488–1500. DOI: 10.1016/j.tibtech.2023.07.003

Priya, S., Boyd, A. S., Torres-Bautista, A., Sawadkar, P., Mosahebi, A., Narayan Rath, S., & Kalaskar, D. M. (2024). Accelerating vascular graft development: Adipose-derived stem cells and PODS® (Polyhedrin delivery system with tissue-specific growth factors) – Enhanced 3D bioprinting for functional blood vessels. *Chemical Engineering Journal*, *494*, 152934. DOI: 10.1016/j.cej.2024.152934

Pw, F. (2023). Longitudinal scRNA-seq analysis in mouse and human informs optimization of rapid mouse astrocyte differentiation protocols. *Nature Neuroscience*, *26*(10), 1726–1738. Advance online publication. DOI: 10.1038/s41593-023-01424-2

Qi, J., Wang, K., Han, Y., Huang, X., Liu, X., Xu, J., Duan, R., Xu, J., Chen, J., Wang, J., Chen, G., Kang, M., Chen, Y., Li, Q., Wang, S., & Chen, J. (2026). Computed Tomography and Whole Slide Image-Based Prediction of Pathological Responses to Neoadjuvant Chemotherapy in Locally Advanced Gastric Cancer: A Multi-Omics Study. *Biomedical Signal Processing and Control*, *112*, 108489. DOI: 10.1016/j.bspc.2025.108489

Raghavan, R., Moger, S. S., Umesh, S., & Bhuvana, G. N. (2025). Revolutionizing Pain Management. *Unveiling Technological Advancements and Interdisciplinary Solutions for Pain Care*, 129–164. DOI: 10.4018/979-8-3693-9501-1.ch005

Ramnauth, A. D., Tippani, M., Divecha, H. R., Papariello, A. R., Miller, R. A., Pattie, E. A., Kleinman, J. E., Maynard, K. R., Collado-Torres, L., Hyde, T. M., Martinowich, K., Hicks, S. C., & Page, S. C. (2023). Spatially-resolved transcriptomics of human dentate gyrus across postnatal lifespan reveals heterogeneity in markers for proliferation, extracellular matrix, and neuroinflammation. *bioRxiv*, *4*(12–01), 2023.11.20.567883. DOI: 10.1101/2023.11.20.567883

Rana, D., Rangel, V. R., Padmanaban, P., Trikalitis, V. D., Kandar, A., Kim, H.-W., & Rouwkema, J. (2025). Bioprinting of Aptamer-Based Programmable Bioinks to Modulate Multiscale Microvascular Morphogenesis in 4D. *Advanced Healthcare Materials*, *14*(1), 2402302. DOI: 10.1002/adhm.202402302

Ruan, X., Hu, K., & Zhang, X. (2023). PIE-seq: Identifying RNA-binding protein targets by dual RNA-deaminase editing and sequencing. *Nature Communications*, *14*(1), 3275. DOI: 10.1038/s41467-023-39054-8

Sadraei, A., Naghib, Seyed Morteza, & and Rabiee, N. (2025). 4D printing biological stimuli-responsive hydrogels for tissue engineering and localized drug delivery applications – part 1. *Expert Opinion on Drug Delivery, 22*(4), 471–490. DOI: 10.1080/17425247.2025.2466772

Sherif, Z. A., Ogunwobi, O. O., & Ressom, H. W. (2025). Mechanisms and technologies in cancer epigenetics. *Frontiers in Oncology*, *14*, 1513654. Advance online publication. DOI: 10.3389/fonc.2024.1513654

Shi, Y., Huang, L., Dong, H., Yang, M., Ding, W., Zhou, X., Lu, T., Liu, Z., Zhou, X., Wang, M., Zeng, B., Sun, Y., Zhong, S., Wang, B., Wang, W., Yin, C., Wang, X., & Wu, Q. (2024). Decoding the spatiotemporal regulation of transcription factors during human spinal cord development. *Cell Research*, *34*(3), 193–213. DOI: 10.1038/s41422-023-00897-x

Spiegel, B. M. R., Rizzo, A., Persky, S., Liran, O., Wiederhold, B., Woods, S., Donovan, K., Sarkar, K., Xiang, H., Joo, S., Jotwani, R., Lang, M., Paul, M., Senter-Zapata, M., Widmeier, K., & Zhang, H. (2024). What Is Medical Extended Reality? A Taxonomy Defining the Current Breadth and Depth of an Evolving Field. *Journal of Medical Extended Reality*, *1*(1), 4–12. DOI: 10.1089/jmxr.2023.0012

Valcz, G., Buzás, E. I., Gatenby, R. A., Újvári, B., & Molnár, B. (2024). Small extracellular vesicles from surviving cancer cells as multiparametric monitoring tools of measurable residual disease and therapeutic efficiency. *Biochimica et Biophysica Acta (BBA) -. Biochimica et Biophysica Acta. Reviews on Cancer*, *1879*(2), 189088. DOI: 10.1016/j.bbcan.2024.189088

Verstegen, M. M. A., Coppes, R. P., Beghin, A., De Coppi, P., Gerli, M. F. M., de Graeff, N., Pan, Q., Saito, Y., Shi, S., Zadpoor, A. A., & van der Laan, L. J. W. (2025). Clinical applications of human organoids. *Nature Medicine*, *31*(2), 409–421. DOI: 10.1038/s41591-024-03489-3

Wess, M., Andersen, M. K., Midtbust, E., Guillem, J. C. C., Viset, T., Størkersen, Ø., Krossa, S., Rye, M. B., & Tessem, M.-B. (2025). Spatial integration of multi-omics data from serial sections using the novel Multi-Omics Imaging Integration Toolset. *GigaScience*, *14*, giaf035. Advance online publication. DOI: 10.1093/gigascience/giaf035

Williams, C. G., Lee, H. J., Asatsuma, T., Vento-Tormo, R., & Haque, A. (2022). An introduction to spatial transcriptomics for biomedical research. *Genome Medicine*, *14*(1), 1. Advance online publication. DOI: 10.1186/s13073-022-01075-1

Wu, X., Xu, W., Lin, D., Sun, L., Loo, L.-H., Dai, J., & Cao, G. (2025). The application and prospects of spatial omics technologies in clinical medical research and molecular diagnostics. *Journal of Genetics and Genomics*. Advance online publication. DOI: 10.1016/j.jgg.2025.09.003

Xu, P., Peng, J., Yuan, T., Chen, Z., He, H., Wu, Z., Li, T., Li, X., Wang, L., Gao, L., Yan, J., Wei, W., Li, C. T., Luo, Z.-G., & Chen, Y. (2024). High-throughput mapping of single-neuron projection and molecular features by retrograde barcoded labeling. *eLife*, *13*, e85419. DOI: 10.7554/eLife.85419

Xu, X., Su, J., Zhu, R., Li, K., Zhao, X., Fan, J., & Mao, F. (2025). From morphology to single-cell molecules: High-resolution 3D histology in biomedicine. *Molecular Cancer*, *24*(1), 1. Advance online publication. DOI: 10.1186/s12943-025-02240-x

Yang, F., Zhao, Z., Zhang, D., Xiong, Y., Dong, X., Wang, Y., Yang, M., Pan, T., Liu, C., Liu, K., Lin, Y., Liu, Y., Tu, Q., Dang, Y., Xia, M., Mi, D., Zhou, W., & Xu, Z. (2024). Single-cell multi-omics analysis of lineage development and spatial organization in the human fetal cerebellum. *Cell Discovery*, *10*(1), 22. DOI: 10.1038/s41421-024-00656-1

Yao, Z., van Velthoven, C. T. J., Kunst, M., Zhang, M., McMillen, D., Lee, C., Jung, W., Goldy, J., Abdelhak, A., Aitken, M., Baker, K., Baker, P., Barkan, E., Bertagnolli, D., Bhandiwad, A., Bielstein, C., Bishwakarma, P., Campos, J., Carey, D., & Zeng, H. (2023). A high-resolution transcriptomic and spatial atlas of cell types in the whole mouse brain. *Nature*, *624*(7991), 317–332. DOI: 10.1038/s41586-023-06812-z

Zhang, Q., Li, T., Xu, M., Islam, B., & Wang, J. (2024). Application of Optogenetics in Neurodegenerative Diseases. *Cellular and Molecular Neurobiology*, *44*(1), 57. DOI: 10.1007/s10571-024-01486-1

Zhang, R., Li, Y., Fu, H., Zhao, C., Li, X., Wang, Y., Sun, Y., & Li, Y. (2025). Nanomedicine strategies for cuproptosis: Metabolic reprogramming and tumor immunotherapy. *Acta Pharmaceutica Sinica. B*, *15*(9), 4582–4613. Advance online publication. DOI: 10.1016/j.apsb.2025.07.007

Zhang, S., Deshpande, A., Verma, B. K., Wang, H., Mi, H., Yuan, L., Ho, W. J., Jaffee, E. M., Zhu, Q., Anders, R. A., Yarchoan, M., Kagohara, L. T., Fertig, E. J., & Popel, A. S. (2024). Integration of Clinical Trial Spatial Multiomics Analysis and Virtual Clinical Trials Enables Immunotherapy Response Prediction and Biomarker Discovery. *Cancer Research*, *84*(16), 2734–2748. DOI: 10.1158/0008-5472.CAN-24-0943

Chapter 8
Spatial Multi-Omics in Cardiovascular Diseases

Vijaya Anand Arumugam
https://orcid.org/0000-0001-7485-1586
Bharathiar University, India

Kamalesh Marudhachalam
https://orcid.org/0009-0001-1906-0498
Bharathiar University, India

Kumar Ebenezar Kesavarao
SRM Institute of Science and Technology, India

Jerline Babu Maria
Dr. R.A.N.M. Arts and Science College, India

Manoj Mani
https://orcid.org/0009-0008-6601-0339
Bharathiar University, India

Akilandeswari Govindraj
CMS College of Science and Commerce, India

Mokesh Kadhirvel
Oxymed Hospital, India

Velayuthaprabhu Shanmugam
Bharathiar University, India

ABSTRACT

Spatial multi-omics has become a transformative frontier of cardiovascular research, offering integrative understandings of cellular heterogeneity and spatially organized cardiac tissues. Through integration of omics by high-resolution spatial profiling, these technologies empower mapping of interactions underlying cardiac development, disease, and remodelling. Developments for atherosclerosis, myocardial infarction, and cardiomyopathies have revealed dynamic networks of communication, metabolic reprogramming, and evolution of the fibrotic niche. Next-generation computational platforms, including graph neural networks, variational autoencoders, and digital pathology guided by AI, support multi-layered data integration and predictive modelling. Novel human heart spatial atlases additionally connect molecular architecture and clinical phenotypes, advancing precision diagnostics

DOI: 10.4018/979-8-3373-4849-0.ch008

and targeted therapeutics. Although data harmonization and standardization are key challenges, spatial multi-omics represents untold promise for transformative precision cardiology and regenerative medicine.

1. INTRODUCTION

Diseases are spreading faster in the industrialized world and these urge proper diagnosis and therapeutic measures (Velayuthaprabhu & Archunan, 2005; Velayuthaprabhu *et al.*, 2007; Alagendran *et al.*, 2010; Velayuthaprabhu *et al.*, 2011; Shanmugam Velayuthaprabhu *et al.*, 2013; Velayuthaprabhu *et al.*, 2016; Bhotla *et al.*, 2020; Meyyazhagan *et al.*, 2020; Arumugam *et al.*, 2020; Mohd *et al.*, 2021; Bhotla *et al.*, 2021; Mohd *et al.*, 2021; Lee *et al.*, 2021; Pushparaj *et al.*, 2022; Sangeetha *et al.*, 2022; Gundappa *et al.*, 2022; Sangeetha *et al.*, 2022; Meyyazhagan *et al.*, 2022; Ramya *et al.*, 2023; Paranitharan *et al.*, 2025). Cardiovascular diseases (CVDs) are the primary cause of deaths all over the world, responsible for 17.6 million deaths in 2012, i.e., 31.3% of total mortality, and ischemic heart disease (IHD) alone accounts for 7.4 million deaths (Mensah *et al.*, 2014). Prevalence of CVDs is far higher in the low- and middle-income nations (LMICs), which are responsible for more than 80% of CVD deaths globally. Though interventional treatments like percutaneous coronary intervention have emerged, long-term morbidity and mortality are still prominent because of post-ischemic issues such as cardiac remodelling and ischemic heart failure. Economic impact is also devastating, with direct healthcare costs and loss of productive years estimated to be US$20 trillion by 2030 (Mensah *et al.*, 2014; McAloon *et al.*, 2016). Elucidation of the intricate pathophysiology of CVDs, therefore, requires high-resolution molecular profiling of the cellular and molecular heterogeneity of the progressive phases of the disorder.

The field of cardiovascular research has been revolutionized by multi-omic technologies to conduct detailed molecular profiling of cardiac tissues. Bulk omics strategies, for example, bulk RNA sequencing, have offered insightful observations of broad-based gene expression and molecular networks. Bulk-based strategies, though, pool signals of heterogeneous cell populations and hide the cellular heterogeneity and complex molecular interplay responsible for the core cardiac function and pathology (Hegenbarth *et al.*, 2022). This drawback limits the resolution power needed to untangle the coordinate cell activities of the varied cell composition of cardiomyocytes, fibroblasts, endothelial cells, and leukocytes, which sync the physiology and pathology of the heart.

To break through these constraints, single-cell RNA sequencing (scRNA-seq) has become an efficacious resource, allowing for the high-resolution interrogation of cellular heterogeneity and dynamic transcriptional variability of the heart.

scRNA-seq allows the uncovering of sparse cardiac cell populations and gives us unprecedented views of cell–cell communication networks (Wang *et al.,* 2021; Forte *et al.,* 2021). However, various challenges remain, including the integration of data across platforms, limited capacity for spatial contextualization, and the necessity of better methodologies to interrogate sparse cell populations. Combining scRNA-seq by means of spatially resolved omics, therefore, remains crucial to the attainment of a better understanding of cardiovascular pathophysiology (Tang *et al.,* 2025).

Spatial multi-omics technologies have lately come forth as paradigm-shifting instruments for the exploration of tissue heterogeneity and spatially resolved molecular interactions within CVDs. Integrating various omic layers of the transcriptomic, epigenomic, proteomic, and metabolomic varieties within a spatial map, these technologies allow the delineation of molecular processes through the cell and organism scores and the identification of cell populations responsible for CVDs (Kiessling & Kuppe, 2024; Liu *et al.,* 2024a). Spatial multi-omics are capable of resolving 10,000 or fewer individual molecules, providing the molecular circuitry of cardiac cell biology and CVD progression through high-definition resolution. Integrating spatial multi-omics with single-cell transcriptomics gives even broader delineation of cellular heterogeneity and intercellular crosstalk and facilitates the discovery of new biomarkers and therapeutic targets through personalized cardiovascular medicine (*Figure 1*) (Tang *et al.,* 2025; Kiessling & Kuppe, 2024).

Figure 1. Spatial omics in cellular heterogeneity

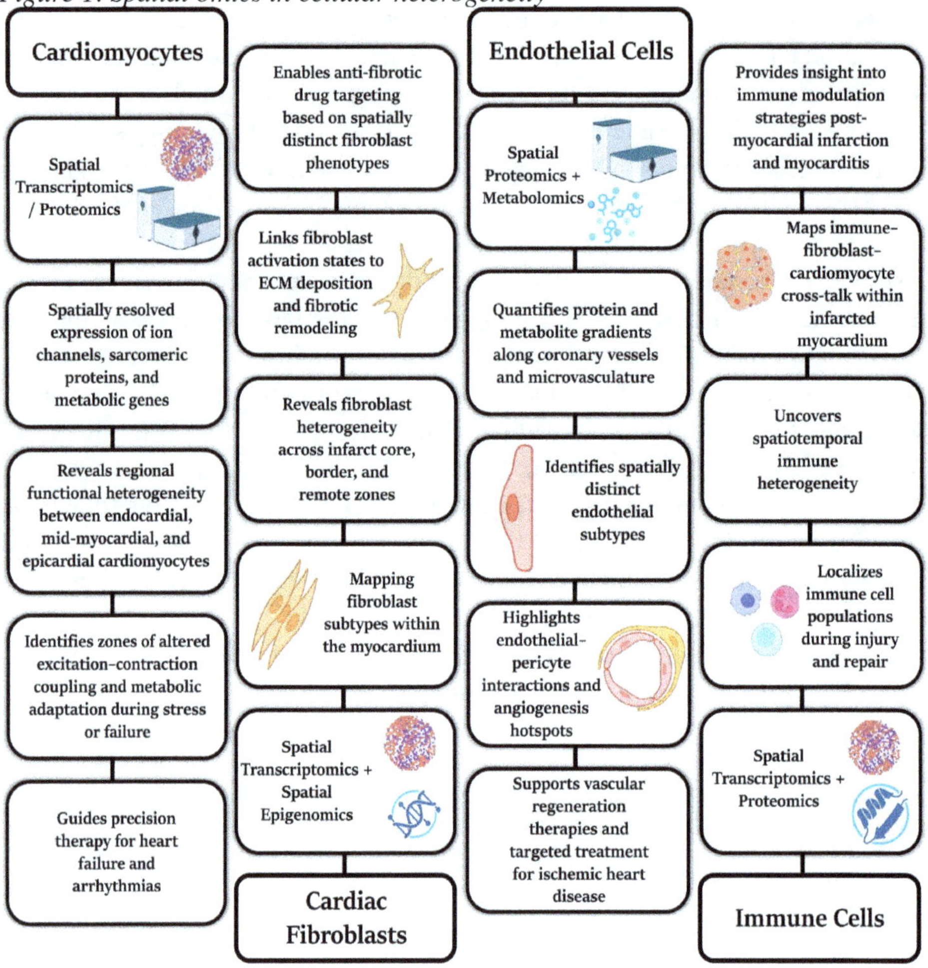

Network medicine complements these approaches by providing systems-level methods to integrate multi-omics information and uncover functional relationships within complex biological networks. This integrative model makes it possible to undertake an unbiased exploration of the pathophysiology of the disease and open new avenues of precision diagnosis, prognosis, and treatment of CVD (Wang *et al.,* 2023; Liu *et al.,* 2024b). Spatial resolution, on the other hand, continues to be indispensable for mechanistic insight into CVDs, because it makes it possible to visualize molecular gradients, cell organization, and cell communication throughout cardiac tissues, which are crucial components of tissue homeostasis, cardiomyopathic remodelling, and the progression of the disease (Han *et al.,* 2024; Schumacher & Kramann, 2023).

2. PRINCIPLES OF SPATIAL MULTI-OMICS

The spatial multi-omics paradigm of cardiology is an integrative model combining multiple omics layers, such as genomics, transcriptomics, proteomics, metabolomics, and epigenomics, to understand CVDs comprehensively at the molecular level. This multi-faceted methodology allows for the concurrent profiling of numerous molecular modalities of the same tissue section, retaining the spatial organization and molecular interactions at the level of the subcellular. Resolving as many as 10,000 individual molecules, spatial multi-omics gives unprecedented access to the intricate interplay of genetic, environmental, and lifestyle drivers of cardiometabolic disorders (Kiessling & Kuppe, 2024; Zhan *et al.*, 2023). Modalities integration allows the reporting of molecular circuits controlling the cell biology of the heart, and consequently, the discovery of new biomarkers and therapeutic targets of major importance for translational cardiovascular and precision medicine (Kiessling & Kuppe, 2024). In myocardial infarction, one of the most frequent CVDs, multi-omics strategies have been decisive for the discovery of pathologically operative signaling networks, molecular controllers, and putative prognosis markers, ultimately benefiting the diagnosis, the treatment, and the prognosis prediction (Liu *et al.*, 2023b). However, current descriptive models are challenged by data harmonization and integration across platforms, highlighting the urgency of sophisticated computational tools and standardized pipelines to fully exploit spatial multi-omics data sets (Huan *et al.*, 2025).

Cardiovascular studies have experienced methodological evolution from bulk omics to single-cell and, most recently, spatial omics technologies. Early bulk omics approaches-including genomics, proteomics, and metabolomics-provided broad molecular portraits of cardiovascular pathophysiology of large patient cohorts, facilitating biomarker discovery and target identification (Shah & Gerszten, 2024). However, the bias toward averaging inherent in bulk tests obscured cellular heterogeneity and the spatial organization of molecular processes of the myocardium. Single-cell omics, and particularly scRNA-seq, revolutionized cardiovascular studies by providing cell-type–specific resolution of transcriptomic, genome, and proteome alterations. These approaches revealed rich transcription networks and cellular heterogeneity of the cardiac tissues, uncovering new insight into lineage specification, disease remodelling, and intercellular communication (Kim *et al.*, 2025; Long *et al.*, 2023a).

Using the foundation of single-cell technologies, spatial omics technologies have filled the void between tissue architecture and molecular profiling by embedding molecular data into spatial coordinates. This innovation makes it possible to reconstruct cellular microenvironments, detect spatially limited signaling domains, and map intercellular communication networks of key importance to cardiac physiology and pathology (Palmer *et al.,* 2024; Tan *et al.,* 2024). An integration of spatial and single-cell omics renders an integrative paradigm to recapture both the physical localization and molecular identity of cells of key importance to the comprehension of the spatial dynamics of cardiac remodelling, inflammation, and fibrosis after injury.

Spatially resolved multi-omics technologies extend the resolution of the mapping of cardiac tissue even further to reveal the detailed structural and functional organization of the heart. High-resolution spatial profiling has outlined defined cellular niches-the compartmentalization of the sinoatrial node and the immune niches of the epicardium-itself responsible for key aspects of cardiac electrophysiology and immune defense (Kanemaru *et al.,* 2023). In addition, spatial omics has revealed important developmental processes by delineating temporally and spatially separated cell states responsible for the formation of specialized cardiac structures, including the pacemaker–conduction system and the heart valves (Lázár *et al.,* 2024). Through these kinds of integrative analyses, spatial multi-omics provides a multidimensional view of the cardiac microenvironment, and this, in turn, allows understanding of the mechanistic details of tissue homeostasis, cellular heterogeneity, and pathological remodelling of CVD (Schumacher & Kramann, 2023; Palmer *et al.,* 2024).

3. TECHNOLOGICAL PLATFORMS

Multi-omics spatial platforms have revolutionized cardiovascular research by enabling the simultaneous interrogation of various molecular layers within the spatial context of cardiac tissues. These technologies, as systems, allow for an integrated understanding of gene expression, protein localization, metabolic dynamics, and chromatin accessibility and, through these, an understanding of the cardiac physiology and pathology. Spatial transcriptomics technologies, such as Slide-seqV2, Stereo-seq, MERFISH, and seqFISH+, have been indispensable resources in the field of cardiology for the delineation of transcriptomic heterogeneity with spatial resolution. These technologies permit the mapping of the expression of genes spatially within intact cardiac tissues, providing molecular gradients and cellular interactions obscured by classical transcriptomic studies. They have been instrumental for the identification of networks of expression of the genes controlling cardiac development, electro-anatomy, and ischemic heart disease (Nguyen *et al.,* 2021). For example, MERFISH was utilized for the identification of an aneurysm-specific male-specific smooth muscle cell lineage responsible for medial calcification, the signature of the pathology of aneurysms, in thoracic aortic aneurysms (Yurdagul Jr & Aikawa, 2023). In addition, the integration of spatial and single-cell transcriptomics has provided new insights into cardiovascular development and CVD progression by identifying dynamic cell–cell communication and cell and tissue remodelling processes (Han *et al.,* 2024; Long *et al.,* 2023b). Extensions of spatial multi-omics platforms integrating transcriptome data and proteome and epigenome layers have further advanced the understanding of molecular circuitry controlling CVDs, providing the stimulus for the identification of biomarkers and the discovery of therapeutic targets (Kiessling & Kuppe, 2024).

Complementing spatial transcriptomics, spatial proteomics technologies such as Imaging Mass Cytometry (IMC), CODEX, and MIBI-TOF provide high-dimensional protein mapping within cardiac tissues. IMC enables multiplexed imaging of dozens of protein markers simultaneously, offering detailed visualization of immune infiltration, cellular crosstalk, and pathological remodelling after myocardial infarction. Using IMC, researchers have elucidated mechanisms of neutrophil-mediated tissue injury and identified key molecular signatures such as endothelial cell–specific upregulation of *H3K9me3* following ischemia-reperfusion injury, which may serve as therapeutic targets to mitigate post-infarction remodelling (Shi *et al.,* 2022; Yao *et al.,* 2023). CODEX utilizes DNA-barcoded antibodies to visualize over 50 cellular markers, enabling simultaneous spatial mapping of RNA and proteins in cardiac tissue and providing critical insights into the complex microenvironment of the diseased myocardium (Kuswanto *et al.,* 2023). Similarly, MIBI-TOF integrates metal-tagged antibodies with time-of-flight detection to achieve high-dimensional spatial proteomics, facilitating immune cell profiling and the study of host–pathogen interactions in cardiac disease models (Hu *et al.,* 2024).

For spatial metabolomics and lipidomics, Matrix-Assisted Laser Desorption/Ionization Mass Spectrometry Imaging (MALDI-MSI) and Desorption Electrospray Ionization Mass Spectrometry Imaging (DESI-MSI) have played key roles in understanding the metabolic and lipidomic profiles of cardiovascular tissues. MALDI-MSI provides excellent spatial resolution and sensitivity for the imaging of various metabolites and lipid species in situ, and it serves as an excellent technique for the investigation of diabetic cardiomyopathy and atherosclerosis. It has allowed the detection of carbohydrates, amino acids, and lipid metabolites, and the demonstration of unique spatial lipid dynamics of the arterial walls following angioplasty and of restenotic lesions (Liu *et al.,* 2023c; Shi *et al.,* 2019). DESI-MSI, on the other hand, provides complementary lipid imaging by increased ion yields and refined resolution of the captured images, favoring neutral lipid species like cholesteryl esters and triacylglycerides (Slijkhuis *et al.,* 2025). Both methods have been utilized to map metabolic disturbances inside the spaces of the atherosclerotic plaques, and they have offered mechanistic explanations of metabolic reprogramming and plaque instability (Seeley *et al.,* 2023; Shi *et al.,* 2019).

Spatial epigenomics advances have brought forth high-resolution mapping approaches, including spatial ATAC-seq and scCUT&Tag, which allow for the delineation of chromatin accessibility and histone modification maps within their native tissue landscapes. Spatial ATAC-seq combines transposase-accessible chromatin profiling and barcoded spatial capture, providing epigenomic maps of the regulatory landscapes and lineage differentiation programs of developmental and disease processes (Llorens-Bobadilla *et al.,* 2022). This technique has been utilized for both mouse embryonic and human tonsil tissues and defined tissue-specific chromatin accessibility and spatial organization of the immune cells (Deng *et al.,* 2021). scCUT&Tag, albeit largely a single-cell profiling technique, supplements spatial ATAC-seq by conducting histone modification profiling at cellular resolution, providing insight into the epigenetic regulation of gene expression within the cardiac tissues (Xie *et al.,* 2024). These technologies, together, present key understandings of the spatial nature of the gene regulation of cardiac development, remodelling, and disease.

The fusion of multimodal spatial data has been accelerated by computation-centered systems designed for aligning and interpreting vast datasets of omic layers. Such software as SpaMosaic applies contrastive learning and graph neural networks to aggregate transcriptomic, proteomic, and epigenomic data and allows for exact delineation of spatial domains and imputation of modalities unobserved (Yan *et al.,* 2024). MaxFuse also addresses challenges of weak cross-modal links by applying iterative co-embedding and data smoothing strategies to achieve strong multimodal integration at the level of single-cell resolution (Chen *et al.,* 2023). These computation-centered innovations, along with experimentally driven innovations, have significantly boosted the interpretability and translational relevance of spatial multi-omics to cardiovascular illness investigations (Kiessling & Kuppe, 2024).

4. CARDIOVASCULAR TISSUE ARCHITECTURE AT THE SPATIAL OMICS SCALE

Spatial omics technologies have revolutionized the comprehension of the architecture of cardiac tissue by offering high-resolution visions of the spatial organisation, cellular heterogeneity, and molecular dynamics of the heart. Through transcriptomic, proteomic, metabolomic, and epigenomic mapping, these technologies allow for the demarcation of microanatomical domains and cellular niches underlying cardiac physiology and pathology. Spatial transcriptomic and proteomic profiling of the cardiac regions, such as the endocardium, the mid-myocardium, and the epicardium, has unveiled biomolecular signatures of specific cellular composition and molecular programs underlying their dedicated functions. The epicardium harbours an original immunological signature of enrichment of IgG^+ and IgA^+ plasma cells, shaping the formation of the immune niches, potentially implicated in infection defence mechanisms (Kanemaru *et al.,* 2023). By contrast, the myocardial-stress niche, richly enriched by activated fibroblasts and stress cardiomyocytes, has been observed throughout the cardiomyopathies. The endocardium, though poorly described by the few spatial studies, harbours the functionality of the conduction and contractility of the heart, implying the existence of a differential cellular and molecular environment of the exterior myocardial regions. Spatial transcriptomics has, moreover, delineated the injury, repair, and remodelling zones corresponding to myocardial infarctions, highlighting the spatial heterogeneity and the region-specific molecular reaction of the heart (Kuppe *et al.,* 2022). These observations are consolidated by integrated space omics studies correlating the expression of the genome along the chromatin accessibilities, and uncovering region-specific regulatory networks and cellular interactions (Cranley *et al.,* 2023; Mohenska *et al.,* 2022).

The use of spatial omics has also revealed the complex microanatomical niches specifying cardiac function. Spatial transcriptome mapping has outlined specialized niches like the sinoatrial node, consisting of pacemaker cells, fibroblasts, and glial-like cells, which regulate the cardiac rhythm (Kanemaru *et al.,* 2023). The immunomodulatory capacity of the heart and defense mechanisms are underscored by the epicardial immune niche, which is enriched by plasma cells and resident macrophages. Single-cell dual-omics protocols have fine-tuned the molecular signature of the molecular classification of cardiac non-myocytes, identifying specific fibroblast subtypes by unique transcriptional signatures associated with the stress response, ECM remodelling, and the regulation of the immune system (Wang *et al.,* 2022). In addition, integrative single-cell and spatial omics approaches have revealed the functional and spatial heterogeneity of cardiomyocytes, and of the endothelial cells, the fibroblasts, and the cells of the immune system, distinguishing the atrial and ventricular subpopulations, which have divergent developmental fates and the ability of the cells to signal (Litviňuková *et al.,* 2020).

Spatial omics technologies have proved most useful in dissecting fibroblast subpopulations and ECM niches, key players of cardiac repair and fibrosis. Automated spatially targeted optical microproteomics (autoSTOMP), for example, has facilitated proteomic profiling of subsets of fibroblasts, e.g., SMA^+ fibroblasts of the myocardial infarction scar region, and illuminated their contributions to ECM remodelling and scar maturation (Mallikarjun *et al.,* 2024). Single-cell and spatial omics have unraveled sex-determined fibroblast gene signatures of fibrotic remodelling, most pronounced under hypertension (Farrugia *et al.,* 2024). Ventricular myocardial-stress niches enriched by cardiomyocytes under stress have localized the activation of fibroblasts by spatial transcriptomics, substantiating their pathogenic significance in cardiomyopathies (Kanemaru *et al.,* 2023). High-resolution spatiotemporal profiling of cardihiac repair has restored the dynamics of the fibrotic niche, delineating the macrophage-mediated balance of fibroblast proliferation to avoid excessive fibrosis and the identification of the dedifferentiating cardiomyocytes sustained by localized paracrine signals (Soliman *et al.,* 2021; Chan *et al.,* 2024).

Spatial multi-omics integrative strategies have expanded these results by outlining, at the single-cell level, the interactions between cardiomyocytes, fibroblasts, endothelial cells, and immune cells. By integrating gene expression, chromatin access, and spatial information, these works have outlined the molecular principles governing the assembly of the heart tissue and intercellular communication (Tan *et al.,* 2024; Kanemaru *et al.,* 2023). For myocardial infarction, spatial multi-omics has defined selective zones of necrosis, inflammation, repair, and remodelling and specified fibroblast-to-myofibroblast transitions governing fibrogenesis (Kuppe *et al.,* 2022; Kramann & Simmons, 2022). With this technology, the compartmentalization of the sinoatrial node and supporting cells has been defined, specifying the structural and paracrine mechanisms of conduction within the cardiovascular system (Kanemaru *et al.,* 2023). In aggregate, these data portray the image of spatial omics as providing an integrated, multidimensional view of cardiovascular tissue structure, stipulating the cell-cell and cell-molecular interactions governing cardiac homeostasis and the processes of CVD (Palmer *et al.,* 2024).

5. SPATIAL MULTI-OMICS IN CARDIOVASCULAR PATHOLOGIES

Spatial multi-omics technologies have substantially accelerated the understanding of cardiovascular pathologies by providing the multi-dimensional observation of cellular organization, molecular signals, and metabolic modifications of the pathological tissues. These interdisciplinary approaches-the integration of spatial transcriptomics, proteomics, metabolomics, and epigenomics can trace complex cellular interactions responsible for inflammation, remodelling, and degeneration of CVD (**Figure 2**). Spatial lipid–immune–stromal cell interactions are an enduring hallmark of the pathogenesis of atherosclerosis. Atherosclerosis lesions are formed by chronic inflammatory and lipid build-up within the vascular wall, where the immunoactive cells, including monocytes, macrophages, dendritic cells, and lymphocytes, associate with the lipids to regulate the progression of the lesion (Annink *et al.,* 2024; Herrero-Fernandez *et al.,* 2019). These interactions are governed through mechanisms of immunometabolism and trained immunity, which maintain the inflammatory profile of the atherosclerotic plaques (Pirillo *et al.,* 2018). Lipid metabolism not only regulates the activation of the immune cell but also the function of lipoproteins, thereby impacting the formation and stability of the lesion (Getz & Reardon, 2014). Spatial transcriptome-based studies have, in the recent past, outlined the structure of the atherosclerotic plaques, identifying the spatially organized populations of the immunoactive, stromal, and vascular cells and the communication networks between them (Bleckwehl *et al.,* 2023). These reports have unraveled the interaction of the endothelial and the vascular smooth muscle cells with the immuno-infiltrates and the coordinated communication of the plaque remodelling and stability.

Figure 2. Spatial multi-omics landscape of cardiovascular pathologies

Within this pro-inflammatory lipid-rich milieu, macrophage reprogramming exerts significant influence on the balance of plaque constituents. Spatially epigenomic and metabolic profiling of the atherosclerotic plaque has emerged as a means of uncovering divergent macrophage phenotypes governed by microenvironmental signals. Epigenetic modulators including TET2, DNMT3A, and HDACs, for instance, regulate the polarization of the macrophage by variably regulating chromatin access and transcriptional potency (Jin *et al.,* 2021). Metabolic rewiring, influenced by cytokines, lipids, ROS, and hypoxia, further governs the macrophage phenotype and function (Koelwyn *et al.,* 2018; Xiao *et al.,* 2023). The interplay between metabolic flux and epigenetic modulation prevents polarization of the macrophage, whereby metabolites dictate patterns of histone modification and gene expression governing pro- or anti-inflammatory states (Yang *et al.,* 2025). Oxygen tension within plaques likewise regulates the behavior of the macrophage, whereby aerobic conditions induce M2 phenotypes through the action of oxidative phosphorylation, whilst hypoxic niches induce glycolytic M1 macrophages (Shen *et al.,* 2023).

Moving on from vascular inflammation to myocardial injury, spatial omics of myocardial infarction (MI) revealed the dynamic patterns of immune cell infiltration and their contribution to the repair of the heart. Spatial mapping demonstrated the new leukocyte accumulation mode by the endocardium through von Willebrand Factor (vWF), supporting the recruitment of $CCR2^+$ monocytes necessary for the subsequent post-MI cardiac function (Wünnemann *et al.,* 2024). Time-profiling of the myocardial transcriptome and proteome indicated the prevalence of Trem2hi macrophages as the foremost immune responders during the subacute stage, and their correction results in better functional recovery (Jung *et al.,* 2022). Longitudinally, the immune microenvironment shifts towards the dominance of the lymphocytes, increasing the numbers of B and T helper cells by day 28, concomitant with the transition towards the secretion of the immunomodulatory cytokines necessary for the maturation of the scar tissues (Mitre *et al.,* 2023). Spatial multi-omic data sets further define the myocardial zones of injury, repair, and remodelling, and reveal fibroblast-to-myofibroblast differentiation as the key driver of the fibrosis (Kuppe *et al.,* 2022).

Extending this understanding, spatially integrated transcriptomic and proteomic profiling has given insight into fibrosis and scar microenvironment remodelling after MI. These strategies allow for detailed mapping of cardiac cell-type composition and gene regulatory states within necrotic, repair, and remodelling zones (Kuppe *et al.,* 2022). Automated spatially targeted optical microproteomics (autoSTOMP) has also defined the proteomes of subsets of fibroblasts and localized macrophages within scar tissue, correlating these populations with extracellular matrix restructuring and energy metabolism of maturing scar tissue (Mallikarjun *et al.,* 2024). Combining such spatial datasets identifies the molecular interplay of resolving inflammation and fibrotic remodelling, and identifies therapeutic targets for the treatment of heart failure (Schumacher & Kramann, 2023).

For heart failure, spatial proteogenomic modifications of sarcomeric and ion channel proteins are the substrates for contractile dysfunction and electrical instability. Remodelling of sarcomeres involves the abnormal phosphorylation of structural proteins, disturbing calcium sensitivity, maximal force generation, and passive stiffness (Hamdani *et al.,* 2008). Ischemic cardiomyopathy proteome analyses reveal dysregulation of the cardiac troponin I, enigma homologue 2, muscle LIM protein, and calsarcin-1, suggesting disturbed kinase-phosphatase balance (Chapman *et al.,* 2023). At the same time, ion channel remodelling, defined by transcriptional and spatial modifications of sodium, potassium, and calcium channel expression, causes conduction system failure and arrhythmias (Yanni *et al.,* 2011; Reisqs *et al.,* 2024). Hypertrophic cardiomyopathy represents an example of how mutations within the sarcomeric genome increase calcium sensitivity, favoring the growth of the myocardium and pathological transcriptional reprogramming (Frey & Olson, 2000).

Spatial omics has also revealed conduction system heterogeneity and arrhythmogenic remodelling by delineating transcriptional and structural alterations that shift impulse conduction. Electrical heterogeneity, the basis for coordinated contraction, turns pro-arrhythmic when compromised by fibrosis or differential gene expression (Boukens & Christoffels, 2012). Spatial measurements reveal how microscopic heterogeneities-are they conduction slowing localized in space, wavefront collision because of fibrosis, or redistribution of the sodium channels?-are manifested as macroscopic arrhythmogenic phenomena (Gokhale *et al.,* 2018; Lammers *et al.,* 1990). Gap junction remodelling, particularly of connexin 43 (CX43), gives rise to reduced plaque size, lateralization, and dephosphorylation, which compromises the electrical coupling of cardiomyocytes and promotes arrhythmogenic susceptibility (Yu & Sheng, 2011; Severs, 2001; Severs, 2003; Hayashi, 2010; Qu *et al.,* 2025).

Spatially resolved multi-omics have further outlined genetic and epigenetic cardiomyopathy determinants. In arrhythmogenic cardiomyopathy (ACM), spatial transcriptomics has outlined *ZBTB11* as the key cardiomyocyte apoptosis and fibro-fatty replacement regulator, and in hypertrophic cardiomyopathy (HCM), single-nucleus RNA sequencing demonstrates vascular lineage expansion and cardiomyocyte contraction as the prominent remodelling processes (Liu *et al.,* 2023b). Differential expression of HCM identifies mitochondrial energetics and ECM remodelling genes like *FGF12* and *AEBP1* as prospects for therapeutic interventions (Laird *et al.,* 2023). Polygenic genetic analysis indicates overlapping polygenic networks propelling hypertrophic and dilated cardiomyopathies, controlling left ventricular traits by shared molecular mechanisms (Tadros *et al.,* 2021).

Spatial epigenetic control of cardiomyopathies focuses the molecular view even further by integrating DNA methylation, histone modification, and chromatin topology into spatial models of disease. Phenotypically variable inherited cardiomyopathies are affected by environmental and metabolic stimuli-reactive epigenetic modifiers (Hajdarpašić *et al.,* 2025). In acquired cardiomyopathies, including dilated cardiomyopathy (DCM), spatial chromatin remodelling through transcription factors like *HAND1* reorganizes enhancer-promoter interactions and induces transcriptional reprogramming (Feng *et al.*, 2022). Dysmorphic methylation by *DNMT3B* also disrupts transcript splicing and is involved in maladaptive remodelling (Vujic *et al.*, 2012).

Spatial multi-omics has unveiled the endothelial-pericyte dynamics and angiogenic niches of CVD remodelling. In the context of heart failure, integration of spatial transcriptomics and fate-mapping has delineated endothelial cell states circumscribed by hypoxia and proliferative reaction transcriptional programs necessary for post-injury angiogenesis (Peisker *et al.*, 2022). Aging induces endothelial dysfunction by promoting nitrative and oxidative stress, perturbing microvascular remodelling and tissue blood flow (Ungvari *et al.*, 2018). Spatial profiling of atherosclerosis lesions has delineated lesion-selective endothelial phenotypes and relationships of these cells and leukocytes, and smooth muscle cells, clarifying mechanisms of vascular inflammation and lesion progression (Nurminen *et al.*, 2023). In the context of myocardial infarction, high-resolution spatial mapping delineates injury and repair zone endothelial-pericyte remodelling (Kuppe *et al.*, 2022). Integrative multiomic and CRISPR-based screen studies have, moreover, delineated therapeutic targets and the coronary artery disease endothelial transcriptional program (Gupta *et al.*, 2023).

6. COMPUTATIONAL AND ANALYTICAL APPROACHES

Integration of multi-layered spatial omics data in the field of cardiology necessitates sophisticated computational strategies capable of handling the complexity, heterogeneity, and high dimensionality of cardiovascular data. Spatial multi-omics enables simultaneous interrogation of the transcriptome, epigenome, proteome, and metabolome of the cardiac tissue, and hence the identification of the molecular circuits and processes controlling CVD progression (Kiessling & Kuppe, 2024). Computational strategies of disease subtyping gain invaluable utility in this area, for they permit the identification of unique patient subsets through the integration of multi-omics and clinical data. Stratification of the type enables precision medicine interventions, though the latter remains limited by challenges within the realm of feature selection, data integration, and the optimization of clustering algorithms (Maiorino & Loscalzo, 2023). In an effort to mitigate the limitations of single-omics analyses, multi-dimensional integrative genomics approaches have been developed, pooling genomics, epigenomics, transcriptomics, metabolomics, proteomics, and microbiome data from both human and model systems (Arneson *et al.*, 2017). To complement these efforts, cardiovascular informatics leverages machine learning (ML) and artificial intelligence (AI) to manage the vast-scale data, pool heterogeneous data modalities, and permit the interrogation of both structured and unstructured information by means of scalable cloud computing platforms (**Figure 3**) (Caufield *et al.*, 2022; Cui & Zhang, 2021)

Figure 3. Computational pipeline for spatial multi-omics analysis in cardiology

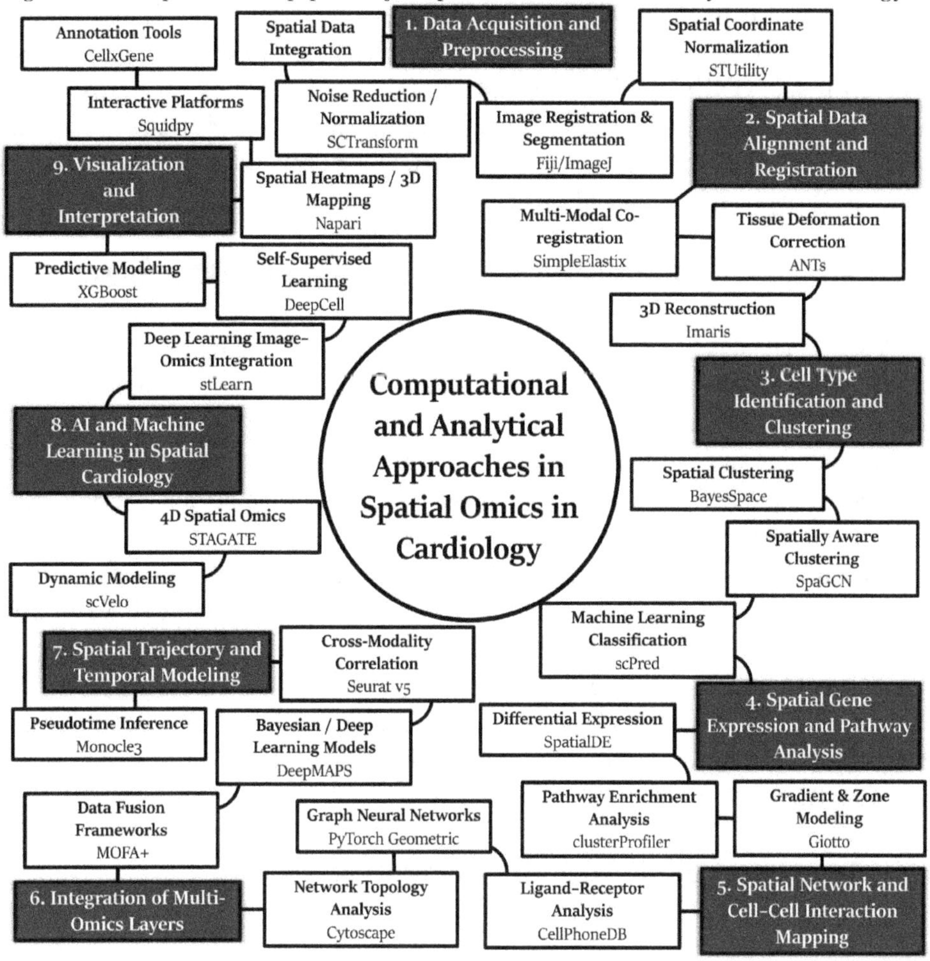

6.1 Graph Neural Networks in Spatial and Functional Modelling

Graph neural networks (GNNs) have become increasingly useful tools in the computational exploration of cardiac tissue microenvironments, due to their ability to retain spatially structured relationships and molecular dependencies within high-dimensional omics data. In cardiovascular simulation, reduced-order models based on GNNs are utilized to estimate hemodynamic parameters-like pressure and flow rates-within intricate cardiac geometries, allowing detailed fine-grained examination of the pathological stress of the cardiac microenvironment (Rangisetti *et al.*, 2024). In addition, GNNs have been incorporated into spatial transcriptomics pipelines, including the STdGCN model, where scRNA-seq data and spatial localization are integrated to reveal endothelial–cardiomyocyte communication networks and dynamic microenvironmental remodelling of the developing heart (Li & Luo, 2023). At the level of tissue, GNNs are likewise applied in predictive analytics towards patient stratification and personalized risk estimation, while, in the case of electrocardiogram (ECG) analysis, they make use of spatial correlations between ECG leads towards refined CVD classification and higher diagnostic accuracy (Zhang *et al.,* 2023a).

6.2 Variational Autoencoders and Cross-Omics Data Integration

Variational autoencoders (VAEs) are instrumental in integrating and denoising complex, multi-modal datasets within spatial cardiology. The STGMVA toolkit, based on a spatiotemporal Gaussian mixture VAE, exemplifies this by integrating gene expression and spatial localization data. It employs Gaussian mixture pretraining and variational graph embeddings to achieve high performance in clustering, gene imputation, and multi-sample integration while correcting batch effects across tissue slices (Liu *et al.,* 2023a). In cardiology applications, multi-modal VAEs have been adapted to integrate physiological and anatomical data-such as ECG waveforms and 3D cardiac structural models-revealing correlations between electrophysiological and morphological features. Applied to large-scale datasets such as the UK Biobank, these models outperform traditional methods in ECG generation and anatomy–physiology coupling, offering a robust framework for modelling cardiac function (Beetz *et al.,* 2022).

6.3 Modelling Cellular Communication and Microenvironmental Interactions

CellPhoneDB is an invaluable computational resource for the identification of the intercellular communication networks of the cardiac tissues by integrating the data of ligand–receptor interactions and single-cell and spatial omics. Latest editions of the resource take into account the multimeric structure of the signalling complex and cell spatial locations and, as such, enable us to dissect cardiac cellular crosstalk by precision (Troulé *et al.,* 2025; Efremova *et al.,* 2020). CellPhoneDB has been instrumental for the description of the cellular niches and the conduction systems' structure of the heart and the delineation of the immune microdomains of the epicardium participating in homeostatic adaptations and repair processes (Kanemaru *et al.,* 2023).

6.4 Spatial Fluxomics and Metabolic Network Reconstruction

Spatial fluxomics integrates spatially resolved metabolomics and metabolic flux analysis to yield a quantitative system for mapping the regional metabolic activity of cardiac tissues. Through MSI, metabolites are profiled throughout the sections of the heart spatially, delineating metabolic states of vulnerable and stable atherosclerosis plaques-a parameter crucial to identifying the vulnerability of plaques and CVD progression (Seeley *et al.,* 2023). Through the complementary mode, isotope-assisted metabolic flux analysis (iMFA) assesses intracellular and extracellular fluxes, and hence dynamic mapping of metabolic pathways, under physiological and stress conditions (Moiz *et al.,* 2022). Alongside systems biology modelling, these approaches have unraveled network-level metabolic remodelling of the failing heart, where differential substrate use and energy dynamics play roles in the pathophysiology of the disease (Aon & Cortassa, 2015; Xie *et al.,* 2023).

6.5 3D and 4D Spatial Multi-Omics: Integrative Computational Reconstruction

Recent advances extend spatial omics to three- and four-dimensional spaces, aligning molecular profiling with volumetric imaging and temporal dynamics. Molecular Cartography and imaging FISH provide volumetric maps of mRNA at the subcellular resolution, while spatial barcoding transcriptomics documents transcriptional heterogeneity between cardiac timepoints (Adusumalli *et al.,* 2023; Wünnemann *et al.,* 2024). Merging this with single-cell and single-nucleus RNA/ATAC-seq facilitates the mapping of transcriptional and epigenetic states of the spatial tissue architecture (Kuppe *et al.,* 2022; Cranley *et al.,* 2024). Multiplexed immunofluorescence, deep visual proteomics, and mass spectrometry imaging validate protein and lipid distributions, and tissue clearing and light-sheet microscopy allow volumetric imaging of the intact myocardium (Ding *et al.,* 2023; Lai & Chan, 2024; Fox *et al.,* 2025).

Computationally, SPACEL and ST-GEARS reconstruct 3D volumes of tissues by deconvolution, registration, and correction of the elastic field of serial sections (Xu *et al.,* 2023; Xia *et al.,* 2024). SpaTrio and COSMOS use probabilistic alignment and graph-based modelling to estimate multimodal heterogeneity and segment 3D/4D spatial domains (Yang *et al.,* 2023; Zhou *et al.,* 2025). These have outlined infarct-border-remodelling zones, fibrotic-niche organization, and conduction-node compartmentalization and temporally dynamic processes of cardiomyocyte dedifferentiation and repair by macrophages (Kuppe *et al.,* 2022; Wünnemann *et al.,* 2024; Adusumalli *et al.,* 2023.

Despite remarkable progress, spatial multi-omics is limited by imaging depth, tissue preparation variability, and availability of standardized computational benchmarks (Lai & Chan, 2024; Xia *et al.,* 2024; Kiessling & Kuppe, 2024). Emerging initiatives involve scalable volumetric reconstruction of the human cardiac tissues, coupled with 4D pipelines of molecular and imaging modalities, and the assembly of open-access atlases of the cardiovascular system for cross-study replication (Christodoulou *et al.,* 2018; Yuan *et al.,* 2023; Zhang *et al.,* 2023b). Integrating these advancing computer algorithms and clinical imaging and molecular mapping remains forefront of advancing precision cardiology, spanning molecular elaboration and functional cardiovascular physiology (Kanemaru *et al.,* 2023; Schumacher & Kramann, 2023).

7. CLINICAL AND TRANSLATIONAL APPLICATIONS

Detection of spatial biomarkers of vulnerable plaque rupture and progression of heart failure represents a key objective of translational cardiovascular medicine, by reason of the overwhelming importance of plaque instability as a trigger of acute coronary syndromes. Vulnerable plaques, most frequently delineated by the characteristics of a thin fibrous cap and an extensive necrotic core, have detectable metabolic and structural signatures amenable to spatial mapping by current spatial metabolomics and imaging technologies. Unstable plaques by the technique of MALDI-MSI were found to have networks significantly enriched for the production of reactive oxygen species, lipid peroxides, and tryptophan metabolism as molecular biomarkers of vulnerability (Luo & Leeper, 2023). Complementary imaging technologies, such as intravascular ultrasound-based iPALP and iMOD, permit quantitative measurements of the dimensions and distribution of the fibrous cap and the necrotic core and distribution of mechanical stress, measurements corresponding to susceptibility to rupture (Gómez *et al.,* 2021). Furthermore, circulating biochemistries and inflammatory indices, including those of the matrix metalloproteinases corresponding to degradation of the extracellular matrix and oxidant stress-related mediators, would be integrated with imaging-derived spatial data to enhance predictive modelling of the risk of rupture of the plaque (Shah, 2014; Prasad & Tsimikas, 2008; Wang & Connolly, 2010). Despite conspicuous improvements, the clinical utility of spatial biomarkers of plaque vulnerability remains constrained by lesion heterogeneity, interpatient variability, and technological limitations of broad-scale application. Overcoming these challenges requires standardization and multimodal integration of molecular and imaging data to permit precision risk stratification and precision therapeutic targeting (Luo & Leeper, 2023; Shah, 2014).

7.1 Spatial Heterogeneity and Patient Stratification in Precision Cardiology

Spatial heterogeneity is the key predictor of patient stratification and therapeutic response of CVDs. Single-cell RNA sequencing and spatial transcriptomics technologies have become capable of super-resolution profiling of the cardiac tissues and have unveiled cell-type–specific transcriptional programs and molecular microenvironments responsible for the variability of the disease (Han *et al.,* 2023). In tandem, cardiovascular magnetic resonance imaging (CMR) yields noninvasive biomarkers of myocardial heterogeneity-including strain mapping and tissue characterization metrics-that enhance early and refined stages of diagnosis and risk prediction beyond the realms of the conventional functional evaluations (Hesse *et al.,* 2024). Fusing these spatially resolved data records with multimodal electronic health records by the power of deep learning algorithms harbors the ability to identify the patient subgroup and tailor the therapy individually (Yin *et al.,* 2025). This integrative paradigm has become instrumental in the fine-tuning of the management of acute coronary disorders (Andreotti *et al.,* 2021) and the refinement of device-based treatments like implantable cardioverter-defibrillators and pacemakers by spatially informed modelling of electrophysiological heterogeneity (Verrier *et al.,* 2022).

7.2 Pharmacoproteomics and Drug Distribution Mapping in Cardiac Tissues

Pharmacoproteomics combined with spatial omics provides an excellent toolbox for the description of the penetration, distribution, and efficacy of drugs within the tissues of the heart. Quantitative targeted proteomics allows the detection and measurement of proteins involved in the interactions of drugs, and provides mechanisms of cardiotoxicity and therapeutic modulation (Ohtsuki, 2012; Nguyen *et al.,* 2021). Spatially resolved proteomics and imaging technologies delineate the localization of the drug molecules and their metabolites throughout defined cell types of the cardiovascular system, avoiding the shortcomings of the bulk pharmacokinetic measurements (Pang & Ye, 2022). This has been useful particularly for the description of anthracycline-induced cardiotoxicity, where the spatial mapping describes cell susceptibility to the injury induced by the drugs (Nguyen *et al.,* 2021). Furthermore, the mapping of the surfaceome of the heart and the receptor networks related thereto identifies new druggable sites, creating the groundwork for the rationalization of the construction of drugs having reduced off-target effects and increased therapeutic potency (Inocencio *et al.,* 2025; Kanemaru *et al.,* 2023).

7.3 Spatial Multi-Omics in Regenerative and Reparative Cardiology

Spatial multi-omics is a paradigm shift for understanding and refining regenerative therapies in cardiology, including stem cell–mediated and engineered heart tissue (EHT)-controlled strategies. By applying spatial transcriptomics, the transcriptional landscapes of transplanted cardiovascular progenitors (CVPs) within the myocardium of infarctions have been characterized, revealing cellular processes associated with engraftment, maturation, and paracrine communication (Adusumalli *et al.,* 2024). These data, integrated with cell–cell communication maps, have provided a significant understanding of host–graft interactions and regenerative niches (Kanemaru *et al.,* 2023). Meanwhile, spatial and electrophysiological signals are also being employed in tissue engineering to promote cardiomyocyte alignment and self-assembly, improving the electrophysiological integration and functional maturation of the engineered cardiac construct (Pahnke *et al.,* 2014; Liau *et al.,* 2012). These integrative multi-omics platforms permit the refinement of the regenerative technique through correlation of molecular signatures and tissue-level function, promoting the rise of clinically feasible cardiac repair modalities (Vollert *et al.,* 2015).

7.4 Spatial Validation of Cardiac Organoids and Heart-on-Chip Models

Cardiac organoid and heart-on-chip platforms' validation by spatial multi-omics is a key innovation in preclinical cardiovascular modelling. An integration of scRNA-seq and spatial transcriptomics provides fine mapping of cellular architecture, lineage specification, and signal microenvironments of organoid systems (Lv *et al.,* 2025). These technologies have revealed congenital heart illness-related regulatory networks, providing organoid-based illness models with mechanistic fidelity (Tirgar *et al.,* 2025; Cimenci, 2025). Cardiac organoid microenvironments are additionally defined by spatially resolved multi-omics, where structural and functional heterogeneity is unveiled and reflects in vivo cardiac physiology (Palmer *et al.,* 2024; Kanemaru *et al.,* 2023). This spatial validation procedure adds translational credibility to organoids and microphysiological systems for screening of drugs, toxicity, and personalized application, ensuring the proper recapitulation of human cardiac biology by in vitro cardiac models.

8. CHALLENGES AND LIMITATIONS

Within the area of spatial multi-omics for cardiology, technical and methodological limitations nevertheless hinder broad progress. Perhaps the most significant hurdle involves the normalization of tissue preservation, sectioning, and analytical methods, which affects the reproduction and interstudy comparability. Attainment of detailed molecular imaging with the concomitant preservation of the spatial and anatomical content of the cardiac tissues represents the current technical limitation, given the prevailing approaches' trade of molecular detail for spatial fidelity (Kiessling & Kuppe, 2024). Complementarity of multimodal omics data-the transcriptomics, proteomics, and metabolomics-together requires ultra-advanced computer frameworks capable of handling the vast, heterogeneous data scales (Kiessling & Kuppe, 2024; Liu *et al.,* 2024a). Solutions-related limitations of the compatibility of the sample, the throughput, and the processing power limit further the applicability of the most advanced technologies of single-cell and spatial profiling (Deng *et al.,* 2023). These latter are crucial for the proper spatial mapping of the pathophysiology of the cardioneurovascular system and for the application of these technologies to the clinical models of interest.

Computational complexity of the inherent high-dimensional spatial omics data of CVD represents another major challenge. Integrating data of various omics modalities gathered from the same tissue section to gain insight into molecular circuits of the disease progression represents the key to unraveling the complexities of the data, but this requires strong data harmonization and advanced analytical pipelines (Kiessling & Kuppe, 2024). Feature selection and biomarker identification are frustrated by the high dimensionality of the spatial omics data, rendering the application of advanced statistical inference, dimensionality reduction, and machine learning algorithms indispensable (Maiorino & Loscalzo, 2023). Again, the vast data volumes generated myriad of which reach terabyte sizes, are computation-intensive for data storing and data scalability, and data mining, necessitating the application of optimized deep learning architectures and big-data infrastructure (Shah & Gerszten, 2024; Kaur *et al.,* 2021). Lack of standardization of the experimental protocol and data analytical pipeline still represents an obstacle, limiting the ability to reproduce the results by laboratories and studies (Kiessling & Kuppe, 2024). Integrating the spatial omics data and clinical phenotypic parameters requires additional sophistication, and this necessitates new computation strategies of aligning molecular features and patient-level endpoints (Shah & Gerszten, 2024).

At the translational level, workflow normalization and clinical validation of cardiology-based spatial omics are still challenged by biological and operational issues. These include the intrinsic heterogeneity of cardiovascular illness and the requirement of tightly annotated phenotypic and multi-omics data, which make the definition of clinically tractable subtypes difficult (Dezem *et al.,* 2024; Maiorino & Loscalzo, 2023). Furthermore, financial and technological limitations make the frequent application of spatial omics throughout the workflow of diagnostics or prognosis undesirable, thus limiting their clinical scalability (Pucciarelli *et al.,* 2023; Palmer *et al.,* 2024). These limitations individually and cumulatively deter the incorporation of spatial omics into precision paradigms of cardiology and the validation of the technologies as clinical-grade tools.

Third, the economic and access constraints inherent to the use of spatial multi-omics technologies are key equity barriers to broad application. Computer-intensive technologies of the caliber of spatial transcriptomics and imaging mass spectrometry are costly to deploy, posing access barriers both to clinicians and to researchers, and especially so to LMICs (Francisco *et al.,* 2024; Nguyen *et al.,* 2021). Use of personalized medicine strategies founded on these technologies favors the high-income nations disproportionately, thus entrenching inequities of the global sphere towards access to advanced cardiovascular therapeutic and diagnostic instruments (Francisco *et al.,* 2024). The instruments of too sophisticated in operational complexity and too hefty in maintenance and data handling expenses, thus limiting applicability on the broader front (Schumacher & Kramann, 2023; Nguyen *et al.,* 2021). Overcoming these fiscal and infrastructural inequities by strategic investment and shared platforms remains key to the scaling of use of spatial multi-omics technologies the world over, particularly towards cardiovascular exploration (Varghese *et al.,* 2021; Schonrock *et al.,* 2022).

9. FUTURE PERSPECTIVES

Integrating AI-driven digital pathology with spatial multi-omics represents a transformative advancement in cardiology, offering unprecedented insights into CVD mechanisms at the molecular and tissue levels. Spatial multi-omics encompasses transcriptomic, epigenomic, proteomic, and metabolomic profiling within intact tissue architecture, enabling comprehensive cellular mapping and contextual understanding of disease progression (Kiessling & Kuppe, 2024). The incorporation of AI technologies, particularly deep learning, enhances this integration by enabling high-throughput, automated analysis of spatial data and correlating molecular signatures with histopathological morphology (Fu & Chen, 2024; Yamada *et al.,* 2022). Tools such as Pix2Path, which utilize conditional generative adversarial networks, facilitate predictive modelling between histological patterns and gene expression landscapes, allowing the exploration of disease mechanisms without predefined conditions (Fu & Chen, 2024). Furthermore, the emerging digitalomics framework unites AI/ML methodologies with multi-omics datasets, integrating information from medical imaging and electronic health records to advance precision medicine and individualized cardiovascular risk prediction (Nomura *et al.,* 2025). Complementing this, SpatialGlue, a graph neural network-based approach, efficiently aggregates cell-type data into spatial domains, providing high-resolution visualization of cross-modality correlations critical for decoding cardiac tissue complexity (Long *et al.,* 2023b). Although challenges persist in data standardization, harmonization, and computational scalability, these convergent approaches promise to revolutionize cardiac diagnostics and therapeutic development through the identification of novel biomarkers and molecular targets (Kiessling & Kuppe, 2024; Long *et al.,* 2023).

Integration of multi-scale systems biology solidifies the promise of spatial multi-omics for the field of cardiology. By integrating genomics, transcriptomics, proteomics, and metabolomics, investigators are able to untangle the complex biological networks of CVDs, including cardiomyopathies and atherosclerosis (Arneson *et al.,* 2017; Doran *et al.,* 2021). These integrative approaches allow for the discovery of molecular circuitry of disease and new therapeutic targets and provide the foundation upon which precision interventions, matched to patient-specific molecular and phenotypic signatures, may be developed and deployed (Reitz *et al.,* 2023; Ghram & Lavie, 2025). Nevertheless, interrogation of these heterogeneous, higher-dimensional data requires next-generation computation paradigms capable of integrating multi-omics data along biological hierarchies-from molecular through organ scales-a challenge of translational systems biology today (Joshi *et al.,* 2021; Doran *et al.,* 2021). Future developments of algorithmic modelling and data interpretation through AI are the key to the full realization of these multi-scale strategies to enhance CVD management and outcomes (Ghram & Lavie, 2025).

The construction of the human heart spatial atlas represents a key resource for the translation of spatial multi-omics observations to clinical cardiology. This atlas combines multi-omic data to reveal the molecular and cellular structure of the heart, creating the foundation for understanding mechanisms of CVD and identifying novel therapeutic targets (Kiessling & Kuppe, 2024). Spatially resolved atlases, as illustrated by Cranley et al., reveal gene regulatory networks and cellular niches of importance for congenital heart disease and cardiadothelial cell developmental processes (Cranley *et al.,* 2023). Furthermore, combining these atlases with experimental systems involving primary human cardiomyocytes raises predictive precision for the effects of drugs, increasing the translational fidelity of preclinical tests of the heart (Abi-Gerges *et al.,* 2020). Lombaert et al. produced the cardiac fiber architecture atlas, and these datasets are supplemented by those detailing myocardial orientation of the myocardial fibers, which are key for the fine-tuning of the computation-based cardiac models and simulation of electrophysiological propensities (Lombaert *et al.,* 2012). In addition, spatial multi-omic maps of myocardial infarction, e.g., those illustrated by Kuppe et al., define post-injury processes of remodelling, providing new perspectives on understanding the processes of fibrosis and repair, key towards developing regenerative treatments. These resources, individually, build the foundation of integrative, data-based translational cardiology (Kuppe *et al.,* 2022).

Integration of spatial multi-omics and the most current imaging modalities, such as MRI, CT, and PET, can redefine the therapeutic and diagnostic strategies of cardiovascular medicine. Combining molecular-level omics data and high-resolution imaging enables a detailed description of the pathophysiology of the heart and the correlating structural and metabolic abnormalities and genetic predispositions of the same (Gupta *et al.,* 2025). PET/MRI hybrid modalities, for example, permit improved spatial and temporal co-registration, leading to improved fidelity of the image and quantitative precision (Stegger *et al.,* 2013; Nekolla *et al.,* 2015). Multimodal integration of this kind enables real-time correlation of tissue-level molecular modifications and physiologic parameters, such as myocardial metabolism and perfusion, and, hence, precision-guided interventions (Manabe *et al.,* 2020). Incorporation of AI and machine learning algorithms into imaging protocols further enhances imaging reconstruction, segmentation, and predictive analytics, eventually increasing the dynamic treatment optimization and the diagnostic accuracy of the field of cardiology (Subhalakshmi *et al.,* 2025).

Personalized mapping of CVD through spatial multi-omics will permit real-time clinical judgment and precision stratification of patients. Single-cell and spatial genomics technologies, aligned, permit detailed spatial mapping of myocardial injury, repair, and remodelling, outlining cell populations and pathways propelling the disease and typically overlooked by classical analyses (Schumacher & Kramann, 2023). Integrating these data through broad clinical phenotyping facilitates genomically guided judgment of risk and personalized therapeutic design, consonant with the paradigm of precision cardiology (Leopold, 2022; Suchkov *et al.,* 2024). Moreover, the application of systems biology paradigms to the interpretation of multi-omics data expands the ability to model cardiovascular function in dynamic terms, driving the field toward an integrated, network-based perspective on cardiopulmonary health and disease (Björnson *et al.,* 2016; Kirchhof *et al.,* 2014). Through these new technologies, the integration of AI, spatial multi-omics, and clinical data will construct the foundation of predictive, preventive, and personalized cardiology in the immediate future.

10. CONCLUSION

Spatial multi-omics is transforming cardiovascular inquiry by offering unrivaled molecular and spatial resolution of the cardiac tissues. By integrating the multi-omics of genomics, transcriptomics, proteomics, metabolomics, and epigenomics, within an anatomical foundation, it gives a systems-level understanding of the mechanisms of cardiovascular illness. This systems methodology has unraveled cellular heterogeneity, interactions of the immune system and stroma, and metabolic modifications underlying atherosclerosis, myocardial infarction, and cardiomyopathies. Combining the strength of AI-enabled digital pathology and spatial omics accelerates data interpretation, enabling predictive modelling and biomarker discovery for personalized medicine. Human heart spatial atlases, along with multimodal imaging such as MRI and PET, provide an integrative foundation to translational cardiology, correlating molecular phenotypes and clinical symptoms. Challenges exist, however, in the areas of data standardization, scalability, and computation intensity, but advancing graph-based models and higher-order imaging integration are continuously expanding analytical precision. With the growth of spatial multi-omics, predictive, preventive, and personalized cardiovascular care shall become a reality by bridging the distance between the discovery of molecular mechanisms and clinical application to transform diagnostics, therapeutics, and regenerative medicine.

ACKNOWLEDGEMENT

All the authors are thankful to their respective Universities and Institutions for their valuable supports.

DATA AVAILABILITY

The authors confirm that the data supporting the findings of this study are available within the book/chapter.

REFERENCES

Abi-Gerges, N., Miller, P. E., & Ghetti, A. (2020). Human heart cardiomyocytes in drug discovery and research: New opportunities in translational sciences. *Current Pharmaceutical Biotechnology*, *21*(9), 787–806. DOI: 10.2174/1389201021666191210142023 PMID: 31820682

Adusumalli, S., Lim, S., Ren, V., Chong, L. Y., Tan, C., Tham, R., & Yap, L. (2023). Spatiotemporal transcriptomes of pig hearts reveal midkine-mediated vascularization in a chronic myocardial infarcted model. bioRxiv, 2023-06.

Adusumalli, S., Lim, S., Ye, L., lim, E., Ren, S., Wang, Y., Petretto, E., Tryggvason, K., & Yap, L. (2024). Abstract We008: Hallmark maturation of the human pluripotent stem cells-derived cardiovascular progenitors in myocardial infarcted large animal model. *Circulation Research*, *135*(Suppl_1), AWe008–AWe008. DOI: 10.1161/res.135.suppl_1.We008

Alagendran, S., Archunan, G., Prabhu, S. V., Orozco, B. E., & Guzman, R. G. (2010). Biochemical evaluation in human saliva with special reference to ovulation detection. *Indian Journal of Dental Research*, *21*(2), 165–168. DOI: 10.4103/0970-9290.66625 PMID: 20657081

Andreotti, F., Iervolino, A., Navarese, E. P., Maggioni, A. P., Crea, F., & Scambia, G. (2021). Precision phenomapping of acute coronary syndromes to improve patient outcomes. *Journal of Clinical Medicine*, *10*(8), 1755. DOI: 10.3390/jcm10081755 PMID: 33919478

Annink, M. E., Kraaijenhof, J. M., Stroes, E. S., & Kroon, J. (2024). Moving from lipids to leukocytes: Inflammation and immune cells in atherosclerosis. *Frontiers in Cell and Developmental Biology*, *12*, 1446758. DOI: 10.3389/fcell.2024.1446758 PMID: 39161593

Aon, M. A., & Cortassa, S. (2015). Systems biology of the fluxome. *Processes (Basel, Switzerland)*, *3*(3), 607–618. DOI: 10.3390/pr3030607

Arneson, D., Shu, L., Tsai, B., Barrere-Cain, R., Sun, C., & Yang, X. (2017). Multidimensional integrative genomics approaches to dissecting cardiovascular disease. *Frontiers in Cardiovascular Medicine*, *4*, 8. DOI: 10.3389/fcvm.2017.00008 PMID: 28289683

Arumugam, V. A., Thangavelu, S., Fathah, Z., Ravindran, P., Sanjeev, A. M. A., Babu, S., Arun, M., Mohd. Iqbal, Y., Khan, S., Ruchi, T., Megha Katare, P., Ranjit, S., Chandra, R., & Dhama, K. (2020). COVID-19 and the world with co-morbidities of heart disease, hypertension and diabetes. *Journal of Pure & Applied Microbiology*, *14*(3), 1623–1638. DOI: 10.22207/JPAM.14.3.01

Beetz, M., Banerjee, A., Sang, Y., & Grau, V. (2022). Combined generation of electrocardiogram and cardiac anatomy models using multi-modal variational autoencoders. In *2022 IEEE 19th International Symposium on Biomedical Imaging (ISBI)* (pp. 1-4). IEEE. DOI: 10.1109/ISBI52829.2022.9761590

Bhotla, H. K., Balasubramanian, B., Meyyazhagan, A., Pushparaj, K., & Easwaran, M. (2021). Opportunistic mycoses in COVID-19 patients/survivors: Epidemic inside a pandemic. *Journal of Infection and Public Health*, *14*(11), 1720–1726. DOI: 10.1016/j.jiph.2021.10.010 PMID: 34700291

Björnson, E., Borén, J., & Mardinoglu, A. (2016). Personalized cardiovascular disease prediction and treatment review of existing strategies and novel systems medicine tools. *Frontiers in Physiology*, *7*, 2. DOI: 10.3389/fphys.2016.00002 PMID: 26858650

Bleckwehl, T., Maryam, S., Babler, A., Nyberg, M., Bosteen, M., Halder, M., & Hayat, S. (2023). Encompassing view of spatial and single-cell RNA-seq renews the role of the microvasculature in human atherosclerosis. bioRxiv, 2023-12.

Boukens, B. J., & Christoffels, V. M. (2012). Electrophysiological patterning of the heart. *Pediatric Cardiology*, *33*(6), 900–906. DOI: 10.1007/s00246-012-0237-4 PMID: 22367553

Caufield, J. H., Sigdel, D., Fu, J., Choi, H., Guevara-Gonzalez, V., Wang, D., & Ping, P. (2022). Cardiovascular informatics: Building a bridge to data harmony. *Cardiovascular Research, 118*(3), 732–745. DOI: 10.1093/cvr/cvab067 PMID: 33751044

Chan, A. S. F., Greiner, J., Brennan, T., Agrawal, A., Hemmer, H., Sinning, K., & Grün, D. (2024). Spatio-temporal dynamics of the fibrotic niche in cardiac repair. bioRxiv, 2024-11.

Chapman, E. A., Aballo, T. J., Melby, J. A., Zhou, T., Price, S. J., Rossler, K. J., & Ge, Y. (2023). Defining the sarcomeric proteoform landscape in ischemic cardiomyopathy by top-down proteomics. *Journal of Proteome Research, 22*(3), 931–941. DOI: 10.1021/acs.jproteome.2c00729 PMID: 36800490

Chen, S., Zhu, B., Huang, S., Hickey, J. W., Lin, K. Z., Snyder, M., & Ma, Z. (2023). Integration of spatial and single-cell data across modalities with weak linkage. *BioRxiv*.

Christodoulou, A. G., Shaw, J. L., Nguyen, C., Yang, Q., Xie, Y., Wang, N., & Li, D. (2018). Magnetic resonance multitasking for motion-resolved quantitative cardiovascular imaging. *Nature Biomedical Engineering, 2*(4), 215–226. DOI: 10.1038/s41551-018-0217-y PMID: 30237910

Cimenci, C. E. (2025). *Engineering the beating heart: cardiac organoid models for tissue regeneration*. IntechOpen.

Cranley, J., Bayraktar, S., Kanemaru, K., Knight-Schrijver, V., Pett, J. P., Davaapil, H., & Teichmann, S. (2023). A spatially-resolved multiomic cell atlas reveals gene regulatory networks underlying cell specification in the developing human heart. *European Heart Journal*, *44*(Supplement_2), ehad655–3042. DOI: 10.1093/eurheartj/ehad655.3042

Cranley, J., Kanemaru, K., Bayraktar, S., Knight-Schrijver, V., Pett, J. P., Polanski, K., & Teichmann, S. A. (2024). Multiomic analysis reveals developmental dynamics of the human heart in health and disease. BioRxiv, 2024-04.

Cui, M., & Zhang, D. Y. (2021). Artificial intelligence and computational pathology. *Laboratory Investigation*, *101*(4), 412–422. DOI: 10.1038/s41374-020-00514-0 PMID: 33454724

Deng, Y., Bai, Z., & Fan, R. (2023). Microtechnologies for single-cell and spatial multi-omics. *Nature Reviews Bioengineering*, *1*(10), 769–784. DOI: 10.1038/s44222-023-00084-y

Deng, Y., Bartosovic, M., Ma, S., Zhang, D., Liu, Y., Qin, X., & Fan, R. (2021). Spatial-ATAC-seq: spatially resolved chromatin accessibility profiling of tissues at genome scale and cellular level. bioRxiv, 2021-06.

Dezem, F. S., Arjumand, W., DuBose, H., Morosini, N. S., & Plummer, J. (2024). Spatially resolved single-cell omics: Methods, challenges, and future perspectives. *Annual Review of Biomedical Data Science*, *7*(1), 131–153. DOI: 10.1146/annurev-biodatasci-102523-103640 PMID: 38768396

Ding, Y., Almasian, M., Zhang, X., Kadam, V., & Yuan, J. (2023). Abstract P1180: Multi-scale imaging and analysis to elucidate cardiac structure and contractile function. *Circulation Research*, *133*(Suppl_1), AP1180–AP1180. DOI: 10.1161/res.133.suppl_1.P1180

Doran, S., Arif, M., Lam, S., Bayraktar, A., Turkez, H., Uhlen, M., & Mardinoglu, A. (2021). Multi-omics approaches for revealing the complexity of cardiovascular disease. *Briefings in Bioinformatics*, *22*(5), bbab061. DOI: 10.1093/bib/bbab061 PMID: 33725119

Efremova, M., Vento-Tormo, M., Teichmann, S. A., & Vento-Tormo, R. (2020). CellPhoneDB: Inferring cell–cell communication from combined expression of multi-subunit ligand–receptor complexes. *Nature Protocols*, *15*(4), 1484–1506. DOI: 10.1038/s41596-020-0292-x PMID: 32103204

Farrugia, G., Harper, R., Krstevski, C., Lelliott, P., Hsu, I., Dona, M., & Pinto, A. (2024). Abstract P355: Multiomic analyses reveal sex-specific mechanisms governing cardiac fibrosis. *Hypertension*, *81*(Suppl_1), AP355–AP355. DOI: 10.1161/hyp.81.suppl_1.P355

Feng, Y., Cai, L., Hong, W., Zhang, C., Tan, N., Wang, M., & Jiang, L. (2022). Rewiring of 3D chromatin topology orchestrates transcriptional reprogramming and the development of human dilated cardiomyopathy. *Circulation*, *145*(22), 1663–1683. DOI: 10.1161/CIRCULATIONAHA.121.055781 PMID: 35400201

Forte, E., McLellan, M. A., Skelly, D. A., & Rosenthal, N. A. (2021). Ex uno, plures–from one tissue to many cells: A review of single-cell transcriptomics in cardiovascular biology. *International Journal of Molecular Sciences*, *22*(4), 2071. DOI: 10.3390/ijms22042071 PMID: 33669808

Fox, A. G., Buonodono, K. E., Jones, A. R., Thomas, M., Schepers, L. E., Earl, C. C., & Crouch, A. C. (2025). Multimodal study of murine cardiovascular remodeling: Four-dimensional ultrasound and mass spectrometry imaging. *Journal of Visualized Experiments*, (215), e67347. DOI: 10.3791/67347 PMID: 39878950

Francisco, K. K. Y., Apuhin, A. E. C., Tan, M. J. T., Byers, M. C., Maravilla, N. M. A. T., Karim, H. A., & AlDahoul, N. (2024). Can personalized medicine coexist with health equity? Examining the cost barrier and ethical implications. *arXiv preprint arXiv:2411.02307*.

Frey, N., & Olson, E. N. (2000). Making matters worse for a broken heart. *The Journal of Clinical Investigation, 106*(12), 1437–1439. DOI: 10.1172/JCI11733 PMID: 11120749

Fu, X., & Chen, Y. (2024). Pix2Path: integrating spatial transcriptomics and digital pathology with deep learning to score pathological risk and link gene expression to disease mechanisms. bioRxiv, 2024-08.

Getz, G. S., & Reardon, C. A. (2014). The mutual interplay of lipid metabolism and the cells of the immune system in relation to atherosclerosis. *Clinical Lipidology, 9*(6), 657–671. DOI: 10.2217/clp.14.50 PMID: 25705263

Ghram, A., & Lavie, C. J. (2025). Integrating omics to optimize precision cardiac rehabilitation. *Metabolism and Target Organ Damage, 5*(3), N-A.

Gokhale, T. A., Asfour, H., Verma, S., Bursac, N., & Henriquez, C. S. (2018). Microheterogeneity-induced conduction slowing and wavefront collisions govern macroscopic conduction behavior: A computational and experimental study. *PLoS Computational Biology, 14*(7), e1006276. DOI: 10.1371/journal.pcbi.1006276 PMID: 30011279

Gómez, A., Tacheau, A., Le Floc'h, S., Pettigrew, R. I., Cloutier, G., Finet, G., & Ohayon, J. (2021). Intravascular ultrasound imaging of human coronary atherosclerotic plaque: novel morpho-elastic biomarkers of instability. In *Biomechanics of coronary atherosclerotic plaque* (pp. 465–489). Academic Press. DOI: 10.1016/B978-0-12-817195-0.00020-2

Gundappa, M., Arumugam, V. A., Hsieh, H. L., Balasubramanian, B., & Shanmugam, V. (2022). Expression of tissue factor and TF-mediated integrin regulation in HTR-8/SVneo trophoblast cells. *Journal of Reproductive Immunology, 150*, 103473. DOI: 10.1016/j.jri.2022.103473 PMID: 35030354

Gupta, R. M., Schnitzler, G. R., Fang, S., Lee-Kim, V. S., & Barry, A. (2023). Multiomic analysis and CRISPR perturbation screens identify endothelial cell programs and novel therapeutic targets for coronary artery disease. *Arteriosclerosis, Thrombosis, and Vascular Biology, 43*(5), 600–608. DOI: 10.1161/ATVBAHA.123.318328 PMID: 36994731

Gupta, T., Aggarwal, S. K., & Mathur, T. (2025). Integrative imaging and molecular profiling: unveiling the nexus of genomics and precision medicine. In *Computer-Assisted Analysis for Digital Medicinal Imagery* (pp. 183–208). IGI Global.

Hajdarpašić, A., Tukker, M., Rijdt, W. T., Mohamedhoesein, S., Meijers, W. C., & Caliskan, K. (2025). Epigenetics of cardiomyopathies: The next frontier. *Heart Failure Reviews*, *30*(1), 257–270. DOI: 10.1007/s10741-024-10460-4 PMID: 39586986

Hamdani, N., Kooij, V., Van Dijk, S., Merkus, D., Paulus, W. J., Remedios, C. D., & van der Velden, J. (2008). Sarcomeric dysfunction in heart failure. *Cardiovascular Research*, *77*(4), 649–658. DOI: 10.1093/cvr/cvm079 PMID: 18055579

Han, S., Xu, Q., Du, Y., Tang, C., Cui, H., Xia, X., & Shang, H. (2024). Single-cell spatial transcriptomics in cardiovascular development, disease, and medicine. *Genes & Diseases*, *11*(6), 101163. DOI: 10.1016/j.gendis.2023.101163 PMID: 39224111

Hayashi, H. (2010). Does remodelling of gap junctions and connexin expression contribute to arrhythmogenesis?–study in an immobilization rat model–. *Circulation Journal*, *74*(12), 2558–2559. DOI: 10.1253/circj.CJ-10-0983 PMID: 21088334

Hegenbarth, J. C., Lezzoche, G., De Windt, L. J., & Stoll, M. (2022). Perspectives on bulk-tissue RNA sequencing and single-cell RNA sequencing for cardiac transcriptomics. *Frontiers in Molecular Medicine*, *2*, 839338. DOI: 10.3389/fmmed.2022.839338 PMID: 39086967

Herrero-Fernandez, B., Gomez-Bris, R., Somovilla-Crespo, B., & Gonzalez-Granado, J. M. (2019). Immunobiology of atherosclerosis: A complex net of interactions. *International Journal of Molecular Sciences*, *20*(21), 5293. DOI: 10.3390/ijms20215293 PMID: 31653058

Hesse, K., Khanji, M. Y., Aung, N., Dabbagh, G. S., Petersen, S. E., & Chahal, C. A. A. (2024). Assessing heterogeneity on cardiovascular magnetic resonance imaging: A novel approach to diagnosis and risk stratification in cardiac diseases. *European Heart Journal Cardiovascular Imaging*, *25*(4), 437–445. DOI: 10.1093/ehjci/jead285 PMID: 37982176

Hu, K., Harman, A., & Baharlou, H. (2024). Imaging mass cytometry for in situ immune profiling. In *Flow Cytometry Protocols* (pp. 407–423). Springer US. DOI: 10.1007/978-1-0716-3738-8_19

Huan, C., Li, J., Li, Y., Zhao, S., Yang, Q., Zhang, Z., & Zhou, L. (2025). Spatially resolved multiomics: data analysis from monoomics to multiomics. *BME frontiers*, *6*, 0084.

Inocencio, I., Rai, A., Donner, D., & Greening, D. W. (2025). The proteomic landscape of the coronary accessible heart cell surfaceome. *Proteomics*, *25*(7), e202400320. DOI: 10.1002/pmic.202400320 PMID: 39790063

Jin, F., Li, J., Guo, J., Doeppner, T. R., Hermann, D. M., Yao, G., & Dai, Y. (2021). Targeting epigenetic modifiers to reprogramme macrophages in non-resolving inflammation-driven atherosclerosis. *European Heart Journal Open*, *1*(2), oeab022. DOI: 10.1093/ehjopen/oeab022 PMID: 35919269

Joshi, A., Rienks, M., Theofilatos, K., & Mayr, M. (2021). Systems biology in cardiovascular disease: A multiomics approach. *Nature Reviews. Cardiology*, *18*(5), 313–330. DOI: 10.1038/s41569-020-00477-1 PMID: 33340009

Jung, S. H., Hwang, B. H., Shin, S., Park, E. H., Park, S. H., Kim, C. W., & Chung, Y. J. (2022). Spatiotemporal dynamics of macrophage heterogeneity and a potential function of Trem2hi macrophages in infarcted hearts. *Nature Communications*, *13*(1), 4580. DOI: 10.1038/s41467-022-32284-2 PMID: 35933399

Kanemaru, K., Cranley, J., Muraro, D., Miranda, A. M., Ho, S. Y., Wilbrey-Clark, A., & Teichmann, S. A. (2023). Spatially resolved multiomics of human cardiac niches. *Nature*, *619*(7971), 801–810. DOI: 10.1038/s41586-023-06311-1 PMID: 37438528

Kaur, P., Singh, A., & Chana, I. (2021). Computational Techniques and Tools for Omics Data Analysis: State-of-the-Art, Challenges, and Future Directions. *Archives of Computational Methods in Engineering*, *28*(7), 4595–4631. DOI: 10.1007/s11831-021-09547-0 PMID: 33519179

Kiessling, P., & Kuppe, C. (2024). Spatial multi-omics: Novel tools to study the complexity of cardiovascular diseases. *Genome Medicine*, *16*(1), 14. DOI: 10.1186/s13073-024-01282-y PMID: 38238823

Kim, H., Choi, S., Heo, H., Cho, S. H., Lee, Y., Kim, D., & Rhee, S. (2025). Applications of single-cell omics technologies for induced pluripotent stem cell-based cardiovascular research. *International Journal of Stem Cells*, *18*(1), 37–48. DOI: 10.15283/ijsc23183 PMID: 39129179

Kirchhof, P., Sipido, K. R., Cowie, M. R., Eschenhagen, T., Fox, K. A., Katus, H., Schroeder, S., Schunkert, H., Priori, S., Alonso, A., Chezaubernard, C., Doevendans, P., Eschenhagen, T., Fox, K., Katus, H., Khder, Y., Kirchhof, P., Kramer, F., Kristensen, S., & Herttuala, S. Y. (2014). The continuum of personalized cardiovascular medicine: A position paper of the european society of cardiology. *European Heart Journal*, *35*(46), 3250–3257. DOI: 10.1093/eurheartj/ehu312 PMID: 25148837

Koelwyn, G. J., Corr, E. M., Erbay, E., & Moore, K. J. (2018). Regulation of macrophage immunometabolism in atherosclerosis. *Nature Immunology*, *19*(6), 526–537. DOI: 10.1038/s41590-018-0113-3 PMID: 29777212

Kuppe, C., Ramirez Flores, R. O., Li, Z., Hayat, S., Levinson, R. T., Liao, X., & Kramann, R. (2022). Spatial multi-omic map of human myocardial infarction. *Nature, 608*(7924), 766–777. DOI: 10.1038/s41586-022-05060-x PMID: 35948637

Kuswanto, W., Nolan, G., & Lu, G. (2023). Highly multiplexed spatial profiling with CODEX: Bioinformatic analysis and application in human disease. []). Berlin/ Heidelberg: Springer Berlin Heidelberg.]. *Seminars in Immunopathology*, *45*(1), 145–157. DOI: 10.1007/s00281-022-00974-0 PMID: 36414691

Lai, H. M., & Chan, S. L. (2024). A scalable 3D spatial multi-omics method for high-plex morpho-molecular imaging of millimeter-thick tissues. *Journal of Clinical Oncology*, *42*(23), 162–162. DOI: 10.1200/JCO.2024.42.23_suppl.162

Laird, J., Perera, G., Batorsky, R., Wang, H., Arkun, K., & Chin, M. T. (2023). Spatial transcriptomic analysis of focal and normal areas of myocyte disarray in human hypertrophic cardiomyopathy. *International Journal of Molecular Sciences*, *24*(16), 12625. DOI: 10.3390/ijms241612625 PMID: 37628806

Lammers, W. J., Schalij, M. J., Kirchhof, C. J., & Allessie, M. A. (1990). Quantification of spatial inhomogeneity in conduction and initiation of reentrant atrial arrhythmias. *American Journal of Physiology. Heart and Circulatory Physiology*, *259*(4), H1254–H1263. DOI: 10.1152/ajpheart.1990.259.4.H1254 PMID: 1699438

Lázár, E., Mauron, R., Andrusivová, Ž., Foyer, J., He, M., Larsson, L., & Lundeberg, J. (2024). Spatial dynamics of the developing human heart. bioRxiv, 2024-03.

Lee, T. H., Liu, P. S., Wang, S. J., Tsai, M. M., Shanmugam, V., & Hsieh, H. L. (2021). Bradykinin, as a reprogramming factor, induces transdifferentiation of brain astrocytes into neuron-like cells. *Biomedicines*, *9*(8), 923. DOI: 10.3390/biomedicines9080923 PMID: 34440126

Leopold, J. A. (2022). Personalizing treatments for patients based on cardiovascular phenotyping. *Expert Review of Precision Medicine and Drug Development*, *7*(1), 4–16. DOI: 10.1080/23808993.2022.2028548 PMID: 36778892

Li, Y., & Luo, Y. (2023). Spatial transcriptomic cell-type deconvolution using graph neural networks. *bioRxiv*.

Liau, B., Zhang, D., & Bursac, N. (2012). Functional cardiac tissue engineering. *Regenerative Medicine*, *7*(2), 187–206. DOI: 10.2217/rme.11.122 PMID: 22397609

Litviňuková, M., Talavera-López, C., Maatz, H., Reichart, D., Worth, C. L., Lindberg, E. L., & Teichmann, S. A. (2020). Cells of the adult human heart. *Nature*, *588*(7838), 466–472. DOI: 10.1038/s41586-020-2797-4 PMID: 32971526

Liu, T., Li, X., Pan, X., Zhang, L., Liao, J., & Yin, M. (2023a). STGMVA: clustering, imputation, and integration for spatial resolved transcriptomics using spatio-temporal gaussian mixture variational autoencoder. DOI: 10.21203/rs.3.pex-2322/v1

Liu, X., Peng, T., Xu, M., Lin, S., Hu, B., Chu, T., & Wu, P. (2024a). Spatial multi-omics: Deciphering technological landscape of integration of multi-omics and its applications. *Journal of Hematology & Oncology*, *17*(1), 72. DOI: 10.1186/s13045-024-01596-9 PMID: 39182134

Liu, X., Wang, L., Wang, Y., Qiao, X., Chen, N., Liu, F., & Shen, H. (2024b). Myocardial infarction complexity: A multi-omics approach. *Clinica Chimica Acta*, *552*, 117680. DOI: 10.1016/j.cca.2023.117680 PMID: 38008153

Liu, X., Yin, K., Chen, L., Chen, W., Li, W., Zhang, T., & Zhou, Z. (2023b). Lineage-specific regulatory changes in hypertrophic cardiomyopathy unraveled by single-nucleus RNA-seq and spatial transcriptomics. *Cell Discovery*, *9*(1), 6. DOI: 10.1038/s41421-022-00490-3 PMID: 36646705

Liu, Y., Zhang, X., Yang, S., Zhou, Z., Tian, L., Li, W., & Wang, Z. (2023c). Integrated mass spectrometry imaging reveals spatial-metabolic alteration in diabetic cardiomyopathy and the intervention effects of ferulic acid. *Journal of Pharmaceutical Analysis*, *13*(12), 1496–1509. DOI: 10.1016/j.jpha.2023.08.011 PMID: 38223449

Llorens-Bobadilla, E., Zamboni, M., Marklund, M., Bhalla, N., Chen, X., Hartman, J., & Ståhl, P. L. (2022). Chromatin accessibility profiling in tissue sections by spatial ATAC. bioRxiv, 2022-07.

Lombaert, H., Peyrat, J. M., Croisille, P., Rapacchi, S., Fanton, L., Cheriet, F., & Ayache, N. (2012). Human atlas of the cardiac fiber architecture: Study on a healthy population. *IEEE Transactions on Medical Imaging*, *31*(7), 1436–1447. DOI: 10.1109/TMI.2012.2192743 PMID: 22481815

Long, X., Yuan, X., & Du, J. (2023a). Single-cell and spatial transcriptomics: Advances in heart development and disease applications. *Computational and Structural Biotechnology Journal*, *21*, 2717–2731. DOI: 10.1016/j.csbj.2023.04.007 PMID: 37181659

Long, Y., Ang, K. S., Liao, S., Sethi, R., Heng, Y., Zhong, C., & Chen, J. (2023b). Integrated analysis of spatial multi-omics with SpatialGlue. bioRxiv, 2023-04.

Luo, L., & Leeper, N. J. (2023). Spatial metabolomics and the vulnerable atherosclerotic plaque. *Arteriosclerosis, Thrombosis, and Vascular Biology*, *43*(9), 1636–1638. DOI: 10.1161/ATVBAHA.123.319739 PMID: 37534466

Lv, H., Sun, F., & Chen, Y. (2025). Decoding congenital heart disease: A multi-omic framework for cardiac lineage and regulatory dysfunction. *Frontiers in Cell and Developmental Biology*, *13*, 1659884. DOI: 10.3389/fcell.2025.1659884 PMID: 41049289

Maiorino, E., & Loscalzo, J. (2023). Phenomics and robust multiomics data for cardiovascular disease subtyping. *Arteriosclerosis, Thrombosis, and Vascular Biology*, *43*(7), 1111–1123. DOI: 10.1161/ATVBAHA.122.318892 PMID: 37226730

Mallikarjun, V., Yin, B., Caggiano, L. R., Blimbaum, S., Pavelec, C. M., Holmes, J. W., & Ewald, S. E. (2024). Automated spatially targeted optical micro proteomics identifies fibroblast-and macrophage-specific regulation of myocardial infarction scar maturation in rats. *Journal of Molecular and Cellular Cardiology*, *186*, 1–15. DOI: 10.1016/j.yjmcc.2023.10.005 PMID: 37951204

Manabe, O., Oyama-Manabe, N., & Tamaki, N. (2020). Positron emission tomography/MRI for cardiac diseases assessment. *The British Journal of Radiology*, *93*(1113), 20190836. DOI: 10.1259/bjr.20190836 PMID: 32023123

McAloon, C. J., Osman, F., Glennon, P., Lim, P. B., & Hayat, S. A. (2016). Global epidemiology and incidence of cardiovascular disease. In *Cardiovascular Diseases* (pp. 57–96). Academic Press. DOI: 10.1016/B978-0-12-803312-8.00004-5

Mensah, G. A., Moran, A. E., Roth, G. A., & Narula, J. (2014). The global burden of cardiovascular diseases, 1990–2010. *Global Heart*, *9*(1), 183. DOI: 10.1016/j.gheart.2014.01.008 PMID: 25432128

Meyyazhagan, A., Balasubramanian, B., Easwaran, M., Alagamuthu, K. K., Shanmugam, S., Kuchi Bhotla, H., Pappusamy, M., Arumugam, V. A., Thangaraj, A., Kaul, T., Keshavarao, S., & Cacabelos, R. (2020). Biomarker study of the biological parameter and neurotransmitter levels in autistics. *Molecular and Cellular Biochemistry*, *474*(1-2), 277–284. DOI: 10.1007/s11010-020-03851-2 PMID: 32740790

Meyyazhagan, A., Pushparaj, K., Balasubramanian, B., Kuchi Bhotla, H., Pappusamy, M., Arumugam, V. A., Easwaran, M., Pottail, L., Mani, P., Tsibizova, V., & Di Renzo, G. C. (2022). COVID-19 in pregnant women and children: Insights on clinical manifestations, complexities, and pathogenesis. *International Journal of Gynaecology and Obstetrics: the Official Organ of the International Federation of Gynaecology and Obstetrics*, *156*(2), 216–224. DOI: 10.1002/ijgo.14007 PMID: 34735717

Mitre, L., Burkard, A., Hansen, S., Li, X., Ehret, F., Heitlinger, S., & Westermann, D. (2023). Changes in the immune cell landscape in the infarcted mouse heart. *European Heart Journal, 44*(Supplement_2), ehad655–3212. DOI: 10.1093/eurheartj/ehad655.3212

Mohd, Y., Balasubramanian, B., Meyyazhagan, A., Kuchi Bhotla, H., Shanmugam, S.K., Ramesh Kumar, M.K., Pappusamy, M., Alagamuthu, K.K., Keshavarao, S., & Arumugam, V.A. (2021). Extricating the association between the prognostic factors of colorectal cancer. *J Gastrointest Cancer, 52*(3), 022-1028.

Mohd, Y., Kumar, P., Kuchi Bhotla, H., Meyyazhagan, A., Balasubramanian, B., Ramesh Kumar, M. K., Pappusamy, M., Alagamuthu, K. K., Orlacchio, A., Keshavarao, S., Sampathkumar, P., & Arumugam, V. A. (2021). Transmission jeopardy of adenomatosis polyposis coli and methylenetetrahydrofolate reductase in colorectal cancer. *Journal of the Renin-Angiotensin-Aldosterone System : JRAAS, 2021*, 2021. DOI: 10.1155/2021/7010706 PMID: 34956401

Mohenska, M., Tan, N. M., Tokolyi, A., Furtado, M. B., Costa, M. W., Perry, A. J., & Ramialison, M. (2022). 3D-cardiomics: A spatial transcriptional atlas of the mammalian heart. *Journal of Molecular and Cellular Cardiology, 163*, 20–32. DOI: 10.1016/j.yjmcc.2021.09.011 PMID: 34624332

Moiz, B., Li, A., Padmanabhan, S., Sriram, G., & Clyne, A. M. (2022). Isotope-assisted metabolic flux analysis: A powerful technique to gain new insights into the human metabolome in health and disease. *Metabolites, 12*(11), 1066. DOI: 10.3390/metabo12111066 PMID: 36355149

Nekolla, S. G., Rischpler, C., & Kunze, K. P. (2015). PET/MRI for cardiac imaging: Technical considerations and potential applications. In *Molecular and Multimodality Imaging in Cardiovascular Disease* (pp. 29–48). Springer International Publishing. DOI: 10.1007/978-3-319-19611-4_3

Nguyen, N., Souza, T., Verheijen, M. C., Gmuender, H., Selevsek, N., Schlapbach, R., & Jennen, D. (2021). Translational proteomics analysis of anthracycline-induced cardiotoxicity from cardiac microtissues to human heart biopsies. *Frontiers in Genetics, 12*, 695625. DOI: 10.3389/fgene.2021.695625 PMID: 34211507

Nomura, A., Takeji, Y., Shimojima, M., & Takamura, M. (2025). Digitalomics: Towards artificial intelligence/machine learning-based precision cardiovascular medicine. *Circulation Journal*, •••, CJ-24. DOI: 10.1253/circj.CJ-24-0865 PMID: 39894532

Nurminen, V., Ravindran, A., Valkonen, M., Örd, T., Mikkola, L., Lönnberg, T., & Kaikkonen, M. (2023). Disease-associated cell states in atherosclerosis defined by spatial transcriptomics. *Atherosclerosis*, *379*, S15. DOI: 10.1016/j.atherosclerosis.2023.06.725

Ohtsuki, S. (2012). Pharmacoproteomic approach by quantitative targeted proteomics. *Yakugaku Zasshi*, *132*(4), 479–487. DOI: 10.1248/yakushi.132.479 PMID: 22465926

Pahnke, A., Montgomery, M., & Radisic, M. (2014). Spatial and electrical factors regulating cardiac regeneration and assembly. *Biomaterials for Cardiac Regeneration*, 71-92.

Palmer, J. A., Rosenthal, N., Teichmann, S. A., & Litvinukova, M. (2024). Revisiting cardiac biology in the era of single cell and spatial omics. *Circulation Research*, *134*(12), 1681–1702. DOI: 10.1161/CIRCRESAHA.124.323672 PMID: 38843288

Pang, Z., & Ye, L. (2022). Illuminating spatial pharmacology with in situ drug imaging. *Clinical and Translational Medicine*, *12*(8), e1024. DOI: 10.1002/ctm2.1024 PMID: 36030504

Paranitharan, N., Kataria, S., Arumugam, V. A., Hsieh, H., Muthukrishnan, S., & Velayuthaprabhu, S. (2025). Integrin α1 upregulation by tf:fviia complex promotes cervical cancer migration through par2-dependent mek1/2 activation. *Biochemical and Biophysical Research Communications*, *742*, 151151. DOI: 10.1016/j.bbrc.2024.151151 PMID: 39657349

Peisker, F., Halder, M., Nagai, J., Ziegler, S., Kaesler, N., Hoeft, K., & Kramann, R. (2022). Mapping the cardiac vascular niche in heart failure. *Nature Communications*, *13*(1), 3027. DOI: 10.1038/s41467-022-30682-0 PMID: 35641541

Pirillo, A., Bonacina, F., Norata, G. D., & Catapano, A. L. (2018). The interplay of lipids, lipoproteins, and immunity in atherosclerosis. *Current Atherosclerosis Reports*, *20*(3), 12. DOI: 10.1007/s11883-018-0715-0 PMID: 29445885

Prasad, A., & Tsimikas, S. (2008). Candidate biomarkers for the detection of coronary plaque destabilization and rupture. *Current Atherosclerosis Reports*, *10*(4), 309–317. DOI: 10.1007/s11883-008-0048-5 PMID: 18618976

Pucciarelli, D. M., Lu, B. Y., Zlobec, I., & DiStasio, M. (2023). Beyond the lab and into the hospital: An outlook on the clinical utility of spatial omics technologies. *GEN Biotechnology*, *2*(5), 360–371. DOI: 10.1089/genbio.2023.0030

Pushparaj, K., Kuchi Bhotla, H., Arumugam, V. A., Pappusamy, M., Easwaran, M., Liu, W. C., Issara, U., Rengasamy, K. R. R., Meyyazhagan, A., & Balasubramanian, B. (2022). Mucormycosis (black fungus) ensuing COVID-19 and comorbidity meets - magnifying global pandemic grieve and catastrophe begins. *The Science of the Total Environment, 805*, 150355. DOI: 10.1016/j.scitotenv.2021.150355 PMID: 34818767

Qu, Z., Hanna, P., Ajijola, O. A., Garfinkel, A., & Shivkumar, K. (2025). Ultrastructure and cardiac impulse propagation: Scaling up from microscopic to macroscopic conduction. *The Journal of Physiology, 603*(7), 1887–1901. DOI: 10.1113/JP287632 PMID: 39612369

Ramya, S., Poornima, P., Jananisri, A., Geofferina, I. P., Bavyataa, V., Divya, M., Priyanga, P., Vadivukarasi, J., Sujitha, S., Elamathi, S., Anand, A. V., & Balamuralikrishnan, B. (2023). Role of hormones and the potential impact of multiple stresses on infertility. *Stresses, 3*(2), 454–474. DOI: 10.3390/stresses3020033

Rangisetti, L. S., Shaik, F., Bhagavatula, A., & Kommareddy, L. S. (2024). Heart disease detection using graph neural networks (GNNs). In *2024 5th International Conference on Electronics and Sustainable Communication Systems (ICESC)* (pp. 1436-1441). IEEE.

Reisqs, J. B., Qu, Y. S., & Boutjdir, M. (2024). Ion channel trafficking implications in heart failure. *Frontiers in Cardiovascular Medicine, 11*, 1351496. DOI: 10.3389/fcvm.2024.1351496 PMID: 38420267

Reitz, C. J., Kuzmanov, U., & Gramolini, A. O. (2023). Multi-omic analyses and network biology in cardiovascular disease. *Proteomics, 23*(21-22), 2200289. DOI: 10.1002/pmic.202200289 PMID: 37691071

Sangeetha, T., Anand, A. V., & Begum, T. N. (2022). Assessment of inter-relationship between anemia and COPD in accordance with altitude. *The Open Respiratory Medicine Journal, 16*(1), e187430642206270. DOI: 10.2174/18743064-v16-e2206270 PMID: 37273961

Sangeetha, T., Nargis Begum, T., Balamuralikrishnan, B., Arun, M., Rengasamy, K. R. R., Senthilkumar, N., Velayuthaprabhu, S., Saradhadevi, M., Sampathkumar, P., & Vijaya Anand, A. (2022). Influence of SERPINA1 gene polymorphisms on anemia and chronic obstructive pulmonary disease. *Journal of the Renin-Angiotensin-Aldosterone System : JRAAS, 2022*, 2022. DOI: 10.1155/2022/2238320 PMID: 36320441

Schonrock, N., Callis, T., Hatchell, K., Esplin, E., Morales, A., Garcia, J., & Nussbaum, R. (2022). Precision Medicine Opportunities for Familial Arrhythmias and Cardiomyopathies Identified When Cost of Genetic Testing is Removed as a Barrier. *Heart Lung and Circulation*, *31*, S19. DOI: 10.1016/j.hlc.2022.04.041

Schumacher, D., & Kramann, R. (2023). Multiomic spatial mapping of myocardial infarction and implications for personalized therapy. *Arteriosclerosis, Thrombosis, and Vascular Biology*, *43*(2), 192–202. DOI: 10.1161/ATVBAHA.122.318333 PMID: 36579644

Seeley, E. H., Liu, Z., Yuan, S., Stroope, C., Cockerham, E., Rashdan, N. A., & Yurdagul, A.Jr. (2023). Spatially resolved metabolites in stable and unstable human atherosclerotic plaques identified by mass spectrometry imaging. *Arteriosclerosis, Thrombosis, and Vascular Biology*, *43*(9), 1626–1635. DOI: 10.1161/ATVBAHA.122.318684 PMID: 37381983

Severs, N. J. (2001). Gap junction remodelling and cardiac arrhythmogenesis: Cause or coincidence? *Journal of Cellular and Molecular Medicine*, *5*(4), 355–366. DOI: 10.1111/j.1582-4934.2001.tb00170.x PMID: 12067469

Severs, N. J. (2003). Communicating junctions, connexins and the cardiomyocyte: from cell biology to cardiology. In *Cardiac Remodelling and Failure* (pp. 417–434). Springer US. DOI: 10.1007/978-1-4419-9262-8_29

Shah, P. K. (2014). Biomarkers of plaque instability. *Current Cardiology Reports*, *16*(12), 547. DOI: 10.1007/s11886-014-0547-7 PMID: 25326730

Shah, S. H., & Gerszten, R. E. (2024). Sifting for Gold in Terabytes of Data: Illuminating Cardiovascular Biology in the 'Omics Age. *Circulation*, *149*(17), 1323–1325. DOI: 10.1161/CIRCULATIONAHA.123.066988 PMID: 38648274

Shen, L., Chen, W., Ding, J., Shu, G., Chen, M., Zhao, Z., & Ji, J. (2023). The role of metabolic reprogramming of oxygen-induced macrophages in the dynamic changes of atherosclerotic plaques. *The FASEB Journal*, *37*(3), e22791. DOI: 10.1096/fj.202201486R PMID: 36723768

Shi, H., El Kazzi, M., Liu, Y., Gao, A., Schroder, A. L., Vuong, S., & Witting, P. K. (2022). Multiplex analysis of mass imaging data: Application to the pathology of experimental myocardial infarction. *Acta Physiologica (Oxford, England)*, *235*(2), e13790. DOI: 10.1111/apha.13790 PMID: 35080155

Shi, Y., Johnson, J., Wang, B., Chen, B., Fisher, G. L., Urabe, G., & Li, L. (2019). Mass spectrometric imaging reveals temporal and spatial dynamics of bioactive lipids in arteries undergoing restenosis. *Journal of Proteome Research, 18*(4), 1669–1678. DOI: 10.1021/acs.jproteome.8b00941 PMID: 30784274

Slijkhuis, N., Towers, M., Claude, E., & Van Soest, G. (2025). MALDI versus DESI mass spectrometry imaging of lipids in atherosclerotic plaque. *Rapid Communications in Mass Spectrometry, 39*(1), e9927. DOI: 10.1002/rcm.9927 PMID: 39435741

Soliman, H., Tung, L. W., & Rossi, F. M. (2021). Fibroblast and myofibroblast subtypes: single cell sequencing. In *Myofibroblasts: Methods and Protocols* (pp. 49–84). Springer US. DOI: 10.1007/978-1-0716-1382-5_4

Stegger, L., Schülke, C., Wenning, C., Rahbar, K., Kies, P., Schober, O., & Schäfers, M. (2013). Cardiac PET/MRI. *Current Cardiovascular Imaging Reports, 6*(2), 169–178. DOI: 10.1007/s12410-012-9189-6

Subhalakshmi, R. T., Nivaashini, M., Ganesh, G. R., & Sathishkumar, A. (2025). Integrating AI into PET/CT and PET/MRI: A paradigm shift in hybrid imaging. In *AI Insights on Nuclear Medicine* (pp. 311-332). IGI Global Scientific Publishing.

Suchkov, S., Scherman, D., Wu, A., Borisova, A., Murphy, S., Smith, D., & Abe, H. (2024). Personalized and precision medicine (ppm) through the view of biomarker-driven applications to support their global impact in the upgraded healthcare-related services. *Journal of Biomedical and Engineering Research, 2*(2), 1–12. DOI: 10.64030/3065-8780.02.02.04

Tadros, R., Francis, C., Xu, X., Vermeer, A. M., Harper, A. R., Huurman, R., & Bezzina, C. R. (2021). Shared genetic pathways contribute to risk of hypertrophic and dilated cardiomyopathies with opposite directions of effect. *Nature Genetics, 53*(2), 128–134. DOI: 10.1038/s41588-020-00762-2 PMID: 33495596

Tan, S., Yang, J., Hu, S., & Lei, W. (2024). Cell-cell interactions in the heart: Advanced cardiac models and omics technologies. *Stem Cell Research & Therapy, 15*(1), 362. DOI: 10.1186/s13287-024-03982-z PMID: 39396018

Tang, Z., Alrumaihi, F., Alwanian, W. M., Alharbi, H. O., Allemailem, K. S., Alissa, M., & Chen, X. (2025). The future of cardiology: Integrating single-cell transcriptomics with multi-omics for enhanced cardiac disease insights. *Current Problems in Cardiology, 50*(4), 103005. DOI: 10.1016/j.cpcardiol.2025.103005 PMID: 39894239

Tirgar, P., Vikstrom, A., Sepúlveda, J. M. R., Srivastava, L. K., Amini, A., Tabata, T., & Ehrlicher, A. (2025). Heart-on-a-Miniscope: A miniaturized solution for electrophysiological drug screening in cardiac organoids. *Small*, *21*(6), 2409571. DOI: 10.1002/smll.202409571 PMID: 39937454

Troulé, K., Petryszak, R., Cakir, B., Cranley, J., Harasty, A., Prete, M., & Vento-Tormo, R. (2025). CellPhoneDB v5: Inferring cell–cell communication from single-cell multiomics data. *Nature Protocols*, *20*(12), 1–29. DOI: 10.1038/s41596-024-01137-1 PMID: 40133495

Ungvari, Z., Tarantini, S., Kiss, T., Wren, J. D., Giles, C. B., Griffin, C. T., & Csiszar, A. (2018). Endothelial dysfunction and angiogenesis impairment in the ageing vasculature. *Nature Reviews. Cardiology*, *15*(9), 555–565. DOI: 10.1038/s41569-018-0030-z PMID: 29795441

Varghese, M. S., Liu, C. L., & Kazi, D. S. (2021). The price of progress: Cost, access, and adoption of novel cardiovascular drugs in clinical practice. *Current Cardiology Reports*, *23*(11), 163. DOI: 10.1007/s11886-021-01598-w PMID: 34599393

Velayuthaprabhu, S., & Archunan, G. (2005). Evaluation of anticardiolipin antibodies and antiphosphatidylserine antibodies in women with recurrent abortion. *Indian Journal of Medical Sciences*, *59*(8), 347–352. DOI: 10.4103/0019-5359.16651 PMID: 16129928

Velayuthaprabhu, S., Archunan, G., & Balakrishnan, K. (2007). Placental thrombosis in experimental anticardiolipin antibodies-mediated intrauterine fetal death. *American Journal of Reproductive Immunology (New York, N.Y.)*, *57*(4), 270–276. DOI: 10.1111/j.1600-0897.2007.00474.x PMID: 17362388

Velayuthaprabhu, S., Matsubayashi, H., & Archunan, G. (2016). Beta-2 GPI induced tissue factor and placental apoptosis for the pathophysiology of pregnancy loss in antiphospholipid syndrome. *International Journal of Research in Medical Sciences*, *4*(8), 3109–3113. DOI: 10.18203/2320-6012.ijrms20162267

Velayuthaprabhu, S., Matsubayashi, H., Sugi, T., Nakamura, M., Ohnishi, Y., Ogura, T., & Archunan, G. (2013). Expression of apoptosis in placenta of experimental antiphospholipid syndrome (APS) mouse. *American Journal of Reproductive Immunology (New York, N.Y.)*, *69*(5), 486–494. DOI: 10.1111/aji.12084 PMID: 23398188

Velayuthaprabhu, S., Matsubayashi, H., Sugi, T., Nakamura, M., Ohnishi, Y., Ogura, T., Tomiyama, T., & Archunan, G. (2011). A unique preliminary study on fetal resorption and placental apoptosis in mice with passive immunization of anti-phosphatidylethanolamine antibodies and anti-Factor XII antibodies. *American Journal of Reproductive Immunology (New York, N.Y.)*, *66*(5), 373–384. DOI: 10.1111/j.1600-0897.2011.01008.x PMID: 21623987

Verrier, R. L., Nearing, B. D., & Zimetbaum, P. J. (2022). *U.S. Patent Application No. 17/607,711.*

Vollert, I., Eder, A., Hansen, A., & Eschenhagen, T. (2015). Engineering cardiovascular regeneration. *Current Stem Cell Reports*, *1*(2), 67–78. DOI: 10.1007/s40778-015-0010-8

Vujic, A., Robinson, E., Ito, M., Haider, S., Skepper, J., Methner, C., & Foo, R. (2012). An altered epigenomic mark predisposes to dilated cardiomyopathy through alternative splicing of sarcomeric gene expression in mice. *Circulation Research*, *111*(suppl_1), A100–A100. DOI: 10.1161/res.111.suppl_1.A100

Wang, L., Yang, Y., Ma, H., Xie, Y., Xu, J., Near, D., & Qian, L. (2022). Single-cell dual-omics reveals the transcriptomic and epigenomic diversity of cardiac non-myocytes. *Cardiovascular Research*, *118*(6), 1548–1563. DOI: 10.1093/cvr/cvab134 PMID: 33839759

Wang, M., Gu, M., Liu, L., Liu, Y., & Tian, L. (2021). Single-cell RNA sequencing (scRNA-seq) in cardiac tissue: Applications and limitations. *Vascular Health and Risk Management*, *17*, 641–657. DOI: 10.2147/VHRM.S288090 PMID: 34629873

Wang, R. S., Maron, B. A., & Loscalzo, J. (2023). Multiomics network medicine approaches to precision medicine and therapeutics in cardiovascular diseases. *Arteriosclerosis, Thrombosis, and Vascular Biology*, *43*(4), 493–503. DOI: 10.1161/ATVBAHA.122.318731 PMID: 36794589

Wang, X., & Connolly, T. M. (2010). Biomarkers of vulnerable atheromatous plaques: Translational medicine perspectives. *Advances in Clinical Chemistry*, *50*, 1–22. DOI: 10.1016/S0065-2423(10)50001-5 PMID: 20521438

Wünnemann, F., Sicklinger, F., Bestak, K., Nimo, J., Thiemann, T., Amrute, J., & Schapiro, D. (2024). Spatial omics of acute myocardial infarction reveals a novel mode of immune cell infiltration. BioRxiv, 2024-05.

Xia, T., Hu, L., Zuo, L., Cao, L., Zhang, Y., Xu, M., & Fang, S. (2024). ST-GEARS: Advancing 3D downstream research through accurate spatial information recovery. *Nature Communications*, *15*(1), 7806. DOI: 10.1038/s41467-024-51935-0 PMID: 39242563

Xiao, Q., Hou, R., Xie, L., Niu, M., Pan, X., & Zhu, X. (2023). Macrophage metabolic reprogramming and atherosclerotic plaque microenvironment: Fostering each other? *Clinical and Translational Medicine*, *13*(5), e1257. DOI: 10.1002/ctm2.1257 PMID: 37151159

Xie, H., Zhang, B., Xie, M., & Li, T. (2023). Circulating metabolic signatures of heart failure in precision cardiology. *Precision Clinical Medicine*, *6*(1), pbad005. DOI: 10.1093/pcmedi/pbad005 PMID: 37020642

Xie, Y., Ruan, F., Li, Y., Luo, M., Zhang, C., Chen, Z., & Tang, C. (2024). Spatial chromatin accessibility sequencing resolves high-order spatial interactions of epigenomic markers. *eLife*, *12*, RP87868. DOI: 10.7554/eLife.87868 PMID: 38236718

Xu, H., Wang, S., Fang, M., Luo, S., Chen, C., Wan, S., & Qu, K. (2023). SPACEL: Deep learning-based characterization of spatial transcriptome architectures. *Nature Communications*, *14*(1), 7603. DOI: 10.1038/s41467-023-43220-3 PMID: 37990022

Yamada, S., Ko, T., Hatsuse, S., Nomura, S., Zhang, B., Dai, Z., & Komuro, I. (2022). Spatiotemporal transcriptome analysis reveals critical roles for mechanosensing genes at the border zone in remodeling after myocardial infarction. *Nature Cardiovascular Research*, *1*(11), 1072–1083. DOI: 10.1038/s44161-022-00140-7 PMID: 39195917

Yan, X., Ang, K. S., van Olst, L., Edwards, A., Watson, T., Zheng, R., & Chen, J. (2024). Mosaic integration of spatial multi-omics with SpaMosaic. BioRxiv, 2024-10.

Yang, P., Jin, L., Liao, J., Jin, K., Shao, X., Li, C., & Fan, X. (2023). Revealing spatial multimodal heterogeneity in tissues with SpaTrio. *Cell Genomics*, *3*(12), 100446. DOI: 10.1016/j.xgen.2023.100446 PMID: 38116121

Yang, P., Rong, X., Gao, Z., Wang, J., & Liu, Z. (2025). Metabolic and epigenetic regulation of macrophage polarization in atherosclerosis: Molecular mechanisms and targeted therapies. *Pharmacological Research*, *212*, 107588. DOI: 10.1016/j.phrs.2025.107588 PMID: 39778637

Yanni, J., Tellez, J. O., Mączewski, M., Mackiewicz, U., Beresewicz, A., Billeter, R., & Boyett, M. R. (2011). Changes in ion channel gene expression underlying heart failure-induced sinoatrial node dysfunction. *Circulation: Heart Failure*, *4*(4), 496–508. DOI: 10.1161/CIRCHEARTFAILURE.110.957647 PMID: 21565973

Yao, L., He, F., Zhao, Q., Li, D., Fu, S., Zhang, M., & Wang, L. (2023). Spatial multiplexed protein profiling of cardiac ischemia-reperfusion injury. *Circulation Research, 133*(1), 86–103. DOI: 10.1161/CIRCRESAHA.123.322620 PMID: 37249015

Yin, Q., Zhong, L., Song, Y., Bai, L., Wang, Z., Li, C., & Yang, X. (2025). A decision support system in precision medicine: Contrastive multimodal learning for patient stratification. *Annals of Operations Research, 348*(1), 579–607. DOI: 10.1007/s10479-023-05545-6

Yu, Z. B., & Sheng, J. J. (2011). Remodelling of cardiac gap junctions and arrhythmias. *Sheng li xue bao:[Acta physiologica Sinica], 63*(6), 586-592.

Yuan, X., Liu, C., & Wang, Y. (2023). 4D myocardium reconstruction with decoupled motion and shape model. In *Proceedings of the IEEE/CVF International Conference on Computer Vision* (pp. 21252-21262). DOI: 10.1109/ICCV51070.2023.01943

Yurdagul, A.Jr, & Aikawa, E. (2023). Discovering a rare smooth muscle cell population specific to men in ascending aortic aneurysm using spatial transcriptomics. *Arteriosclerosis, Thrombosis, and Vascular Biology, 43*(12), 2298–2300. DOI: 10.1161/ATVBAHA.123.320235 PMID: 37916413

Zhan, C., Tang, T., Wu, E., Zhang, Y., He, M., Wu, R., & Shen, B. (2023). From multi-omics approaches to personalized medicine in myocardial infarction. *Frontiers in Cardiovascular Medicine, 10*, 1250340. DOI: 10.3389/fcvm.2023.1250340 PMID: 37965091

Zhang, H., Liu, W., Chang, S., Wang, H., He, J., & Huang, Q. (2023a). ST-ReGE: A novel spatial-temporal residual graph convolutional network for CVD. *IEEE Journal of Biomedical and Health Informatics, 28*(1), 216–227. DOI: 10.1109/JBHI.2023.3327025 PMID: 37871091

Zhang, X., Chen, Y., Hua, K., Xu, S., You, R., Hao, M., & Jiang, R. (2023b). uniHEART: an ensemble atlas of cardiac cells provides multifaceted portraits of the human heart. DOI: 10.21203/rs.3.rs-3215038/v1

Zhou, Y., Xiao, X., Dong, L., Tang, C., Xiao, G., & Xu, L. (2025). Cooperative integration of spatially resolved multi-omics data with COSMOS. *Nature Communications, 16*(1), 27. DOI: 10.1038/s41467-024-55204-y PMID: 39747840

ABBREVIATIONS

ACM: - Arrhythmogenic Cardiomyopathy
AI: - Artificial Intelligence

CMR: - Cardiovascular Magnetic Resonance
CVDs: - Cardiovascular Diseases
CVPs: - Cardiovascular Progenitors
CX43: - Connexin 43
DCM: - Dilated Cardiomyopathy
DESI-MSI: - Desorption Electrospray Ionization Mass Spectrometry Imaging
ECG: - Electrocardiogram
EHT: - Engineered Heart Tissue
GNNs: - Graph Neural Networks
HCM: - Hypertrophic Cardiomyopathy
IHD: - Ischemic Heart Disease
IMC: - Imaging Mass Cytometry
iMFA: - Isotope-Assisted Metabolic Flux Analysis
LMICs: - Low- And Middle-Income Nations
MALDI-MSI: - Matrix-Assisted Laser Desorption/Ionization Mass Spectrometry Imaging
ML: - Machine Learning
MSI: - Mass Spectrometric Imaging
scRNA-seq: - Single-Cell RNA Sequencing
VAEs: - Variational Autoencoders
vWF: - Von Willebrand Factor

Chapter 9
Spatial Multi-Omics in Kidney Diseases

Odangowei Inetiminebi Ogidi
https://orcid.org/0000-0002-9961-893X
Bayelsa Medical University, Yenagoa, Nigeria

Deghinmotei Alfred-Ugbenbo
Bayelsa Medical University, Yenagoa, Nigeria

Mao Ebimobotei Bunu
https://orcid.org/0009-0002-0370-7258
Bayelsa Medical University, Yenagoa, Nigeria

ABSTRACT

Kidney diseases encompass a heterogeneous group of disorders with complex molecular underpinnings and diverse spatial manifestations within renal tissue. This chapter explores the transformative potential of spatial multi-omics technologies in unraveling the cellular and molecular architecture of kidney diseases. By integrating genomics, transcriptomics, proteomics, and metabolomics with spatial resolution, spatial multi-omics enables high-resolution mapping of disease-associated alterations within distinct renal compartments. The chapter discusses recent advances in spatial transcriptomics, and single-cell omics that have deepened our understanding of kidney pathophysiology. Additionally, the chapter addresses the challenges in data integration, computational analysis, and clinical translation. Ultimately, spatial multi-omics holds promise for redefining kidney disease diagnostics, identifying novel biomarkers, and guiding targeted therapies through a more precise, tissue-contextual understanding of renal pathology.

DOI: 10.4018/979-8-3373-4849-0.ch009

INTRODUCTION

The human kidney is crucial for metabolic functions and waste elimination. Acute kidney injury (AKI) may occur due to its substantial blood flow and robust excretory function, rendering it vulnerable to many insults such as toxins, ischaemia, infections, metabolic disturbances, and immune responses (Oe, 2024). Previous investigations by our group indicated that high-risk inpatients exhibited an incidence of acute kidney injury (AKI) between 20% and 50% attributable to various aetiologies (Liu et al., 2024). Hospitalised patients with acute kidney injury (AKI) exhibited a mortality rate of approximately 25%, with 50% of these individuals failing to achieve full recovery of renal function, ultimately progressing to chronic kidney disease (CKD) (Liu et al., 2024).

This is mostly due to the absence of efficient treatment agents and the low incidence of early diagnosis, both of which complicate the clinical management of AKI significantly. To enhance the prognosis of AKI, early diagnosis, complication prevention, and appropriate treatment of pathogenic insults are essential. Nonetheless, the advancement of precision medicine for acute kidney injury (AKI) is significantly impeded by the challenges associated with early diagnosis and thorough assessment of renal damage, attributable to the intricate tissue architecture and physiological functions, alongside the diverse aetiology across a broad spectrum. Therefore, to improve clinical diagnosis and create new pharmaceuticals, it is crucial to establish a thorough and complete renal assessment system (Chebib & Torres, 2016).

Recent advancements in computational analysis and extensive data generation have significantly impacted clinical medicine and biological research. Precision medicine in these "big science" initiatives has been mostly driven by cancer research, but other fields have fallen behind (Ogidi, 2024). Nonetheless, nephrology possesses the potential to take the lead due to recent developments (Saez-Rodriguez et al., 2019). A great deal of molecular data, including the genome, epigenome, transcriptome, proteome, and metabolome, has emerged in the age of big data thanks to the advent of high-throughput omics technology. Using this multimodal depiction, the processes involved in human health and illness can be better understood (Picard et al., 2021; Ogidi & Ajoko, 2024).

It is crucial to use an integrated method that examines many datasets in order to address biological concerns thoroughly. Data at different resolutions can be used for multi-omics integrative analysis, depending on the downstream technique. It is commonly used in bulk omics to classify patients, subtype diseases, and derive biological insights (Wang et al., 2022). To better understand tissue heterogeneity, detect cell subpopulations, and conduct comparative analysis, single-cell data integration provides improved resolution (Ryu et al., 2023). In addition, by combining geographical context with data collected at the single-cell or spot-level, spatial data can help in identifying spatial functional areas, analysing cell-cell interactions, and improving our understanding of cell and tissue composition (Kiessling & Kuppe, 2024). The integration of massive multi-omics data sets presents a number of issues, including data heterogeneity, complexity, interpretability, and validation. In this chapter, we will look at the most recent developments in nephrology-related omics data integration, highlighting the main obstacles, potential solutions, and future directions.

OMICS TECHNOLOGIES IN KIDNEY RESEARCH

Omics technologies have emerged as powerful tools in kidney research (Table 1 & 2), offering deep insights into the molecular mechanisms underlying renal physiology and disease. Genomics, transcriptomics, proteomics, and metabolomics each provide unique layers of information, from genetic variations and gene expression patterns to protein networks and metabolic alterations. When integrated, these approaches allow researchers to map complex biological pathways, identify biomarkers for early diagnosis, and discover novel therapeutic targets for conditions such as chronic kidney disease, diabetic nephropathy, and acute kidney injury (Ogidi et al., 2024a). By capturing a holistic view of kidney biology, omics technologies are driving precision medicine approaches that can improve patient outcomes and enhance our understanding of renal pathophysiology.

Table 1. Omics in kidney research and precision medicine applications

Omics Type	Technologies / Approaches	Applications in Kidney Research	Applications in Precision Medicine
Genomics	Whole-genome/exome sequencing, GWAS, targeted sequencing	Identification of genetic variants, mutations, hereditary kidney disease	Risk stratification (*APOL1* in donors), diagnosis of ADPKD (*PKD1/2*), *COL4A* in Alport syndrome
Transcriptomics	Bulk RNA-seq, single-cell RNA-seq, spatial transcriptomics	Cell-type mapping, biomarker discovery, disease mechanism identification	Patient stratification by expression signatures, discovery of therapeutic targets in lupus nephritis, CKD
Epigenomics	DNA methylation profiling, histone modification mapping (ChIP-seq), ATAC-seq, chromatin conformation	Understanding regulation of fibrosis and disease pathways	Epigenetic biomarkers (DNA methylation, histone modifications), lncRNAs (*MALAT1*) for DKD progression

continued on following page

Table 1. Continued

Omics Type	Technologies / Approaches	Applications in Kidney Research	Applications in Precision Medicine
Proteomics	Mass spectrometry (LC-MS/MS, DIA), urinary proteomics, exosome proteomics, PTM-specific proteomics	Protein-level biomarkers, signaling pathways, post-translational modifications	CKD273 urinary classifier predicting CKD and CVD outcomes; exosomal miRNAs distinguishing DKD vs. FSGS
Metabolomics	NMR spectroscopy, LC-MS/MS, metabolic fingerprinting	Identification of altered metabolic pathways in CKD and DKD	Fumarate and amino acid metabolites as biomarkers; urine metabolomics predicting disease progression
Multi-Omics / Systems Biology	Integrative genomics, transcriptomics, proteomics, metabolomics, ML/AI-based integration	Comprehensive molecular understanding, pathway mapping, atlas creation	Precision medicine models, kidney multi-omics atlas identifying fibrosis pathways, AI/ML models predicting CKD progression

Table 2. Case studies of omics technologies in kidney research

Omics Type	Objective / Focus	Technology / Method	Key Findings	Study/Authors
Genomics	Identify genetic causes of kidney disease in adults	Whole-exome sequencing (WES) on 3,315 patients	9.3% of cases had a monogenic cause identified; reclassification of clinical diagnoses improved patient management	Groopman et al. (2019)
Transcriptomics	Define cellular diversity in healthy and diseased kidneys	Single-cell RNA sequencing (scRNA-seq) of mouse and human kidneys	Identified 16 distinct cell types and novel cell states linked to fibrosis and injury	Park et al. (2018)
Proteomics	Discover urinary biomarkers for diabetic kidney disease (DKD)	Mass spectrometry-based urinary proteomics	Identified and validated several peptide markers associated with DKD progression	Gajjala et al. (2019)
Metabolomics	Explore metabolomic profiles predicting CKD progression	Capillary electrophoresis–mass spectrometry (CE-MS) on urine samples	Specific peptide signatures correlated with CKD stages and outcomes	Nkuipou-Kenfack et al. (2014)

continued on following page

Table 2. Continued

Omics Type	Objective / Focus	Technology / Method	Key Findings	Study/Authors
Epigenomics	Study DNA methylation in kidney fibrosis	Reduced representation bisulfite sequencing (RRBS)	Identified hypermethylated genes (e.g., RASAL1) contributing to fibroblast activation	Marumo et al. (2020)
Multi-omics Integration	Integrate proteomics, metabolomics, and transcriptomics in lupus nephritis	Multi-omics network analysis	Revealed novel molecular subtypes and pathways predictive of therapy response	Steenbergen et al. (2021)

Genomics

Genomic studies have uncovered genetic variations that cause renal illness, which has greatly enhanced our knowledge of the disease. Using high-throughput sequencing techniques, many variants linked to different kidney disorders have been discovered. Monogenic kidney diseases are caused by mutations in only one gene. A disorder known as autosomal dominant polycystic kidney disease (ADPKD) develops when cysts grow and renal function gradually declines due to mutations in the PKD1 and PKD2 genes. For around 85% of ADPKD instances, the underlying mutation is in PKD1, while for approximately 15%, it is in PKD2. New mutations in these genes have recently been discovered, which has improved the accuracy of diagnoses (Chebib & Torres, 2016; Kim et al., 2021).

Newer studies have increased diagnostic precision by broadening the spectrum of potential mutations in these genes. A person with Alport syndrome will have a mutation in one of three genes: COL4A3, COL4A4, or COL4A5. Issues with vision, hearing, and glomerulonephritis are symptoms of this illness. Mutations that affect the generation and assembly of type IV collagen $\alpha3\alpha4\alpha5$ can lead to problems with the basement membranes of the kidney, cochlea, and eye. We now understand the links between genotype and phenotype better, and diagnostic accuracy has grown, thanks to the recent identification of more deleterious variations in these genes (Kashtan et al., 2018). Thanks to advancements in genome sequencing, these mutations can now be more easily detected, enabling early diagnosis and therapy. Complex kidney diseases, beyond those caused by a single gene, can arise from a complex interplay of several genetic variants and environmental factors. Multiple loci connected with chronic kidney disease susceptibility have been revealed by genome-wide association studies (GWAS). There is a strong correlation between uromodulin gene mutations and an increased risk of chronic kidney disease (CKD) (Wang et al., 2924).

Further evidence that uromodulin is involved in renal pathology points to its potential role in the rare genetic disorder autosomal dominant tubulointerstitial kidney disease (ADTKD). These mutations, linked to increased uromodulin expression, influence renal function and the development of illness (Kottgen et al., 2022). Certain hereditary differences increase the risk of kidney illness in some groups. Nephropathy caused by hypertension and focal segmental glomerulosclerosis (FSGS) are more common in African populations due to shared mutations in the APOL1 gene. The G1 and G2 risk alleles are the most common causes of renal illness in these populations, as supported by strong data. Certain kidney problems are substantially more likely to occur in individuals with two APOL1 risk alleles (Freedman et al., 2021; Itoku et al., 2024).

The importance of genetic research in West African nations is underscored by new data showing one-third of the population contains APOL1 risk genes. The field of functional genomics has helped us understand how these genetic variants impact certain circuits. Organoids of the kidney made from induced pluripotent stem cells (iPSCs) allowed scientists to study molecular pathways impacted by mutations, including intercellular junctions and apicobasal polarity. These findings pave the way for investigations into the consequences of genetic alterations that do not alter the fundamental sequence in further investigations (Forbes et al., 2018). The CKD GWAS identified many genes, one of which was Dab2. These genes may hold therapeutic promise because to their involvement in tubular function and the pathogenesis of CKD (Zaiden et al., 2020).

Better prognoses and treatment options are made possible by genetic testing for dangerous mutations. Diagnosing APOL1 risk polymorphisms can impact the management of hypertensive nephropathy and kidney transplantation. Donor and recipient APOL1 genotypes are critical for kidney transplant success. The likelihood of kidney transplant survival was lower for donors who possessed two APOL1 risk alleles. Donor screening and post-transplant care are two areas where the results indicate APOL1 genotyping can be useful. A higher prevalence of hypertension-related chronic kidney disease (CKD) is observed in individuals with APOL1 risk polymorphisms. Treatment tailored according to APOL1 status has the potential to improve hypertensive nephropathy prognoses (Freedman et al., 2018; Malone, 2021).

Research on inaxaplin (VX-407) for the treatment of acute diabetic nephropathy in individuals with certain PKD1 mutations suggests that a better understanding of the genetic basis of renal diseases should lead to the creation of tailored treatments (Doshi et al., 2021; Egbuna et al., 2023; Giudicelli et al., 2022; Hosseinpour et al., 2022). Combining the impacts of many genetic variations is a new and promising strategy called polygenic risk scores (PRS). The results may lead to more accurate risk assessments and stronger advocacy for early interventions in populations prone to certain diseases (Clark et al., 2019).

Transcriptomics

Transcriptomics, which involves analysing RNA transcripts in cells or tissues, can be a significant tool in the search for molecular causes of kidney disorders. This area of study allows researchers to examine cellular relationships and activities in healthy and sick states by measuring thousands of genes at once. Using the transcriptome researchers have discovered novel biomarkers for kidney disease, genes displaying differential expression, and complex metabolic pathways. In inflammatory kidney illnesses such as lupus nephritis and IgA nephropathy, novel molecular biomarkers have been discovered by transcriptome analysis. These findings pave the way for future research into novel approaches to kidney disease detection and therapy and offer crucial new information about the causes of renal sickness (Schena et al., 2024; Si et al., 2024).

Use single-cell RNA sequencing (scRNA-seq), a cutting-edge transcriptomics technique, to investigate the cellular diversity of the kidney tissue. This method evaluates the expression of genes on a single cell level. The ability to detect nuanced differences between cell types and the discovery of new cell populations linked to kidney development, function, and disease are two advantages of single-cell RNA sequencing over traditional bulk RNA sequencing. Using ScRNA-seq, three populations of unidentified cells and thirteen distinct cell types are used to categorise adult mouse kidney cells into sixteen groups. Using scRNA-seq, researchers were able to create a detailed cellular atlas of human kidneys, which showed thirteen groups of healthy renal cells (Lindstrom et al., 2019).

In addition, scRNA-seq has improved our understanding of the aetiology of disorders like lupus nephritis by revealing specific subsets of immune cells involved in autoimmunity and inflammation. A rise in pro-inflammatory CD163+ dendritic cells in the kidneys is one symptom of lupus nephritis, which in turn activates the local immune system through interactions with CD4+ T cells (Chen et al., 2024). Lupus nephritis patients have an active extrafollicular B cell response, which could be a factor in the disease's progression, since atypical B cells and cells that secrete antibodies have been found in renal tissue using single-cell RNA sequencing (Wu et al., 2018). One of the most useful ways to analyse cellular interactions and identify the cells that cause kidney illness is via single-cell transcriptomics, which can show changes in transcription that are exclusive to individual cells (Benjamin et al., 2022).

One of the pathological processes associated with miRNAs is fibrosis in diabetic kidney disease (DKD), making them a promising therapeutic target. As a result of its role in promoting renal fibrosis through elevating the production of extracellular matrix and the epithelial-mesenchymal transition, miR-21 expression is substantially upregulated in DKD. In animals suffering from diabetic kidney disease, antisense oligonucleotides directed at miR-21 improved renal function and reduced fibrosis. In

DKD, there is an inverse relationship between the expression of the miR-29 family and increased collagen synthesis. Restoring miR-29 levels is one potential therapy strategy for reducing renal fibrosis (Dhas et al., 2023; Gu et al., 2021). There are new avenues for diagnosis and prognosis thanks to the identification of miRNAs in kidney tissues that have not been previously studied. This is particularly true for cancers such as clear cell renal cell carcinoma (ccRCC) (Ogidi et al., 2025g). Researchers found that ccRCC tissues showed nine dysregulated miRNAs, which could be signs of disease, when compared to normal renal parenchyma. There is encouraging evidence that certain microRNAs, such as Knm22_2209 and Knm6_2419, could serve as biomarkers for diseases (Gilyazova et al., 2023).

Axon guidance, VEFG signalling, and MAPK signalling are three significant signalling pathways that microRNAs are involved in, according to research on their functions in target prediction. By combining geographical information with gene expression patterns, spatial transcriptomics (ST) maintains the kidney tissue's architecture throughout analysis. Using spatial transcriptomics, researchers have been able to study cellular organisation within nephron segments and examine disease-associated alterations in gene expression at the local level in kidney studies. Thanks to ST, researchers can examine how genetic and environmental variations impact cellular communication inside the tubule microenvironment. Spatial transcriptomics enhances our comprehension of the impact of tissue architecture on renal function and pathology by visualising the spatial distribution of cellular responses (Raghubar et al., 2022).

The ability to produce human kidney atlases that are geographically rooted in both healthy and diseased tissues has been greatly enhanced by ST. The understanding of the molecular and cellular mechanisms that contribute to kidney diseases has been substantially enhanced as a result of this (Isnard et al., 2024). Risk of renal disease knowledge also requires transcriptome diversity that is distinctive to a population. People of African descent are more likely to have APOL1-associated kidney illness, according to a transcriptome analysis that targeted this ethnic group specifically (Yusuf et al., 2021). It is clear from these results that more population-based research is required to pinpoint possible risk factors that may be inherited differently from one generation to the next (Ranek et al., 2022).

Proteomics

Proteome analysis may uncover molecular insights regarding diseases that genomes and transcriptomics could overlook by examining temporal protein alterations. The primary focus has been on examining protein localisation, post-translational modifications (PTMs), protein-protein interactions, and expression levels. Protein-tissue interactions, such as glycosylation and phosphorylation, can alter protein functionality, localisation, and inter-protein interactions, hence influencing disease progression. An increasing amount of evidence in health and disease indicates that post-translational modifications (PTMs) significantly influence cellular signalling pathways and protein-protein interactions (Elangovan et al., 2022).

Quantitative proteomics, using mass spectrometry and antibody-based assays, has facilitated the identification of proteins differently expressed in diverse kidney diseases. Protein expression patterns associated with oxidative stress, inflammation, and fibrosis have been correlated with chronic kidney disease (CKD). These profiles elucidate the aetiology of chronic kidney disease (CKD) and indicate potential treatment targets that may mitigate or even reverse the condition (Ling & Kuo, 2018). Rising research indicates that proteome patterns linked to the complement cascade and ubiquitin-proteasome pathways are pivotal in the progression of chronic kidney disease (CKD). Chronic inflammation and fibrosis are hallmarks of chronic kidney disease (CKD), which is characterised by decreased proteasome activity and elevated complement activation (RamirezMedina et al., 2023).

Important biomarkers for these processes include transthyretin, complement component C6 and biglycan. Therapeutic alternatives to improve proteasome activity or regulate complement activation are also presented by these biomarkers, which may lead to efficient means of reducing the progression of disease (RamirezMedina et al., 2023). Comparable to transcriptomics, studies of spatial and temporal proteomics have played an essential role in tracking the development of disease throughout different parts of the kidney and different stages of time. Researchers have been able to map protein distributions inside kidney tissue sections using spatially resolution proteomics, which has allowed them to explore localised pathophysiological changes in response to damage and pinpoint the important intervention window. Regardless of the amounts of individual proteins, it is vital to understand functional networks and protein-protein interactions in renal disorders. The synthesis of extracellular matrix and tissue fibrosis are induced when CTGF interacts with signalling molecules like TGF-β. The role of inflammation and fibrosis in the development of CKD can be better understood by conducting studies based on networks (Panizo et al., 2021).

Investigating these relationships has led to the discovery of key proteins and processes, like inflammatory cascades, which have improved our understanding of the molecular basis of DKD [56]. In renal disorders, the preservation of cellular homeostasis relies on protein breakdown mechanisms such as autophagy and the ubiquitin-proteasome system. To prevent cytotoxicity and buildup, the ubiquitin-proteasome system eliminates unfolded or damaged proteins selectively. Autophagy allows larger cellular components, like organelles, to be broken down through lysosomal pathways. In the event that either pathway is compromised, the accumulation of damaged proteins and organelles could worsen renal impairment. Decreased ubiquitin-proteasome activity is linked to renal fibrosis and a number of renal diseases. Several kidney diseases have been associated with autophagy dysfunction, despite autophagy's central role in maintaining healthy kidneys. Autophagy and the ubiquitin-proteasome system work together to control cellular quality; a breakdown in either system can hasten the development of kidney disease (Meyer-Schwesinger, 2019;Takabatake et al., 2014; Kocaturk & Gozuacik, 2018).

Conditions like DKD and CKD are associated with cellular stress and protein accumulation in tubular and glomerular cells when these pathways are disrupted (Dubin & Rhee, 2020). Alterations to proteins can occur after they have been translated into a different form, and these changes are known as post-translational modifications (PTMs). These changes impact biological processes and events by altering the protein's stability, localisation, activity, and interactions with other molecules. Alterations linked with pathological states in kidney disease have been uncovered through proteomic research that aims to characterise post-translational modifications in renal tissues and fluids. Fibrogenesis relies on abnormal phosphorylation events and is made possible by EGFR signalling hyperactivation. Alterations in glycosylation, especially of uromodulin, have also been linked to chronic kidney disease (CKD). Important glycosylation patterns are found on uromodulin, the primary protein in normal urine. The renal function of uromodulin may be impacted by changes in these glycosylation patterns, which have been associated with various illnesses (Zhuang & Liu, 2014; Devuyst & Bochud, 2015).

The number of glycosylation and phosphorylation sites on important proteins that regulate cellular transport and membrane integrity exceeds seven hundred and twenty. Phosphoproteins and glycoproteins may have many modification sites, just like rat kidney tissues. Proteins such as fibronectin undergo both glycosylation and concurrent phosphorylation, which suggests that these post-translational modifications may be associated with the processes of kidney disease (Hao et al., 2011; Pagel et al., 2015; Tran et al., 2013). Thanks to proteomic methods, novel kidney disease biomarkers can be developed. Urine proteomics is an intriguing non-invasive method for the early detection of kidney disease. Cardiovascular events (fatal and non-fatal) occurred at a rate more than 10 times greater in the highest CKD273 tertile compared to the lowest tertile, according to a cohort study (Verbeke et al., 2021).

This was very helpful for patients with early chronic renal failure who did not have a history of cardiovascular disease. The distinctiveness of exosomal components as biomarkers for renal illnesses was further demonstrated by microRNA profiling of urine exosomes, which differentiated diabetic kidney disease from focal segmental glomerulosclerosis (Chen et al., 2018; Knepper & Pisitkun, 2007; Trabulus et al., 2024; Du et al., 2023). The emerging discipline of proteogenomics has been crucial in the identification of novel protein variants and mutations that exacerbate renal disorders. These include changes influencing alternative splicing and protein coding. An important mechanism for transcriptome and proteome variability, alternative splicing influences a broad range of biological processes and disorders. Mistakes or alterations in the splicing process, which is responsible for the production of mature messenger RNA and functioning proteins, can lead to many illnesses. To express genes and ensure proteins work as intended in eukaryotic organisms, messenger RNAs must undergo alternative splicing (Kim et al., 2018; Ren et al., 2021).

Metabolomics

Identification of metabolic alterations associated with kidney illness has been made possible through metabolomics, the study of small molecules in living organisms. By facilitating the discovery of potential treatment targets, biomarkers, and disease origins, this method enhances renal diagnosis, management, and measurement. Novel immunotherapies and personalised approaches are now within reach, thanks to this discovery. In diabetic kidney disease, adenine is one of many signs of renal fibrosis. The basic pathogenic variables that cause chronic kidney disease (CKD) to progress are believed to be exacerbated by these metabolic alterations, which include renal damage and fibrosis. Mitochondrial dysfunction, which can happen in renal tubular cells in stressful environments like hypoxia, is a factor in chronic kidney disease (CKD). This is because the kidneys have a high concentration of mitochondria and require a lot of energy (Ogidi, 2025). Fumarate is one indicator of mitochondrial failure, and NAD precursors are another. Fumarate links oxidative stress to tissue fibrosis in particular (Morevati et al., 2022; Patera et al., 2024).

By increasing concentrations of nicotinamide adenine dinucleotide (NAD+), niacinamide supplementation is a potential therapeutic alternative with protective benefits. Chronic kidney disease (CKD) caused by a NAD+ deficit manifests itself in cellular senescence and mitochondrial malfunction. By increasing NAD+ levels, niacinamide treatment has the potential to delay the course of disease (Morevati et al., 2022). By delving into these disrupted pathways, we can uncover the root causes of disease and pave the road for the creation of innovative, tailored treatments that can block these pathways and delay the course of the disease (Kalim & Rhee, 2017). People with chronic kidney disease (CKD) clearly have major anomalies in amino acid metabolism, as seen by reduced levels of essential amino acids and elevated levels of uremic toxins. As an example of how chronic kidney disease impacts protein metabolism and clearance, this imbalance can lead to the accumulation of toxic chemicals, which in turn accelerates the progression of the disease (Kang et al., 2023).

In addition, alterations in lipid metabolism, particularly elevated levels of certain sphingolipids, have been linked to renal fibrosis and decreased kidney function in diabetic kidney disease (DKD). Bioactive sphingolipids regulate cell growth, inflammation, and cell death. Ceramides and sphingosine-1-phosphate (S1P) are the components of sphingolipids. Damage to podocytes, glomerular sclerosis, and tubulointerstitial fibrosis are hallmarks of chronic kidney disease (DKD) development and may be due to deregulation of sphingolipid metabolism. Various variables contribute to renal failure, including oxidative stress, inflammatory reactions, and fibrotic processes. Recent research have linked variations in sphingolipid levels to these processes. For instance, increased ceramide buildup exacerbates the relationship between renal cell death and mitochondrial dysfunction. Red blood cell enlargement and elevated S1P signalling are associated with profibrotic pathway activation. The findings raise the possibility of novel therapeutic strategies for diabetic kidney disease (DKD) that aim to decrease kidney fibrosis while preserving renal function through regulation of sphingolipid metabolism (Nicholson et al., 2021; Sakic et al., 2024; Schwalm et al., 2024).

In detecting illnesses early and monitoring DKD, these findings emphasise the significance of metabolomic profiling (Pereira et al., 2022). There is some evidence that specific compounds in urine may have harmful effects, including exacerbation or even mortality, in those with chronic kidney disease (CKD). Timely intervention and the development of customised treatment methods are made possible by these markers, which are especially helpful for forecasting which patients are most prone to rapid health decline (Steinbrenner et al., 2021). Individual metabolites have been identified, and specific pathways impacted by kidney illness have been discerned through the analysis of clusters of related compounds. Metabolomics has provided new insight into metabolic changes in lipids and amino acids; in particular, branched-chain amino acids and their metabolites are strongly associated with chronic renal disease. These findings provide credence to the idea that targeting pathways could be a useful treatment strategy (Hasegawa & Inagi, 2021).

Lipidomics is a branch of metabolomics that focusses on lipid patterns; it has shown that GD and CKD share issues with lipid metabolism. Damage to the kidneys, inflammation, and fibrosis are all symptoms of chronic kidney disease, and these three main reasons are associated with changes in ceramides, phosphatidylcholines, and cholesterol esters, among other lipid species. Fibrosis in the kidneys occurs when ceramide levels are elevated, leading to increased oxidative stress and cellular apoptosis. Likewise, aberrant phosphatidylcholine metabolism can aggravate renal damage by compromising cell membrane integrity and disrupting signalling pathways. Evidence indicates a correlation between inflammation, kidney injury, and the accumulation of cholesterol esters. By examining these lipidomic alterations, potential therapy targets for mitigating disease development can be discovered, revealing insights into the molecular mechanisms underlying CKD (Sakic et al., 2024). This discovery offers more evidence that lipidomics may facilitate the development of lipid-targeted treatments aimed at potentially modulating critical pathways that drive inflammation and fibrosis (Davies, 2018).

Epigenomics

The term "epigenomics" is used to characterise the scientific investigation of these alterations, which are variations in gene expression that are passed down across generations instead of new DNA sequences (Ogidi et al., 2025a). Understanding the distinct epigenetic patterns of podocytes and tubular cells which are the two types of renal cells and they has been crucial in elucidating their roles in disease states (Chan & Li, 2021). Some of the post-translational alterations that histone proteins can experience are acetylation and methylation. Gene accessibility and chromatin architecture are affected by these alterations. The activation of profibrotic genes like TGF-β1 is associated with elevated levels of Histone H3 lysine 9 acetylation (H3K9ac), which is observed in renal biopsies taken from patients with diabetic kidney disease (DKD). In mesangial cells, histone acetyltransferases, like p300/CBP, are activated by TGF-β1, leading to histone acetylation in the promoters of target genes, most notably plasminogen activator inhibitor-1 (PAI-1) and p21 (Lu et al., 2021; Yuan et al., 2013).

A potential future approach to treating renal disease could centre on enzymes that change histones (Wanner & Bechtel-Walz, 2017). Methylation of cytosine residues in DNA is an important process that controls gene expression. Scientists have discovered that different kidney diseases are associated with different patterns of methylation. The promoter region of the Ras-association domain family 1 isoform A (RASSF1A) gene may be hypermethylated, which could speed up the evolution of renal cell carcinoma. Patients with RASSF1A promoter methylation in RCC had a worse chance of cancer-specific survival (Tezval et al., 2008; Zhuang et al., 2018; Ogidi & Tawariowei, 2023). Chronic kidney disease (CKD) progression has been associated with hypomethylation of particular genes, which may provide markers for early detection (Ding et al., 2021).

The development of kidney disease is affected by histone alterations because these changes affect gene expression and chromatin characteristics. The H2A.Z histone variant is involved in transcription activation and repression in opposite ways. Since this element can change the structure of chromatin, its incorporation into nucleosomes can affect gene expression. H2A.Z is essential for several physiological processes, such as DNA repair, transcriptional control, and centromeric heterochromatin modulation. Several diseases, including cancer, have been linked to H2A.Z dysregulation (Ogidi et al., 2025e). Gene expression, particularly in response to stress or illness, may be impacted by H2A.Z-induced changes in chromatin architecture in renal cells (Ogidi et al., 2025h). However, more evidence connecting changes in H2A.Z dynamics to changes in kidney cell gene expression is needed. Determining H2A.Z's exact function in renal aetiology calls for additional investigation (Chan & Li, 2021; Giaimo et al., 2019).

The spatial layout of chromatin influences gene expression in kidney illness, according to recent research. This is because it enhances the interaction between enhancers and promoters. Chromosome loops link genes specific to the kidney to distant regulatory elements, which in turn affect gene regulation in CKD and nephropathy. Researchers have used chromatin conformation capture methods to identify the transcription start sites of 304 genes that are associated with chronic kidney disease (CKD). Many of these genes are essential for keeping the kidneys in a state of equilibrium. The fibrotic environment in renal sickness has been better understood because to a human kidney multi-omics single-cell atlas that sheds light on the significance of spatial chromatin arrangement in disease development (Martire & Banaszynski, 2020; Brandt et al., 2018).

Renal cells can undergo epigenetic modifications in response to environmental variables such as pollutants, poisons, and lifestyle choices (Ogidi et al., 2024b). These alterations, such as elevated DNA methylation or histone acetylation, may accumulate in susceptible individuals as a result of exposure and impact the course of the disease, such as diabetic kidney disease (DKD) or chronic kidney disease (CKD) (Wilson et al., 2020). Different kinds of non-coding RNAs affect gene expression after transcription. Some examples of these are microRNAs and short non-coding RNAs (lncRNAs). There is mounting evidence that a disruption in microRNA regulation is associated with renal illness. Chronic kidney disease (CKD) patients have elevated levels of miR-21, a gene inhibitor that inhibits genes that prevent fibrosis. Kidney fibrosis is the eventual outcome of MiR-21's downregulation of antifibrotic genes, such as PPAR-α. The attachment of miR-21 to renal epithelial cells activates Toll-like receptor 7, which in turn triggers pro-inflammatory responses, further aggravating the fibrosis (Chau et al., 2012; Kim et al., 2023).

Long non-coding RNAs (lncRNAs) like MALAT1, which are also linked to DKD, influence inflammatory reactions and the production of extracellular matrix components. The metastasis-associated lung adenocarcinoma transcript 1 (MALAT1) gene is more prevalent in diabetic kidney disease (DKD). The regulation of ECM synthesis and the immune system's response to it is a key function of this gene in disease progression. In HK-2 cells, which are triggered by high glucose levels, the data showed that MALAT1 promotes DKD via the miR-15b-5p/TLR4 signalling pathway by reducing cell viability, apoptosis, and inflammatory responses. A miR-15b-5p inhibitor or an increase of TLR4 might mitigate the effect. Increased levels of MALAT1 in the urine of individuals suffering from diabetic ketoacidosis (DKD) have been found to be an indicator of possible type 1 diabetes mellitus (Yang et al., 2022; Dieter et al., 2023). In order to rectify abnormal epigenetic changes and return normal gene expression patterns, researchers are investigating the possibility of using inhibitors of DNAmethyltransferase and histone deacetylase. New treatment options may soon be available as a result of this.

INTEGRATION OF MULTI-OMICS DATA IN KIDNEY DISEASES

Genomic, transcriptomic, proteomic, metabolomics, and epigenomics datasets must be combined for a complete understanding of renal disorders (Figure 1). By expanding our knowledge of intricate biological processes, creating novel biomarkers, and discovering potential treatment targets, this all-encompassing strategy promotes advancements in precision nephrology. Genomic and transcriptomic data integration has uncovered unique genetic variations in CKD patients linked to altered gene expression. These analyses uncover regulatory networks and mechanisms that impact the course of the disease, in addition to potential intervention sites. New genes associated with kidney function were successfully identified by the CKDGen Consortium using GWAS and transcriptome data (Duan et al., 2021).

Tubule interstitial fibrosis and estimated glomerular filtration rate (eGFR) are kidney health indicators. TENM2, COL4A3, and DCLK1 are among the genes linked to this. This study intends to provide insight into the mechanisms that drive the progression of kidney disease by documenting the association between fibrosis and genes like DCLK1, as well as the positive correlation between improved eGFR and reduced fibrosis levels and increased expression of protective genes like TENM2 and SNX30. The regulatory components of gene expression can be revealed, and possible treatment targets identified, by combining kidney-specific eQTL data with GWAS data. To understand the progression of renal disease across time, it is essential to integrate omics data at various points in the process (Liu et al., 2022). A comparison of spatial omics platforms relevant to kidney research is shown in Table 3.

Researchers have shown through time-series analyses of DKD that molecular profiles change across the course of the disease, with changes in gene expression associated with metabolic failure, inflammation, and fibrosis occurring at distinct stages. Delaying disease progression may be possible by identifying early biomarkers and intervention targets by stage-specific analysis of these active pathways (Liu et al., 2022). Molecular abnormalities inside the various kidney compartments, including the tubules and glomeruli, can be better understood with the use of spatial multi-omics, which could improve our knowledge of compartment-specific pathology. Using structure-specific spatial omics data, molecular alterations linked to fibrosis and inflammation have been discovered in glomerulonephritis (Duan et al., 2021).

Proteins and metabolites associated with various stages of chronic kidney disease (CKD) have been discovered using proteomics and metabolomics studies; these biomarkers may help in the diagnosis and monitoring of the illness at an earlier stage. Investigations using Mendelian randomisation (MR) and proteome-wide Mendelian randomisation (PWMR) have linked inflammatory proteins and metabolites like 2-hydroxyoctanoate and N-acetylalliin to chronic kidney disease (CKD), kidney function, and possible treatment options (Lu & Wang, 2024).

The likelihood of finding new genetic correlations and molecular markers will increase as a result. In the field of nephrology, this method is essential for creating drugs specifically for underserved groups (Wilson et al., 2020). The repurposing of current drugs is made possible by combining multiple omics data sets that link established pharmacological pathways with distinct biological targets in renal illness. Medications with potential nephroprotective effects can be more easily identified with the use of omics-based approaches that correlate genomic and proteomic features with pathways in renal disease. By combining cutting-edge research methods like molecular docking and route mapping, this integration speeds up the process of finding new treatments for kidney disease. The goal is to find drugs that can have a greater impact on important disease mechanisms (Pulley et al., 2020).

Machine learning's (ML) advent has paved the way for the examination of massive multi-omics datasets, pattern recognition, and the creation of early warning systems for renal illness (Ogidi et al., 2025f). Personalised treatment plans have been developed through the integration of clinical and omics data, and machine learning methods have been used to enhance the risk prediction of chronic renal disease, for example, random forest survival analysis (Aoki et al., 2023). Computational approaches that extend beyond machine learning, including causal inference models, deep learning, and Bayesian inference, enhance the detection of patterns and the study of molecular interactions in renal illness (Hasanzadeh et al., 2022; Ogidi & Emaikwu, 2024).

Figure 1. Integration of multi-omics data in kidney diseases

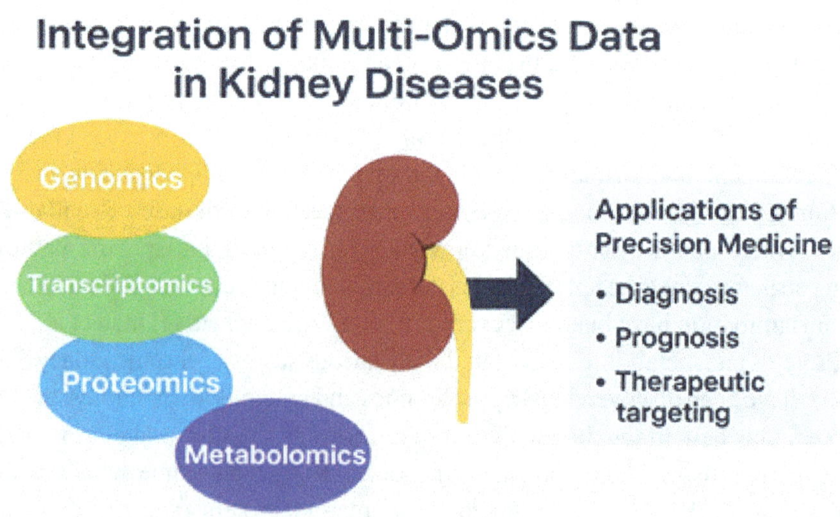

Table 3. Comparison of spatial omics platforms relevant to kidney research

Platform / Method	Technology Type	Spatial Resolution	Throughput (Genes/Proteins per assay)	Compatibility with Kidney Tissue	Key Strengths / Limitations	Reference
10x Genomics Visium	Spatial transcriptomics (barcoded array)	~55 µm spot diameter (~1–10 cells)	~18,000 genes	Widely used for mouse & human kidneys	High throughput and reproducibility; limited single-cell resolution	(10x Genomics, 2021)
Slide-seqV2	Bead-based spatial transcriptomics	10 µm	~10,000–15,000 genes	Compatible with cryosectioned kidney tissue	Near single-cell resolution; requires frozen tissue	(Stickels et al., 2021)
MERFISH (Multiplexed Error-Robust Fluorescence In Situ Hybridization)	Spatial transcriptomics (in situ hybridization)	~100 nm (subcellular)	100s–1,000s of genes (predefined panels)	Works with fixed kidney tissue	Ultra-high resolution; limited gene panel flexibility	(Xia et al., 2019).
seqFISH+	Sequential fluorescence in situ hybridization	~100 nm (subcellular)	Up to 10,000 genes	Complex due to autofluorescence in kidney	Single-molecule resolution; high imaging burden	(Eng et al., 2019).
GeoMx DSP (NanoString)	Spatial	~10–600 µm ROI (region-dependent)	1,800+ genes or 100+ proteins	FFPE-compatible and validated for kidney biopsies	Good for clinical samples; moderate throughput	(Merritt et al., 2020)
MALDI Imaging Mass Spectrometry (IMS)	Spatial	10–50 µm	1,000s of metabolites/ proteins	Excellent for kidney metabolic mapping	Broad molecular coverage; lower spatial resolution than FISH	(Patterson & Caprioli, 2022)
CODEX (CO-Detection by Indexing)	Spatial proteomics (multiplexed antibody imaging)	~700 nm	50–100 proteins	Used in kidney immune microenvironment studies	Multiplexed protein detection; antibody validation required	(Goltse et al., 2018)

APPLICATIONS OF PRECISION MEDICINE USING OMICS IN KIDNEY DISEASES

Nephrology precision medicine stands to benefit greatly from genomic technology's capacity to enhance diagnostic accuracy and pave the way for the development of individualised treatment regimens. Drugs like VX-407 are designed to target particular types of polycystic kidney disease. Genomic research has recently uncovered subtypes of the disease, each with its own distinct genetic pathways. For example, APOL1 mutations impact transplant decisions (Table 1). By revealing various gene expression profiles linked to diseases, including glomerulonephritis, transcriptomics improves the accuracy of diagnostic testing. As a factor in glomerulosclerosis, podocyte failure is explained, and therapy options for preserving podocyte integrity are offered (Hall, 2024).

Proteomics allows us to differentiate between glomerular illnesses like DKD by the identification of unique protein signatures, including biomarkers linked to disease progression such laminin and fibronectin (Barutta et al., 2021). Among the metabolomics-identified early signs of renal impairment is fumarate, which highlights oxidative stress and mitochondrial dysfunction as potential biomarkers and treatment targets in diabetic ketoacidosis (You et al., 2016). Chronic kidney disease (CKD) treatments must take the spatial context of molecular expression within renal tissues into account. Disease manifestation in certain kidney regions can be better understood with the use of spatial omics. Chromatin interaction studies have recently expanded our understanding of 3D regulatory networks and their relationship to disease risk (Abedini et al., 2022). When it comes to kidney diseases like lupus nephritis, single-cell omics sheds light on specific cells and pathways involved, which improves our knowledge of renal development and the biological diversity of kidney illness (Chen et al., 2024; Wei et al., 2024).

More targeted and maybe reversible treatments for disorders like inflammation and renal fibrosis are made possible by epigenomics, which reveals how epigenetic alterations affect gene expression (Fontecha-Barriuso et al., 2018). Predictive models driven by machine learning boost healthcare by accurately measuring the course of chronic kidney disease (CKD) and the success of treatment (Gogoi & Valan, 2024; Isaza-Ruget et al., 2024). Pharmacogenomics tailors pharmacological treatments to each patient's specific genetic composition in an effort to improve the safety and effectiveness of medications. This method is quite helpful for patients using immunosuppressants, since their window of opportunity for treatment is very small. No matter how advanced the technology gets, omics-driven biomarkers still need to undergo validation in the form of trials and longitudinal studies before they can be used in clinical settings (Qiu et al., 2020).

Recently, genomics in precision medicine has shown promise in bettering chronic renal disease care. Reznichenko et al. (2024) created a molecular classification that is particular to the kidneys, which led to significant advancements in this area. Four molecular categories (CKD-Blue, CKD-Gold, CKD-Olive, and CKD-Plum) were identified by examining the transcriptome of kidney biopsies using a self-organising map (SOM) machine learning technique. Along with well-established CKD staging methods and histological diagnoses, these groups give light on the basic biological mechanisms that drive development and encompass a wide range of illnesses. These educated molecular classifications suggest possible therapeutic targets, in line with other biological processes. There is a correlation between protein synthesis and vesicular transport in CKD-Plum, increased mitochondrial activity in CKD-Olive, and pro-inflammatory and profibrotic pathways in CKD-Gold (Reznichenko et al., 2024).

The predictive and therapeutic utility of CKD-Gold was demonstrated by the fact that it had the greatest incidence of progression to renal failure among different patient groups. By determining unique patterns in the urine proteome linked to several molecular classes, the study made it possible to classify using non-invasive biomarkers. This approach can replace the current systemic and indirect CKD classifications with a more accurate, tissue-specific molecular method. This would lead to greater projected accuracy and more personalised therapy. The data is consistent across several cohorts, which means it can be scaled and used for therapeutic reasons. This study demonstrates the potential of omics technology to classify chronic kidney disease, which, when applied to the field of precision medicine, can provide valuable information on the causes, progression, and potential treatments for this condition. More people can benefit from advances in nephrology if we all pitch in (Qiu et al., 2020).

Spatial multi-omics offers a promising pathway toward precision diagnostics and prognostics in kidney disease by linking molecular signatures to tissue microarchitecture. In a clinical framework, spatial transcriptomic or proteomic profiling of kidney biopsies could be integrated into existing diagnostic pipelines to complement histopathology, enabling the identification of cell type–specific injury patterns, molecular subtypes, and therapeutic targets. Standardized workflows using FFPE-compatible platforms such as NanoString GeoMx DSP or 10x Visium could generate spatially resolved molecular profiles from routine clinical samples, which, when coupled with AI-based image and omics integration, may allow automated classification of disease states and prediction of progression or treatment response. Longitudinal incorporation of these molecular maps into electronic health records and biobank datasets would support risk stratification models and facilitate biomarker validation across diverse patient cohorts, ultimately establishing spatial multi-omics as a bridge between molecular discovery and clinical nephrology practice (Reznichenko et al., 2024).

CHALLENGES AND FUTURE DIRECTIONS

The application of omics technology in nephrology has greatly expanded our understanding of kidney disease. However, there is still a long way to go before it can be effectively utilised in clinical practice. The lack of established formats and standards for integrating data across many omics platforms is a major obstacle. The enormous variety in data kinds, quantities, and formats among omics datasets makes integration a formidable obstacle. These datasets include genomics, transcriptomics, proteomics, and metabolomics, among others. For accurate and useful analysis, it is necessary to use well-established techniques and up-to-date computer tools for efficient integration. The necessity for advanced analytical methods to manage data complexity and heterogeneity has been brought to light by the utilisation of multi-omics data in renal medulla studies (Hodgin et al., 2024).

Robust omics research that is both statistically powerful and reproducible requires large, high-quality datasets. However, in most cases, these limitations are not severe, especially when it comes to marginalised populations or certain kidney diseases. In order to facilitate research, collaboration, and data-sharing, massive repositories must be supported. An excellent illustration of the significance of big databases for in-depth study is the German Chronic Kidney Disease (GCKD) program (Zacharias et al., 2018). Renal disease can be better understood with the use of the high-resolution cellular data produced by single-cell and spatial omics approaches. Data analysis becomes even more difficult as a result of these technologies because of the more complex data structures. Precisely defining cellular diversity in renal tissues requires ongoing work on improving methods for integrating bulk omics with single-cell and spatial data (Duan et al., 2021).

Machine learning and artificial intelligence are state-of-the-art computing tools that are essential for the analysis and management of complicated multi-omics data (Ogidi & Bunu, 2025). The use of these models in real-world clinical settings, however, makes their comprehension all the more difficult. It is critical to use explainable AI approaches that make model predictions and decisions more clear in order to increase trust in omics-based models among doctors and patients (Ogidi et al., 2025b,c,d). Healthcare providers will find this helpful for comprehending the reasoning behind these results. It is still challenging to convert the many possible kidney disease indicators uncovered by omics technologies into tools that are useful in clinical practice. The prevalence of renal illnesses makes longitudinal data essential for understanding the temporal dynamics of gene expression. On the other hand, time-series omics data collection and analysis necessitates substantial resources and cross-site cooperation. Logistical and economical limitations prevent longitudinal studies from being implemented, despite their ability to identify crucial intervention areas. To allow for earlier and more successful treatments, there has to be an increase in funding for longitudinal omics research in nephrology (Sheng et al., 2020).

To determine the reliability and validity of biomarkers in various settings and populations, large-scale validation studies are necessary. Potentially hindering the clinical implementation of promising candidates are regulatory constraints that could slow down the approval of omics-based biomarkers. This technology may be out of reach for healthcare organisations with restricted budgets due to the high expense of omics analysis. To make scientific discoveries available to more people, it is crucial to find affordable omics solutions, which is the focus of this topic. Before they are integrated into regular nephrology practice, expenses must be reduced and accessibility must be improved so that equitable access to omics technology breakthroughs may occur (Zacharias et al., 2018).

Particularly weighty are the ethical concerns of privacy, data ownership, and genetic prejudice. When dealing with genetic and epigenetic data, for example, there are ethical concerns because of the unknown effects it could have on the patient's family. In order to use data responsibly, patients need to know what omics testing entails and how it will affect them and their loved ones. A change in legislation, such as the elimination of algorithms based on race in kidney function tests, highlights the need for ethical frameworks that promote equity and protect patients' rights. Even with all the obstacles we're facing, new therapeutic advances might emerge from integrating omics data. As a result of genomic techniques' ability to shed light on the molecular pathways linked to kidney illnesses, new therapeutic targets can be found and individualised treatments can be developed. The integration of multi-omics has greatly improved our understanding and management of renal illness, which is a complex condition (Hodgin et al., 2024).

Emerging spatial omics technologies are rapidly transforming kidney research by enabling molecular mapping of renal architecture at unprecedented resolution. However, challenges persist in integrating multi-omics datasets such as genomic, transcriptomic, proteomic, and metabolomics due to differences in data scales, modalities, and spatial resolution. Recent advances in AI-driven data integration, including deep learning and graph-based models, now facilitate cross-platform harmonization, while new spatial frameworks such as Tangram, Seurat v5, and SpaceM allow alignment of single-cell data with spatial context. Simultaneously, innovations in tissue preparation, such as expansion microscopy, tissue clearing, and hydrogel embedding, are improving compatibility with complex kidney structures. New platforms like Stereo-seq and Seq-Scope offer subcellular resolution, bridging the gap between cellular localization and molecular identity within renal compartments.

Looking forward, translating spatial omics insights into clinical nephrology requires addressing reproducibility, cost, and data management barriers. The establishment of standardized reference maps, such as the Human Kidney Atlas and HuBMAP, promotes consistency across studies and facilitates integration with clinical-grade assays like NanoString GeoMx DSP, which are compatible with routine kidney biopsies. Cloud-based computational frameworks and interoperable data standards (e.g., OME-NGFF, FAIR principles) are improving data accessibility and reproducibility. As these innovations converge, kidney pathology is shifting toward precision nephrology, where spatially resolved molecular profiles guide diagnosis, prognosis, and therapeutic targeting. Ultimately, the fusion of high-resolution spatial omics with computational intelligence promises to move kidney research from descriptive observation to predictive, patient-tailored intervention.

CONCLUSION

Spatial multi-omics has emerged as a transformative approach in nephrology, offering unprecedented insight into the molecular, cellular, and spatial complexities of kidney diseases. By integrating spatial transcriptomics, proteomics, and metabolomics with histopathology, this technique allows researchers to map disease-specific molecular alterations within the anatomical context of kidney tissues. Such integration overcomes the limitations of bulk omics data, which lack spatial resolution, thereby enabling precise identification of pathological changes across different nephron segments and microenvironments. Research on kidney diseases, such as CKD, glomerulonephritis, and acute kidney damage, has started to use spatial multi-omics to uncover important pathways. Tissue injury, immunological infiltration, fibrosis, and vascular changes are correlated with regional differences in gene and protein expression, which have been uncovered by this method. Our knowledge of renal pathophysiology has been enhanced by these results, which also reveal molecular targets for diagnostics and treatment that were previously unknown.

More significantly, spatial multi-omics supports the development of personalized medicine approaches in nephrology. By identifying patient-specific molecular signatures within affected kidney regions, clinicians can potentially stratify patients based on disease progression and therapeutic response. This opens the door to targeted interventions, reducing unnecessary treatments and improving outcomes. Moreover, spatial omics can guide renal biopsy interpretation by adding molecular depth to traditional histological assessments. Despite its promise, several challenges remain. Spatial multi-omics is still limited by high costs, technical complexity, and the need for advanced bioinformatics tools. Standardization of data acquisition, integration, and interpretation is essential for widespread clinical adoption. Additionally, more extensive validation in larger patient cohorts is required to translate early discoveries into actionable clinical tools.

Future research will likely focus on refining spatial multi-omics platforms to achieve higher resolution, broader coverage, and integration with longitudinal patient data. Combining spatial omics with artificial intelligence and machine learning may further accelerate discovery and aid in real-time diagnostics. Such efforts will be critical in bridging the gap between bench and bedside, bringing precision nephrology closer to clinical reality. Spatial multi-omics represents a significant leap forward in the study and management of kidney diseases. Its ability to contextualize molecular information within tissue architecture provides a powerful framework for understanding disease mechanisms, identifying therapeutic targets, and advancing personalized treatment. Continued innovation, collaboration, and investment will be vital to harnessing the full potential of this emerging field in kidney health.

REFERENCES

Abedini, A., Levinsohn, J., Klötzer, K. A., Dumoulin, B., Ma, Z., Frederick, J., Dhillon, P., Balzer, M. S., Shrestha, R., & Liu, H., (2022). Spatially resolved human kidney multi-omics single cell atlas highlights the key role of the fibrotic microenvironment in kidney disease progression. *bioRxiv*. https://doi.org/DOI: 10.1101/2022.09.29.510141

Aoki, J., Kaya, C., Khalid, O., Kothari, T., Silberman, M. A., Skordis, C., Hughes, J., Hussong, J., & Salama, M. E. (2023). CKD progression prediction in a diverse US population: A machine-learning model. *Kidney Medicine*, *5*(3), 100692. DOI: 10.1016/j.xkme.2023.100692 PMID: 37637863

Barutta, F., Bellini, S., Canepa, S., Durazzo, M., & Gruden, G. (2021). Novel biomarkers of diabetic kidney disease: Current status and potential clinical application. *Acta Diabetologica*, *58*(7), 819–830. DOI: 10.1007/s00592-020-01656-9 PMID: 33528734

Benjamin, K., Bhandari, A., Shang, Z., Xing, Y., An, Y., Zhang, N., Hou, Y., Tillmann, U., Bull, K. R., & Harrington, H. A. (2022). Multiscale topology classifies and quantifies cell types in subcellular spatial transcriptomics. *arXiv*. https://arxiv .org/abs/2212.06505

Brandt, M. M., Meddens, C. A., Louzao-Martinez, L., Van Den Dungen, N. A. M., Lansu, N. R., Nieuwenhuis, E. E. S., Duncker, D. J., Verhaar, M. C., Joles, J. A., Mokry, M., & Cheng, C. (2018). Chromatin conformation links distal target genes to CKD loci. *Journal of the American Society of Nephrology*, *29*(2), 462–476. DOI: 10.1681/ASN.2016080875 PMID: 29093029

Chan, K., & Li, X. (2021). Current epigenetic insights in kidney development. *Genes*, *12*(8), 1281. DOI: 10.3390/genes12081281 PMID: 34440455

Chau, B. N., Xin, C., Hartner, J., Ren, S., Castano, A. P., Linn, G., Li, J., Tran, P. T., Kaimal, V., Huang, X., Chang, A. N., Li, S., Kalra, A., Grafals, M., Portilla, D., MacKenna, D. A., Orkin, S. H., & Duffield, J. S. (2012). MicroRNA-21 promotes fibrosis of the kidney by silencing metabolic pathways. *Science Translational Medicine*, *4*(121), 121ra18. Advance online publication. DOI: 10.1126/scitranslmed.3003205 PMID: 22344686

Chebib, F. T., & Torres, V. E. (2016). Autosomal dominant polycystic kidney disease: Core curriculum 2016. *American Journal of Kidney Diseases*, *67*(5), 792–810. DOI: 10.1053/j.ajkd.2015.07.037 PMID: 26530876

Chen, L., Su, W., Chen, H., Chen, D.-Q., Wang, M., Guo, Y., & Zhao, Y.-Y. (2018). Proteomics for biomarker identification and clinical application in kidney disease. In Makowski, G. S. (Ed.), *Advances in Clinical Chemistry* (Vol. 85, pp. 91–113). Elsevier., DOI: 10.1016/bs.acc.2018.02.004

Chen, W., Jin, B., Cheng, C., Peng, H., Zhang, X., Tan, W., Tang, R., Lian, X., Diao, H., Luo, N., Li, X., Fan, J., Shi, J., Yin, C., Wang, J., Peng, S., Yu, L., Li, J., Wu, R.-Q., & Jiang, X. (2024). Single-cell profiling reveals kidney CD163+ dendritic cell participation in human lupus nephritis. *Annals of the Rheumatic Diseases*, *83*(5), 608–623. DOI: 10.1136/ard-2023-224788 PMID: 38290829

Clark, D. J., Dhanasekaran, S. M., Petralia, F., Pan, J., Song, X., Hu, Y., Da Veiga Leprevost, F., Reva, B., Lih, T.-S. M., Chang, H.-Y., Ma, W., Huang, C., Ricketts, C. J., Chen, L., Krek, A., Li, Y., Rykunov, D., Li, Q. K., Chen, L. S., & Tu, Z. (2019). Integrated proteogenomic characterization of clear cell renal cell carcinoma. *Cell*, *179*(4), 964–983.e31. DOI: 10.1016/j.cell.2019.10.007 PMID: 31675502

Davies, R. (2018). The metabolomic quest for a biomarker in chronic kidney disease. *Clinical Kidney Journal*, *11*(6), 694–703. DOI: 10.1093/ckj/sfy037 PMID: 30288265

Devuyst, O., & Bochud, M. (2015). Uromodulin, kidney function, cardiovascular disease, and mortality. *Kidney International*, *88*(5), 944–946. DOI: 10.1038/ki.2015.267 PMID: 26579681

Dhas, Y., Arshad, N., Biswas, N., Jones, L., & Ashili, S. (2023). MicroRNA-21 silencing in diabetic nephropathy: Insights on therapeutic strategies. *Biomedicines*, *11*(9), 2583. DOI: 10.3390/biomedicines11092583 PMID: 37761024

Dieter, C., Lemos, N. E., Girardi, E., Ramos, D. T., Corrêa, N. R. D. F., Canani, L. H., Bauer, A. C., Assmann, T. S., & Crispim, D. (2023). The lncRNA MALAT1 is upregulated in urine of type 1 diabetes mellitus patients with diabetic kidney disease. *Genetics and Molecular Biology*, *46*(2), e20220291. DOI: 10.1590/1678-4685-gmb-2022-0291 PMID: 37272835

Ding, H., Zhang, L., Yang, Q., Zhang, X., & Li, X. (2021). Epigenetics in kidney diseases. In Makowski, G. S. (Ed.), *Advances in Clinical Chemistry* (Vol. 104, pp. 233–297). Elsevier., DOI: 10.1016/bs.acc.2020.09.001

Doshi, M. D., Gordon, E. J., Freedman, B. I., Glover, C., Locke, J. E., & Thomas, C. P. (2021). Integrating APOL1 kidney-risk variant testing in live kidney donor evaluation: An expert panel opinion. *Transplantation*, *105*(10), 2132–2134. DOI: 10.1097/TP.0000000000003641 PMID: 33534524

Du, S., Zhai, L., Ye, S., Wang, L., Liu, M., & Tan, M. (2023). In-depth urinary and exosome proteome profiling analysis identifies novel biomarkers for diabetic kidney disease. *Science China. Life Sciences*, *66*(12), 2587–2603. DOI: 10.1007/s11427-022-2348-0 PMID: 37405567

Duan, A., Wang, H., Zhu, Y., Wang, Q., Zhang, J., Hou, Q., Xing, Y., Shi, J., Hou, J., Qin, Z., Chen, Z., Liu, Z., & Yang, J. (2021). Chromatin architecture reveals cell type-specific target genes for kidney disease risk variants. *BMC Biology*, *19*(1), 38. DOI: 10.1186/s12915-021-00977-7 PMID: 33627123

Dubin, R. F., & Rhee, E. P. (2020). Proteomics and metabolomics in kidney disease, including insights into etiology, treatment, and prevention. *Clinical Journal of the American Society of Nephrology; CJASN*, *15*(3), 404–411. DOI: 10.2215/CJN.07420619 PMID: 31636087

Egbuna, O., Zimmerman, B., Manos, G., Fortier, A., Chirieac, M. C., Dakin, L. A., Friedman, D. J., Bramham, K., Campbell, K., Knebelmann, B., Barisoni, L., Falk, R. J., Gipson, D. S., Lipkowitz, M. S., Ojo, A., Bunnage, M. E., Pollak, M. R., Altshuler, D., & Chertow, G. M. (2023). Inaxaplin for proteinuric kidney disease in persons with two APOL1 variants. *The New England Journal of Medicine*, *388*(11), 969–979. DOI: 10.1056/NEJMoa2202396 PMID: 36920755

Elangovan, A., Li, Y., Pires, D. E. V., Davis, M. J., & Verspoor, K. (2022). Large-scale protein-protein post-translational modification extraction with distant supervision and confidence calibrated BioBERT. *BMC Bioinformatics*, *23*(1), 4. DOI: 10.1186/s12859-021-04504-x PMID: 34983371

Eng, C. H. L., Lawson, M., Zhu, Q., Dries, R., Koulena, N., Takei, Y., Yun, J., Cronin, C., Karp, C., Yuan, G.-C., & Cai, L. (2019). Transcriptome-scale super-resolved imaging in tissues by RNA seqFISH+. *Nature*, *568*(7751), 235–239. DOI: 10.1038/s41586-019-1049-y PMID: 30911168

Fontecha-Barriuso, M., Martin-Sanchez, D., Ruiz-Andres, O., Poveda, J., Sanchez-Niño, M. D., Valiño-Rivas, L., Ruiz-Ortega, M., Ortiz, A., & Sanz, A. B. (2018). Targeting epigenetic DNA and histone modifications to treat kidney disease. *Nephrology, Dialysis, Transplantation*, *33*(11), 1875–1886. DOI: 10.1093/ndt/gfy009 PMID: 29534238

Forbes, T. A., Howden, S. E., Lawlor, K., Phipson, B., Maksimovic, J., Hale, L., Wilson, S., Quinlan, C., Ho, G., Holman, K., Bennetts, B., Crawford, J., Trnka, P., Oshlack, A., Patel, C., Mallett, A., Simons, C., & Little, M. H. (2018). Patient-iPSC-derived kidney organoids show functional validation of a ciliopathic renal phenotype and reveal underlying pathogenetic mechanisms. *American Journal of Human Genetics*, *102*(5), 816–831. DOI: 10.1016/j.ajhg.2018.03.014 PMID: 29706353

Freedman, B. I., Kopp, J. B., Sampson, M. G., & Susztak, K. (2021). APOL1 at 10 years: Progress and next steps. *Kidney International*, *99*(6), 1296–1302. DOI: 10.1016/j.kint.2021.03.013 PMID: 33794228

Freedman, B. I., Limou, S., Ma, L., & Kopp, J. B. (2018). APOL1-associated nephropathy: A key contributor to racial disparities in CKD. *American Journal of Kidney Diseases*, *72*(5, Suppl 1), S8–S16. DOI: 10.1053/j.ajkd.2018.06.020 PMID: 30343724

Gajjala, P. R., Fliser, D., Speer, T., Jankowski, V., & Jankowski, J. (2019). Proteomic biomarkers of diabetic kidney disease: Progress and challenges. *Current Diabetes Reports*, *19*(9), 87. DOI: 10.1007/s11892-019-1198-0

Genomics. (2021). *Visium Spatial Gene Expression User Guide*. 10x Genomics. https://www.10xgenomics.com (Access on 22nd October 2025)

Giaimo, B. D., Ferrante, F., Herchenröther, A., Hake, S. B., & Borggrefe, T. (2019). The histone variant H2A.Z in gene regulation. *Epigenetics & Chromatin*, *12*(1), 37. DOI: 10.1186/s13072-019-0274-9 PMID: 31200754

Gilyazova, I., Ivanova, E., Izmailov, A., Sharifgaliev, I., Karunas, A., Pudova, E., Kobelyatskaya, A., Gilyazova, G., Izmailova, A., Pavlov, V., & Khusnutdinova, E. (2023). MicroRNA expression signatures in clear cell renal cell carcinoma: High-throughput searching for key miRNA markers in patients from the Volga-Ural region of Eurasian continent. *International Journal of Molecular Sciences*, *24*(13), 6909. DOI: 10.3390/ijms24086909 PMID: 37108073

Giudicelli, G. C., De Souza, C. M. B., Veronese, F. V., Pereira, L. V., Hünemeier, T., & Vianna, F. S. L. (2022). Precision medicine implementation challenges for APOL1 testing in chronic kidney disease in admixed populations. *Frontiers in Genetics*, *13*, 1016341. DOI: 10.3389/fgene.2022.1016341 PMID: 36588788

Gogoi, P., & Valan, J. A. (2024). Machine learning approaches for predicting and diagnosing chronic kidney disease: Current trends, challenges, solutions, and future directions. *International Urology and Nephrology*, *57*(4), 1245–1268. Advance online publication. DOI: 10.1007/s11255-024-04281-5 PMID: 39560857

Groopman, E. E., Marasa, M., Cameron-Christie, S., Petrovski, S., Aggarwal, V. S., Milo-Rasouly, H., Li, Y., Zhang, J., Nestor, J., Krithivasan, P., Lam, W. Y., Mitrotti, A., Piva, S., Kil, B. H., Chatterjee, D., Reingold, R., Bradbury, D., DiVecchia, M., Snyder, H., & Gharavi, A. G. (2019). Diagnostic utility of exome sequencing for kidney disease. *The New England Journal of Medicine*, *380*(2), 142–151. DOI: 10.1056/NEJMoa1806891 PMID: 30586318

Gu, Y.-Y., Lu, F.-H., Huang, X.-R., Zhang, L., Mao, W., Yu, X.-Q., Liu, X.-S., & Lan, H.-Y. (2021). Non-coding RNAs as biomarkers and therapeutic targets for diabetic kidney disease. *Frontiers in Pharmacology*, *11*, 583528. DOI: 10.3389/fphar.2020.583528 PMID: 33574750

Hall, G. (2024). Genetic insights into the mechanisms of proliferative glomerulonephritis. *The Journal of Clinical Investigation*, *134*(2), e183090. DOI: 10.1172/JCI183090 PMID: 39087471

Hao, P., Guo, T., & Sze, S. K. (2011). Simultaneous analysis of proteome, phospho- and glycoproteome of rat kidney tissue with electrostatic repulsion hydrophilic interaction chromatography. *PLoS One*, *6*(2), e16884. DOI: 10.1371/journal.pone.0016884 PMID: 21373199

Hasanzadeh, A., Hajiramezanali, E., Duffield, N., & Qian, X. (2022). MoReL: Multi-omics relational learning. *arXiv*. https://arxiv.org/abs/2203.08149

Hasegawa, S., & Inagi, R. (2021). Harnessing metabolomics to describe the pathophysiology underlying progression in diabetic kidney disease. *Current Diabetes Reports*, *21*(7), 21. DOI: 10.1007/s11892-021-01390-8 PMID: 33974145

Hodgin, J. B., Smith, C., & Kretzler, M. (2024). Multi-omics data integration shines a light on the renal medulla. *Kidney International*, *105*(2), 242–244. DOI: 10.1016/j.kint.2023.11.014 PMID: 38245213

Hosseinpour, M., Ardalani, F., Mohseni, M., Beheshtian, M., Arzhangi, S., Ossareh, S., Najmabadi, H., Nobakht, A., Kahrizi, K., & Broumand, B. (2022). Targeted next generation sequencing revealed novel variants in the PKD1 and PKD2 genes of Iranian patients with autosomal dominant polycystic kidney disease. *Archives of Iranian Medicine*, *25*(9), 600–608. DOI: 10.34172/aim.2022.95 PMID: 37543885

Isaza-Ruget, M. A., Yomayusa, N., González, C. A., De Oro, V. F. A., Cely, A., Murcia, J., Gonzalez-Velez, A., Robayo, A., Colmenares Mejía, C. C., & Castillo, A.. (2024). Predicting chronic kidney disease progression with artificial intelligence. *BMC Nephrology*, *25*(1), 148. DOI: 10.1186/s12882-024-03545-7 PMID: 38671349

Isnard, P., Li, D., Xuanyuan, Q., Wu, H., & Humphreys, B. D. (2024). Histopathological-based analysis of human kidney spatial transcriptomics data. *American Journal of Pathology*. Advance online publication. DOI: 10.1016/j.ajpath.2023.12.001

Itoku, A., Isaac, J., Wilson, S., Reidy, K., & Kaskel, F. (2024). APOL1 nephropathy risk variants through the life course: A review. *American Journal of Kidney Diseases*, *84*(1), 102–110. DOI: 10.1053/j.ajkd.2023.12.014 PMID: 38341125

Kalim, S., & Rhee, E. P. (2017). An overview of renal metabolomics. *Kidney International*, *91*(1), 61–69. DOI: 10.1016/j.kint.2016.08.021 PMID: 27692817

Kang, J., Guo, X., Peng, H., Deng, Y., Lai, J., Tang, L., Aoieong, C., Tou, T., Tsai, T., & Liu, X. (2023). Metabolic implications of amino acid metabolites in chronic kidney disease progression: A metabolomics analysis using OPLS-DA and MBRole2.0 database. *International Urology and Nephrology*, *56*(6), 1173–1184. DOI: 10.1007/s11255-023-03779-8 PMID: 37728808

Kashtan, C. E., Ding, J., Garosi, G., Heidet, L., Massella, L., Nakanishi, K., Nozu, K., Renieri, A., Rheault, M., Wang, F., & Gross, O. (2018). Alport syndrome: A unified classification of genetic disorders of collagen IV A345: A position paper of the Alport Syndrome Classification Working Group. *Kidney International*, *93*(5), 1045–1051. DOI: 10.1016/j.kint.2017.12.018 PMID: 29551517

Kiessling, P., & Kuppe, C. (2024). Spatial multi-omics: Novel tools to study the complexity of cardiovascular diseases. *Genome Medicine*, *16*(1), 14. DOI: 10.1186/s13073-024-01282-y PMID: 38238823

Kim, H., Kim, H.-H., Chang, C. L., Song, S. H., & Kim, N. (2021). Novel PKD1 mutations in patients with autosomal dominant polycystic kidney disease. *Laboratory Medicine*, *52*(2), 174–180. DOI: 10.1093/labmed/lmaa047 PMID: 32816041

Kim, H. K., Pham, M. H. C., Ko, K. S., Rhee, B. D., & Han, J. (2018). Alternative splicing isoforms in health and disease. *Pflügers Archiv*, *470*(7), 995–1016. DOI: 10.1007/s00424-018-2136-x PMID: 29536164

Kim, J., Ha, S., Son, M., Kim, D., Kim, M.-J., Kim, B., Kim, D., Chung, H. Y., & Chung, K. W. (2023). TLR7 activation by miR-21 promotes renal fibrosis by activating the pro-inflammatory signaling pathway in tubule epithelial cells. *Cell Communication and Signaling : CCS*, *21*(1), 215. DOI: 10.1186/s12964-023-01234-w PMID: 37596656

Knepper, M. A., & Pisitkun, T. (2007). Exosomes in urine: Who would have thought...? *Kidney International*, *72*(9), 1043–1045. DOI: 10.1038/sj.ki.5002510 PMID: 17943150

Kocaturk, N. M., & Gozuacik, D. (2018). Crosstalk between mammalian autophagy and the ubiquitin-proteasome system. *Frontiers in Cell and Developmental Biology*, *6*, 128. DOI: 10.3389/fcell.2018.00128 PMID: 30333975

Köttgen, A., Cornec-Le Gall, E., Halbritter, J., Kiryluk, K., Mallett, A. J., Parekh, R. S., Rasouly, H. M., Sampson, M. G., Tin, A., Antignac, C., Ars, E., Bergmann, C., Bleyer, A. J., Bockenhauer, D., Devuyst, O., Florez, J. C., Fowler, K. J., Franceschini, N., Fukagawa, M., & Gharavi, A. G. (2022). Genetics in chronic kidney disease: Conclusions from a Kidney Disease: Improving Global Outcomes (KDIGO) controversies conference. *Kidney International*, *101*(6), 1126–1141. DOI: 10.1016/j.kint.2022.03.019 PMID: 35460632

Lindström, N. O., De Sena Brandine, G., Ransick, A., & McMahon, A. P. (2019). Single-cell RNA sequencing of the adult mouse kidney: From molecular cataloging of cell types to disease-associated predictions. *American Journal of Kidney Diseases*, *73*(1), 140–142. DOI: 10.1053/j.ajkd.2018.07.002 PMID: 30241960

Ling, X. C., & Kuo, K.-L. (2018). Oxidative stress in chronic kidney disease. *Renal Replacement Therapy*, *4*(1), 53. DOI: 10.1186/s41100-018-0195-2

Liu, F., Chen, J., Li, Z., & Meng, X. (2022). Recent advances in epigenetics of age-related kidney diseases. *Genes*, *13*(5), 796. DOI: 10.3390/genes13050796 PMID: 35627181

Liu, X., Shi, J., Jiao, Y., An, J., Tian, J., Yang, Y., & Zhuo, L. (2024). Integrated multi-omics with machine learning to uncover the intricacies of kidney disease. *Briefings in Bioinformatics*, *25*(5), bbae364. Advance online publication. DOI: 10.1093/bib/bbae364 PMID: 39082652

Lu, H.-C., Dai, W.-N., & He, L.-Y. (2021). Epigenetic histone modifications in the pathogenesis of diabetic kidney disease. *Diabetes, Metabolic Syndrome and Obesity*, *14*, 329–344. DOI: 10.2147/DMSO.S288500 PMID: 33519221

Lu, Y.-Q., & Wang, Y. (2024). Multi-omic analysis reveals genetic determinants and therapeutic targets of chronic kidney disease and kidney function. *International Journal of Molecular Sciences*, *25*(11), 6033. DOI: 10.3390/ijms25116033 PMID: 38892221

Malone, A. F. (2021). APOL1 risk variants in kidney transplantation: A modulation of immune cell function. *The Journal of Clinical Investigation*, *131*(9), e154676. DOI: 10.1172/JCI154676 PMID: 34779415

Martire, S., & Banaszynski, L. A. (2020). The roles of histone variants in fine-tuning chromatin organization and function. *Nature Reviews. Molecular Cell Biology*, *21*(9), 522–541. DOI: 10.1038/s41580-020-0262-8 PMID: 32665685

Marumo, T., Yamasaki, Y., Ueda, S., & Kawachi, H. (2020). Epigenetic regulation of RASAL1 promotes kidney fibrosis. *Nature Communications*, *11*(1), 4469. DOI: 10.1038/s41467-020-18241-9 PMID: 32901013

Merritt, C. R., Ong, G. T., Church, S. E., Barker, K., Danaher, P., Geiss, G., Hoang, M., Jung, J., Liang, Y., McKay-Fleisch, J., Nguyen, K., Norgaard, Z., Sorg, K., Sprague, I., Warren, C., Warren, S., Webster, P. J., Zhou, Z., Zollinger, D. R., & Beechem, J. M. (2020). Multiplex digital spatial profiling of proteins and RNA in fixed tissue. *Nature Biotechnology*, *38*(8), 1000–1008. DOI: 10.1038/s41587-020-0472-9 PMID: 32393914

Meyer-Schwesinger, C. (2019). The ubiquitin–proteasome system in kidney physiology and disease. *Nature Reviews. Nephrology*, *15*(7), 393–411. DOI: 10.1038/s41581-019-0148-1 PMID: 31036905

Morevati, M., Fang, E. F., Mace, M. L., Kanbay, M., Gravesen, E., Nordholm, A., Egstrand, S., & Hornum, M. (2022). Roles of NAD+ in acute and chronic kidney diseases. *International Journal of Molecular Sciences*, *24*(1), 137. DOI: 10.3390/ijms24010137 PMID: 36613582

Nicholson, R. J., Pezzolesi, M. G., & Summers, S. A. (2021). Rotten to the cortex: Ceramide-mediated lipotoxicity in diabetic kidney disease. *Frontiers in Endocrinology*, *11*, 622692. DOI: 10.3389/fendo.2020.622692 PMID: 33584550

Nkuipou-Kenfack, E., Zürbig, P., & Mischak, H. (2014). The use of urine proteomic classifiers in diagnosis and prediction of kidney disease. *Nature Reviews. Nephrology*, *10*(3), 153–162. DOI: 10.1038/nrneph.2013.282

Oe, Y. (2024). Editorial for the special issue: Pathophysiology of chronic kidney disease and its complications. *Biomedicines*, *12*(416), 416. Advance online publication. DOI: 10.3390/biomedicines12020416 PMID: 38398018

Ogidi, O. I. (2024). Recent advances in anticancer activity and Bioinformatics Approach from Potential Plants. In: Computational Approaches in Biomaterials, Bioinformatics and Biomedical Engineering Applications. Published *CRC Press, Taylor and Francis Group, Florida, USA*. DOI: 10.1201/9781003354437-3

Ogidi, O. I. (2025). Natural Products Modulating Mitophagy: Implications for Cancer Therapy. In Pandurangan, A. (Ed.), *Roles of Mitophagy in Cancer Regulation* (pp. 325–358). IGI Global Scientific Publishing., DOI: 10.4018/979-8-3693-3836-0.ch012

Ogidi, O. I., & Ajoko, I. (2024). Neuroprotective Potentials of Phytochemicals. In: M.S. Rajput, T. Sarachana, P.D. Sarkar, M. Raval (eds) NeuroPhytomedicine. Published by *CRC Press, Routledge Taylor and Francis Group.*

Ogidi, O. I., Alfred-Ugbenbo, D., & Ebitimitula, O. (2025f). Drug Discovery and Development with Generative Artificial Intelligence. In Gupta, S., Gupta, U., Sharma, M., & Malik, K. (Eds.), *Generative Intelligence in Healthcare: Transforming Patient Care with AI Creativity* (1st ed.). CRC Press., DOI: 10.1201/9781003539483-1

Ogidi, O. I., Alfred-Ugbenbo, D., & Poripo, B. E. (2025d). AI-Driven Drug Discovery, Design, and Development in Immunological Disorders. In Chen, J. (Ed.), *AI-Assisted Computational Approaches for Immunological Disorders* (pp. 143–180). IGI Global Scientific Publishing., DOI: 10.4018/979-8-3693-9725-1.ch005

Ogidi, O. I., Aye, T., & Felagha, I. (2025c). Antiviral Therapeutics for Emerging Viruses. In Aslam, M., Yingngam, B., Javed, T., & Riaz, S. (Eds.), *Global Perspectives on Antiviral Drug Development* (pp. 217–260). IGI Global Scientific Publishing., DOI: 10.4018/979-8-3693-9276-8.ch006

Ogidi, O. I., Barwant, M. M., & Okara, E. (2024a). Biomarker Discovery: Protein Phosphorylation State. In *Spatially Variable Genes in Cancer: Development, Progression, and Treatment Response.* IGI Global., DOI: 10.4018/979-8-3693-7728-4.ch005

Ogidi, O. I., & Bunu, E. M. (2025). Blockchain Technology and Genomic Research. In Chaudhary, A., Singh, R., & Agarwal, G. (Eds.), *Blockchain in Health.* John Wiley & Sons, Inc., DOI: 10.1002/9781394336043.ch19

Ogidi, O. I., Christopher, E., & Orlu, H. A. (2025a). Environmental DNA and RNA Metabarcoding: Advancing Insights in Cancer Research. In Raghavan, R. (Ed.), *Cancer Exposomics and Environmental Influences on Carcinogenesis* (pp. 279–322). IGI Global Scientific Publishing., DOI: 10.4018/979-8-3373-2165-3.ch007

Ogidi, O. I., Chukwunonso, O. A., Afakwu-Adimoha, H. C., & Barwant, M. M. (2025b). Ethical Considerations in Generative Artificial Intelligence in Healthcare. In: Generative Artificial Intelligence in Healthcare: Current Practice and Future Development. *CRC Press, Taylor and Francis Group, Florida, USA.* https://doi.org/ DOI: 10.1201/9781003488255-5

Ogidi, O. I., & Emaikwu, N. G. (2024). Plant Phenolic compound isolation and its Bioinformatics Approaches of Molecular Mechanisms in Antimicrobial activities and Resistance. In: Computational Approaches in Biomaterials, Bioinformatics and Biomedical Engineering Applications. Publishered *CRC Press, Taylor and Francis Group, Florida, USA.* DOI: 10.1201/9781003354437-8

Ogidi, O. I., Erho, S. A., & Omu, O. (2025e). Reactive Oxygen Species in Cancer. In Prabhakar, P. (Ed.), *The Role of Reactive Oxygen Species in Human Health and Disease* (pp. 339–376). IGI Global Scientific Publishing., DOI: 10.4018/979-8-3693-7919-6.ch010

Ogidi, O. I., Ijere, N. D., & Sogoi, P. B. (2025g). Neural and Endocrine Regulation of Immunity. In Reddy, K. (Ed.), *Research Methodologies and Practical Applications in Psychoneuroimmunology* (pp. 17–42). IGI Global Scientific Publishing., DOI: 10.4018/979-8-3693-7432-0.ch002

Ogidi, O. I., Ogregade, I., & Joshua, S. (2024b). Investigation of Renal Protective Effects of Bryophyllum pinnatum (Lam) Extract in Acetaminophen Induced Toxicity in Albino Rats. *Medical and Pharmaceutical Journal, 3*(2), 82–93. DOI: 10.55940/medphar202485

Ogidi, O. I., Orlu, H. A., & Poripo, B. E. (2025h). Evaluation of Hepatoprotective Activities of *Bryophyllum pinnatum* Leaf Extract in Paracetamol Induced Toxicity In Wistar Rats. *Trends in Pharmaceutical Sciences, 11*(1), -. DOI: 10.30476/tips.2025.105055.1272

Ogidi, O. I., & Tawariowei, A. M. (2023). Endocrine Disrupting Chemicals and their Role in Cancer-A review. *Molecular Sciences And Applications., 3*, 5–14. DOI: 10.37394/232023.2023.3.2

Pagel, O., Loroch, S., Sickmann, A., & Zahedi, R. P. (2015). Current strategies and findings in clinically relevant post-translational modification-specific proteomics. *Expert Review of Proteomics, 12*(3), 235–253. DOI: 10.1586/14789450.2015.1042867 PMID: 25955281

Panizo, S., Martínez-Arias, L., Alonso-Montes, C., Cannata, P., Martín-Carro, B., Fernández-Martín, J. L., Naves-Díaz, M., Carrillo López, N., & Cannata-Andía, J. B. (2021). Fibrosis in chronic kidney disease: Pathogenesis and consequences. *International Journal of Molecular Sciences, 22*(8), 408. DOI: 10.3390/ijms22010408 PMID: 33401711

Park, J., Shrestha, R., Qiu, C., Kondo, A., Huang, S., Werth, M., Li, M., Barasch, J., & Susztak, K. (2018). Single-cell transcriptomics of the mouse kidney reveals potential cellular targets of kidney disease. *Science, 360*(6390), 758–763. DOI: 10.1126/science.aar2131 PMID: 29622724

Patera, F., Gatticchi, L., Cellini, B., Chiasserini, D., & Reboldi, G. (2024). Kidney fibrosis and oxidative stress: From molecular pathways to new pharmacological opportunities. *Biomolecules, 14*(1), 137. DOI: 10.3390/biom14010137 PMID: 38275766

Patterson, N. H., & Caprioli, R. M. (2022). Integrated molecular imaging of kidney tissue by mass spectrometry. *Nature Reviews. Nephrology*, *18*(1), 23–38. DOI: 10.1038/s41581-021-00487-2 PMID: 36253509

Pereira, P. R., Carrageta, D. F., Oliveira, P. F., Rodrigues, A., Alves, M. G., & Monteiro, M. P. (2022). Metabolomics as a tool for the early diagnosis and prognosis of diabetic kidney disease. *Medicinal Research Reviews*, *42*(3), 1518–1544. DOI: 10.1002/med.21883 PMID: 35274315

Picard, M., Scott-Boyer, M. P., Bodein, A., Périn, O., & Droit, A. (2021). Integration strategies of multi-omics data for machine learning analysis. *Computational and Structural Biotechnology Journal*, *19*, 3735–3746. DOI: 10.1016/j.csbj.2021.06.030 PMID: 34285775

Pulley, J. M., Rhoads, J. P., Jerome, R. N., Challa, A. P., Erreger, K. B., Joly, M. M., Lavieri, R. R., Perry, K. E., Zaleski, N. M., Shirey-Rice, J. K., & Aronoff, D. M. (2020). Using what we already have: Uncovering new drug repurposing strategies in existing omics data. *Annual Review of Pharmacology and Toxicology*, *60*(1), 333–352. DOI: 10.1146/annurev-pharmtox-010919-023537 PMID: 31337270

Qiu, X., Wu, Z., Xu, Q., Sheng, C., & Jiao, Z. (2020). Pharmacogenomics of immunosuppressants. In Cai, W., Liu, Z., Miao, L., & Xiang, X. (Eds.), *Pharmacogenomics in precision medicine* (pp. 83–106). Springer., DOI: 10.1007/978-981-15-3895-7_5

Raghubar, A. M., Pham, D. T., Tan, X., Grice, L. F., Crawford, J., Lam, P. Y., Andersen, S. B., Yoon, S., Teoh, S. M., Matigian, N. A., Stewart, A., Francis, L., Ng, M. S. Y., Healy, H. G., Combes, A. N., Kassianos, A. J., Nguyen, Q., & Mallett, A. J. (2022). Spatially resolved transcriptomes of mammalian kidneys illustrate the molecular complexity and interactions of functional nephron segments. *Frontiers in Medicine*, *9*, 873923. DOI: 10.3389/fmed.2022.873923 PMID: 35872784

Ramírez Medina, C. R., Ali, I., Baricevic-Jones, I., Odudu, A., Saleem, M. A., Whetton, A. D., Kalra, P. A., & Geifman, N. (2023). Proteomic signature associated with chronic kidney disease (CKD) progression identified by data-independent acquisition mass spectrometry. *Clinical Proteomics*, *20*(1), 19. DOI: 10.1186/s12014-023-09405-0 PMID: 37076799

Ranek, J. S., Stanley, N., & Purvis, J. E. (2022). Integrating temporal single-cell gene expression modalities for trajectory inference and disease prediction. *Genome Biology*, *23*(1), 186. DOI: 10.1186/s13059-022-02749-0 PMID: 36064614

Ren, P., Lu, L., Cai, S., Chen, J., Lin, W., & Han, F. (2021). Alternative splicing: A new cause and potential therapeutic target in autoimmune disease. *Frontiers in Immunology*, *12*, 713540. DOI: 10.3389/fimmu.2021.713540 PMID: 34484216

Reznichenko, A., Nair, V., Eddy, S., Fermin, D., Tomilo, M., Slidel, T., Ju, W., Henry, I., Badal, S. S., Wesley, J. D., Liles, J. T., Moosmang, S., Williams, J. M., Quinn, C. M., Bitzer, M., Hodgin, J. B., Barisoni, L., Karihaloo, A., Breyer, M. D., & Kretzler, M. (2024). Unbiased kidney-centric molecular categorization of chronic kidney disease as a step towards precision medicine. *Kidney International, 105*(6), 1263–1278. DOI: 10.1016/j.kint.2024.01.012 PMID: 38286178

Ryu, Y., Han, G. H., Jung, E., & Hwang, D. (2023). Integration of single-cell RNA-seq datasets: A review of computational methods. *Molecules and Cells, 46*(2), 106–119. DOI: 10.14348/molcells.2023.0009 PMID: 36859475

Saez-Rodriguez, J., Rinschen, M. M., Floege, J., & Kramann, R. (2019). Big science and big data in nephrology. *Kidney International, 95*(6), 1326–1337. DOI: 10.1016/j.kint.2018.11.048 PMID: 30982672

Schena, F. P., Chiurlia, S., Abbrescia, D. I., & Cox, S. N. (2024). Kidney and urine cell transcriptomics in IgA nephropathy and lupus nephritis: A narrative review. *Clinical Kidney Journal, 17*(1), sfad121. Advance online publication. DOI: 10.1093/ckj/sfad121 PMID: 38186900

Schwalm, S., Manaila, R., Oftring, A., Schaefer, L., Von Gunten, S., & Pfeilschifter, J. (2024). The contribution of the sphingosine 1-phosphate signaling pathway to chronic kidney diseases: Recent findings and new perspectives. *Pflügers Archiv, 476*(10), 1845–1861. DOI: 10.1007/s00424-024-03029-5 PMID: 39384640

Sheng, X., Qiu, C., Liu, H., Gluck, C., Hsu, J. Y., He, J., Hsu, C., Sha, D., Weir, M. R., Isakova, T., Raj, D., Rincon-Choles, H., Feldman, H. I., Townsend, R., Li, H., & Susztak, K. (2020). Systematic integrated analysis of genetic and epigenetic variation in diabetic kidney disease. *Proceedings of the National Academy of Sciences of the United States of America, 117*(47), 29013–29024. DOI: 10.1073/pnas.2005905117 PMID: 33144501

Si, S., Liu, H., Xu, L., & Zhan, S. (2024). Identification of novel therapeutic targets for chronic kidney disease and kidney function by integrating multi-omics proteome with transcriptome. *Genome Medicine, 16*(1), 84. DOI: 10.1186/s13073-024-01356-x PMID: 38898508

Steenbergen, E. J., Troyanov, S., Rovin, B. H., & Parikh, S. V. (2021). Multi-omics integration in lupus nephritis identifies molecular subgroups and therapeutic targets. *Kidney International, 100*(5), 1013–1026. DOI: 10.1016/j.kint.2021.05.012

Steinbrenner, I., Schultheiss, U. T., Kotsis, F., Schlosser, P., Stockmann, H., Mohney, R. P., Schmid, M., Oefner, P. J., Eckardt, K.-U., Köttgen, A., Sekula, P., Eckardt, K.-U., Meiselbach, H., Schneider, M. P., Schiffer, M., Prokosch, H.-U., Bärthlein, B., Beck, A., Reis, A., & Nadal, J. (2021). Urine metabolite levels, adverse kidney outcomes, and mortality in CKD patients: A metabolome-wide association study. *American Journal of Kidney Diseases*, *78*(5), 669–677.e1. DOI: 10.1053/j.ajkd.2021.01.018 PMID: 33839201

Stickels, R. R., Murray, E., Kumar, P., Li, J., Marshall, J. L., Di Bella, D. J., & Macosko, E. Z. (2021). Highly sensitive spatial transcriptomics at near-cellular resolution with Slide-seqV2. *Nature Biotechnology*, *39*(3), 313–319. DOI: 10.1038/s41587-020-0739-1 PMID: 33288904

Takabatake, Y., Kimura, T., Takahashi, A., & Isaka, Y. (2014). Autophagy and the kidney: Health and disease. *Nephrology, Dialysis, Transplantation*, *29*(9), 1639–1647. DOI: 10.1093/ndt/gft535 PMID: 24520117

Tezval, H., Merseburger, A. S., Matuschek, I., Machtens, S., Kuczyk, M. A., & Serth, J. (2008). RASSF1A protein expression and correlation with clinicopathological parameters in renal cell carcinoma. *BMC Urology*, *8*(1), 12. DOI: 10.1186/1471-2490-8-12 PMID: 18822131

Trabulus, S., Zor, M. S., Alagoz, S., Dincer, M. T., Meşe, M., Yilmaz, E., Tahir Turanli, E., & Seyahi, N. (2024). Profiling of five urinary exosomal miRNAs for the differential diagnosis of patients with diabetic kidney disease and focal segmental glomerulosclerosis. *PLoS One*, *19*(1), e0312470. DOI: 10.1371/journal.pone.0312470 PMID: 39471136

Tran, T., Park, J., Kim, O., Kim, B., Choi, D., Lee, J., Kim, K., Oh, B., & Lee, H. (2013). Combined phospho- and glycoproteome enrichment in nephrocalcinosis tissues of phytate-fed rats. *Rapid Communications in Mass Spectrometry*, *27*(23), 2767–2776. DOI: 10.1002/rcm.6742 PMID: 24214862

Verbeke, F., Siwy, J., Van Biesen, W., Mischak, H., Pletinck, A., Schepers, E., Neirynck, N., Magalhães, P., Pejchinovski, M., & Pontillo, C.. (2021). The urinary proteomics classifier CKD273 predicts cardiovascular outcome in patients with chronic kidney disease. *Nephrology, Dialysis, Transplantation*, *36*(5), 811–818. DOI: 10.1093/ndt/gfz242 PMID: 31837226

Wang, D., Pan, M., Li, H., Li, M., Li, P., Xiong, F., & Xiao, H. (2024). Four novel mutations identified in the COL4A3, COL4A4 and COL4A5 genes in 10 families with Alport syndrome. *BMC Medical Genomics*, *17*(1), 181. DOI: 10.1186/s12920-024-01953-0 PMID: 38978054

Wang, Y., Tang, S., Ma, R., Zamit, I., Wei, Y., & Pan, Y. (2022). Multi-modal intermediate integrative methods in neuropsychiatric disorders: A review. *Computational and Structural Biotechnology Journal*, *20*, 6149–6162. DOI: 10.1016/j.csbj.2022.11.008 PMID: 36420153

Wanner, N., & Bechtel-Walz, W. (2017). Epigenetics of kidney disease. *Cell and Tissue Research*, *369*(1), 75–92. DOI: 10.1007/s00441-017-2588-x PMID: 28286899

Wei, S., Shen, H., Zhang, Y., Liu, C., Li, S., Yao, J., Jin, Z., & Yu, H. (2024). Integrative analysis of single-cell and bulk transcriptome data reveal the significant role of macrophages in lupus nephritis. *Arthritis Research & Therapy*, *26*(1), 84. DOI: 10.1186/s13075-024-03311-y PMID: 38610007

Wilson, P. C., Ledru, N., & Humphreys, B. D. (2020). Epigenomics and the kidney. *Current Opinion in Nephrology and Hypertension*, *29*(3), 280–285. DOI: 10.1097/MNH.0000000000000602 PMID: 32235270

Wu, H., Malone, A. F., Donnelly, E. L., Kirita, Y., Uchimura, K., Ramakrishnan, S. M., Gaut, J. P., & Humphreys, B. D. (2018). Single-cell transcriptomics of a human kidney allograft biopsy specimen defines a diverse inflammatory response. *Journal of the American Society of Nephrology*, *29*(8), 2069–2080. DOI: 10.1681/ASN.2018020125 PMID: 29980650

Xia, C., Fan, J., Emanuel, G., Hao, J., & Zhuang, X. (2019). Spatial transcriptome profiling by MERFISH reveals subcellular RNA compartmentalization and cell cycle-dependent gene expression. *Nature Methods*, *16*(10), 989–993. DOI: 10.1038/s41592-019-0548-y PMID: 31501331

Yang, Z., Song, D., Wang, Y., & Tang, L. (2022). lncRNA MALAT1 promotes diabetic nephropathy progression via miR-15b-5p/TLR4 signaling axis. *Journal of Immunology Research*, *8098001*, 1–13. Advance online publication. DOI: 10.1155/2022/8098001 PMID: 35910856

You, Y.-H., Quach, T., Saito, R., Pham, J., & Sharma, K. (2016). Metabolomics reveals a key role for fumarate in mediating the effects of NADPH oxidase 4 in diabetic kidney disease. *Journal of the American Society of Nephrology*, *27*(2), 466–481. DOI: 10.1681/ASN.2015030302 PMID: 26203118

Yuan, H., Reddy, M. A., Sun, G., Lanting, L., Wang, M., Kato, M., & Natarajan, R. (2013). Involvement of P300/CBP and epigenetic histone acetylation in TGF-β1-mediated gene transcription in mesangial cells. *American Journal of Physiology. Renal Physiology*, *304*(5), F601–F613. DOI: 10.1152/ajprenal.00523.2012 PMID: 23235480

Yusuf, A. A., Govender, M. A., Brandenburg, J.-T., & Winkler, C. A. (2021). Kidney disease and APOL1. *Human Molecular Genetics*, *30*(R1, R2), R129–R137. DOI: 10.1093/hmg/ddab024 PMID: 33744923

Zacharias, H. U., Altenbuchinger, M., Solbrig, S., Schäfer, A., Buyukozkan, M., Schultheiß, U. T., Kotsis, F., Köttgen, A., Krumsiek, J., & Theis, F. J., (2018). Fully integrative data analysis of NMR metabolic fingerprints with comprehensive patient data: A case report based on the German Chronic Kidney Disease (GCKD) study. *arXiv*. https://arxiv.org/abs/1810.04281

Zaidan, M., Burtin, M., Zhang, J. D., Blanc, T., Barre, P., Garbay, S., Nguyen, C., Vasseur, F., Yammine, L., Germano, S., Badi, L., Gubler, M.-C., Gallazzini, M., Friedlander, G., Pontoglio, M., & Terzi, F. (2020). Signaling pathways predisposing to chronic kidney disease progression. *JCI Insight*, *5*(10), e126183. DOI: 10.1172/jci.insight.126183 PMID: 32376805

Zhuang, Q., Chen, Z., Shen, J., Fan, M., Xue, D., Lu, H., Xu, R., & He, X. (2018). RASSF1A promoter methylation correlates with development, progression, and poor cancer-specific survival of renal cell carcinoma: Trial sequential analysis. *OncoTargets and Therapy*, *12*, 119–134. DOI: 10.2147/OTT.S183142 PMID: 30588036

Zhuang, S., & Liu, N. (2014). EGFR signaling in renal fibrosis. *Kidney International Supplements*, *4*(1), 70–74. DOI: 10.1038/kisup.2014.13 PMID: 26312153

Chapter 10
Spatial Multi-Omics in Veterinary Infectious Diseases

Dragos Constantin Anita
Regional Center of Advanced Research for Emerging Diseases, Zoonoses, and Food Safety, Iasi University of Life Sciences, Romania

Adriana Elena Anita
Iasi University of Life Sciences, Romania

Irina Negut
https://orcid.org/0000-0003-4038-7548
National Institute for Lasers, Plasma, and Radiation Physics, Romania

ABSTRACT

Spatial multi-omics is transforming veterinary infectious disease research by integrating genomic, transcriptomic, proteomic, and metabolomic data with spatial and histological context. This approach reveals host–pathogen interactions, cellular heterogeneity, and localized immune or metabolic responses that shape disease outcomes. Emerging technologies such as SM-Omics, SpatialGlue, and MOSAIK enable multi-layered data integration across imaging and molecular scales. In veterinary medicine, spatial multi-omics provides insights into pathogen localization, immune microenvironments, and species-specific mechanisms in bacterial, viral, and parasitic infections. Combined with artificial intelligence (AI) and digital pathology, it enhances diagnostic accuracy, biomarker discovery, and outbreak prediction. Despite challenges in cost and data standardization, the convergence of spatial omics, AI, and One Health strategies is paving the way for precision veterinary medicine and improved surveillance of zoonotic diseases.

DOI: 10.4018/979-8-3373-4849-0.ch010

1. INTRODUCTION

Spatial multi-omics has emerged as a transformative approach in biomedical and veterinary sciences, enabling researchers to map the molecular architecture of tissues while preserving spatial context. Unlike conventional bulk or single-cell omics, which lose positional information during cell dissociation, spatial multi-omics integrates transcriptomic, proteomic, metabolomic, and epigenomic data with their precise tissue localization, revealing how cellular microenvironments and molecular networks shape disease processes (Bressan et al., 2023).

The integration of multiple molecular modalities, such as spatial transcriptomics, proteomics, and metabolomics, has revolutionized our understanding of tissue heterogeneity, intercellular communication, and host–pathogen interactions. In infectious diseases, these technologies allow for the precise identification of infection niches, immune cell localization, and metabolic alterations within infected tissues, facilitating a more comprehensive understanding of disease progression and pathogen dynamics.

Recent breakthroughs in spatial omics technologies, including automated high-throughput systems such as SM-Omics and computational frameworks like SpatialGlue, have accelerated the integration of imaging, sequencing, and mass spectrometry data across biological scales (Vickovic et al., 2022).

In the veterinary context, the integration of omics technologies is still in its early stages but holds significant promise for advancing diagnostics, vaccine development, and comparative pathology. Omics approaches such as proteomics, metabolomics, and lipidomics are being applied to unravel host-specific responses and pathogen adaptations across animal species, providing insights that enhance One Health surveillance and translational research.

Spatial multi-omics is particularly powerful for investigating infectious diseases with strong spatial heterogeneity, such as tuberculosis, leishmaniasis and numerous tick-borne infections that affect both domestic animals and wildlife. By integrating imaging mass spectrometry, in situ sequencing, and AI-driven spatial analysis, researchers can visualize host–pathogen interactions and immune microenvironments at single-cell resolution. This is especially valuable in veterinary medicine, where species-specific immune architectures, diverse pathogen reservoirs, and complex tissue responses often complicate traditional diagnostic and research approaches (Rodríguez-Durán et al, 2025).

Ultimately, the convergence of spatial biology, advanced bioinformatics, and artificial intelligence is propelling a new generation of infectious disease research in both human and veterinary health. For veterinary medicine, this cross-disciplinary integration offers major advantages: deeper insights into host–pathogen ecosystems across species, improved identification of clinically relevant biomarkers, enhanced translational precision for vaccine and therapy development, and stronger preparedness for emerging and zoonotic diseases.

2. TECHNOLOGICAL LANDSCAPE OF SPATIAL MULTI-OMICS

Multi-omics refers to the integrated application of multiple high-throughput screening technologies, such as genomics, transcriptomics, single-cell transcriptomics, proteomics, metabolomics, and spatial transcriptomics. In veterinary research, these approaches play an important role in advancing our understanding of animal health and disease, enabling deeper insights into disease mechanisms, host–pathogen interactions, and species-specific biological processes.

The technological landscape of spatial multi-omics in veterinary infectious disease research is rapidly evolving, offering detailed insights into the spatial organization of biological molecules and their roles in host–pathogen interactions. Below, we present the main technological modalities and their integration in this context.

2.1. Spatial transcriptomics

Spatial transcriptomics (ST) has emerged as a transformative tool in understanding host-pathogen interactions within the tissue microenvironment, allowing researchers to map gene expression while preserving the spatial architecture of the sample.

ST combines traditional histology with next-generation sequencing (NGS). The process generally involves several key steps:

a) Sample preparation:

Tissue samples are collected and preserved via freezing or formalin-fixation followed by paraffin embedding (FFPE). Fresh-frozen tissues typically yield higher RNA quality but are more difficult to handle, whereas FFPE offers ease of storage and is compatible with clinical samples. Protocols like those developed by Wehrle et al. provide detailed guidance for FFPE spatial transcriptomics [REMOVED HYPERLINK FIELD](Wehrle et al., 2024). For example, dual spatial transcriptomics has been applied to COVID-19-infected human lungs, revealing co-localized expression of viral and host genes, and identifying key modulators of infection through unbiased, transcriptome-wide analysis (Sounart et al., 2023).

b) Tissue sectioning and mounting:

Thin tissue sections (usually 10 μm) are mounted onto specialized glass slides pre-coated with barcoded oligonucleotide arrays (e.g., Visium platform by 10x Genomics). These barcodes uniquely label mRNA molecules based on their spatial origin within the tissue section (Haber et al, 2025).

c) Tissue fixation, permeabilization, and staining:

After fixation, permeabilization enables RNA release from cells. Hematoxylin and eosin staining is often used to preserve morphological context for alignment between histological features and gene expression data.

d) Reverse transcription and cDNA synthesis:

RNA molecules hybridize to spatially barcoded probes on the slide, followed by reverse transcription to generate spatially encoded cDNA.

e) Library preparation and sequencing:

Spatially barcoded cDNA is amplified, fragmented, and sequenced using NGS platforms. The spatial barcodes allow for re-mapping of sequencing reads to their original tissue coordinates (Baptista et al., 2025)

f) Data analysis:

Advanced computational tools such as stPipe, SpaRTaCo, and MOSAIK are used to align gene expression profiles to tissue architecture, perform clustering and deconvolution, and identify spatial expression patterns.

This is particularly critical in veterinary infectious disease research, where animal models are essential for exploring zoonotic transmission, immune responses, and pathogen localization.

In veterinary contexts, studies on *Mycobacterium abscessus* infections in murine models demonstrate the utility of custom bacterial probes for mapping bacterial transcripts such as *rpoB* and virulence factors like *lsr2*, enabling dual visualization of bacterial and host responses in lung tissues under oxidative stress. Similarly, in murine models of ocular infection with *Pseudomonas aeruginosa*, spatial transcriptomics was used to discover a novel bacterial virulence factor (*PA2590*), as well as host inflammation patterns that predict disease severity.

The integration of spatial and single-cell transcriptomics further enhances resolution. For example, spatial mapping in *Plasmodium berghei*-infected mouse livers revealed "inflammatory hotspots" and gene expression changes related to lipid metabolism and immune zonation, improving our understanding of liver-stage malaria pathogenesis.

Current work in companion animals highlights the promise of ST technologies, for example, generating new insights into cancer immunotherapy in dogs (Ammons et al., 2024). In addition, companion animal research has begun applying ST to the study of equine neurodegeneration, offering valuable insights into the spatial organization of neural injury and inflammatory responses. These advances illustrate the broader potential to develop innovative disease models across additional companion species and to improve livestock health, productivity, and welfare.

Overall, spatial transcriptomics offers a powerful platform for examining complex host-pathogen dynamics, identifying disease biomarkers, and informing therapeutic strategies in veterinary infectious diseases.

2.2. Spatial proteomics

Spatial proteomics is a cutting-edge technique that enables the mapping of protein expression and localization within tissue architecture, thereby revealing complex interactions between host and pathogen at the protein level. In veterinary infectious disease research, it is particularly valuable for understanding how pathogens alter host cellular machinery and identifying potential protein biomarkers for diagnostics or therapeutic targeting.

Spatial proteomics typically employs high-resolution mass spectrometry (MS) or multiplexed imaging approaches such as immunofluorescence or mass cytometry to study protein distributions. Recent advances have enabled the use of FFPE tissue samples for spatial proteomic profiling. A notable contribution comes from Makhmut et al., who developed a protocol for ultra-low input proteomics using single-cell laser capture microdissection combined with MS-based analysis. Their approach allowed the quantification of over 2,000 proteins from minimal FFPE material, making it feasible for routine use in archived veterinary samples (Makmut et al., 2023).

Broadly, proteomics has been used to study various veterinary conditions, from bovine mastitis to equine joint diseases, by identifying disease biomarkers such as cathelicidin and haptoglobin (Eckersall et al., 2020), and it holds potential for improving diagnostics, therapeutics, and surveillance in both livestock and companion animals [20].

In veterinary medicine, proteomic studies have been instrumental in advancing our understanding of pathogenesis and in identifying host response markers. Proteomic analysis of tissues and biological fluids from animals is increasingly being applied to questions of health and disease in veterinary medicine, mirroring similar advances in human biomedical research. Comparative proteomics, focused on evaluating protein profiles across different species or sample types, adds further scientific value by enabling cross-species insights in cattle, dogs, poultry, cats, pigs, horses, fish, and other domesticated or farmed animals. For instance, spatial proteomics has enabled the differentiation of infection-specific protein profiles in murine kidney tissue infected with *Staphylococcus aureus*, pinpointing key proteins involved in immune defense and tissue remodeling (Ceciliani et al., 2014).

Proteomic studies in veterinary species have identified disease-specific protein biomarkers in blood, saliva, and tissues, aiding early diagnosis. Examples include serum protein changes in dogs with leishmaniosis or mitral valve degeneration, and bronchoalveolar proteins in terriers with pulmonary fibrosis.

Spatial proteomics supports subcellular localization of both host and pathogen proteins during infection, informing studies on zoonotic agents. For example, machine learning-enhanced tools like Spatial-DC have been proposed to reconstruct proteomic profiles at the cell-type level from complex tissues, useful in characterizing immune cell niches in infected organs.

Together, these innovations highlight the growing power of spatial proteomics to dissect infection biology with subcellular precision, bridging histopathology and systems biology for more accurate diagnosis, prognosis, and therapeutic design in veterinary medicine.

2.3. Spatial metabolomics

Spatial metabolomics is an advanced analytical approach that enables the visualization and quantification of small-molecule metabolites within tissue sections while preserving their spatial context. By combining sophisticated computational and statistical methods with state-of-the-art analytical technologies and chemical profiling, it represents a highly multidisciplinary field. In contrast to bulk metabolomics, which averages metabolic signals across entire tissues or fluids, spatial metabolomics pinpoints localized metabolic changes linked to infection, inflammation, tissue injury, or microbial colonization. This spatial precision makes it especially valuable for veterinary infectious disease research, where metabolic heterogeneity across tissues and species can profoundly influence disease progression and host responses. The primary engine of spatial metabolomics is MSI, especially matrix-assisted laser desorption/ionization (MALDI-MSI), desorption electrospray ionization (DESI), and secondary ion mass spectrometry (SIMS). These techniques allow label-free, high-resolution visualization of metabolite distributions at cellular or even subcellular levels, and are critical for untargeted discovery of infection-related metabolic signatures. MSI-based spatial metabolomics has already been applied across diverse research fields, including drug discovery, toxicology, nutrition, microbiology, and biomedical science. In each of these areas, it has proven especially effective for identifying disease-associated biomarkers and uncovering mechanistic metabolic pathways.

In veterinary settings, spatial metabolomics has begun to demonstrate value in characterizing infection-induced metabolic remodeling. For example, targeted metabolomic analysis in poultry infected with *Salmonella Enteritidis* showed significant changes in plasma metabolites before and after inoculation, pointing to its utility in early infection diagnostics[REMOVED HYPERLINK FIELD] (Lu et al., 2023).

Beyond diagnostics, spatial metabolomics provides insights into host–pathogen interactions by uncovering shifts in host metabolic pathways, such as glycolysis, lipid metabolism, and amino acid biosynthesis, which are often exploited by pathogens for survival and replication. These approaches have revealed "metabolic signatures" unique to specific infections, offering a route to biomarker discovery and even host-targeted therapies. Moreover, MSI approaches have uncovered host–pathogen metabolic interactions by highlighting shifts in key metabolic pathways. For example, pathogens are known to hijack host lipid metabolism or alter glycolytic flux to favor their replication. These interactions can be visualized and quantified using MSI techniques such as high-resolution MALDI-MS or SIMS (Chen et al., 2023)].

An exciting development in spatial metabolomics is the combination of fluorescence in situ hybridization (FISH) with MSI, exemplified by the metaFISH pipeline, which links microbial identity with localized metabolic outputs on the same tissue section. This dual approach was used to study deep-sea mussel symbioses but is adaptable to veterinary infectious models, especially in complex microbial communities (Greier et al., 2020).

The recent integration of artificial intelligence and DL tools into spatial metabolomics is further enhancing the accuracy and interpretability of MSI data, offering automated metabolite annotation and biological inference at unprecedented scale.

Despite these advantages, spatial metabolomics faces several challenges. These include metabolite annotation bottlenecks, sample preparation artifacts, and limitations in detection sensitivity for low-abundance metabolites. Metabolite identification remains difficult due to signal overlap, limited spectral databases, and ion suppression effects. Moreover, sample preparation methods, particularly for veterinary tissues with varied compositions, can introduce artifacts or lead to analyte degradation.

2.4. Imaging and spatial data integration

Imaging and spatial data integration form the critical connective tissue between spatial omics modalities, such as transcriptomics, proteomics, and metabolomics, allowing researchers to visualize and contextualize complex biological interactions in situ. Imaging informatics focuses on collecting detailed information at the tissue level, capturing not only the physical structure of organs but, in some cases, their functional activity. In veterinary infectious disease research, these integrative approaches enable comprehensive, multilayered characterization of host–pathogen interactions across tissue microenvironments.

2.4.1. Imaging technologies

Advanced imaging techniques provide spatial reference frameworks onto which molecular data can be mapped:

- H&E staining remains a gold standard for histopathology and is often used to guide spatial transcriptomic analyses by aligning morphological features with gene expression maps.
- Multiplexed Immunofluorescence (mIF) and immunohistochemistry (IHC) enable simultaneous visualization of multiple protein markers in tissue sections. These are vital for co-registering proteomic findings with cell phenotypes and local immune responses.
- Mass spectrometry imaging (MSI), including MALDI and DESI, facilitates untargeted spatial metabolomics, while imaging mass cytometry supports highly multiplexed protein imaging with subcellular resolution (Alexandrov et al., 2020).
- Fluorescence in situ hybridization (FISH), including metaFISH, has been combined with spatial metabolomics to link microbial identity with metabolite distribution within the same tissue section.

2.4.2. Spatial data integration platforms

By merging these imaging data with spatial omics layers requires computational alignment and multi-modal integration:

- Seurat and Giotto are widely used for integrating spatial transcriptomics with high-resolution images and other omics data, offering capabilities for clustering, trajectory inference, and spatially aware cell–cell interaction modeling (Dries et al., 2021).
- SpaGCN integrates gene expression, spatial proximity, and histological features using graph convolutional networks to produce more accurate tissue segmentation and functional zone detection

- Visium platform (10x Genomics) combines H&E imaging with barcoded transcript capture, enabling precise alignment between spatial gene expression and histology.

    MOSAIC and SpaRTaCo are newer platforms tailored for veterinary and biomedical research that align multiple omics layers to histological maps and facilitate cross-modality inference .
- MIIT (Multi-omics Imaging Integration Toolset) is a Python framework for integrating spatially resolved multi-omcis data. The main spatial-omics technologies that we focus on are Spatial Transcriptomics (through Visium) and MSI (ImzML). MIIT is designed for integrating spatial omics data, but it also includes tools for processing these datasets and supports formats like pointset data and tissue annotation masks. The integration of spatial omics data is particularly important in cancer research, as it helps in identifying novel tissue biomarkers (Wess et al., 2025).

2.4.3. Veterinary applications

In veterinary pathology, these integrative imaging approaches allow for:

- Mapping immune niches within infected tissues, identifying spatial zones of inflammation, immune suppression, or pathogen persistence.
- Detecting co-localization of pathogens and host responses, particularly in zoonotic diseases like brucellosis or mycoplasmosis.
- Visualizing disease progression in complex tissues such as lung, liver, or lymphoid organs in livestock and companion animals.

Integrated imaging and spatial transcriptomics are increasingly being used in veterinary and translational models to map the microanatomy of infection. These approaches allow researchers to simultaneously detect spatial gene expression, immune activation, and pathogen distribution in complex tissues such as lungs, eyes, or lymphoid organs.

Recent studies provide strong examples of such approaches. A murine model of ocular infection with *Pseudomonas aeruginosa* used dual spatial transcriptomics to simultaneously profile host and bacterial transcripts. This uncovered spatially enriched immune response zones and predicted bacterial burden with high accuracy. Importantly, the study identified novel virulence genes such as *PA2590*, highlighting the utility of integrated spatial omics in revealing host-pathogen interactions (Zhou et al., 2025). In a mouse lung model of *P. aeruginosa* infection, spatial lipidomics using mass spectrometry imaging (MSI) showed distinct phospholipid remodeling in airway versus parenchymal regions, correlating with neutrophilic inflammation and metabolic pathway shifts. This spatial resolution provided insights into local host-pathogen dynamics that would be missed by bulk tissue analysis (Scott et al., 2023). A related study applied TOF-SIMS imaging to lungs of mice infected with *P. aeruginosa*, identifying spatial accumulation of cholesteryl sulfate in bronchial epithelium and demonstrating bacterial effects on lipid distribution at the tissue level. These methods, though initially developed in human or rodent systems, are directly transferable to veterinary infectious disease research. For example, pathogens like *Brucella abortus* or *Mycoplasma bovis* could be investigated using similar approaches to identify immune microenvironments, pathogen niches, and tissue-specific immune evasion strategies in livestock.

2.5. Bioinformatics tools and pipelines

Spatial multi-omics integration represents a transformative approach in infectious disease research, particularly in veterinary and translational contexts. By co-registering spatial transcriptomics, proteomics, metabolomics, and epigenomics datasets onto tissue architecture, scientists can investigate the precise localization of cellular responses, immune niches, and pathogen-induced molecular alterations.

To enable comprehensive spatial multi-omics analyses in infectious disease research, a range of specialized bioinformatics tools and pipelines have been developed. These platforms facilitate the integration of transcriptomic, proteomic, metabolomic, and epigenomic data with spatial resolution, as summarized in the table below (*Table 1*).

Table 1. Key bioinformatics tools and pipelines for spatial multi-omics integration in infectious disease research

Tool / Pipeline	Functionality	Relevant studies
SpaMTP	Integrates spatial metabolomics and transcriptomics; provides automated metabolite annotation and visualization	Causer et al., 2024
AtlasXbrowser	GUI for processing DBiT-seq data; image alignment, spot calling, integration	Barnet et al., 2022
MOFA (Multi-Omics Factor Analysis)	Unsupervised integration of multi-modal data; discovers latent variables linking omics layers	Argelaguet et al, 2018
Deep Learning Pipelines	Uses ML/DL to classify cell types, spatial patterns, and regulatory states	Erfanian et al., 2023
Custom Python/R Pipelines	Combines Seurat, Cardinal, and CellProfiler for user-defined workflows	Balzer et al., 2021
stLearn	Integrates spatial transcriptomics with histology and tissue morphology features	Pham et al., 2020

continued on following page

Table 1. Continued

Tool / Pipeline	Functionality	Relevant studies
Giotto	Comprehensive spatial analysis platform; includes segmentation, visualization, and statistics	Dries et al., 2021
Seurat	Adds spatial-aware analysis tools to Seurat; allows multimodal data integration	
Squidpy	Framework for spatial transcriptomics and cell-cell communication in Python	Palla et al., 2022

3. VETERINARY PATHOGEN-HOST INTERACTIONS THROUGH A SPATIAL LENS

Host–pathogen interactions are defined as the complex biological processes through which bacteria, parasites, or viruses adapt and behave in order to survive, replicate, and disseminate within a host organism. These interactions also encompass the host's physiological and immune responses to invasion, which determine the outcome of infection. Depending on the scale of observation, host–pathogen interactions can be studied at the population level (epidemiological dynamics), the organismal level (clinical manifestation and immune response), or at the molecular and cellular levels (mechanisms of infection, replication, and immune evasion). Obligate pathogens replicate exclusively inside host cells, whereas facultative pathogens are capable of surviving outside the host as well.

These interactions are broadly categorized into three main types: (I) mutualism, in which both the host and the microorganism derive a net benefit from the association. A prominent example is the intestinal microbiota, composed primarily of commensal bacteria that protect the host against pathogenic species, contribute to nutrient absorption and metabolism, and play a critical role in immune system maturation and regulation; (II) commensalism, where one organism benefits from the relationship while the other remains largely unaffected, experiencing neither harm nor advantage; (III) parasitism, in which one organism benefits at the expense of the other, typically exploiting host resources for its own growth and reproduction.

In parasitic interactions, pathogens have evolved highly sophisticated strategies to ensure their survival and persistence within the host. These mechanisms often involve subverting or co-opting host cellular machinery to facilitate replication, while simultaneously inhibiting or evading immune defenses. Such molecular and cellular tactics can include interference with signaling pathways, modulation of apoptosis, and suppression of antigen presentation, all of which enable pathogens to establish chronic or latent infections.

3.1. Cellular microenvironments in animal infections

Host–pathogen interactions at the cellular level are central to the outcome of animal infections, shaping both pathogen survival and host defense. After breaching physical barriers such as the skin, mucosa, or epithelial linings, pathogens, including viruses, bacteria, fungi, and intracellular parasites, target specific host cells capable of supporting their replication. Once inside, pathogens hijack cellular machinery to replicate, disrupt normal cellular processes, and manipulate host signaling pathways. These interactions can lead to diverse cellular outcomes, including cytopathic effects and cell death, latent or asymptomatic infection, or pathological cellular transformations that contribute to disease progression (Omotade et al., 2019).

The cellular environment plays an important role in pathogen replication and survival. Pathogens rely on host cell machinery to access energy, biosynthetic components, and molecular pathways required for growth and propagation. The type and physiological state of the host cell strongly influence the pathogen's replication strategy, efficiency, and the overall course of infection.

Key aspects of the cellular environment in pathogen replication include: (I) Exploitation of host cell machinery: many pathogens cannot complete their life cycles independently. Viruses, obligate intracellular bacteria, and intracellular parasites rely on host processes, including protein synthesis, metabolic pathways, and energy production, to replicate and assemble new infectious particles or progeny. (II) Variability of replication dynamics: replication efficiency and strategies vary according to pathogen type and host cell characteristics. A given virus, bacterium, or parasite may replicate differently in epithelial cells, macrophages, or neurons, reflecting differences in cellular resources, innate defenses, and signaling pathways. The interplay between pathogens and the host cellular environment determines the success of infection, the kinetics of replication, and ultimately the severity and progression of disease (Li et al., 2021).

When tissue macrophages and dendritic cells detect a pathogen via pattern recognition receptors (PRRs), immune signaling cascades activate transcription factors such as NF-κB, driving the production of a broad array of inflammatory mediators. These mediators encompass a diverse range of molecules, including vasoactive amines like histamine, bradykinin, and serotonin; vasoactive peptides such as substance P and fibrin degradation products; pro-inflammatory cytokines including TNF-α, IL-1, IL-6, IL-10, IL-12, IL-15, IL-18, and IFN-γ; chemokines such as IL-8, MCP-1, and CXCL5; lipid mediators like platelet-activating factor (PAF) and eicosanoids; complement-derived anaphylatoxins C3a and C5a; and inducible nitric oxide synthase (iNOS), which increases nitric oxide production, promoting vascular smooth muscle relaxation and vasodilation. Collectively, these molecules act in concert to initiate and sustain the local inflammatory response by enhancing blood flow, directing immune cell migration through chemotaxis, promoting adhesion and diapedesis of neutrophils, and activating both humoral defenses, such as natural antibodies and complement, and cellular mechanisms, including phagocytosis and bactericidal activity [52].

Pathogens have evolved a wide array of strategies to manipulate host defenses. Intracellular pathogens may inhibit phagosome maturation, escape into the cytoplasm, block apoptotic pathways, or downregulate antigen presentation to evade detection. Extracellular pathogens may secrete toxins, form biofilms, or manipulate host signaling to dampen inflammatory responses. Host cells counter these tactics through intrinsic defense mechanisms such as autophagy, inflammasome activation, and the production of interferons and other antimicrobial mediators.

The key steps in cellular interactions with an intracellular pathogen depend on cellular susceptibility, penetration and replication, and host modulation. For a cell to be infected, it must first be capable of binding to the pathogen, a process often mediated by specific receptors on the cell surface. The cell must also be permeable, allowing the pathogen's genetic material to enter and replicate within it. Infection can subsequently alter host cell functions in various ways: lytic infection when the pathogen replicates rapidly, ultimately destroying the host cell to release new infectious particles; asymptomatic infection: The pathogen replicates within the cell without causing apparent damage, as observed with certain viruses; or cellular transformation: In some cases, infection can lead to alterations in the host cell, such as immortalization or other pathological modifications.

The resulting cellular microenvironment is highly dynamic and context-dependent, reflecting the continuous interplay between pathogen virulence strategies and host immune responses. Communication between immune and non-immune cells, combined with molecular and biochemical signaling, establishes a spatially and temporally organized defense landscape. This landscape governs the efficiency of pathogen clearance, the extent of tissue damage, and the overall progression of disease.

3.2. Host immune response mapping

In vertebrate organisms, the host response to infection involves a diverse and coordinated network of defense mechanisms that can be broadly divided into two major biological systems: innate immunity and adaptive immunity.

Innate immunity constitutes the first line of defense against invading pathogens. It is responsible for the rapid and largely nonspecific inflammatory response that occurs almost immediately after infection. The main goals of this system are to eliminate the invading microorganism, limit tissue damage, and initiate and shape the subsequent adaptive immune response. Innate immune mechanisms are evolutionarily conserved and present across virtually all eukaryotic organisms, highlighting their fundamental role in host defense.

The effectors of the innate immune system can be categorized into humoral and cellular components: (I) The *humoral effectors* include antimicrobial peptides such as defensins, soluble lectins like collectins, and the complement system, all of which can bind to pathogen surfaces, promote opsonization, or directly mediate pathogen lysis. (II) The *cellular effectors* comprise various immune cells, including monocytes/macrophages, neutrophilic granulocytes, and natural killer cells, which are responsible for phagocytosis, cytotoxicity, and the secretion of pro-inflammatory cytokines (Degn et al., 2013).

A key event in the innate immune response is the recognition of conserved molecular signatures expressed by pathogens, known as pathogen-associated molecular patterns (PAMPs). These molecular motifs are detected by specialized receptors called pattern-recognition receptors (PRRs). PRRs can be either soluble (circulating in bodily fluids) or cell-associated. In the latter case, they may be expressed on the cell surface or located within intracellular compartments of innate immune cells (such as macrophages and dendritic cells) and certain epithelial cells.

Several types of pattern-recognition receptors (PRRs) exist, including soluble receptors, membrane-bound receptors, and cytoplasmic receptors. Soluble PRRs are proteins that can bind directly to pathogens and initiate a cascade of events leading to immune activation. The main families of soluble PRRs include: Collectins (C-type lectins): these proteins recognize specific sugar residues, such as fucose and mannose, commonly found in the cell walls of bacteria and fungi. Upon binding, collectins can activate the complement pathway and/or facilitate opsonization, enhancing phagocytosis by immune cells.

Ficolins: structurally similar to collectins, ficolins bind to N-acetylated compounds, such as N-acetylglucosamine present in bacterial and fungal cell walls. Functionally, ficolins contribute to complement activation and opsonization in a manner analogous to collectins (Cedzyński et al., 2023).

Cellular PRRs can be located on the cell surface (membrane-bound) or inside the cell (cytoplasmic). These receptors detect PAMPs directly within infected cells or in extracellular compartments, initiating intracellular signaling pathways that induce cytokine and chemokine production, inflammasome activation, and other defense mechanisms.

Together, these PRRs form a highly integrated network that allows the innate immune system to rapidly detect and respond to invading pathogens. By discriminating between self and non-self-molecular patterns, PRRs provide the foundation for both immediate defense and the subsequent activation of adaptive immunity (Sellge et al., 2015). These receptors include Toll-like receptors (TLRs), NOD-like receptors (NLRs), and C-type lectin receptors (CLRs). TLRs, in particular, are capable of recognizing pathogen-associated molecular patterns (PAMPs) from bacteria, viruses, and fungi, thereby initiating signaling pathways that lead to the transcription of target genes involved in the immune response. PRRs are expressed on immune cells such as macrophages and dendritic cells, enabling them to detect pathogens and orchestrate appropriate defense mechanisms.

The engagement of PRRs with PAMPs triggers a cascade of signaling events that culminates in the activation of inflammatory pathways. This process leads to the production of cytokines, chemokines, and other mediators that recruit immune cells to the site of infection, enhance phagocytic activity, and establish the characteristic inflammatory response that defines innate immunity.

Adaptive immunity is subsequently engaged, providing highly specific, long-lasting defense. Antigen-presenting cells, primarily dendritic cells and macrophages, process pathogen-derived antigens and present them to T and B lymphocytes in lymphoid tissues. This activation leads to pathogen-specific cytotoxic T cell responses, helper T cell-mediated cytokine production, and B cell-mediated antibody generation. The integration of innate and adaptive responses ensures effective clearance of the pathogen while establishing immunological memory for future protection.

3.3. Zoonotic potential and comparative pathology

Zoonotic pathogens are infectious agents capable of crossing species barriers, causing disease in both animals and humans. Understanding the zoonotic potential of these pathogens is critical for predicting, preventing, and controlling emerging infectious diseases. The likelihood of cross-species transmission depends on multiple factors, including pathogen genetics, host susceptibility, environmental conditions, and the frequency and type of human–animal interactions. Viruses, bacteria, fungi, and parasites can all possess zoonotic potential, and some pathogens may undergo adaptive changes that enhance their ability to infect novel hosts.

Comparative pathology is the integrative study of disease biology in humans and animals, aimed at understanding pathological mechanisms by comparing processes across species. This approach not only enriches our fundamental knowledge of pathology but also helps develop strategies to improve both human and animal health. By systematically comparing disease manifestations, cellular tropism, immune responses, and tissue pathology across species, it can be identified the conserved pathogenic mechanisms as well as species-specific susceptibilities (Wekesa et al., 2023). For example, studies of influenza viruses in birds, pigs, and humans reveal both shared pathways of viral replication and host-specific responses that influence disease severity, while comparative analyses of bacterial zoonoses such as Salmonella or Brucella highlight the impact of host immune defenses, tissue architecture, and microbiota on infection outcomes.

To evaluate similarities and differences in pathological processes across species, open-source bioinformatics tools are increasingly applied in comparative and molecular pathology. These tools allow for the systematic analysis of multispecies datasets, including histopathology, imaging, and 'omics data, facilitating the integration of complex biological information. By combining molecular, cellular, histological, and clinical data, researchers can gain a comprehensive understanding of disease mechanisms (Gutleben et al., 2018). Comparative analyses reveal conserved pathogenic pathways, species-specific variations, and potential therapeutic targets, bridging basic research with translational applications.

Incorporating histopathology, molecular assays, and omics approaches within a comparative framework provides a holistic view of pathogen–host interactions. This approach helps identify key determinants of virulence, tissue tropism, and immune evasion, generating critical insights for the development of cross-species vaccines, therapeutics, and surveillance strategies. By linking the study of zoonotic potential with comparative pathology, researchers can inform both veterinary and human medicine, supporting proactive measures against emerging infectious diseases and mitigating their pandemic risk.

4. APPLICATIONS IN KEY VETERINARY INFECTIOUS DISEASES

Veterinary medicine is essential for the health and protection of livestock, domestic animals, companion animals, and wildlife. Animal health is therefore central to human–environment interactions. More than three-quarters of emerging human infectious diseases originate from animals, and it is estimated that two-thirds of all human pathogens have an animal source or natural reservoir. The term zoonosis encompasses all transmissible diseases (including viruses, bacteria, prions, and parasites) capable of infecting both humans and animals. Transmission mechanisms are diverse, including wild or domestic vertebrates (livestock or companion animals), animal products, excreta, and vectors. Multiple factors underlie the risk of human infection, such as hunting, farming, vector population dynamics, and human–wildlife ecotones.

4.1. Bacterial infections

Brucellosis is a zoonotic disease caused by bacteria of the genus *Brucella*, with a global incidence estimated at 500,000 human cases per year (www.who.int). The primary reservoir is animal, predominantly affecting livestock species. The *Brucella* species most relevant to humans are *B. melitensis*, *B. abortus*, and *B. suis*, which share over 90% sequence identity (Figure 1).

Figure 1. Zoonotic potential of the various Brucella species. The colors and arrow sizes represent the zoonotic risk displayed by each species isolated from its preferred host

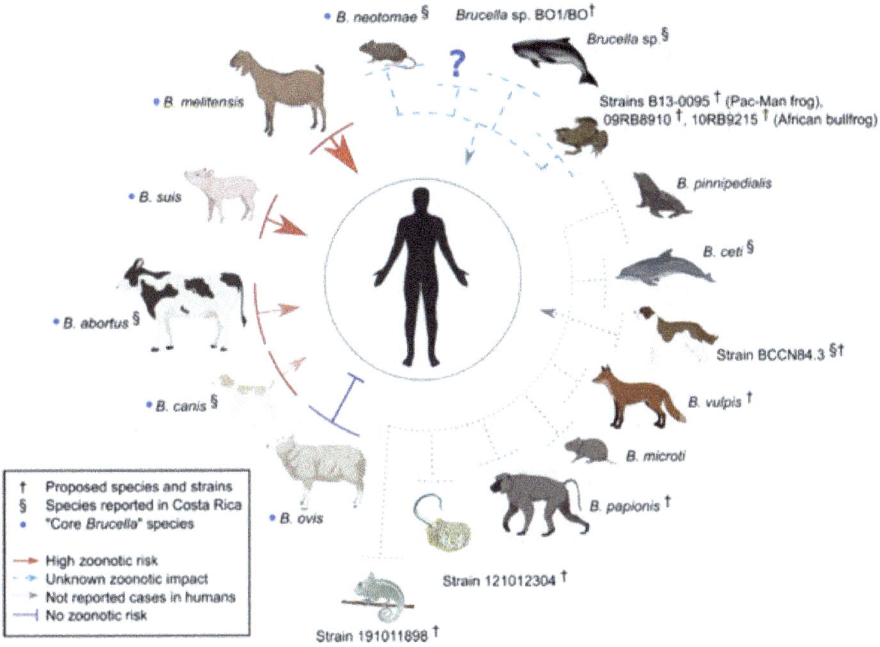

(Whatmore et al., 2021)

Although these species are highly related genetically, they exhibit diversity in phenotypic characteristics, host preference, and pathogenicity. The genetic homogeneity of these species can be advantageous for the development of robust universal diagnostic tools. Conversely, it poses a challenge for differentiating them, complicating the precise characterization of isolates from the same outbreak. Thus, omics analyses in brucellosis may provide a comprehensive understanding of the disease, the *Brucella* bacterium, and host responses through genomics, transcriptomics, proteomics, and metabolomics. These techniques may reveal the molecular basis of bacterial virulence, antibiotic resistance, and host immunity, offering critical insights for diagnostics, therapeutic interventions, and vaccine development.

The pathogenicity of *Brucella* is closely linked to its capacity to express a diverse array of virulence factors that modulate both its extracellular and intracellular life cycle stages. Extracellularly, *Brucella* can evade host defenses through mechanisms such as resistance to complement-mediated lysis and avoidance of opsonization. Once inside host cells, particularly macrophages and dendritic cells, the bacteria exploit intracellular survival strategies, including inhibition of phagosome-lysosome fusion, modulation of autophagic pathways, and resistance to reactive oxygen and nitrogen species (Roop et al., 2021). These virulence mechanisms enable *Brucella* to establish chronic infections by persisting within host cells, avoiding immune clearance, and manipulating host signaling pathways to its advantage. The coordinated action of these factors underlies the bacterium's ability to cause systemic disease, highlighting the complex interplay between pathogen adaptation and host immune responses. The lipopolysaccharide (LPS) of *Brucella* plays an important role in the bacterium's pathogenicity, influencing both its intracellular survival and trafficking within host cells. Beyond its contribution to bacterial life cycle dynamics, *Brucella* LPS is also a key modulator of the host immune response, affecting the recognition and activation of immune cells. Its unique structure allows the bacterium to evade detection by certain pattern recognition receptors, dampen pro-inflammatory signaling, and manipulate host defenses, thereby promoting persistent infection and contributing to the overall virulence of *Brucella* (smith et al., 2018).

Advantages of genomics and transcriptomic applied in the study of *Brucella* bacterium may contribute to identification of virulence factors, antibiotic resistance genes, and genomic adaptations that allow survival in diverse host species. Comparative genomic analyses across *Brucella* species and strains may reveal both conserved mechanisms and host-specific adaptations, highlighting factors that drive zoonotic transmission. Moreover, gene expression studies elucidate how *Brucella* adapts to intracellular environments and interacts with host cells. These analyses may shed light on bacterial strategies for evading immune detection and persisting in different host tissues, including macrophages and reproductive organs in livestock (Whatmore et al., 2021). Profiling the complete set of bacterial and host proteins allows identification of infection biomarkers and clarifies the molecular mechanisms underlying pathogenesis, immune modulation, and tissue tropism. Analysis of small metabolites provides insights into the metabolic interplay between *Brucella* and its host. Metabolic adaptations enable bacterial survival within host cells, while changes in host metabolites reflect immune activation and tissue responses.

4.2. Viral infections

Foot-and-mouth disease (FMD) is one of the most contagious viral diseases affecting animals. Despite being known since antiquity, FMD remains a persistent global concern and poses a continual threat of re-emergence in countries previously declared free of the disease. The causative agent, foot-and-mouth disease virus (FMDV), belongs to the family *Picornaviridae*, *genus Aphthovirus*, and primarily affects domestic and wild artiodactyls, including cattle, sheep, goats, pigs, camelids, and deer. FMD has a major socio-economic impact on livestock production and agricultural trade, particularly in endemic regions and during incursions into previously FMD-free zones, as observed during the 2001 and 2025 outbreak in Europe (WOAH). Historically, FMDV has been one of the most extensively studied animal viruses. Its high genetic and antigenic variability, the absence of cross-immunity among its seven serotypes, its environmental resilience, extreme contagiousness, broad host range, and distinctive biological characteristics make it a continuing priority in animal health research.

The RNA genome of the FMD disease virus (FMDV) consists of a single open reading frame (ORF) that encodes a polyprotein of approximately 2,300 amino acids. This polyprotein undergoes a series of proteolytic cleavages, primarily mediated by the viral proteases Lpro and 3Cpro. These successive cleavages generate several precursor intermediates and ultimately yield 15 mature structural and nonstructural proteins that are essential for viral replication, assembly, and host interaction. The low replication fidelity of the viral RNA-dependent RNA polymerase, together with the short replication cycle of the foot-and-mouth disease virus, drives the generation of a heterogeneous population of closely related viral variants, collectively referred to as quasi-species, within the infected host (Carrillo et al., 2005). This high mutation rate enables the virus to rapidly adapt to selective pressures, including host immune responses and environmental constraints. As a result, FMDV exhibits remarkable genetic and antigenic diversity, complicating both vaccine development and disease control efforts. The quasi-species dynamics also facilitate viral persistence and cross-species transmission by allowing the selection of variants with enhanced fitness, altered receptor affinity, or resistance to neutralizing antibodies. Immunologically, FMDV comprises seven distinct serotypes: A, O, C, Asia-1, and the Southern African Territories (SAT1-3), with over 65 genetic lineages and topotypes (Jamal et al., 2013).

Despite decades of research on the FMDV, there remains a critical gap in the development of robust predictive models capable of accurately forecasting viral emergence, spread, and evolution. The high mutation rate of FMDV, coupled with its quasi-species dynamics and broad host range, generates significant genetic and antigenic variability that challenges traditional epidemiological and molecular modeling approaches. Existing models often fail to integrate the complex interplay between viral evolution, host immune responses, and environmental factors influencing transmission dynamics. Thus, an integrated predictive framework combining bioinformatics, systems biology, and spatiotemporal epidemiology represents a major tool to proactive disease control. Developing such models is essential for improving surveillance, guiding vaccination strategies, and mitigating the risk of transboundary FMD incursions in both endemic and disease-free regions.

4.3. Parasitic infections

Toxoplasmosis, a parasitic disease for which cats and other felids serve as the definitive hosts, is a well-known zoonosis primarily feared for the developmental disorders it can cause in fetuses when contracted by pregnant women. Until recently, its pathogenic role in humans was considered limited to this specific context, with only a few exceptional cases reported outside of congenital infection. However, emerging and converging studies now suggest that *Toxoplasma gondii* infection may have far broader implications for human health, potentially influencing a range of physiological and neurological processes.

In intermediate hosts, the route of *Toxoplasma gondii* transmission primarily depends on dietary behavior. Herbivorous animals become infected through the ingestion of sporulated oocysts excreted in the feces of parasitized cats, whereas carnivorous species acquire the infection by consuming prey or raw meat containing tissue cysts harboring bradyzoites (Figure 2). In both cases, the parasite invades intestinal epithelial cells, initiating a complex intracellular cycle that leads to the formation of tachyzoites, rapidly dividing forms responsible for systemic dissemination. These tachyzoites subsequently differentiate into bradyzoites, establishing long-lived cysts within neural and muscular tissues (Blader et Saeij, 2009). Furthermore, transplacental transmission can occur in both herbivores and carnivores, enabling vertical passage of the parasite to the fetus during gestation, which may result in congenital toxoplasmosis. *T. gondii*'s dual ability to infect a wide range of intermediate hosts and to persist in latent tissue cysts underlies its extraordinary epidemiological success.

Figure 2. Life cycle of Toxoplasma gondii (Sanchez et Besteiro, 2021)

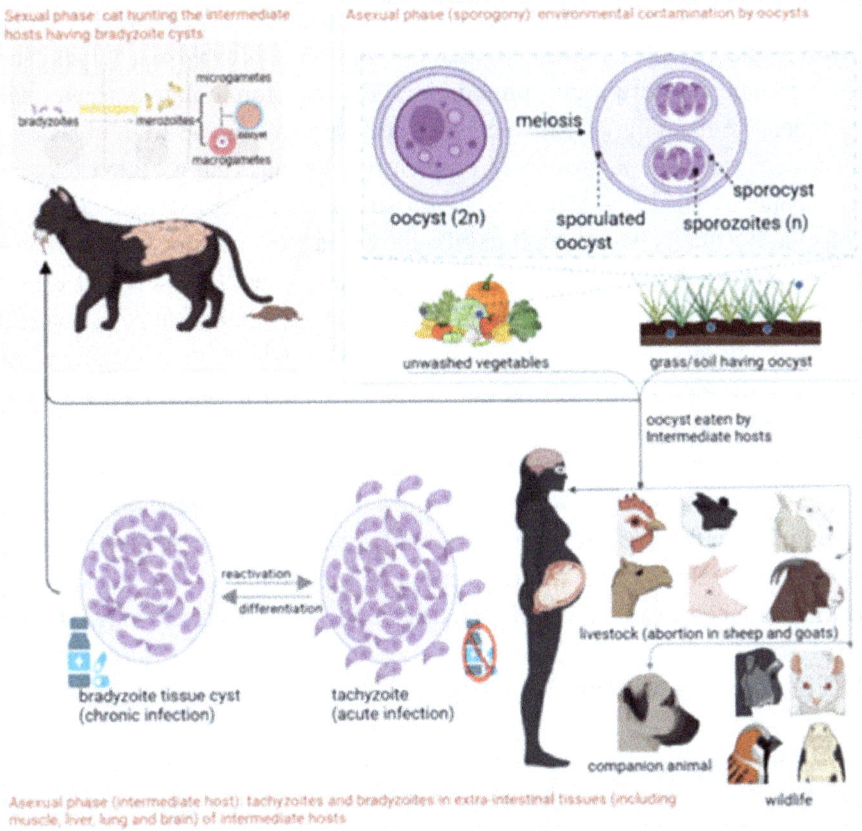

The interaction between *Toxoplasma gondii* and the host immune system is characterized by a finely tuned balance between parasite survival and immune control. Upon infection, innate immune cells such as macrophages and dendritic cells recognize *T. gondii* through pattern recognition receptors (PRRs), including Toll-like receptors (TLR2, TLR4, and TLR11/12 in rodents), which detect parasite-derived molecules such as profilin. This recognition triggers the activation of intracellular signaling cascades leading to the production of pro-inflammatory cytokines, notably interleukin-12, which in turn stimulates interferon-gamma (IFN-γ) secretion by natural killer cells and T lymphocytes. However, *T. gondii* has evolved multiple mechanisms to evade or subvert these immune defenses. . The parasite secretes a variety of effector proteins from its rhoptries and dense granules that interfere with host cell signaling pathways. Understanding these molecular interactions provides crucial insight into the mechanisms of parasite survival, immune regulation, and the establishment of long-term latent infection.

In recent years, multi-omics technologies have provided powerful tools for unraveling the complex molecular dialogue between *Toxoplasma gondii* and its hosts. Genomic, transcriptomic, proteomic, and metabolomic analyses have enabled comprehensive mapping of parasite genes and pathways involved in invasion, immune modulation, and persistence. Comparative genomics has revealed key genetic determinants of virulence across *T. gondii* strains, while transcriptomic profiling of infected host cells has highlighted dynamic changes in immune and metabolic pathways during infection. Proteomic studies have identified secreted effectors, such as rhoptry and dense granule proteins, that manipulate host cell signaling and gene expression to favor parasite survival. Meanwhile, metabolomic analyses have uncovered significant reprogramming of host energy metabolism and nutrient acquisition by the parasite.

The integration of these multi-omics datasets provides a systems-level understanding of host–parasite interactions, offering new perspectives for the identification of molecular biomarkers of infection and potential therapeutic or vaccine targets. Such approaches also support the development of predictive models for parasite adaptation and immune evasion, enabling the design of more effective control strategies.

4.4. Fungal infections

Aspergillus species belong to the class *Ascomycetes*, family *Aspergillaceae*. These fungi proliferate abundantly in damp environments, on decaying organic matter, and particularly on moldy cereals. They are also opportunistic exosaprophytes commonly present in the atmosphere, and approximately thirty species are associated with diseases in humans and animals. Infection typically occurs via inhalation of airborne conidia, ingestion of contaminated feed, or, less commonly, via wounds. In susceptible hosts, *Aspergillus* can cause a spectrum of diseases ranging from superficial infections to invasive systemic mycoses. In animals, infections are particularly relevant in poultry, ruminants, and companion animals, leading to respiratory, gastrointestinal, or disseminated disease. In humans, immunocompromised individuals are at greatest risk, with *Aspergillus* spp. causing conditions such as allergic bronchopulmonary aspergillosis, chronic pulmonary aspergillosis, or invasive aspergillosis (Paulussen et al., 2017). *Aspergillus fumigatus* is the primary species responsible for most cases of aspergillosis. Its spores are extremely small, enabling them to readily enter the host via inhalation or ingestion. In veterinary medicine, *A. fumigatus* and related species are considered opportunistic pathogens with a global distribution, capable of causing mycoses across a wide range of animal species. These infections are particularly severe in immunocompromised animals, young livestock, and birds, often manifesting as respiratory, systemic, or gastrointestinal disease (Earle et al., 2023).

The primary route of *Aspergillus fumigatus* infection is the inhalation of spores, which represent the fungus's infectious morphotype. In healthy individuals, however, the innate immune system can efficiently eliminate the fungus. The epithelium of the upper respiratory tract, through the mucociliary clearance mechanism, serves as the first line of defense against inhaled spores. The host's innate immune system serves as the first line of defense against *Aspergillus fumigatus* infections. Fungal cells are surrounded by a thick cell wall composed primarily of polysaccharides, pigments, and proteins. As the outermost structure, the cell wall is the first fungal component to engage with the host immune system. Its composition changes with the fungus's morphological stage (van de Veerdonk et al., 2017). Therefore, the roles of different fungal morphotypes and their cell wall components in complement activation, as well as in interactions with collectins and antimicrobial peptides, are of major importance for understanding host–pathogen interactions.

Once the pathogen is inhaled, alveolar macrophages constitute a highly efficient first line of defense, capable of phagocytosing and eliminating up to 90% of *Aspergillus fumigatus* spores. The second line of immune defense involves polymorphonuclear neutrophils, a large population of intravascular phagocytes that are rapidly recruited to the infection site to bolster host defenses.

These neutrophils are essential for controlling *Aspergillus fumigatus* hyphae, releasing the contents of their granules into the extracellular space to directly attack the fungus [70]. Furthermore, respiratory epithelial cells secrete a diverse array of antimicrobial molecules that help maintain pulmonary sterility. These molecules function either directly as endogenous antibiotics or indirectly by enhancing pathogen clearance by phagocytes. Until roughly two decades ago, *Aspergillus fumigatus* was regarded as a minor pathogen due to the effectiveness of the innate immune response. However, the rising number of immunocompromised individuals and the extensive use of potent immunosuppressive therapies have dramatically shifted this perspective. Today, *Aspergillus fumigatus* is recognized as the predominant airborne fungal pathogen, posing a significant threat to vulnerable populations. The integration of modern genetic approaches offers promising avenues for improving the control of *Aspergillus* infections. Genomic and transcriptomic analyses enable the identification of genes and regulatory networks essential for fungal virulence, stress adaptation, and antifungal resistance, while proteomics provide insight into the functional expression of these pathways during host infection. Such comprehensive datasets can reveal biomarkers for early detection, highlight potential therapeutic targets, and improve understanding of host-pathogen interactions at the molecular level. Additionally, advances in CRISPR-based functional genomics and comparative genomics allow precise characterization of virulence factors and immune evasion mechanisms, facilitating the development of targeted antifungal therapies and more effective vaccines.

5. CASE STUDIES AND EMERGING INSIGHTS

5.1. Spatial omics in respiratory pathologies in livestock

Spatial omics offer a powerful lens for dissecting complex respiratory infections in livestock, especially where pathogens like *Mycoplasma*, *Pasteurella*, or coronaviruses interact with heterogeneous lung microenvironments. By integrating transcriptomic, proteomic, and metabolomic data with tissue histology, researchers can now spatially map inflammation, immune cell infiltration, and microbial localization.

One relevant example includes efforts to assess the role of camels in the transmission of Middle East respiratory syndrome coronavirus (MERS-CoV). A review by Roess et al. (2015) emphasized the need for integrating ecological and sociocultural data to understand camel movement and human-animal contact, which influence MERS outbreaks in endemic regions.

In the context of bovine respiratory disease (BRD), transcriptomic studies have revealed immune dysregulation in lung tissues of infected cattle, and spatial mapping approaches are beginning to localize expression of key cytokines and bacterial virulence factors within lung lesions (behura et al., 2017)

Moreover, systems immunology approaches incorporating spatial omics have been used to analyze the distribution and function of alveolar macrophages and neutrophils in pneumonia-infected swine lungs, revealing compartmentalized immune microenvironments that influence disease outcome (Li et al., 2023).

Advanced imaging combined with spatial transcriptomics is now used to assess coinfections and disease severity in livestock, highlighting tissue-specific host-pathogen dynamics, especially relevant for co-infections involving viral and bacterial agents such as influenza and *Haemophilus parasuis*.

5.2. Companion animals: canine and feline infectious diseases

Companion animals like dogs and cats are susceptible to infectious diseases from various pathogens, and recent research using spatial multi-omics has shown they harbor diverse microbial communities with potential implications for both animal and human health. Studies of meta-transcriptomic analysis of infectomes in cats and dogs found a wide range of viruses, bacteria, and fungi, with differences in composition sometimes linked to the animal's health status. Unlike traditional diagnostics, which target specific pathogens, meta-transcriptomics captures the entire RNA landscape of an infection, allowing detection of co-infections, rare or novel pathogens, and dynamic interactions between microbes and the host. This approach can identify transcriptionally active virulence factors, antimicrobial resistance genes, and host immune responses, offering a holistic view of pathogen activity and disease mechanisms.

In a study analyzing 239 individual swabs from dogs and cats across China (Wu et al., 2024), researchers identified a wide spectrum of both known and previously unrecognized pathogens, including 27 potential zoonotic agents. Several of these zoonotic pathogens were detected even in apparently healthy pets, underscoring the need for continuous surveillance and monitoring. The total infectome unveiled through meta-transcriptomic analysis provides a robust framework for understanding infectious diseases in domestic animals, extending far beyond mere pathogen discovery and characterization. It offers insights into pathogen diversity, host-pathogen interactions, and potential disease dynamics, serving as a comprehensive model for studying animal health and zoonotic risks.

Recent findings on the companion animal gut microbiome highlight the advantages of metagenomic analysis, which enables high-resolution profiling of microbial communities. Using 2,639 shotgun metagenomes, 730 previously published and 1,909 newly generated, the study identified 2,320 taxa, including 561 absent from the human gut and 19 newly described (Branck et al., 2024). These results provide key insights into the evolutionary relationships between companion animal and human microbiomes, shaped by shared environments, diets, and disease profiles. Moreover, host-specific genetic and functional adaptations, including differences in metabolism and antibiotic resistance, underscore the value of metagenomics in advancing our understanding of microbiome diversity and evolution.

5.3. Wildlife and one health applications

The severe impact of African swine fever virus (ASFV) on global domestic and wild pig populations has driven integrated multi-omics research aimed at elucidating the complex regulatory mechanisms governing host–pathogen interactions. Through multi-omics analyses, researchers have identified a concurrent upregulation of innate immune effectors alongside the suppression of apoptosis inhibitors, suggesting a coordinated antiviral response coupled with regulated apoptotic activity (Wang et al., 2025). These findings provide valuable insights into the molecular dynamics of ASFV infection and support the development of genetic and immunological strategies to enhance disease control in swine populations.

Similarly, transcriptomic analyses of rabies virus (RABV) have revealed complex gene expression patterns and immune signaling networks activated during infection. To elucidate the role of the salivary glands in RABV propagation, mRNA expression profiles were analyzed in parotid tissue from infected mice (Guo et al., 2024). This analysis identified thousands of differentially expressed genes involved in pathways such as salivary secretion and PI3K-Akt signaling. The study provides a comprehensive transcriptomic map of RABV-infected parotid tissue, offering new insights into viral pathogenic mechanisms and potential targets for rabies control. Furthermore, these findings contribute to the identification of biomarkers for early diagnosis and highlight molecular targets for antiviral therapies, underscoring the value of transcriptomics in advancing both animal health and disease control strategies. In Europe, wildlife plays a critical role in rabies transmission, with red foxes (*Vulpes vulpes*) serving as the main reservoirs of the virus (Backer et al., 2020). Spillover from wildlife hosts can lead to outbreaks in domestic animals and, rarely, humans. Understanding the molecular and cellular mechanisms of RABV infection, including the role of the salivary glands in viral shedding, is therefore essential not only for controlling rabies in livestock but also for mitigating the risk of wildlife-mediated transmission to domestic animals and humans.

6. CHALLENGES AND LIMITATIONS

The main challenges and limitations of applying multi-omics approaches to veterinary infectious disease research stem from multiple factors. These include the complexity of multi-layered datasets, which require advanced computational tools and robust statistical frameworks for effective integration and interpretation; experimental constraints, such as high operational costs, sample heterogeneity, and the necessity for carefully standardized study designs; and analytical challenges, including the risk of data overfitting or misinterpretation, the difficulty in distinguishing correlation from causation, and the limited reproducibility or validation of findings across diverse animal species and populations.

6.1. Sample preservation and spatial resolution

Due to the high cost and technical complexity of multi-omics approaches, substantial variation often exists among different high-throughput technologies with respect to sample collection timing, storage conditions, and experimental grouping. In veterinary research, samples obtained from animals, particularly those from rare species, valuable livestock, or clinical cases are often scarce and difficult to obtain, necessitating meticulous experimental planning [79]. Therefore, rigorous study design is important in veterinary multi-omics research to ensure the reproducibility and reliability of results, especially when working with large datasets, multiple comparison groups, or intricate analytical frameworks.

When investigating infectious diseases, the biological characteristics of the disease must be central to the design of multi-omics studies. Unlike disorders caused by single-gene mutations, infectious diseases result from complex interactions among the pathogen (bacterial, viral, parasitic, or fungal), the host immune response, and various environmental influences. Consequently, sample collection becomes an important aspect of study design, as the choice of tissue, timing of sampling, and storage conditions can have a great impact on the quality and interpretability of multi-omics data [80]. For example, in viral respiratory infections of cattle or swine, collecting samples from different regions of the respiratory tract at multiple time points is essential to capture the temporal dynamics of pathogen replication and local immune responses. Furthermore, studies involving farm or wild animals often face additional challenges, including limited opportunities for repeated sampling, individual physiological variability, and stress induced by handling (van Cise et al., 2024). To overcome these constraints, a well-structured experimental strategy, encompassing careful subject selection, standardized sampling procedures, and close collaboration with experts in bioinformatics and biostatistics, is vital to achieve robust, reproducible, and biologically meaningful multi-omics insights into infectious diseases of animals.

6.2. Cross-species annotation issues

In veterinary research, substantial variability can arise from interspecies biological differences, husbandry practices, dietary factors, microbiome composition, and environmental influences, all of which can profoundly affect the molecular profiles captured in omics analyses. Factors such as diurnal rhythms, diet, and the specific biomolecular composition of different sample types must be carefully considered to avoid introducing artifacts or confounding variables that could obscure true biological signals. Many animal diseases exhibit considerable heterogeneity due to substantial variation in infection progression among affected individuals. Understanding this heterogeneity is essential for elucidating disease etiology, predicting outcomes, and developing targeted interventions. Consequently, identifying underlying disease subtypes or classifying samples into biologically meaningful subgroups is a critical step in veterinary research. Achieving representative and standardized sample collection is therefore essential to generate data that accurately reflect biological processes rather than external variability (Allaband et al., 2024).

In studies of bacterial or viral infections in livestock, multiple factors can significantly influence transcriptomic, proteomic, or metabolomic outcomes. These include the timing of sample collection relative to infection onset, the anatomical site of sampling, and the physiological or immune status of the animals. Challenges are further compounded in wildlife or field-based investigations, where limited access to samples often results in small sample sizes, reducing statistical power and increasing the risk of false-positive or misleading findings. Diet and feeding patterns also exert a strong and reproducible influence on the gastrointestinal environment and luminal microenvironments; however, reported effects frequently vary between studies. One key source of variability is the timing of sample collection, which, if not carefully standardized, can lead to inconsistent or conflicting conclusions. Comprehensive recording of phenotypic and environmental metadata, such as infection severity, clinical signs, immune response markers, and management practices is likewise critical for accurately correlating molecular profiles with biological outcomes.

Profiling proteins in tissues or bodily fluids enables the accurate identification of disease-specific protein signatures. This approach, when applied to livestock health monitoring, provides a comprehensive view of protein expression patterns associated with both normal physiological functions and pathological states. Adding to the complexity of veterinary multi-omics research are cross-species annotation challenges. Many animal species, particularly livestock and wildlife, lack fully annotated genomes, transcriptomes, or proteomes, and reference databases are often incomplete or limited to model organisms. As a result, the identification and functional annotation of genes, proteins, or metabolites can be ambiguous, and downstream analyses, including pathway mapping, molecular network construction, and comparative studies, frequently rely on homology-based inference from other species [85]. These limitations can complicate the interpretation of molecular responses to infection and may obscure species-specific versus conserved mechanisms, emphasizing the need for careful bioinformatic strategies and cross-species validation.

6.3. Cost, accessibility and infrastructure gaps

In addition to annotation challenges, veterinary multi-omics research faces substantial practical limitations related to cost, accessibility, and infrastructure. High-throughput omics technologies, including proteomics, transcriptomics, and metabolomics, require specialized instrumentation, reagents, and skilled personnel, making large-scale studies in livestock or wildlife populations prohibitively expensive for many research groups. Access to advanced analytical platforms is often limited to well-equipped laboratories, creating geographic and institutional disparities that hinder widespread application of multi-omics approaches in veterinary science. Moreover, the handling and storage of biological samples, particularly in field-based or remote settings, pose significant logistical challenges. Maintaining cold chains, preventing sample degradation, and standardizing collection protocols are critical for preserving molecular integrity, yet these requirements often exceed the available infrastructure in rural or resource-limited areas.

Such constraints not only limit the scale and scope of studies but also affect reproducibility and data quality, particularly when integrating multi-omics layers across different species or experimental sites (Schurch et al., 2016). Addressing these gaps requires strategic investment in laboratory infrastructure, training programs for technical personnel, and the development of cost-effective and portable technologies for sample collection and molecular analysis. Collaborative networks that share resources, standardized protocols, and reference datasets can also help mitigate accessibility challenges, enabling more comprehensive and comparable multi-omics studies across diverse veterinary populations.

6.4. Joint analysis, data sharing and standardization

Multi-omics approaches have greatly expanded the potential for genetic and health improvements in veterinary research by enabling the exploration of molecular mechanisms underlying complex traits, the discovery of novel biomarkers, and a better understanding of gene–environment interactions that influence performance, disease resistance, and productivity. Integrating classical genetic models with multi-omics data allows the development of precision breeding strategies that optimize health, productivity, and resilience across diverse species and production systems.

Several analytical strategies have been developed to interpret multi-omics datasets. Concatenation-based integration merges measurements from different omics layers into a single matrix, enabling the application of standard machine learning techniques. While highly predictive in supervised frameworks, this approach can be sensitive to unbalanced feature numbers across omics layers, potentially biasing results toward layers with more abundant data. Recent methodological improvements, such as feature normalization or weighting, help address these limitations. Consensus clustering performs individual clustering on each omics layer before deriving an overall sample classification. Although this method has been successfully applied in several studies, including disease subtyping, its two-step nature can reduce the ability to capture interactions or crosstalk between omics layers. Multiomic pathway enrichment focuses on identifying biological pathways associated with specific phenotypes by integrating data from multiple omics layers. In practice, most current tools still conduct enrichment analyses separately for each layer and then combine the results, similar to conceptual integration. Emerging approaches that integrate pathway-level information directly across omics layers provide more precise identification of key regulatory networks and cross-layer interactions that underlie complex biological processes.

In veterinary multi-omics research, data sharing and standardization are critical for ensuring reproducibility, comparability, and translational relevance. Generating multi-omics datasets from the same samples offers a comprehensive view of biological processes across molecular layers, providing valuable insights into infectious and complex diseases (Elrashedy et al., 2025).

The integration of molecular data with detailed phenotypic and environmental information further enhances interpretation, supporting more accurate diagnoses, informing precision veterinary interventions, and guiding effective management strategies for livestock and wildlife populations. The method widely adopted in multi-omics studies is conceptual integration, which involves analyzing each omics layer independently and then combining the results during the interpretation phase. Its simplicity and the lack of a standardized approach in the field have contributed to its widespread use.

Standardized protocols for sample collection, processing, and molecular profiling are essential to minimize technical variability and batch effects, particularly when comparing data across laboratories, species, or geographic regions. Centralized, accessible databases that store multi-omics datasets can facilitate collaborative research, cross-species comparisons, and meta-analyses (Feldner-Busztin et al., 2023). Adherence to clear data-sharing policies and FAIR principles (Findable, Accessible, Interoperable, and Reusable) maximizes the value of these data, accelerates biomarker discovery, and advances the understanding of disease mechanisms in veterinary pathology.

7. FUTURE PERSPECTIVES

The future of spatial multi-omics in veterinary infectious disease research lies in the convergence of advanced analytical technologies, computational intelligence, and translational applications. As high-dimensional datasets grow in complexity, AI, digital pathology, and integrated surveillance systems will play central roles in transforming how infectious diseases are diagnosed, monitored, and controlled across animal populations. The following subsections outline key emerging directions shaping the next generation of spatial multi-omics research in veterinary and One Health contexts.

7.1. Integration with digital pathology and AI

Digital pathology and AI are revolutionizing the way biological and clinical data are analyzed and interpreted. Whole-slide imaging and high-resolution microscopy enable the digitization of tissue morphology at subcellular levels, which, when combined with spatial omics, produce molecular–morphological maps of infected tissues.

AI-driven computational pathology is increasingly being applied to veterinary science. For example, a pan-species digital pathology atlas demonstrated that convolutional neural networks could accurately classify immune responses and predict spatial immune scores across multiple animal species, laying the foundation for AI-assisted veterinary diagnostics (Elrashedy et al., 2025).

The integration of AI and digital pathology in both human and veterinary medicine has been recognized as inevitable, enhancing automated diagnosis and enabling telepathology-based data sharing across institutions (Zhang et al., 2022). Similarly, visual–omics models such as OmiCLIP and spEMO now link histopathology images with transcriptomic data, demonstrating robust accuracy in predicting gene expression from hematoxylin and eosin slides (Battaza et al., 2024)

Emerging AI approaches for spatial omics are further transforming biological interpretation, integrating data-driven and constraint-based modeling to decipher tissue microenvironments.

7.2. Personalized veterinary medicine

Personalized veterinary medicine is emerging as a transformative paradigm, supported by the convergence of omics technologies, digital pathology, and AI. Just as human precision medicine integrates genomic and spatial data to tailor therapy, veterinary science is beginning to adopt similar individualized strategies for disease prediction, prevention, and treatment.

Spatial multi-omics provides high-resolution maps of host immune responses and pathogen localization, enabling identification of species- or breed-specific disease susceptibilities. AI-assisted multi-omics integration in veterinary systems, such as the MOSAIC multi-omic spatial atlas, demonstrates how molecular signatures and imaging data can be harmonized to define biological subtypes and therapeutic targets.

In animal health, this approach could revolutionize vaccine design and antimicrobial stewardship by integrating pathogen genomics with host immune phenotyping. Moreover, AI-based systems can model infection trajectories and predict therapeutic responses, advancing evidence-based, individualized veterinary care.

Histo-genomics, the integration of histology with genomic biomarkers, has already enhanced human precision medicine and is expected to bring similar benefits to animal diagnostics by linking tissue morphology with molecular pathways (Chen et al., 2025).

7.3. Surveillance and outbreak preparedness in animal populations

Spatial multi-omics technologies are poised to enhance global veterinary disease surveillance and outbreak preparedness. Traditional surveillance systems rely heavily on epidemiological observations, but integrating spatial molecular data adds unprecedented precision in identifying early infection biomarkers, tracking pathogen evolution, and characterizing tissue-specific tropism.

Real-time sequencing and AI-driven analytics have already proven transformative in human infectious disease surveillance, and similar applications are emerging for veterinary contexts. Integrating multi-omics datasets with digital monitoring tools can identify emerging zoonotic threats and elucidate mechanisms of cross-species transmission.

Spatial omics can detect molecular "early warning" signatures, such as immune activation zones or metabolic alterations preceding clinical disease. Combining these insights with AI-powered epidemiological models will enable predictive outbreak management, particularly for high-risk zoonotic agents such as African swine fever virus or *Brucella* spp.

Cloud-based data platforms adhering to FAIR principles will support interoperability among veterinary laboratories, field surveillance programs, and One Health initiatives, fostering global collaboration and improving early detection capabilities. As emphasized in *The Dawn of Spatial Omics*, integrating multi-omics and AI at scale will not only expand biological insight but also improve accessibility in real-world applications.

7.4. One Health and Comparative Multi-Omics

Because many pathogens readily cross species barriers, multi-omics approaches have become essential tools within the *One Health* framework. Metagenomics, transcriptomics, and related techniques provide high-resolution insights into host–pathogen interactions in zoonotic diseases such as influenza, leptospirosis, and coronaviruses, enhancing outbreak detection and mechanistic understanding (Grubaugh et al., 2019). In parallel, comparative oncology and immunology increasingly leverage shared molecular pathways between companion animals and humans, positioning veterinary species as valuable translational models.

At the same time, microbiome-focused multi-omics has expanded rapidly across veterinary science. Integrative studies across cattle, poultry, horses, and companion animals highlight the complex relationships between microbial communities, immune modulation, and metabolic phenotypes (Suchodolski, 2016). Newer layers such as metaproteomics and metabolomics extend these insights by capturing functional microbial activity rather than solely taxonomic composition, offering a more comprehensive view of host–microbiome interactions.

8. RECOMMENDATION AND FUTURE RESEARCH DIRECTIONS

To accelerate the adoption and impact of spatial multi-omics in veterinary medicine, it is essential to establish standardized, species-specific protocols for the collection, preservation, and processing of diverse animal tissues. Parallel to this, the development of robust AI and machine learning tools is important for managing the complexity of multi-layered spatial datasets and enabling cross-species comparisons at scale. Strengthening computational infrastructure, particularly platforms capable of harmonizing data across different species, tissue types, and omics modalities, will support more rigorous analysis and interpretation. Progress also depends on expanding the availability of probes, antibodies, and other molecular reagents tailored to veterinary-relevant pathogens, as these remain a major bottleneck. Integrating advanced imaging modalities, such as high-resolution microscopy, PET, or MSI, with spatial transcriptomic and proteomic data will enhance biological validation and enable richer spatial context. Finally, fostering multi-institutional collaborations to build shared, standardized databases will generate comparative datasets that reveal both conserved and species-specific biological processes. Key priorities include creating validated experimental platforms for both fresh-frozen and FFPE samples, applying advanced imaging for cross-modal verification, leveraging AI-driven pattern discovery, and establishing open, well-annotated repositories to accelerate discoveries in animal health and disease.

9. CONCLUSION

Spatial multi-omics represents a revolutionary advance in the understanding of infectious disease biology, integrating molecular, cellular, and spatial information into a single analytical framework. By capturing the intricate organization of tissues and the spatial relationships between host and pathogen, this approach provides unparalleled insights into disease pathogenesis, immune responses, and microbial ecology.

The convergence of genomics, transcriptomics, proteomics, metabolomics, and imaging technologies allows researchers to uncover cellular heterogeneity and tissue microenvironments that drive infection outcomes. For example, integrating multi-omics datasets has illuminated spatially confined immune responses and metabolic niches that influence the success or failure of host defense mechanisms.

In veterinary infectious disease research, spatial omics has begun to elucidate the complexity of host–pathogen interactions across diverse animal species. Studies using multi-omics approaches in ticks and other disease vectors have revealed molecular mechanisms of feeding, pathogen transmission, and immune evasion, paving the way for novel control strategies.

Artificial intelligence and deep learning are expected to play a central role in managing and interpreting spatial omics datasets. Generative AI and transformer-based models are already improving data integration, enhancing predictive accuracy, and accelerating biomarker discovery across single-cell and tissue-level analyses (Barsoum et al., 2019) .

Despite its immense promise, spatial multi-omics faces challenges related to data standardization, cost, and accessibility, particularly in veterinary research settings. Addressing these barriers will require collaborative data-sharing frameworks, open-source analytical pipelines, and species-adapted reagents. As highlighted by technology developers, the next phase of spatial biology must prioritize reproducibility, standardization, and democratization of analytical tools.

Looking forward, spatial multi-omics will reshape the landscape of infectious disease research through integration with AI, digital pathology, and One Health initiatives. By connecting molecular events to tissue structure, environmental data, and epidemiological trends, this field holds the potential to revolutionize precision diagnostics, therapeutic discovery, and outbreak preparedness. The continued expansion of spatial omics into veterinary and comparative medicine will be essential to achieving holistic understanding of pathogen biology and cross-species transmission dynamics, ultimately strengthening global biosecurity and advancing precision veterinary care.

ACKNOWLEDGEMENT

D.C.A. and A.E.A. acknowledges the support by research grant no. 26ROMD/20.05.2024, code PN-IV-P8-8.3-ROMD2023-0053 from the Executive Unit for Financing Higher Education, Research, Development and Innovation (UE-FISCDI). I.N. acknowledges a grant from the Ministry of Research, Innovation and Digitalization, CCCDI UEFISCDI, project number PN-IV-P8-8.3-ROMD-2023-0060, within PNCDI IV. IN also acknowledges the Romanian Ministry of Research, Innovation and Digitalization under the Romanian National Nucleu Program LAPLAS VII—contract No. 30N/2023.

REFERENCES

Alexandrov, T. (2020). Spatial metabolomics and imaging mass spectrometry in the age of artificial intelligence. *Annual Review of Biomedical Data Science*, *3*(1), 61–87. DOI: 10.1146/annurev-biodatasci-011420-031537 PMID: 34056560

Allaband, C., Lingaraju, A., Flores Ramos, S., Kumar, T., Javaheri, H., Tiu, M. D., Dantas Machado, A. C., Richter, R. A., Elijah, E., Haddad, G. G., Leone, V. A., Dorrestein, P. C., Knight, R., & Zarrinpar, A. (2024). Time of sample collection is critical for the replicability of microbiome analyses. *Nature Metabolism*, *6*(7), 1282–1293. DOI: 10.1038/s42255-024-01064-1 PMID: 38951660

Ammons, D. T., Hopkins, L. S., Cronise, K. E., Kurihara, J., Regan, D. P., & Dow, S. (2024). Single-cell RNA sequencing reveals the cellular and molecular heterogeneity of treatment-naïve primary osteosarcoma in dogs. *Communications Biology*, *7*(1), 496. DOI: 10.1038/s42003-024-06182-w PMID: 38658617

Argelaguet, R., Velten, B., Arnol, D., Dietrich, S., Zenz, T., Marioni, J. C., Buettner, F., Huber, W., & Stegle, O. (2018). Multi-omics factor analysis—A framework for unsupervised integration of multi-omics data sets. *Molecular Systems Biology*, *14*(6), e8124. DOI: 10.15252/msb.20178124 PMID: 29925568

Baker, L., Matthiopoulos, J., Müller, T., Freuling, C., & Hampson, K. (2020). Local rabies transmission and regional spatial coupling in European foxes. *PLoS One*, *15*(5), e0220592. DOI: 10.1371/journal.pone.0220592 PMID: 32469961

Balzer, M. S., Ma, Z., Zhou, J., Abedini, A., & Susztak, K. (2021). How to get started with single-cell RNA sequencing data analysis. *Journal of the American Society of Nephrology*, *32*(6), 1279–1292. DOI: 10.1681/ASN.2020121742 PMID: 33722930

Baptista, A., Nuamah, R., Chiappini, C., & Grigoriadis, A. (2025). *MOSAIK: Multi-origin spatial transcriptomics analysis and integration kit.* https://doi.org//arXiv.2505.11384DOI: 10.48550

Barnett, J., Silverman, J., Wetzel, M., Rao, P., Sotudeh, N., & Wang, L. (2022). AtlasX-browser enables spatial multi-omics data analysis through the precise determination of the region of interest. *bioRxiv*. https://doi.org/DOI: 10.1101/2022.05.11.491526

Barsoum, I., Tawedrous, E., Faragalla, H., & Yousef, G. M. (2019). Histo-genomics: Digital pathology at the forefront of precision medicine. *Diagnosis (Berlin, Germany), 6*(3), 203–212. DOI: 10.1515/dx-2018-0064 PMID: 30827078

Battazza, A., Brasileiro, F. C. D. S., Tasaka, A. C., Bulla, C., Ximenes, P. P., Hosomi, J. E., da Silva, P. F., da Silva, L. F., de Moura, F. B. C., & Rocha, N. S. (2024). Integrating telepathology and digital pathology with artificial intelligence: An inevitable future. *Veterinary World, 17*(8), 1667–1671. DOI: 10.14202/vetworld.2024.1667-1671 PMID: 39328444

Behura, S. K., Tizioto, P. C., Kim, J., Grupioni, N. V., Seabury, C. M., Schnabel, R. D., Gershwin, L. J., Van Eenennaam, A. L., Toaff-Rosenstein, R., Neibergs, H. L., Regitano, L. C. A., & Taylor, J. F. (2017). Tissue tropism in host transcriptional response to members of the bovine respiratory disease complex. *Scientific Reports, 7*(1), 17938. DOI: 10.1038/s41598-017-18205-0 PMID: 29263411

Blader, I. J., & Saeij, J. P. (2009). Communication between *Toxoplasma gondii* and its host. *APMIS, 117*(5-6), 458–476. DOI: 10.1111/j.1600-0463.2009.02453.x PMID: 19400868

Branck, T., Hu, Z., Nickols, W. A., Walsh, A. M., Bhosle, A., Short, M. I., Nearing, J. T., Asnicar, F., McIver, L. J., Maharjan, S., Rahnavard, A., Louyakis, A. S., Badri, D. V., Brockel, C., Thompson, K. N., & Huttenhower, C. (2024). Comprehensive profile of the companion animal gut microbiome integrating reference-based and reference-free methods. *The ISME Journal, 18*(1), wrae201. Advance online publication. DOI: 10.1093/ismejo/wrae201 PMID: 39394961

Bressan, D., Battistoni, G., & Hannon, G. J. (2023). The dawn of spatial omics. *Science, 381*(6657), eabq4964. Advance online publication. DOI: 10.1126/science.abq4964 PMID: 37535749

Carrillo, C., Tulman, E. R., Delhon, G., Lu, Z., Carreno, A., Vagnozzi, A., Kutish, G. F., & Rock, D. L. (2005). Comparative genomics of foot-and-mouth disease virus. *Journal of Virology, 79*(10), 6487–6504. DOI: 10.1128/JVI.79.10.6487-6504.2005 PMID: 15858032

Causer, A., Lu, T., & Fitzgerald, C., (2024). *SpaMTP: Integrative statistical analysis and visualisation of spatial metabolomics and transcriptomics data. bioRxiv*. https://doi.org/DOI: 10.1101/2024.10.31.621429

Ceciliani, F., Eckersall, D., Burchmore, R., & Lecchi, C. (2014). Proteomics in veterinary medicine. *Veterinary Pathology*, *51*(2), 351–362. DOI: 10.1177/0300985813502819 PMID: 24045891

Cedzyński, M., & Świerzko, A. S. (2023). Collectins and ficolins in neonatal health and disease. *Frontiers in Immunology*, *14*, 1328658. DOI: 10.3389/fimmu.2023.1328658 PMID: 38193083

Chen, W., Zhang, P., Tran, T. N., Xiao, Y., Li, S., Shah, V. V., Cheng, H., Brannan, K. W., Youker, K., Lai, L., Fang, L., Yang, Y., Le, N.-T., Abe, J., Chen, S.-H., Ma, Q., Chen, K., Song, Q., Cooke, J. P., & Wang, G. (2025). A visual–omics foundation model to bridge histopathology with spatial transcriptomics. *Nature Methods*, *22*(7), 1568–1582. DOI: 10.1038/s41592-025-02707-1 PMID: 40442373

Roess, A. A., Winch, P. J., Akhter, A., Afroz, D., Ali, N. A., Shah, R., Begum, N., Seraji, H. R., El Arifeen, S., Darmstadt, G. L., & Baqui, A. H.Bangladesh Projahnmo Study Group. (2015). Household animal and human medicine use and animal husbandry practices in rural Bangladesh: Risk factors for emerging zoonotic disease and antibiotic resistance. *Zoonoses and Public Health*, *62*(7), 569–578. DOI: 10.1111/zph.12186 PMID: 25787116

World Health Organization. (2025). *Brucellosis*. https://www.who.int/news-room/fact-sheets/detail/brucellosis

Chapter 11
AI Approaches for Constructing Digital Transcriptomics

Siddhartha Das
https://orcid.org/0000-0002-7482-2555
Centurion University of Technology and Management, Paralakhemundi, India

Sweety Maity
Odisha University of Agriculture and Technology, Bhubaneswar, India

ABSTRACT

The integration of AI in transcriptomics has revolutionized the analysis of complex biological data. This chapter highlights AI's role in digital transcriptomics, especially in spatial transcriptomics, pathology, and toxicity studies. AI techniques like ML and DL tackle high-dimensional data through preprocessing, feature extraction, and modeling. Tools such as HE2RNA predict RNA-Seq data from histology slides, aiding virtual gene expression mapping and tumor analysis. GANs, like TransTox, translate organ-specific profiles to support drug safety. In cancer immunotherapy, AI helps detect immune signatures and tumor heterogeneity. AI also enables multi-omics integration, improving signal clarity and accelerating precision medicine.

INTRODUCTION

A wide range of factors, including cell type, stages of development, environmental conditions, and disease status, can affect the transcriptome's composition, which is temporally dynamic. Because the genome is static, information about how gene expression is regulated by biological conditions, including growth and development,

DOI: 10.4018/979-8-3373-4849-0.ch011

the development of diseases, and exposure to stress, cannot be gleaned from it. Consequently, one of the most researched biological datasets is the human transcriptome. Numerous studies on the human transcriptome have facilitated quick developments in personalized medicine techniques, cancer diagnosis (Sager *et al.*, 2015), and the comprehension of complex diseases (Casamassimi *et al.*, 2017). In the last decade, long-read genome sequencing methods have been created, commercialized, and are currently subject to ongoing enhancement. Oxford Nanopore Innovative Technologies (ONT) and Pacific Biosciences (PacBio) single-molecule real-time (SMRT) decoding are two potential examples of these technologies (Logsdon *et al.*, 2020). Over time, the entire genome or transcriptome sequencing has become more efficient. A human genome sequencing project for cancer research using Illumina's NGS technology was expected to cost $4000 USD by 2020 (Schwarze *et al.*, 2020), which is normally within the budget of most research institutions. It's just a small portion of the Human Genome Project, which spent 2.7billion US dollars over 13 years to sequence the first human genome. The lack of computing tools enabling data processing was the problem brought on by the ongoing developments in the field of DNA and RNA sequencing technology. Modern transcriptome methods, for instance single cell in addition to the spatial transcriptomics, or transcripts, are becoming more computationally demanding and more impressive and challenging (Eraslan *et al.*, 2019). Many transcriptome datasets that are currently available in public sources have not yet been completely utilized. A limited number of characteristics in the full-length transcriptomes that have not been previously sequenced are examined in the many studies that are released each year (Xu Jackson, 2019). Of these databases, only a small portion serves as the foundation for further investigation. Conventional analytical techniques would not be adequate to manage the huge amount and variety of transcriptomic metadata that is currently available, especially since data analysis is not evolving at the same rate as data generation. Machine learning (ML) approaches are being included in the transcriptomic analysis pipeline, which is undergoing a revolution. Transcriptomics using machine learning is not a new idea. Examples of using machine learning techniques to analyze gene expression data date back to the 2000s, despite the fact that they were not part of the standardized analytical pipeline. However, since 2020, the quantity of articles published in the bioinformatics discipline has grown rapidly. A subsection of machine learning (ML) called deep learning (DL) makes use of multiple layers of artificial neural networks (ANN) (Min *et al.*, 2017). Machine learning technologies encompass the computation of transcriptomics data, such as RNA sequences together with data on their epigenetic alteration sites and levels of expression (Libbrecht *et al.*, 2015). Traditional analytical methods often struggle with the volume and complexity of transcriptomics data, posing difficulties in tasks such as managing large data sets, discerning underlying patterns, and integrating diverse data sources; machine learning techniques can offer improved accuracy and efficiency in addressing these challenges (Auslander *et al.*, 2021). Biologists could also use machine learning (ML) in transcriptomics by applying ML techniques to solve new research problems, for instance, selecting or developing a suitable ML model (Le *et al.*, 2020). However, a major obstacle for many bioinformaticians is their lack of knowledge about the state of machine learning's application in this field, particularly for those who are not familiar with the technology in the study of complex biological systems and human transcriptomics. Following an overview of 40 current research studies that use ML methods to address various transcriptomics issues, the review's next sections describe the primary categories of machine learning. Included are significant computational outcomes such as task formulation, data pre-processing, results (ML model performance), and validation techniques. There is a lot of emphasis on published research in the hopes of improving practical relevance. In the past five years, nearly entirely analyzing human transcriptomes and comparing the results with those of conventional non-ML methods.

1. VARIOUS TYPES OF MACHINE LEARNING

Machine learning models are used to foresee the researcher's objective and find underlying patterns in datasets. The three main ML paradigms are reinforcement learning, unsupervised learning, and supervised learning (Sarker, 2021). DL is another well-known subset of ML that uses multiple layers of ANN and is not classified under any of these paradigms.

1.1. Supervised learning

Training the model using labeled data is a form of learning known as supervised learning. Data labeling is the process of giving each data point a meaningful tag. It is the source of supervision, and according to the researcher, it identifies particular dataset aspects. The term "ground truth," which is used in machine learning, statistics, and related fields, refers to the correct answer. Both upfront and following training, it acts as a benchmark for assessing the model's performance. Tasks with clear objectives and readily available labeled data are conducive to this type of learning. Unsupervised learning differs from supervised learning in that sense which marks it does not require labeled data for training. Instead, the model explores the incoming data on its own to find patterns.

1.2. Unsupervised learning

In contrast to supervised learning, unsupervised learning does not use structured information for training. Instead, the model finds rhythms in the data that arrives on its own. When the objective is to identify a structure or pattern in the data that was previously unknown, learning is frequently the preferred method (Parasa *et al.*, 2021). Unsupervised learning models can produce unexpected outcomes that deviate greatly from the problem-solving path since the researcher does not provide labels to guide the algorithm. Sometimes, though, those outcomes can reveal latent structures or patterns in the data that supervised processes might not have picked up on.

1.3. Reinforcement learning

Unlike supervised and unsupervised learning, which rely on fixed datasets, reinforcement learning uses allied model components that interact with its environment dynamically (Sarker, 2021). The agent must be rewarded for choices that further the goal and punished for those that don't, according to a feedback mechanism. This type of learning occurs most effectively when decisions impact one another.

1.4. Deep learning

The integration of different ANN layers distinguishes DL as a subset of ML. Artificial Neural Network (ANN) is a machine learning technique that simulates how the human nervous system and brain function during learning (Montesinos López *et al.*, 2022). It is made up of interconnected nodes that function as the neural network's basic processors and resemble neurons (Wu and Feng, 2018). Nodes are arranged into several levels, and a node's function within the network is determined by the layer to which it belongs. ANNs typically consist of three major layers: input, hidden, and output. A DL model's "deep" quality can be explained by the existence of multiple hidden layers. According to LeCun et al. (2015), learning occurs in the hidden layers of the DL model. Compared to classic machine learning (ML) computational models, which tend to be older and do not incorporate deep neural networks (DNN), deep learning (DL) algorithms are typically more intricate and computationally demanding. Nevertheless, compared to typical ML techniques, they are more appropriate for analyzing larger and more complicated data sets and require less data pre-processing.

2. TRANSCRIPTOMICS WITH MACHINE LEARNING

2.1. DNA microarray-based analysis

Hybridization techniques are employed to provide DNA microarray data that measure RNA quantities and gene expression based on established DNA sequences obtained from RNA to cDNA conversion (Figure 1). They offer details on the RNA transcripts' gene expression profiles (Lowe et al.,2017). Since the 2010s, RNA sequencing (RNA-seq) methods have been seen as a better option than microarrays for researching gene expression profiling (Rao *et al.*, 2019), although microarrays are still carried out since they are less expensive and can be used as a validation tool to support RNA-seq results. Publicly available microarray datasets, such as those in the NCBI archive, represent a rich resource for further analysis. In recent years, machine learning (ML) approaches have been increasingly applied to these datasets to predict disease outcomes and discover potential biomarker genes. Two well-liked conventional techniques, PLIER and RMA, were surpassed by random forest with the condition (Table 1) (Korir *et al.*, 2015), and were the most effective machine learning algorithms for measuring the expression of transcript isoforms. Conditional random forests, an improvement on the classic random forest algorithm, use statistical testing to determine the significance of each characteristic at each node in order to choose whether to split the data. Despite its computational demands and need for extensive data, it offers insights into feature importance and performs well with intricate datasets characterized by interconnected elements. The Genotype Tissue Expression (GTEx) project website was used to obtain RNA–seq and microarray data in order to create the study's training data (Korir *et al.* 2015).

Table 1. Machine learning methods applied to recent transcriptome investigations using DNA microarray data

Research focus	Objective	Machine learning technique
Disease diagnosis and prediction	Make predictions about pulmonary arterial hypertension and pinpoint the genes that are crucial for these predictions. Determine the stage of multiple sclerosis and its prognosis.	Recursive feature removal with sliced inverse regression Using adaptive boosting on functional trees
Calculating the amount of gene expression	Determine the gene and transcriptional levels of expression of a transcript isoform.	Forest with conditional randomness
Identifying genes that serve as disease biomarkers	Determine the gene biomarkers for high-grade gliomas. Determine the genes crucial for predicting and forecasting intervertebral disc degeneration. Determine which genes are involved in the subtypes of cancer prediction.	Operator for selection and the least amount of absolute shrinkage Recursive feature elimination and vector machine support Deep learning with neural networks

For the analysis of microarray data, standard pre-processing steps are background correction and quantile normalization (Korir *et al.*, 2015). After training, the algorithms were evaluated on another dataset to see which one worked best. Numerous machine learning methods have been tried for disease prediction. L1-penalized logistic regression was employed by Warnat-Herresthal *et al.* (Warnat-Herresthal *et al.*, 2020) to find differentially expressed genes and predict acute myeloid leukemia (Table 1). L1-penalized logistic regression combines logistic regression, used for modeling binary outcome probabilities, with L1 normalization, which shrinks the absolute values of model parameters. This combination is used for binary classification and offers the advantage of automatic feature selection, leading to more interpretable models. However, L1-penalization can struggle with small datasets, potentially resulting in inaccurate coefficient estimation and suboptimal feature selection, ultimately hindering performance. Hernandez-Warnat *et al.* (Warnat-Herresthal *et al.*, 2020) 's training data were acquired from many platforms, and each platform's output was normalized separately for pre-processing purposes. The training data was then extracted from data on common genes that were identified across all datasets. Each sample should have a label identifying it to a particular illness condition. The training and test data used to assess the models for generalization may also be designed not to overlap (Warnat-Herresthal *et al.*, 2020). Using the blood transcriptome, multiple sclerosis phases can be determined. Acquaviva *et al.* (Acquaviva *et al.*, 2020) used a pre-processing pipeline that was comparable to that of Warnat-Herresthal *et al.* (Warnat-Herresthal *et al.*, 2020); the datasets were normalized separately, and only genes that were shared by all datasets were curated. Functional trees can have their performance enhanced by adding the adaptive boosting technique, which raises the classification models' predictive accuracy (Sarker, 2021).

To achieve improved outcomes, machine learning algorithms are often combined. For example, in Cui *et al.*'s research (Cui *et al*, 2019), to find the genes required for the classification of pulmonary arterial hypertension, they employed sliced inverse regression-based recursive feature elimination (SIR-RFE) in conjunction with other supervised learning methods such as artificial neural network (ANN), support vector machines (SVM), and linear discriminant analysis. By combining the feature selection and dimensionality reduction methods of SIR and RFE, SIR-RFE chooses the most significant features and eliminates less significant ones (Hilafu & Safo, 2022). When there are many variables in high-dimensional data, like in data analysis when the variables are individual genes, this approach works well. However, feature removal from the original feature space may result in information loss. Using a limited panel of interesting genes found by the ML algorithm, Cui *et al.* (Cui *et al*, 2019) generated very accurate results by combining two approaches. The identification of disease-associated biomarker genes is a work that has a close relationship with disease prediction. As opposed to the former, which is essentially a task of classification, the latter is an attribute selection task that entails prioritizing the relevant features. Shu et al. employed L1 regularization, specifically the least absolute shrinkage and selection operator (LASSO), to identify biomarkers for high-grade glioma. Following data collection and batch effect correction, their algorithm pinpointed differentially expressed genes within the microarray data. Subsequently, they concentrated on genes exhibiting non-zero coefficients and utilized LASSO to select the most informative genes. The prognosticity of the prognostic biomarker gene WEE1 was proven by additional in vitro research that showed its overexpression causes cell invasions (Shu *et al.*, 2018). Another feature selection method is recursive feature elimination, which can be used in conjunction with a classification algorithm, such as SVM, to get ready for illness classification beforehand. SVM classifies points into one of the two classes designated by specifying a hyperplane. SVM is poor at dealing with big data because it is prone to noise, but does well in small, complex data sets. This is due to the use of outliers in the computations by SVM. Data pre-processing procedures in association with illness prediction, such as the detection of differential expression genes, must be performed before the machine learning model's training and testing. Testing of the ML model on several different datasets is recommended in order to validate its ability to recognize overfitting issues (Chang *et al.*, 2020).

Figure 1. DNA microarray operation principle. mRNAs are isolated and purified from the relevant sample. Using fluorescent dyes, they are labeled after being reverse transcribed to cDNAs. After being combined, the labeled RNAs are put onto a suitable DNA microarray chip that has probes—known DNA sequences—that complementary labeled RNAs can attach to. Following chip washing and fluorescence data collection (often in the form of an image), gene expression and further analysis can be carried out

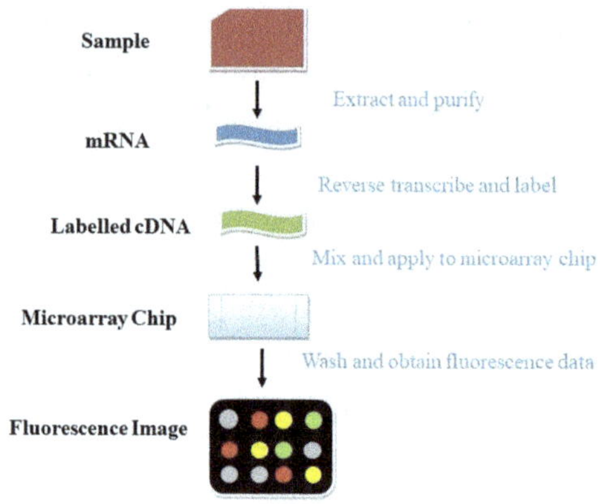

2.2. Methods based on RNA sequencing

Next-generation sequencing (NGS) technology, illustrated in Figure 2, is commonly used to generate RNA-seq data. This data primarily offers insights into the sequences and expression levels of RNA molecules within a sample. Illumina sequencing is a widely used NGS platform for producing RNA-seq data. Recent studies reveal that machine learning (ML) is widely applied to RNA-seq data for tasks such as identifying diseases, pinpointing key transcripts, discovering disease biomarker genes, detecting differentially expressed genes, deconvoluting single-cell data, and normalizing data for cross-platform comparison with microarray or long-read sequencing data. Both with and without DNN are used in varying amounts. The first four of the aforementioned six applications are ultimately categorization tasks. Algorithms that give the input objects class labels are used in machine learning to solve this kind of problem. According to studies that will be covered below, the stages involved in data pre-processing are comparable to those for handling microarray data: Get RNA-seq data, name the samples if you want to, choose two unrelated datasets for training machine learning models and practical applications, and finally normalize the data. Quantile-normalization can be implemented using the basis of reads per kilobase of exons per million reads mapped (RPKM) (Korir *et al.*, 2015). Ridge regression, another name for L2 regularization (Table 2), was a conventional machine learning technique that was employed to forecast disease stages and progression (Pandey *et al.,* 2018). L2 regularization models give the merits of a few features and non-overfitting in analyzing human transcriptome data. DL methods based on recurrent neural networks (RNN), generative adversarial networks (GAN), and DNN were used for some recent articles. (Ahmed *et al.*, 2022). It has been established that GANs can generate a wide variety of excellent simulated data. Their ability to learn from unlabeled input makes them appropriate for unsupervised learning tasks like gene expression and sequencing simulations (Creswell *et al.*,2018). However, because they have two structures—the discriminator and the generator—they are also challenging to develop. RNNs are particularly well-suited for time series research, such as predicting disease symptoms by taking transcriptome fluctuations into account, because of their recurrent connections or feedback loops, which allow information to be retained over time in the network (Salehinejad *et al.*, 2017). Despite being computationally demanding and intended for sequential data analysis, the classical paradigm may experience gradient issues with lengthy sequences (Salehinejad *et al.*, 2017). Data augmentation should be done before training since overfitting is a major problem with DL approaches, particularly if the initial set of data is too small to support diversity. In order to produce variants of the current data, the data must be randomly assigned. Despite having been developed on mouse data, the synthetic gene activity profiles produced by Park et al. (Park *et al.*, 2020) were based on real human data. Through pairwise pairings among several samples, data augmentation was carried out to increase the amount and diversity of the training information (Park *et al.*, 2020). Following data augmentation, the profiles produced by the GAN were examined to learn more about the part genes play in the onset of disease. By using batch normalization and dense connections as optimizers at the input layer level, Ahmed et al. (Ahmed *et al.*, 2022) have demonstrated significant improvements in the error rate and rank order correlation. By using these optimizers, RNN performance can be improved (Gupta *et al.*, 2022). It was found that adaptive moment estimation enhanced the DNN the most among the tested selections. According to the frequency and recentness of each parameter's changes, the adaptive moment estimation method modifies the learning rate for each one (Kingma and Ba, 2014). In their 2019 study, Figgett et al. used multi-fold, repeated double cross-validation. Statistical testing confirmed that the resulting machine learning models effectively distinguished differences in gene regulation across patient samples with varying subtypes of systemic lupus erythematosus. Differentially expressed genes, disease biomarker genes, and transcripts of interest are all related activities that employ the same range of machine learning approaches. RNA-seq data should be obtained, samples should be labeled, and normalization should be performed as in earlier jobs in order to create the training and test data (Khatri and Bhasin,2020). Before applying machine learning, identifying target transcripts and completing the categorization using a non-ML approach offers several advantages. The resulting data can serve as valuable training data or be used to assess and confirm the performance of the ML classifier model. Feature selection models like information gain, correlation feature selection, LASSO, and Relief can minimize errors. Classification models include logistic regression, classification via regression, random forest, logistic model tree, K-nearest neighbor, random subspace, and repeated incremental pruning.

Figure 2. Illumina is a popular next-generation sequencing platform that is used to sequence RNA. The sample's RNAs are taken out and broken up using the Illumina RNA sequencing technology. Following RNA extraction, reverse transcription generates complementary DNAs (cDNAs). Integrated machine learning-artificial intelligence helps in impactful data delivery, interpretation, and model establishment

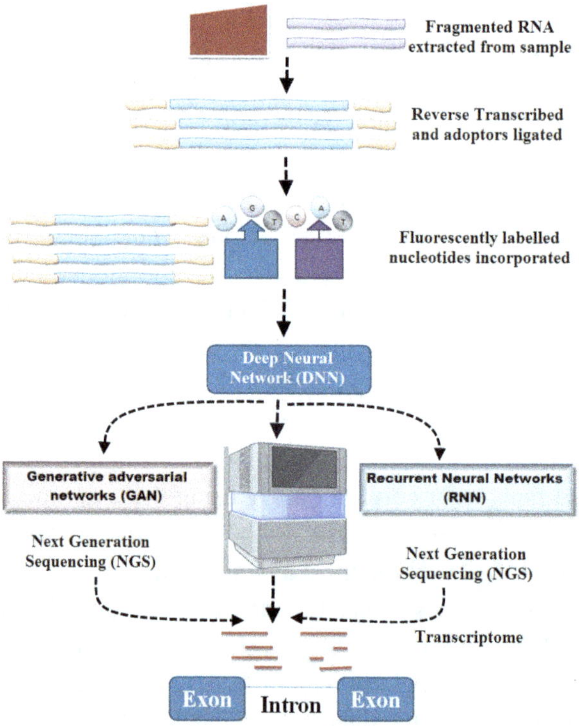

3.3 Long-read sequencing-based approaches

Sequencing methods that are able to sequence long stretches of DNA or RNA without breaking them down into a couple of hundred base pairs are referred to as long-read sequencing data (Pollard et del 2018). ONT and SMRT, the two leading long-read sequencing technology providers, employ distinct sequencing methodologies. Furthermore, genome assembly benefits significantly from long-read data compared to NGS data, primarily because the extended read lengths can effectively span and resolve complex genomic regions, including repetitive sequences (Amarasinghe *et al.,* 2020).

But initially, NGS performed better than several long-read techniques according to accuracy per read (Workman *et al.*, 2019). RNNs are appropriate for jobs involving sequential patterns because of their layer connections, which include feedback to nodes in the layers before them. Due to such a connection, output from a preceding stage may be utilized as part of the input for the current step. In a bid to attain greater base calling accuracy with RNN, Wick et al. (Wick *et al.*, 2019) suggested training their own model with additional hidden layers and training data using ONT's RNN training research software, Sloika. They also noted notable improvement every time the DL model was trained using data whose species belonged to the same taxon as the experimental data (Wick *et al.*, 2019). The process of choosing the best architecture of neural networks for a task was automated by neural architecture search, where sample models were chosen, trained, and then automatically assessed by the search algorithm in the search space (Zoph & Le, 2016). There have been cases when ML models like logistic regression, decision trees, and learning recursive models have been used to attain quicker and more precise alignment (Kalikar *et al.*,2022). By substituting a learnt recursive model index for the conventional hash-table index during the alignment's seeding phase, Kalikar et al. (Kalikar *et al.*,2022) accelerated the Minimap2 program's alignment process (Li,2018). Without converting messenger RNAs into cDNA libraries, the majority of long-read sequencing techniques, including ONT and SMRT, enable direct sequencing of RNA molecules and epigenetic markers (Lucas and Nonoa, 2023). N 6-methyladenosine (m 6 A) has been the most investigated RNA modification, and it can be identified using ML-based workflows (Jiang *et al.*, 2021). Because they fall under the classification task category (Table 4), SVM can be utilized to identify m6A alterations or other forms of RNA changes. Basecalling and mapping information generated by direct RNA-seq technology are obtained according to the manufacturer's guidance, preferably with the Minimap2 and Guppy software, respectively, to obtain the training data. When trained on synthetic data, such a system can achieve an in vivo accuracy of approximately 90% in yeast data (Liu *et al.,* 2019). On the other hand, Pratanwanich et al. relied on the unsupervised probabilistic multi-sample two-Gaussian mixture model to develop machine learning software for the task. Two Gaussian distributions make up the two-Gaussian mixture model, one of which is used to represent the modified or unaltered data points. The term “multi-sample” refers to the way the signal properties of the two data point samples are interlinked while accounting for their differences. Being a kind of multivariate Gaussian mixture hypothesis, it has the advantage of being able to represent Gaussian mixtures with flexibility in input and capturing intricate data distributions. A model used by Pratanwanich *et al.* (Pratanwanich *et al.*,2021) divides continuous base-called sequences into five-nucleotide segments and aligns them with the eventalign function of Nanopolish. A neural network with pooling layers based on numerous instance learning was employed by Hendra et al. (Hendra *et al.,* 2022) to predict the probability of a modification. They observed ROC AUC ratings of about 0.8, indicating robust generalization across datasets. Although the fairly high basecalling error rate invalidates the barcode demultiplexing method, single-cell sequencing is useful for long-read sequencing. Papetti et al. (Papetti *et al.*, 2023) used unsupervised learning techniques such as self-organizing maps (SOMs) for clustering and autoencoders for feature learning. In order to represent the input information in a lower dimension, SOMs employ competitive learning techniques using initial parameters derived from principal component testing. The unprocessed signal data with tags was converted into two-dimensional images using the Python module “pyts” with an accuracy of over 95% on test data and below that on independent biological samples. The study uses random forest models to find regions of double-stranded RNA in PacBio's long-read sequencing data. Features like length and exon-skipping ratio are extracted by the model, and a tree-based pipeline optimization technique is used to tune the hyperparameters. A one-dimensional CNN architecture that is used to identify intron retention. ML is also used to quantify transcript expression, using a unique method that changes according to the transcriptome's dynamic context. Compared to other techniques, this method offers more accurate quantification.

3. ANALYTICAL CHALLENGES

Skepticism about using a device with black-box techniques was the primary barrier to incorporating machine learning (ML) into the transcriptional analytical process as a general strategy, as was the case with many DL algorithms. The phrase “black-box” refers to the model's internal workings, poor interpretability, and transparency from the perspective of the user (Sheu, 2020). One may argue that a model's perceived level of “black-box” status can change based on the user's level of subject-matter competence. Researchers may refrain from employing a tool if they are unaware of its precise working mechanics. One significant obstacle to using machine learning (ML) in transcriptomics is figuring out how to make it more efficient than conventional analytical techniques without abusing resources. Many biologists find machine learning (ML) to be a complex topic that necessitates substantial additional learning, which hinders its application in biology as a whole. Although there are many trained machine learning models and tools for handling typical tasks in transcriptomics that were created by bioinformaticians with ML expertise, they are not necessarily appropriate for the task at hand. This is because transcriptome data, which come from many studies, organisms, and settings, are distinct and dynamic. While it can be taught to adapt to specific conditions, machine learning models often struggle to generalize to other conditions after being taught. An unfamiliar biologist to machine learning does not possess the experience and knowledge required to optimize an ML algorithm to perform better than the established traditional analytical methods. Constructing and training an ML model can also be time- and resource-intensive. Such resources are computer hardware, vast datasets, and machine learning expertise. But pre-trained models can often bypass this limitation, though with the problems introduced by the model's tendency to poor generalization. Additionally, a variety of cloud-based deep learning environments with pertinent infrastructures have been developed that may help new biologists.

4. TRANSCRIPTOMICS' USE IN CANCER RESEARCH IS ENCOURAGED BY ARTIFICIAL INTELLIGENCE

RNA-seq is a crucial component of next-generation sequencing that enables the provision of an abundant amount of transcriptome data of individual cells or of cells in general (Tang *et al.*,2009). In the early days of RNA-seq development, the majority of transcriptome profiles were generated using bulk RNA-seq, which provides average gene expression levels. Arithmetic methods are more frequently used in both upstream and downstream analysis of these data (such as analyses of differentially expressed genes and gene set enrichment). Currently, a number of state-of-the-art AI-based methods have been developed to extract more information from RNA-seq bulk data. Numerous transcriptome analytic domains, such as deconvolution, transcription factor binding, gene expression, and alternative splicing, have made substantial use of AI-based computational tools. Furthermore, using artificial intelligence algorithms in the analysis of transcripts speeds up the advancement of personalized medicine and makes it easier to determine the cancer stage and disease condition. However, even with the use of sophisticated algorithms, bulk RNA-seq is unable to accurately depict the complex ecosystems that make up tumor tissues. Thus, scRNA-seq techniques are developing into effective instruments for high-resolution tumor tissue dissection with the emergence of precision medicine (Suvà et al., 2019). Thus, it makes it possible for researchers to investigate more complex biological issues, like how cells interact inside the tumor's microenvironment, how cancer affects cells over time, and how anticancer therapies affect immune cell infiltration. Oncology research now widely utilizes scRNA-seq technologies. Data analysis techniques have been improved significantly as a result of the widespread adoption of AI techniques, which have also aided in the advancement of single-cell omics technology. With significant implications for cancer immunotherapy, scRNA-seq has made impressive strides in cancer immune research in recent years. Thousands of transcriptome profiles are currently accessible thanks to high-throughput sequencing technology. Transcriptomic data's high variability and complexity are the main obstacles preventing academics from making new scientific discoveries (DelGiudice *et al.*,2021). The demand for AI algorithms grows as transcriptome data volumes increase. AI can distinguish the real biological signal from noise and offer workable ways to decipher the complexity of cancer's complicated transcriptomes.

5. FRESH PERSPECTIVES ON CANCER IMMUNOTHERAPY USING AI-AIDED TRANSCRIPTOME ANALYSIS

Toxicology and resistance to immunotherapy typically arise and limit its use in clinical practice, despite the fact that it has demonstrated long-lasting anticancer effectiveness in individuals (Michot *et al.*,2016). Investigating the factors influencing therapy response is so crucial. Toxicology and resistance to immunotherapy typically arise and limit its use in clinical practice, even though it has demonstrated long-lasting anticancer effectiveness in individuals. Investigating the factors influencing therapeutic response, medication resistance, and side effects is so crucial. AI-assisted transcriptome analysis has become widely employed in recent years to assess the intrinsic mechanisms of medication resistance and side effects, ascertain cellular state, and interpret variability in tumors and the tumor microenvironment.

5.1. Diverseness of Tumors

Intratumor heterogeneity (ITH) is a critically important feature that influences immunotherapy response and patient outcomes (Suphavilai et al., 2021). Sung et al. used transcriptomics data and machine learning techniques to predict ITH and assess its level. They showed that their method could detect transcriptome markers linked to ITH and differentiate between tumor samples that had high and low ITH levels. In order to better analyze the wide spectrum of the ICI response and identify the elements influencing the ICI response in TCGA samples, Zeng et al. developed an algorithm for machine learning in accordance with nonnegative matrix factorization (NMF). According to their findings, ubiquitin E3 ligases might regulate the ICI response (Zeng *et al.*, 2022). Ovarian cancer (OV) cell line transcriptome profiles were subjected to the NMF algorithm by a different group. The research is concerned with the application of scRNA-seq methods to investigate the genetic and molecular features of ovarian tumors (OVs) at different tumor formation. The processing of scRNA-seq data heavily depends on AI algorithms, and the need for precision medicine necessitates speeding up its usage in cancer studies. Various bioinformatics platforms have been built based on MI and DL algorithms in order to accelerate the processing of scRNA-seq data. In lung adenocarcinoma (LUAD), scRNA-seq data were utilized to disentangle the epithelial cell lineages' characterization across different stages of histology. In the early stages of LUAD, a distinct population of cells known as ATD-like cells—which bear a striking resemblance to alveolar type 2 cells (AT2)—emerged and added cellular heterogeneity to the development of LUAD. Researchers used scRNA-seq and genotyping to describe AMI ecosystems in acute myeloid leukemia (AML) and proposed RF-based machine learning models for classifying AML cell types and differentiating between cancerous and healthy cells in AML tumors. Although they are a type of ITH, copy number variants (CNVs) are challenging to identify in transcriptomic data. CNVs have been inferred from scRNA-seq data using a variety of methods, including CopyKAT, inferCNV, and HoneyBADGER. It has been suggested that ITH in cancer is caused by lineage plasticity, which allows cells to change from one phenotypic state to another. Cancer development includes the gain of stem cell-like properties, and mRNAsi can be employed to mark molecular and clinical characteristics in TCCA tumors.

5.2. The Microscopic environment of tumors

The intricate and heterogeneous tumor microenvironment (TME) influences the efficacy of immunotherapy and anticancer activity. Understanding the TME is necessary to develop immunotherapeutic approaches. Computational methods, including gene signature-based and deconvolution-based approaches, offer mcans to estimate cell types from bulk tumor transcriptome profiles. In the case of identifying the cellular composition of the tumor microenvironment, deep neural networks (DNNs) have proven to be outstanding. The prognosis of many malignancies is affected by immune cell infiltration, and this has led to the development of novel immune-related prognostic factors. In patients with esophageal cancer, ICIs improve overall survival and have a better safety record than chemotherapy. ScRNA-seq has illuminated new insights into immune mechanisms and immunotherapy resistance. The content and types of T cells are connected to the efficiency of immunotherapy. Therefore, features of T cells determine how effective immunotherapy will be. Approximately 10,000 T cells from 14 untreated patients were analyzed with scRNA-seq by Guo et al. (Guo *et al.,* 2018) recognized heterogeneity in various T cell types. Additionally, the co-workers discovered a new biomarker for NSCLC patient stratification: an elevated tumor regulatory T-cell-associated gene called IL1R2 (interleukin 1 receptor type 2) was linked to a poor prognosis. Mei et al. (2021) used scRNA seq to find new immune cell compositions in patients with colorectal cancer. In addition to that it was found a lot of T cell clusters, like CXCL1+ AND XCL1/2+ T cells, that help patients respond better to ICIs. Other cell populations have also been found recently to be associated with ICI therapy response, besides T cells, whose types and compositions affect ICI treatment (Helmink *et al.*, 2020). In addition, there is growing evidence that cancer-associated fibrosis (CAF) impacts the TME and immunotherapy sensitivity. Lu et al. developed a CAF score model using the RF and LASSO algorithms, which can be used to precisely forecast the quantity of invasive immune cells and how they would react to immunotherapy (Lu *et al.*, 2022). Moreover, they discovered that a protein related to CAF, fibroblast growth factor receptor, or FGFR4, may be involved in pancreatic cancer. Yang et al. used tumor cell ligand-receptor gene pairs with macrophages to build an ML prediction model. It was connected to the prognosis of LUAD patients. Understanding cell states and multicellular communities is crucial to comprehending cancer. There have been many phenotypic classifications, such as T-cell infiltration, in human tumors defined by previous studies. To find cell types and communities at large through substantial tissue samples, researchers created the machine learning platform EcoTyper to discover a wide variety of newly unidentified cells that had not been previously discovered using scRNA-seq. Also, most cell states have a strong correlation with prognosis. It's interesting to note that AEBP1+ foamy macrophages were linked to a lower survival period, which may indicate that they could be a target for immunotherapy in cancer. EcoTyper is also capable of restoring multicellular communities and identifying cohesive cell populations. They also used it on tumors of diffuse large B-cell lymphoma.

5.3. The toxic effects of immunotherapy

Cancer patients benefit greatly from immunotherapy; however, side effects such as inflammatory responses (irAEs) are common. It is unclear what immunological mechanisms underlie irAEs. scRNA-seq has been used to identify alterations in the effector programs and immune cell compositions that underlie irAEs. When hepatocellular carcinoma (HCC) patients were treated with anti-PD-1, single-cell analysis revealed that CXCR3^{+} CD8^{+} effector memory T (TEM) cells had a role in both response and adverse events. For HCC patients who received ICI therapy, it may be possible to distinguish between response and irAEs by taking advantage of a particular TNF interaction between CXCR3+ CD8+ TEM cells and response-specific and irAE–specific myeloid cell population. Selective TNFR2 inhibition can increase ICI effectiveness without increasing irAEs, highlighting its potential as an innovative strategy to distinguish immunotherapy toxicity from anticancer efficacy. Since microRNA-146a (miR-146a) regulates the autoimmune dysregulation brought on by ICIs, it has been proposed as a potential biological target to stop ICI-induced irAEs.

5.4. Finding novel targets and triumphing over drug resistance

The significance of certain novel targets in the immunotherapy response may also be suggested by AI-based transcriptome analysis (Table 2). A gene expression signature called PredictIO was created by Bareche et al., who demonstrated its higher predictive value than alternative biomarkers (Bareche *et al.*,2022). Additionally, they used PredictIO to find possible targets to get past ICI resistance in conjunction with TIDE, which integrated the expression profiles of T-cell malfunction and exclusion to simulate tumor immune evasion (Jiang *et al.*,2018). Because of their high T-cell dysfunction scores, RBFOX2 (RNA-binding fox-1 homolog 2) and F2RL1 (protease-activated receptor 2-like trypsin receptor 1) were considered novel biomarkers and possible targets. In another study, scientists employed machine learning-based algorithms to confirm the prognostic and therapeutic response importance of complement factor B (CFB) and serum amyloid A2 (SAA2), stating that both proteins are predictive biomarkers and contribute importantly to tumor metastasis. To anticipate ICI reaction, several prediction models have been developed to analyze data utilizing machine learning. The significance of immune-related gene expression in antitumor immunity, whose identification is crucial for targeting, is not well understood, despite the fact that several studies have been conducted to better understand the relationships between immune-related gene expression and immunotherapy responses in different malignancies. By expanding our knowledge of the related immune response mechanism and discovering novel targets implicated in cancer initiation, immune evasion, growth, and metastasis, the ScRNA-seq technology may hasten target discovery. For instance, Somani et al. found a significant correlation between T cell exhaustion and NF-kB activation mediated by interleukin-1 receptor-associated kinase-4 in pancreatic ductal adenocarcinoma (PDAC). IRAK4 inhibition or deletion affected the expression of checkpoint molecules and increased the number and activity of infiltrative CD4+ and CD8+ T cells. IRAK4 inhibition causes myeloid and fibroblast cells to reprogramme to acute inflammatory responses, which results in a T cell-facilitative TME, according to ScRNA-seq research. To increase the survival rate in mouse models, it can be used with ICI treatment. NKG7 reduces he effectiveness of ICIs in NKG7 mutant mice and only exists in $CD8^+$ T cells and NK cells. C-C motif chemokine ligand 5 (CCL5) was also discovered by

ScRNA-seq analysis to be a potential target for promoting immune cell infiltration. The significant role that E74-like ETS transcription factor 3 (ELF3) has in modulating interactions between tumor cells and TME components indicates that it could be a potential therapeutic target in LUAD. SHP2 can be a target for myeloid-targeted immunotherapy, which has been shown in CRC patients as well as in murine tumor models. One of the strategies for bypassing immunotherapy resistance is to combine ICIs with BIRC2 inhibitors.

Table 2. AI techniques applied to transcriptomic analysis

Input Data	Tumor	AI Model	Function	Application
Spatial RNA-seq data and images	Breast cancer	DL:CNN	Connecting visual characteristics and gene expression since morphology	ITH
Drug response data, scRNA-seq and bulk RNA-seq	Ovarian cancer	ML:RF	Helping to identify safe and efficient combination treatment plans	Drug resistance
ScRNA-seqdata	Malignant melanoma	DL:Denoising autoencoder	Dealing with a significant quantity of noise	ITH
ScRNA-seq data	Colorectal cancer	DL:GAN,deep autoencoders	Determining the cell types specific to a batch	ITH,TME
BulkRNA-seq, scRNA-seqand spatialRNA-seq data	16typesofcarcinoma	ML:EcoTyper	Providingaframeworkfor large-scaleprofilingof cellularecosystemsinany tissue	TME,ITH
BulkRNA-seq data	Gastrointestinal cancer	ML:linearSVM	Forecasting a patient's reaction to immunotherapy	TME
Bulk RNA-seq data	Pancreatic cancer	ML: RF, LASSO	Estimating how many immune cells will infiltrate and how well immunotherapy will work	TME

6. COVERAGE OF MAJOR DL MODELS FOR ANALYZING SRT DATA

For varying purposes, SRT exploration employs diverse supervised and unsupervised learning methods. Cell segmentation and gene prediction are tackled by supervised learning, while clustering, gene expression imputation, and dimension reduction are tackled by unsupervised learning. A general overview of the DL models used for SRT data analysis is presented in Table 3. The models are convolutional neural networks (CNNs), graph neural networks (GNNs), variational autoencoders (VAEs), autoencoders (AEs), and deep neural networks (DNNs). Pre-processing methods and reviewed DL models of gene expression matrices, geographical data. The aforementioned models are classified as sequential models, which generate a sequence of hidden states from the previous hidden state. Sequential mechanism problem hinders training examples from parallelizing, which becomes essential at longer sequence lengths for bigger data sets and causes memory constraints (Vaswani *et al.*,2017). Attention mechanisms (HA), which establish dependencies without considering their distance in the input or output sequences, have been newly introduced to reduce the constraint of sequential computation. In addition, another strong model architecture that allows additional parallelization in DNN-based methods is the multi-head attention mechanism (MHA) or transformer. As there are several DL model topologies and loss functions proposed by the reviewed papers.

Table 3. An overview of DL models

Model	Explanation
CNNs	Supervised image processing models. To extract features from multidimensional input data, use pooling and convolutional layers.
GNNs	General models for processing graph data. Combine and modify node data using different network topologies, such as graph attention networks (GAT) and graph convolution networks (GCN).
VAEs	Inputs should be encoded as latent space distributions. Use multi-layer neural networks to discover latent properties.
AEs	Deep generative models to reduce dimensionality. Data input should be reconstructed after being encoded into latent variables.
DNNs	feed-forward neural networks that have several activation functions and hidden layers. Estimate nonlinear transformations for certain objectives

7. EXAMINING DL MODELS FOR SRT EVALUATION

Improvement of Gene Expression Solution (GER), cell-cell relationships, absent gene imputation, spatial area recognition, spatially varying genes (SVGs), and cell-type deconvolution are the six subgroups upon which we separate the 26 DL methods that are employed in examining SRT data. Supplementary Tables S1 and S2 include the measures for the datasets within each group, respectively.

8. DETERMINING THE SPATIAL DOMAIN

Identifying spatially coherent areas with uniform histology and gene expression—also referred to as spatial domain identification—is a crucial step in spatial transcriptomics research. There are several technologies available for spatial transcriptomics, some of which, like slide-seq, generate both tissue pictures and gene expression data (Rodriques *et al.*, 2019). The majority of methods rely on clustering techniques that describe cell types only by gene expression properties (e.g., Seurat). Bayesian methods such as BayesSpace and conventional machine learning techniques like the hidden Markov random field (HMRF) model have been used to incorporate spatial information and tissue heterogeneity. Moreover, Single-Cell Microscopy Empirical Bayes is used for spatial domain identification, which predicts labels using techniques like expectation-maximization (EM) and empirical Bayes. Liu et al. have observed that the conventional process of doing dimension reduction and spatial clustering consecutively is not always accurate. As a result, they presented DR-SC, a technique that uses a single joint model framework to solve both spatial clustering and dimension reduction concurrently. This method derives low-dimensional embeddings effectively by mainly using a probabilistic and hierarchical model that is based on an HMRF model. These ML techniques, however, may not work well with little experimental control since they assume certain aspects of the data-generating process. For the analysis of high-dimensional spatial transcriptomics data, particularly in sequencing, DL models have become more and more common. DL methods with spatial data and histological images for identifying spatial domains are as follows: SpaCell, stLearn, SpaGCN, SEDR, STAGATE, RESEPT, ECNN, JSTA, conST, CCST, GraphST, spatial-MGCN, MGCN, and STGNNKs. The model decouples the preprocessing of images and count matrices and normalizes and divides the images into 299×299 pixel tiles. Each point in the pictures comes with a count of genes associated with it. ResNet50, a CNN that has already been trained, is used by the model to extract significant characteristics from every tile. SpaCell uses the same images and a two-layered DNN model to analyze gene count matrices and pixel attributes in order to provide a likelihood for four phases of sickness. SpaCell is based on a pre-trained model that does not incorporate spatial information in the embeddings and was trained on non-histology images, which can produce uninformative results. The development of stLearn, which maximizes the spatial distance, integration of tissue morphology, and gene expression in spatial transcriptomics, was prompted by SpaCell (Pham *et al.*, 2023).

9. UNDERSTANDING SVGS

The identification of spatial expression patterns in tissue slices is an important task that incorporates SVG detection. Certain statistical methods have been developed (Sun *et al.*, 2020), but they neglect predictors that are associated with morphology and tissue classification. SVG detection methods are classified into two types: cluster-based and whole tissue-based. Cluster-based methods (like STAGATE, SpaGCN, MGCN, and conST) employ statistical tests on spatial regions that are derived from clustering algorithms; however, they cannot detect samples that are resistant to clustering and gradient expression genes. In these situations, whole tissue methods must be used. Deep or spectrum-based machine learning methods are present for SVG detection. Spectral methods, like Rayleigh Selection (Govek *et al.*,2019), calculate each gene's combinatorial Laplacian score (CLS); increased spatial variability is reflected in lower CLS values. However, spectral methods are computationally expensive and do not perform well on large datasets. Computational efficiency of the latest methods to detect SVGs was investigated by Zhang et al. (Zhang *et al.*, 2022). To detect SV genes with accuracy, they suggested ScGCO, which is a new algorithm that relies on a probabilistic graph model to identify statistical and geographical features of the modeled variables. It is better to employ methods that integrate SVG detection with H&E-stained histology images. It has been demonstrated that DL-based methods, including in situ hybridization and in situ sequencing, are better. For SVG detection in cluster and deep-based domains, DL methods like CoSTA, ST-NET, SPADE, HisToGene, CNNTL, and DeepSpaCE have been suggested. SpaGCN is a method designed to find SVGs in certain clusters that was originally introduced by Hu *et al.* To find SVGs, the approach uses the Wilcoxon rank-sum test, focusing on genes referred to as "metagenes" with elevated expression levels across scattered regions. SpaGCN effectively identifies metrics specific to target domains by adjusting threshold values, selecting base genes, and modulating the addition and subtraction of positive and negative genes. The method is more effective in detecting SVG compared to SPARK (Sun *et al.*, 2020) and SpatialDE based on Moran's I statistic (Li *et al.*, 2007), and also provides an option for expressing heterogeneity with sub-clustering. Yet, tissue heterogeneity can be insufficiently represented by the marker genes SpaGCN identifies since its primary goal is to construct deep models for clustering. In a similar vein, Zong et al. (Zong *et al.*,2022) employed conST to discover spatial marker genes within clusters as a subsequent task to verify the correctness of clustering. ConST replicates SpaGCN's approach by applying it to its main algorithm's latent embeddings and does better in Moran's I tests on the spatial LIBD dataset, especially at the white matter border layer. STAGATE is a model that is also attempting to find SVGs in spatial spaces, though not specifically trained for this type of task. Like SpaGCN, it identifies SVGs that are not always associated with tissue shape. STAGATE utilizes the Wilcoxon test through SCANPY to label SVGs in all spatial domains. Applied to the Slide-seqV2 dataset, which was obtained from mouse olfactory bulb tissue, STAGATE identified more genes in smaller-sized tissue structures than the SPARK-X algorithm. Finally, in comparison to both SpaGCN and SpatialDE using Moran's I statistic test, MGCN compares its performance and finds SVGs by adopting the process laid out in SpaGCN. The results indicate that MGCN is better than the above-discussed methods in detecting more SVGs. Instead, based on pixel location data from spatial transcriptomic images, Xu et al. (Xu *et al.*, 2021) introduced CoSTA, a cluster-based approach employing an unsupervised CNN for finding spatial relationships between genes. Pixel binning, gene matrix normalization, and scaling are all components of its pre-processing. The two phases of this process are training neural networks and clustering. Convolution, batch normalization, and max-pooling layers all exist within each of the three convolution boxes that constitute a ConvNet, which is initially provided with normalized images by CoSTA.Labels are constructed for ConvNet training following clustering. A completely linked layer with a function for softmax activation is added in the second phase, which provides the chance that the ith sample is from the ith cluster by a supplementary target distribution. In the datasets for breast, prostate, and olfactory bulb cancer, SPADE effectively identified tissue-specific markers. However, spot density and distance have an impact on marker gene identification. In order to address the shortcomings of ST Net, such as its CNN model's insufficient use of spatial information, Pang et al. (Pang *et al.*, 2021) presented HisTo Gene. HisToGene forecasts gene expression levels while taking histology images and spatial location into consideration using an attention-based algorithm and an AE. The model goes through pre-processing procedures like normalizing UMI counts, removing genes, and converting them to a natural log scale. To create a new matrix that takes into account differences in spot counts within tissues, which are comparable to sentence lengths in natural language processing (NLP), HisToGene separates patches from histology pictures. The final embedding matrix is formed by adding the encoding matrices that the encoder has encoded through a single layer, which includes the new picture matrix and the spatial coordinate. The next step is to apply multi-head attention layers, which have eight layers and sixteen attention heads.When tested on high-resolution SRT data, HisToGene always showed better correlation than ST-Net. Gene expression patterns from human brain pictures can be embedded in the ISH approach using CNNTL, a CNN-based technique with contrastive loss, as suggested by Abed-Esfahani et al. (Abed-Esfahani *et al.*, 2022). Triplet loss was used during training in CNNTL, which was created as an alternative to classification–based methods. It outperformed random and single ResNet models with a rank-1 proficiency of 38.3% on the Cortex dataset. However, CNNTL was pre-trained on an orthogonal dataset, and its use is restricted to a small sample of genes in brain layers. A CNN model called DeepSpaCE was presented by Monjo et al. (Monjo *et al.*, 2022) with an emphasis on oncology and designed especially for in situ capture techniques for spatial transcriptomics. Breast cancer indicators are among the 24 genes whose expression is predicted by DeepSpaCE using a VGG16 network. When tested on a dataset of human breast cancer, DeepSpaCE demonstrated a correlation coefficient of 0.588 between the predicted and measured values. DeepSpaCE's ability to forecast a certain number of genes is its only limitation.

10. GENE IMPUTATION FOR MISSING GENES

Although scRNA-seq offers comprehensive gene expression profiles, it is devoid of spatial context. On the other hand, spatial transcriptomics preserves spatial context but has a low resolution. Integrating scRNA-seq with SRT to predict unmeasured genes was suggested by recent publications as a solution to overcome shortcomings in each method. In machine learning techniques like LIGER and SpaGE, joint dimension reduction methods like NMF and PCA are used, and then linear models are used for joint embedding. Deep generative models are used by GimVI, a joint non-linear model, to adapt to different domains. These techniques, however, frequently rely on common genes across the datasets and potentially deceptive metrics, such as the Spearman correlation coefficient, as Shengquan et al. (Shengquan *et al.*,2021) pointed out. Consequently, they put out the stPlus AE model, which predicts genes using the k-NN method. AMI, ARI, Home, and NMI are the four clustering metrics that stPlus fared better on than SpaGE, Seurat, Liger, and gimVI. Lopez et al. (2019) suggested gimVI, which uses a VAE model to assess gene expression matrices from scRNA-seq and SRT studies. As of December 10, it was in preprint status. Using a binary variable to distinguish between the two creates a latent vector that represents the different sorts of cells. It showed better imputation ability than Liger and Seurat, albcit thc cvaluation was restricted to a subset of genes in the SRT dataset, and the choice of K may have affected the outcomes. Tangram, which applies DL to align histology data to anatomical coordinates and transfer spatial information into scRNA-seq data, was created by Biancalani *et al.* (Biancalani et al.,2021). It updates the alignment using nonconvex optimization according to an objective function that assesses gene expression and compares cell-density distributions. Tangram can effectively do gene imputation even if its primary purpose is the reconstitution of spatial maps. The original paper did not include a quantitative comparison with alternative imputation techniques. Due to its lengthier running time, Tangram was evaluated as the third–best approximation method by recent research, behind stPlus and gimVI.

11. CELL-TYPE DECONVOLUTION

Transcripts are recorded at spatial locations, or "spots," in spatial transcriptomics, which are made up of low-resolution, heterogeneous cells. Due to tissue variability or spatial transcriptomics technology, identifying the cell composition at the spot level is essential (Saiselet *et al.*,2020). Three types of computational approaches can be used for this task: DL-based, multivariate analysis/linear algebra-based, and inference-based approaches. Assuming input data distribution, inference-based techniques such as RCTD, Stereoscope, cell2location DestVI, and STdeconvolve use likelihood-based methodologies. Both machine learning and statistics are included in multivariate analysis and linear algebra-based techniques like SPOTlight and SpatialDWLS. Cell type proportions are estimated using DL models in the meantime by DL-based techniques such as GIST, Tangram, and GraphST. In this work, certain approaches, including VAE, are classified as DL methods even if they are probability-based. These techniques do, however, have limits in terms of practical use. A semi-supervised graph convolution network technique called DSTG was presented by Song *et al.* (Song *et al.*,2021) for breaking down cell mixes in SRT data. In order to estimate cell type proportions, it builds a linked graph, generates pseudo-ST data from scRNA–seq data, and makes use of a GCN network with three convolution layers. On both artificial and real SRT datasets, DSTG performs better than the SPOTlight approach. However, utilizing Euclidean distance to compare pseudo-ST with real-ST might not yield an accurate assessment. The drawbacks of inference-based deconvolution techniques that ignore spatial information and result in inaccurate cell-type detection were discussed by Biancalani et al. (Biancalani *et al.*,2021). Tangram, which they invented, deconvolutionally processes ST/Visium technology, namely, low-resolution SRT data. Tangram maps cell-type ratios consistently across several datasets and computes cell counts through initial segmentation. But depending on past knowledge of cell counts for segmentation could be problematic, particularly in tissues with a high density like malignancies. GIST is a combined model that Zubair et al. (Zubair *et al.*,2022) suggested to improve cell-type deconvolution by combining SRT and image-derived data. Within a Bayesian framework, GIST uses DL on pictures to provide early data for cell type identification. After converting JPEG photos into encoded TIFF format, a CNN model creates probability maps to predict the quantity of cell types, especially immune cells. The weighted summing of overlapping patches is used to compute spot-level probability. When identifying immune cells in the pathophysiology of breast cancer, GIST fared better than the base model that relied solely on expression data. However, in order to ascertain its relative performance, comparisons with Tangram and DSTG are required. In order to determine the cell proportion in each area, Long et al. (Long *et al.*, 2023) seek to create a mapping matrix M. The reconstructed cell gene expression matrix (HC) from SCRNA-seq data and the reconstructed gene expression HS from ST data are first obtained using an AE model.

12. INTERACTIONS BETWEEN CELLS

An essential extracellular communication pathway involving ligand-receptor interactions (LRI) is cell-cell interactions, which are the main focus of research on intercellular communication (Guryanov *et al.*, 2016). Most computational approaches now in use focus on intracellular interactions or are only applicable to small-scale studies. The ability to forecast LRIs using spatial transcriptomics, which provides gene expression patterns in spatial coordinates within cells, is promising. An extensive framework for examining such data, which includes a module on cell-cell interaction, is provided by Giotto. Cellular microenvironment interactions can be found using statistical techniques like Giotto and machine learning frameworks like MISTy (Tanevski *et al.*,2020), which model marker expression and produce pairwise distances, respectively. MISTy aids in the study of marker-interaction interactions by uniquely identifying interactions inside particular regions, although it requires significant computing power. Although increasing SRT data diversity affects the performance of non-DL models, DL techniques are more effective at detecting cell interactions in sizable SRT datasets. stLearn is a technique for examining cell type diversity and locating receptor–ligand interactions (RLIs) in tissue slices that was created by Pham et al. (Pham *et al.*,2023). It entails using CellPhoneDB to compute the co-expression of ligand-receptor pairings and assessing the diversity of cell types. Tissue areas with comparable co-expression are grouped to form a substantial ligand–receptor pair matrix (CCI matrix). Cell density and CCI measurements are combined by stLearn to pinpoint regions with a high likelihood of cell-to-cell interaction. StLearn's validation on a dataset related to breast cancer showed noteworthy linkages between immune cells and tumors. To get around thc difficulties in detecting extracellular interactions from spatial transcriptomics data, Yuan and Bar-Joseph created GCNG, a graph convolutional network for gene expression. GCNG creates an adjacency matrix based on Euclidean distance using spatial cell locations and gene expression pairings as inputs. The approach allows for the examination of interactions between cells that are not physically connected by mapping the matrices' output to an embedding vector. In comparison to other models, GCNG had high AUROC/AUPRC values. Nevertheless, biases could be introduced by its predetermined distance parameters for neighbor cell selection. Finally, the benefits of clustering, SVG detection, and trajectory inference were utilized by conST. The microenvironment of breast cancer cells in IDC regions was analyzed using this approach, which also identified target receptors on the cells. First, the dataset's latent properties are extracted and divided into 20 groups. Trajectory inference is then used for pseudotime ordering after identifying three clusters with notable lesion regions. The tumor microenvironment's marker genes are found using the SVG discovery algorithm. Lastly, TraSig is used for cross-cluster CCI analysis, while label transfer from Seurat is used for within-cluster analysis to find active ligand–receptor couples.

13. USING CONTRASTIVE LEARNING TO ACQUIRE THE SIMILARITY MEASURE

CAWarm-BO is more effective than other optimization methods for optimizing the function (fR) for a given sample R. Our experimental results, however, demonstrate that it is still insufficiently effective to be used directly to find the optimal parameter vector for each new sample; in other words, because of its poor pace, it cannot be completely trusted as the oracle O. As seen in Eq. 3, this restriction emphasizes the necessity of creating a similarity measure S that will allow model M to rapidly produce the specific advisor set for every new sample via nearest neighbor search.

14. DEPICTION OF AN RNA-SEQ SAMPLE

Since every RNA-seq sample contains a large number of sequencing reads that are not well suited for direct neural network input, we create representations for these samples using the MinHash sketch (Ondov *et al.*, 2016). The first step in the MinHash sketch process is breaking down an RNA-seq sample's readings into their component k-mers. A hash value is then obtained by processing each k-mer via a hash function h. In conclusion, MinHash sketch delivers the number of k-mers correlated with these hash values as well as the s shortest hashes generated by h across all k-mers for a given sketch size s. A set input X = xi s i=1 represents each RNA-seq sample R. Each xi is a two-dimensional vector, with the first dimension representing a hash value and the subsequent dimension indicating the total quantity of k-mers in R that correspond to this hash.

15. DATA AUGMENTATION MODULE

There is a limit on the size of the representative sample set because it takes a lot of time to perform Bayesian Optimization (BO) on each RNA-seq sample in order to obtain the ideal parameter vector. The quantity of training data that can be used for contrastive learning is thus constrained. To increase training accuracy in the face of these limitations, we create a brand-new data augmentation technique. The training dataset can be significantly increased using this strategy without requiring more BO compute rounds. Our data augmentation method's fundamental idea is as follows: reads from a previous sample R, are subsampled to create a new RNA-seq sample Rsub. It is expected that the best parameter vector of its corresponding function, fRsub, will be similar to fR. By subsampling reads from the original RNA-seq samples, we can possibly create more training data and infer the similarity scores without additional calculations of Bayesian Optimization (BO). Let Rj sub2 and Ri sub1 be the subsampled counterparts of Rj rep and Ri rep, respectively. In practice, we create seven subsampled instances with sampling proportions of 0.65, 0.7, 0.75, 0.8, 0.85, 0.9, and 0.95 for each representative sample. This approach essentially triples our training dataset without demanding additional computation with Bayesian Optimization (BO).

16. THE ESTABLISHMENT OF NEURAL NETWORKS AND THEIR LOSS FUNCTIONS

Architecture for a neural network that is permutation invariant is necessary since the RNA-seq sample is represented as a set. Drawing inspiration from Zaheer et al. [2017], the encoder Enc is a combination of two sub-networks with trainable parameters. In our situation, the contrastive loss function, which is mean squared error, performs well without requiring the adjustment of hyperparameters like temperature. Training data matching to zi batch and zi batch embeddings can be used with this method, which maintains permutation invariance.

17. APPLICATION OF BIOINFORMATICS IN TRANSCRIPTOMICS

17.1. Parameter tuning in bioinformatics

Parameter adjustment has been explored in various bioinformatics domains, including biological pathway reconstruction (Magnano and Gitter, 2021), transcript assembly (DeBlasio *et al.*, 2020), and multiple sequence alignment (Cedillo *et al.*, 2022). While DeBlasio et al. [2020] represent the first attempt at a data-driven parameter tuning system for transcript assembly by learning an advisor set, this method relies on a fixed advisor set for all new samples. Although it demonstrates improved performance compared to default settings on transcript assemblers like Scallop (Shao and Kingsford, 2017) and StringTie (Pertea *et al.*, 2015), its generalizability is limited due to the small and potentially unrepresentative training set of RNA-seq samples. As a result—a common problem with short training datasets—this method performs poorly when new samples are encountered that are significantly different from those in the training set. AutoTuneX uses a sizable, representative set of RNA-seq samples to overcome these drawbacks. As a result, a more flexible and reliable solution for transcript assembly can be provided by learning a system that can produce sample-specific advisor sets.

17.2. Bayesian optimization

An efficient sample-based algorithm named Bayesian optimization (BO) was developed to optimize a black-box function f(x) globally, which is costly to compute (Frazier, 2018). Bayesian optimization (BO) seeks to find the input point that maximizes (or minimizes) a function f by iteratively sampling points predicted to yield high functional values. Each iteration involves two key components: an acquisition function, which guides the selection of the next point to evaluate, and a Bayesian statistical model, typically a Gaussian Process regressor, which models the function f based on previously observed data (Schulz *et al.,* 2018), to predict f(x). To read an exhaustive account of BO, we refer the reader to Frazier [2018]. Although BO has been effectively used in a variety of contexts, including chemical design, reinforcement learning (Brochu *et al.*, 2010; Marco *et al.*, 2017), and hyper-parameter tuning, its effectiveness in adjusting the parameters of bioinformatics tools has not been thoroughly investigated. To estimate the assembler's performance, we employ a loss function in conjunction with a transcript assembler. Finding the ideal parameters for a particular RNA-seq sample R is, in this case, the same as optimizing the loss function. This loss function appears as a black-box function, as we show below. Because BO is effective at managing these black-box situations, it becomes an ideal tool for this improvement process.

17.3. Contrastive learning

Learning through comparison of input samples is known as Contrastive Learning (CL). CL aims to learn input sample embeddings so that comparable samples (like two cat photos) are embedded near one another and dissimilar samples (like one dog image and one dog image) are inserted farther apart (Le-Khac *et al.*, 2020). A basic CL framework consists of three parts: In the training set, the similarity definition establishes whether two samples are similar (positive pair) or different (negative pair). The specific challenge determines the different similarity criteria. In supervised learning, for example, similarity may be defined by the labels of input data: samples with the same label are deemed similar, while samples that have various labels are considered distinct (Khosla et al., 2020). An encoder for features is an adaptable neural network that creates an embedding space for every input sample. The contrast loss function directs the training process by utilizing the feature encoder's outputs and the specified similarity. The encoding device learns to map samples from a negative pair far apart and those from far apart and those from a positive pair adjacent in the embedding space by minimizing this loss.

18. CONCLUSION

Rapid advancements in transcriptomic technology have significantly advanced the study of cancer immunotherapy. Transcriptome technologies have a wider range of applications now that multi-omics technologies have advanced. Comprehensive molecular profiles are provided by multi-modal data, allowing for a more thorough comprehension of cancer biology. Moreover, spatial transcriptomics of single cells can reveal intercellular communication operating in situ and record spatial distribution data. Furthermore, the incorporation of cutting-edge computational techniques into high-dimensional single-cell analysis advances research on cancer immunotherapy as AI technology advances in biomedical research. The emergence of the precision medicine age will lead to the ongoing development and increased use of AI-assisted transcriptome technologies in immunotherapy research, which will yield important data for the creation of fresh approaches to boost cancer immunotherapy's effectiveness.

REFERENCES

Abed-Esfahani, P., Darwin, B. C., Howard, D., Wang, N., Kim, E., Lerch, J., & French, L. (2022, January 24). Evaluation of deep convolutional neural networks for in situ hybridization gene expression image rep resentation. *PLoS One*, *17*(1), e0262717. DOI: 10.1371/journal.pone.0262717 PMID: 35073334

Acquaviva, M., Menon, R., Di Dario, M., Dalla Costa, G., Romeo, M., Sangalli, F., Colombo, B., Moiola, L., Martinelli, V., Comi, G., & Farina, C. (2020). Inferring multiple sclerosis stages from the blood transcriptome via machine learning. *Cell Reports Medicine*, *1*(4), 100053. Advance online publication. DOI: 10.1016/j.xcrm.2020.100053 PMID: 33205062

Ahmed, S., Komeili, M., & Park, J. (2022). Predictive modelling of Parkinson's disease progression based on RNA-Sequence with densely connected deep recurrent neural networks. *Scientific Reports*, *12*(1), 21469. DOI: 10.1038/s41598-022-25454-1 PMID: 36509776

Amarasinghe, S. L., Su, S., Dong, X., Zappia, L., Ritchie, M. E., & Gouil, Q. (2020). Opportunities and challenges in long-read sequencing data analysis. *Genome Biology*, *21*(1), 30. DOI: 10.1186/s13059-020-1935-5 PMID: 32033565

Auslander, N., Gussow, A. B., & Koonin, E. V. (2021). Incorporating machine learning into established bioinformatics frameworks. *International Journal of Molecular Sciences*, *22*(6), 2903. DOI: 10.3390/ijms22062903 PMID: 33809353

Bareche, Y., Kelly, D., Abbas-Aghababazadeh, F., Nakano, M., Esfahani, P., Tkachuk, D., Mohammad, H., Samstein, R., Lee, C.-H., Morris, L. G., Bedard, P. L., Haibe-Kains, B., & Stagg, J. (2022). Leveraging big data of immune checkpoint blockade response identifies novel potential targets. *Annals of Oncology : Official Journal of the European Society for Medical Oncology*, *33*(12), 1304–1317. DOI: 10.1016/j.annonc.2022.08.084 PMID: 36055464

Biancalani, T., Scalia, G., Buffoni, L., Avasthi, R., Lu, Z., Sanger, A., Tokcan, N., Vanderburg, C. R., Segerstolpe, Å., Zhang, M., Avraham-Davidi, I., Vickovic, S., Nitzan, M., Ma, S., Subramanian, A., Lipinski, M., Buenrostro, J., Brown, N. B., Fanelli, D., & Regev, A. (2021). Deep learning and alignment of spatially resolved single-cell transcriptomes with tangram. *Nature Methods*, *18*(11), 1352–1362. DOI: 10.1038/s41592-021-01264-7 PMID: 34711971

Brochu, E., Cora, V. M., & De Freitas, N. (2010). A tutorial on Bayesian optimization of expensive cost functions, with application to active user modeling and hierarchical reinforcement learning. *arXiv preprint arXiv:1012.2599.*

Casamassimi, A., Federico, A., Rienzo, M., Esposito, S., & Ciccodicola, A. (2017). Transcriptome profiling in human diseases: New advances and perspectives. *International Journal of Molecular Sciences*, *18*(8), 1652. DOI: 10.3390/ijms18081652 PMID: 28758927

Chang, H., Yang, X., You, K., Jiang, M., Cai, F., Zhang, Y., Liu, L., Liu, H., & Liu, X. (2020). Integrating multiple microarray dataset analysis and machine learning methods to reveal the key genes and regulatory mechanisms underlying human intervertebral disc degeneration. *PeerJ*, *8*, e10120. DOI: 10.7717/peerj.10120 PMID: 33083145

Creswell, A., White, T., Dumoulin, V., Arulkumaran, K., Sengupta, B., & Bharath, A. A. (2018). Generative adversarial networks: An overview. *IEEE Signal Processing Magazine*, *35*(1), 53–65. DOI: 10.1109/MSP.2017.2765202

Cui, S., Wu, Q., West, J., & Bai, J. (2019). Machine learning-based microarray analyses indicate low-expression genes might collectively influence PAH disease. *PLoS Computational Biology*, *15*(8), e1007264. DOI: 10.1371/journal.pcbi.1007264 PMID: 31404060

DeBlasio, D., Kim, K., & Kingsford, C. (2020). More accurate transcript assembly via parameter advising. *Journal of Computational Biology*, *27*(8), 1181–1189. DOI: 10.1089/cmb.2019.0286 PMID: 32315544

DelGiudice, M., Peirone, S., Perrone, S., Priante, F., Varese, F., Tirtei, E., Fagioli, F., & Cereda, M. (2021). Artificial Intelligence in Bulk and Single-Cell RNA-Sequencing Data to Foster Precision Oncology. *International Journal of Molecular Sciences*, *22*(9), 4563. DOI: 10.3390/ijms22094563 PMID: 33925407

Eraslan, G., Avsec, Ž., Gagneur, J., & Theis, F. J. (2019). Deep learning: New computational modelling techniques for genomics. *Nature Reviews. Genetics*, *20*(7), 389–403. DOI: 10.1038/s41576-019-0122-6 PMID: 30971806

Figgett, W. A., Monaghan, K., Ng, M., Alhamdoosh, M., Maraskovsky, E., Wilson, N. J., & Mackay, F. (2019). Machine learning applied to whole-blood RNA-sequencing data uncovers distinct subsets of patients with systemic lupus erythematosus. *Clinical & Translational Immunology*, *8*(12), e01093. DOI: 10.1002/cti2.1093 PMID: 31921420

Frazier, P. I. (2018). A tutorial on Bayesian optimization. *arXiv preprint arXiv:1807.02811*. https://doi.org//arXiv.1807.02811DOI: 10.48550

Govek, K. W., Yamajala, V. S., & Camara, P. G. (2019). Clustering-independent analysis of genomic data using spectral simplicial theory. *PLoS Computational Biology*, *15*(11), e1007509. DOI: 10.1371/journal.pcbi.1007509 PMID: 31756191

Guo, X., Zhang, Y., Zheng, L., Zheng, C., Song, J., Zhang, Q., Kang, B., Liu, Z., Jin, L., Xing, R., Gao, R., Zhang, L., Dong, M., Hu, X., Ren, X., Kirchhoff, D., Roider, H. G., Yan, T., & Zhang, Z. (2018). Global characterization of T cells in non-small-cell lung cancer by single-cell sequencing. *Nature Medicine*, *24*(7), 978–985. DOI: 10.1038/s41591-018-0045-3 PMID: 29942094

Gupta, S., Gupta, M. K., Shabaz, M., & Sharma, A. (2022). Deep learning techniques for cancer classification using microarray gene expression data. *Frontiers in Physiology*, *13*, 952709. DOI: 10.3389/fphys.2022.952709 PMID: 36246115

Guryanov, I., Fiorucci, S., & Tennikova, T. (2016). Receptor-ligand inter actions: Advanced biomedical applications. *Materials Science and Engineering C*, *68*, 890–903. DOI: 10.1016/j.msec.2016.07.072 PMID: 27524092

Helmink, B. A., Reddy, S. M., Gao, J., Zhang, S., Basar, R., Thakur, R., Yizhak, K., Sade-Feldman, M., Blando, J., Han, G., Gopalakrishnan, V., Xi, Y., Zhao, H., Amaria, R. N., Tawbi, H. A., Cogdill, A. P., Liu, W., LeBleu, V. S., Kugeratski, F. G., & Wargo, J. A. (2020). B cells and tertiary lymphoid structures promote immunotherapy response. *Nature*, *577*(7791), 549–555. DOI: 10.1038/s41586-019-1922-8 PMID: 31942075

Hendra, C., Pratanwanich, P. N., Wan, Y. K., Goh, W. S., Thiery, A., & Göke, J. (2022). Detection of m6A from direct RNA sequencing using a multiple instance learning framework. *Nature Methods*, *19*(12), 1590–1598. DOI: 10.1038/s41592-022-01666-1 PMID: 36357692

Hilafu, H., & Safo, S. E. (2022). Sparse sliced inverse regression for high dimensional data analysis. *BMC Bioinformatics*, *23*(1), 168. DOI: 10.1186/s12859-022-04700-3 PMID: 35525975

Jiang, P., Gu, S., Pan, D., Fu, J., Sahu, A., Hu, X., Li, Z., Traugh, N., Bu, X., Li, B., Liu, J., Freeman, G. J., Brown, M. A., Wucherpfennig, K. W., & Liu, X. S. (2018). Signatures of T cell dysfunction and exclusion predict cancer immunotherapy response. *Nature Medicine*, *24*(10), 1550–1558. DOI: 10.1038/s41591-018-0136-1 PMID: 30127393

Jiang, X., Liu, B., Nie, Z., Duan, L., Xiong, Q., Jin, Z., Yang, C., & Chen, Y. (2021). The role of m6A modification in the biological functions and diseases. *Signal Transduction and Targeted Therapy*, *6*(1), 74. DOI: 10.1038/s41392-020-00450-x PMID: 33611339

Kalikar, S., Jain, C., Vasimuddin, M., & Misra, S. (2022). Accelerating minimap2 for long-read sequencing applications on modern CPUs. *Nature Computational Science*, *2*(2), 78–83. DOI: 10.1038/s43588-022-00201-8 PMID: 38177520

Khatri, I., & Bhasin, M. K. (2020). A transcriptomics-based meta-analysis combined with machine learning identifies a secretory biomarker panel for diagnosis of pancreatic adenocarcinoma. *Frontiers in Genetics*, *11*, 572284. DOI: 10.3389/fgene.2020.572284 PMID: 33133160

Kingma, D. P., & Ba, J. (2014). Adam: A method for stochastic optimization. *arXiv preprint arXiv:1412.6980.*

Korir, P. K., Geeleher, P., & Seoighe, C. (2015). Seq-ing improved gene expression estimates from microarrays using machine learning. *BMC Bioinformatics*, *16*(1), 1–11. DOI: 10.1186/s12859-015-0712-z PMID: 26338512

Le, T. T., Fu, W., & Moore, J. H. (2020). Scaling tree-based automated machine learning to biomedical big data with a feature set selector. *Bioinformatics (Oxford, England)*, *36*(1), 250–256. DOI: 10.1093/bioinformatics/btz470 PMID: 31165141

Le-Khac, P. H., Healy, G., & Smeaton, A. F. (2020). Contrastive representation learning: A framework and review. *IEEE Access : Practical Innovations, Open Solutions*, *8*, 193907–193934. DOI: 10.1109/ACCESS.2020.3031549

LeCun, Y., Bengio, Y., & Hinton, G. (2015). Deep learning. *nature, 521*(7553), 436-444. DOI: 10.1038/nature14539

Li, H. (2018). Minimap2: Pairwise alignment for nucleotide sequences. *Bioinformatics (Oxford, England)*, *34*(18), 3094–3100. DOI: 10.1093/bioinformatics/bty191 PMID: 29750242

Li, H., Calder, C. A., & Cressie, N. (2007). Beyond moran's i: Testing for spatial dependence based on the spatial autoregressive model. *Geographical Analysis*, *39*(4), 357–375. DOI: 10.1111/j.1538-4632.2007.00708.x

Libbrecht, M. W., & Noble, W. S. (2015). Machine learning applications in genetics and genomics. *Nature Reviews. Genetics*, *16*(6), 321–332. DOI: 10.1038/nrg3920 PMID: 25948244

Liu, H., Begik, O., Lucas, M. C., Ramirez, J. M., Mason, C. E., Wiener, D., Schwartz, S., Mattick, J. S., Smith, M. A., & Novoa, E. M. (2019). Accurate detection of m6A RNA modifications in native RNA sequences. *Nature Communications*, *10*(1), 4079. DOI: 10.1038/s41467-019-11713-9 PMID: 31501426

Logsdon, G. A., Vollger, M. R., & Eichler, E. E. (2020). Long-read human genome sequencing and its applications. *Nature Reviews. Genetics*, *21*(10), 597–614. DOI: 10.1038/s41576-020-0236-x PMID: 32504078

Long, Y., Ang, K. S., Li, M., Chong, K. L. K., Sethi, R., Zhong, C., Xu, H., Ong, Z., Sachaphibulkij, K., Chen, A., Zeng, L., Fu, H., Wu, M., Lim, L. H. K., Liu, L., & Chen, J. (2023). Spatially informed clustering, integration, and deconvolution of spatial transcriptomics with graphst. *Nature Communications*, *14*(1), 1155. DOI: 10.1038/s41467-023-36796-3 PMID: 36859400

Lopez, R., Nazaret, A., Langevin, M., Samaran, J., Regier, J., Jordan, M. I., & Yosef, N. (2019). A joint model of unpaired data from scRNA-seq and spatial transcriptomics for imputing missing gene expression measurements. arXiv preprint arXiv: 1905.02269. https://doi.org//arXiv.1905.02269DOI: 10.48550

Lowe, R., Shirley, N., Bleackley, M., Dolan, S., & Shafee, T. (2017). Transcriptomics technologies. *PLoS Computational Biology*, *13*(5), e1005457. DOI: 10.1371/journal.pcbi.1005457 PMID: 28545146

Lu, S., Hua, J., Xu, J., Wei, M., Liang, C., Meng, Q., Liu, J., Zhang, B., Wang, W., Yu, X., & Shi, S. (2022). Turning towards nonimmunoreactive tumors: Evaluation of cancer-associated fibroblasts enables prediction of the immune microenvironment and treatment sensitivity in pancreatic cancer. *Computational and Structural Biotechnology Journal*, *20*, 3911–3923. DOI: 10.1016/j.csbj.2022.07.029 PMID: 35950187

Lucas, M. C., & Novoa, E. M. (2023). Long-read sequencing in the era of epigenomics and epitranscriptomics. *nature methods, 20*(1), 25-29. DOI: 10.1038/s41592-022-01724-8

Magnano, C. S., & Gitter, A. (2021). Automating parameter selection to avoid implausible biological pathway models. *NPJ Systems Biology and Applications*, *7*(1), 12. DOI: 10.1038/s41540-020-00167-1 PMID: 33623016

Marco, A., Berkenkamp, F., Hennig, P., Schoellig, A. P., Krause, A., Schaal, S., & Trimpe, S. (2017, May). Virtual vs. real: Trading off simulations and physical experiments in reinforcement learning with Bayesian optimization. In *2017 IEEE International Conference on Robotics and Automation (ICRA)* (pp. 1557-1563). IEEE. DOI: DOI: 10.1109/ICRA.2017.7989186

Mei, Y., Xiao, W., Hu, H., Lu, G., Chen, L., Sun, Z., Lü, M., Ma, W., Jiang, T., Gao, Y., Li, L., Chen, G., Wang, Z., Li, H., Wu, D., Zhou, P., Leng, Q., & Jia, G. (2021). Single-cell analyses reveal suppressive tumor microenvironment of human colorectal cancer. *Clinical and Translational Medicine*, *11*(6), e422. DOI: 10.1002/ctm2.422 PMID: 34185431

Michot, J., Bigenwald, C., Champiat, S., Collins, M., Carbonnel, F., Postel-Vinay, S., Berdelou, A., Varga, A., Bahleda, R., Hollebecque, A., Massard, C., Fuerea, A., Ribrag, V., Gazzah, A., Armand, J. P., Amellal, N., Angevin, E., Noel, N., Boutros, C., & Lambotte, O. (2016). Immune-related adverse events with immune checkpoint blockade: A comprehensive review. *European Journal of Cancer (Oxford, England)*, *54*, 139–148. DOI: 10.1016/j.ejca.2015.11.016 PMID: 26765102

Min, S., Lee, B., & Yoon, S. (2017). Deep learning in bioinformatics. *Briefings in Bioinformatics*, *18*(5), 851–869. DOI: 10.1093/bib/bbw068 PMID: 27473064

Montesinos López, O. A., Montesinos López, A., & Crossa, J. (2022). *Multivariate statistical machine learning methods for genomic prediction*. Springer Nature., DOI: 10.1007/978-3-030-89010-0

Ondov, B. D., Treangen, T. J., Melsted, P., Mallonee, A. B., Bergman, N. H., Koren, S., & Phillippy, A. M. (2016). Mash: Fast genome and metagenome distance estimation using MinHash. *Genome Biology*, *17*(1), 1–14. DOI: 10.1186/s13059-016-0997-x PMID: 27323842

Pandey, G., Pandey, O. P., Rogers, A. J., Ahsen, M. E., Hoffman, G. E., Raby, B. A., Weiss, S. T., Schadt, E. E., & Bunyavanich, S. (2018). A nasal brush-based classifier of asthma identified by machine learning analysis of nasal RNA sequence data. *Scientific Reports*, *8*(1), 8826. DOI: 10.1038/s41598-018-27189-4 PMID: 29891868

Pang, M., Su, K., & Li, M. (2021). Leveraging information in spatial tran scriptomics to predict super-resolution gene expression from histology images in tumors. *BioRxiv* pp. 2021–11. DOI: 10.1101/2021.11.28.470212

Papetti, D. M., Spolaor, S., Nazari, I., Tirelli, A., Leonardi, T., Caprioli, C., Besozzi, D., Vlachou, T., Pelicci, P. G., Cazzaniga, P., & Nobile, M. S. (2023). Barcode demultiplexing of nanopore sequencing raw signals by unsupervised machine learning. *Frontiers in Bioinformatics*, *3*, 1067113. DOI: 10.3389/fbinf.2023.1067113 PMID: 37181486

Parasa, N. A., Namgiri, J. V., Mohanty, S. N., & Dash, J. K. (2021). Introduction to unsupervised learning in bioinformatics. *Data analytics in bioinformatics: A machine learning perspective*, 35-49. DOI: 10.1002/9781119785620.ch2

Park, J., Kim, H., Kim, J., & Cheon, M. (2020). A practical application of generative adversarial networks for RNA-seq analysis to predict the molecular progress of Alzheimer's disease. *PLoS Computational Biology*, *16*(7), e1008099. DOI: 10.1371/journal.pcbi.1008099 PMID: 32706788

Pertea, M., Pertea, G. M., Antonescu, C. M., Chang, T. C., Mendell, J. T., & Salzberg, S. L. (2015). StringTie enables improved reconstruction of a transcriptome from RNA-seq reads. *Nature Biotechnology*, *33*(3), 290–295. DOI: 10.1038/nbt.3122 PMID: 25690850

Pham, D., Tan, X., Balderson, B., Xu, J., Grice, L. F., Yoon, S., Willis, E. F., Tran, M., Lam, P. Y., Raghubar, A., Kalita-de Croft, P., Lakhani, S., Vukovic, J., Ruitenberg, M. J., & Nguyen, Q. H. (2023). Robust mapping of spa tiotemporal trajectories and cell–cell interactions in healthy and diseased tissues. *Nature Communications*, *14*(1), 7739. DOI: 10.1038/s41467-023-43120-6 PMID: 38007580

Pollard, M. O., Gurdasani, D., Mentzer, A. J., Porter, T., & Sandhu, M. S. (2018). Long reads: Their purpose and place. *Human Molecular Genetics*, *27*(R2), R234–R241. DOI: 10.1093/hmg/ddy177 PMID: 29767702

Pratanwanich, P. N., Yao, F., Chen, Y., Koh, C. W., Wan, Y. K., Hendra, C., Poon, P., Goh, Y. T., Yap, P. M., Chooi, J. Y., Chng, W. J., Ng, S. B., Thiery, A., Goh, W. S. S., & Göke, J. (2021). Identification of differential RNA modifications from nanopore direct RNA sequencing with xPore. *Nature Biotechnology*, *39*(11), 1394–1402. DOI: 10.1038/s41587-021-00949-w PMID: 34282325

Qu, Y., He, F., & Chen, Y. (2010). Different effects of the probe summarization algorithms PLIER and RMA on high-level analysis of Affymetrix exon arrays. *BMC Bioinformatics*, *11*(1), 1–13. DOI: 10.1186/1471-2105-11-211 PMID: 20426803

Rao, M. S., Van Vleet, T. R., Ciurlionis, R., Buck, W. R., Mittelstadt, S. W., Blomme, E. A., & Liguori, M. J. (2019). Comparison of RNA-Seq and microarray gene expression platforms for the toxicogenomic evaluation of liver from short-term rat toxicity studies. *Frontiers in Genetics*, *9*, 636. DOI: 10.3389/fgene.2018.00636 PMID: 30723492

Rodriques, S. G., Stickels, R. R., Goeva, A., Martin, C. A., Murray, E., Vanderburg, C. R., Welch, J., Chen, L. M., Chen, F., & Macosko, E. Z. (2019). Slide-seq: A scalable technology for measuring genome-wide expression at high spatial resolution. *Science*, *363*(6434), 463–467. DOI: 10.1126/science.aaw1219 PMID: 30923225

Sager, M., Yeat, N. C., Pajaro-Van der Stadt, S., Lin, C., Ren, Q., & Lin, J. (2015). Transcriptomics in cancer diagnostics: Developments in technology, clinical research and commercialization. *Expert Review of Molecular Diagnostics*, *15*(12), 1589–1603. DOI: 10.1586/14737159.2015.1105133 PMID: 26565429

Saiselet, M., Rodrigues-Vitória, J., Tourneur, A., Craciun, L., Spinette, A., Larsimont, D., Andry, G., Lundeberg, J., Maenhaut, C., & Detours, V. (2020). Transcrip tional output, cell-type densities, and normalization in spatial transcriptomics. *Journal of Molecular Cell Biology*, *12*(11), 906–908. DOI: 10.1093/jmcb/mjaa028 PMID: 32573704

Salehinejad, H., Sankar, S., Barfett, J., Colak, E., & Valaee, S. (2017). Recent advances in recurrent neural networks. *arXiv preprint arXiv:1801.01078*. https://doi.org//arXiv.1801.01078DOI: 10.48550

Sarker, I. H. (2021). Machine learning: Algorithms, real-world applications and research directions. *SN Computer Science*, *2*(3), 160. DOI: 10.1007/s42979-021-00592-x PMID: 33778771

Schulz, E., Speekenbrink, M., & Krause, A. (2018). A tutorial on Gaussian process regression: Modelling, exploring, and exploiting functions. *Journal of Mathematical Psychology*, *85*, 1–16. DOI: 10.1016/j.jmp.2018.03.001

Schwarze, K., Buchanan, J., Fermont, J. M., Dreau, H., Tilley, M. W., Taylor, J. M., Antoniou, P., Knight, S. J., Camps, C., Pentony, M. M., Kvikstad, E. M., Harris, S., Popitsch, N., Pagnamenta, A. T., Schuh, A., Taylor, J. C., & Wordsworth, S. (2020). The complete costs of genome sequencing: A microcosting study in cancer and rare diseases from a single center in the United Kingdom. *Genetics in Medicine*, *22*(1), 85–94. DOI: 10.1038/s41436-019-0618-7 PMID: 31358947

Shao, M., & Kingsford, C. (2017). Accurate assembly of transcripts through phase-preserving graph decomposition. *Nature Biotechnology*, *35*(12), 1167–1169. DOI: 10.1038/nbt.4020 PMID: 29131147

Shengquan, C., Boheng, Z., Xiaoyang, C., Xuegong, Z., & Rui, J. (2021). STPLUS: A reference based method for the accurate enhancement of spatial tran scriptomics. *Bioinformatics (Oxford, England)*, *37*(1, Supplement_1), i299–i307. DOI: 10.1093/bioinformatics/btab298 PMID: 34252941

Sheu, Y. H. (2020). Illuminating the black box: Interpreting deep neural network models for psychiatric research. *Frontiers in Psychiatry*, *11*, 551299. DOI: 10.3389/fpsyt.2020.551299 PMID: 33192663

Shu, C., Wang, Q., Yan, X., & Wang, J. (2018). Whole-genome expression microarray combined with machine learning to identify prognostic biomarkers for high-grade glioma. *Journal of Molecular Neuroscience*, *64*(4), 491–500. DOI: 10.1007/s12031-018-1049-7 PMID: 29502292

Song, Q., & Su, J. (2021). Dstg: Deconvoluting spatial transcriptomics data through graph-based artificial intelligence. *Briefings in Bioinformatics*, *22*(5), bbaa414. Advance online publication. DOI: 10.1093/bib/bbaa414 PMID: 33480403

Sun, S., Zhu, J., & Zhou, X. (2021). Statistical analysis of spatial expres sion patterns for spatially resolved transcriptomic studies. *Nature Methods*, *17*(2), 193–200. DOI: 10.1038/s41592-019-0701-7 PMID: 31988518

Suphavilai, C., Chia, S., Sharma, A., Tu, L., Da Silva, R. P., Mongia, A., DasGupta, R., & Nagarajan, N. (2021). Predicting heterogeneity in clone-specific therapeutic vulnerabilities using single-cell transcriptomic signatures. *Genome Medicine*, *13*(1), 189. DOI: 10.1186/s13073-021-01000-y PMID: 34915921

Suvà, M. L., & Tirosh, I. (2019). Single-Cell RNA Sequencing in Cancer: Lessons Learned and Emerging Challenges. *Molecular Cell*, *75*(1), 7–12. DOI: 10.1016/j.molcel.2019.05.003 PMID: 31299208

Tanevski, J., Flores, R. O. R., Gabor, A., Schapiro, D., & Saez-Rodriguez, J. (2020). Explainable multi-view framework for dissecting intercellular signaling from highly multiplexed spatial data. *BioRxiv*, pp. 2020–05. DOI: 10.1101/2020.05.08.084145

Tang, F., Barbacioru, C., Wang, Y., Nordman, E., Lee, C., Xu, N., Wang, X., Bodeau, J., Tuch, B. B., Siddiqui, A., Lao, K., & Surani, M. A. (2009). mRNA-Seq whole-transcriptome analysis of a single cell. *Nature Methods*, *6*(5), 377–382. DOI: 10.1038/nmeth.1315 PMID: 19349980

Vaswani, A., Shazeer, N., Parmar, N., Uszkoreit, J., Jones, L., Gomez, A. N., Kaiser, Ł., & Polosukhin, I. (2017). Attention is all you need. In von Luxburg, U. (Ed.), Advances in neural information processing systems: Vol. 30. *NIPS 2017). Curran Associates Inc* (pp. 5998–6008).

Warnat-Herresthal, S., Perrakis, K., Taschler, B., Becker, M., Baßler, K., Beyer, M., & Schultze, J. L. (2020). Scalable prediction of acute myeloid leukemia using high-dimensional machine learning and blood transcriptomics. *iScience*, *23*(1), 100780. Advance online publication. DOI: 10.1016/j.isci.2019.100780 PMID: 31918046

Wick, R. R., Judd, L. M., & Holt, K. E. (2019). Performance of neural network basecalling tools for Oxford Nanopore sequencing. *Genome Biology*, *20*(1), 1–10. DOI: 10.1186/s13059-019-1727-y PMID: 31234903

Workman, R. E., Tang, A. D., Tang, P. S., Jain, M., Tyson, J. R., Razaghi, R., Zuzarte, P. C., Gilpatrick, T., Payne, A., Quick, J., Sadowski, N., Holmes, N., de Jesus, J. G., Jones, K. L., Soulette, C. M., Snutch, T. P., Loman, N., Paten, B., Loose, M., & Timp, W. (2019). Nanopore native RNA sequencing of a human poly (A) transcriptome. *Nature Methods*, *16*(12), 1297–1305. DOI: 10.1038/s41592-019-0617-2 PMID: 31740818

Wu, Y. C., & Feng, J. W. (2018). Development and application of artificial neural network. *Wireless Personal Communications*, *102*(2), 1645–1656. DOI: 10.1007/s11277-017-5224-x

Xu, C., & Jackson, S. A. (2019). Machine learning and complex biological data. *Genome Biology*, *20*(1), 1–4. DOI: 10.1186/s13059-019-1689-0 PMID: 30992073

Xu, Y., & McCord, R. P. (2021). Costa: Unsupervised convolutional neural network learning for spatial transcriptomics analysis. *BMC Bioinformatics*, *22*(1), 397. DOI: 10.1186/s12859-021-04314-1 PMID: 34372758

Zeng, Z., Gu, S. S., Wong, C. J., Yang, L., Ouardaoui, N., Li, D., Zhang, W., Brown, M., & Liu, X. S. (2022). Machine learning on syngeneic mouse tumor profiles to model clinical immunotherapy response. *Science Advances*, *8*(41), eabm8564. Advance online publication. DOI: 10.1126/sciadv.abm8564 PMID: 36240281

Zhang, K., Feng, W., & Wang, P. (2022). Identification of spatially variable genes with graph cuts. *Nature Communications*, *13*(1), 5488. DOI: 10.1038/s41467-022-33182-3 PMID: 36123336

Zong, Y., Yu, T., Wang, X., Wang, Y., Hu, Z., & Li, Y. (2022). conST:an interpretable multi-modal contrastive learning framework for spatial transcriptomics. bioRxiv 2022, 2022–01. DOI: 10.1101/2022.01.14.476408

Zoph, B., & Le, Q. V. (2016). Neural architecture search with reinforcement learning. *arXiv preprint arXiv:1611.01578*. https://doi.org//arXiv.1611.01578DOI: 10.48550

Zubair, A., Chapple, R., Natarajan, S., Wright, W. C., Pan, M., Lee, H. M., Tillman, H., Easton, J., & Geeleher, P. (2022). Jointly leveraging spatial transcriptomics and deep learning models for image annotation achieves better-than-pathologist performance in cell type identification in tumors. *Cancer Research*, *82*(12), 456–456. DOI: 10.1158/1538-7445.AM2022-456

Compilation of References

Abbott, R. K., & Crotty, S. (2020). Factors in B cell competition and immunodominance. *Immunological Reviews*, *296*(1), 120–131. DOI: 10.1111/imr.12861

Abed-Esfahani, P., Darwin, B. C., Howard, D., Wang, N., Kim, E., Lerch, J., & French, L. (2022, January 24). Evaluation of deep convolutional neural networks for in situ hybridization gene expression image rep resentation. *PLoS One*, *17*(1), e0262717. DOI: 10.1371/journal.pone.0262717 PMID: 35073334

Abedini, A., Levinsohn, J., Klötzer, K. A., Dumoulin, B., Ma, Z., Frederick, J., Dhillon, P., Balzer, M. S., Shrestha, R., & Liu, H., (2022). Spatially resolved human kidney multi-omics single cell atlas highlights the key role of the fibrotic microenvironment in kidney disease progression. *bioRxiv*. https://doi.org/DOI: 10.1101/2022.09.29.510141

Abi-Gerges, N., Miller, P. E., & Ghetti, A. (2020). Human heart cardiomyocytes in drug discovery and research: New opportunities in translational sciences. *Current Pharmaceutical Biotechnology*, *21*(9), 787–806. DOI: 10.2174/1389201021666191210142023 PMID: 31820682

Acquaviva, M., Menon, R., Di Dario, M., Dalla Costa, G., Romeo, M., Sangalli, F., Colombo, B., Moiola, L., Martinelli, V., Comi, G., & Farina, C. (2020). Inferring multiple sclerosis stages from the blood transcriptome via machine learning. *Cell Reports Medicine*, *1*(4), 100053. Advance online publication. DOI: 10.1016/j.xcrm.2020.100053 PMID: 33205062

Adegoke, B. O., Odugbose, T., & Adeyemi, C. (2024). Data analytics for predicting disease outbreaks: A review of models and tools. *International Journal of Life Science Research Updates [Online]*, 2(2), 1–9. DOI: 10.53430/ijlsru.2024.2.2.0023

Adossa, N., Khan, S., Rytkönen, K. T., & Elo, L. L. (2021). Computational strategies for single-cell multi-omics integration. *Computational and Structural Biotechnology Journal*, *19*, 2588–2596. DOI: 10.1016/j.csbj.2021.04.060

Adusumalli, S., Lim, S., Ren, V., Chong, L. Y., Tan, C., Tham, R., & Yap, L. (2023). Spatiotemporal transcriptomes of pig hearts reveal midkine-mediated vascularization in a chronic myocardial infarcted model. bioRxiv, 2023-06.

Adusumalli, S., Lim, S., Ye, L., lim, E., Ren, S., Wang, Y., Petretto, E., Tryggvason, K., & Yap, L. (2024). Abstract We008: Hallmark maturation of the human pluripotent stem cells-derived cardiovascular progenitors in myocardial infarcted large animal model. *Circulation Research, 135*(Suppl_1), AWe008–AWe008. DOI: 10.1161/res.135.suppl_1.We008

Agrebi, S., & Larbi, A. (2020). Use of artificial intelligence in infectious diseases. In *Artificial intelligence in precision health* (pp. 415–438). Elsevier. DOI: 10.1016/B978-0-12-817133-2.00018-5

Ahmed, F. J. (2024). Personalized Medicine: The Dawn of a New Era in Healthcare. *Futuristic Biotechnology, 01*, 01. Advance online publication. DOI: 10.54393/fbt.v4i03.138

Ahmed, S., Komeili, M., & Park, J. (2022). Predictive modelling of Parkinson's disease progression based on RNA-Sequence with densely connected deep recurrent neural networks. *Scientific Reports, 12*(1), 21469. DOI: 10.1038/s41598-022-25454-1 PMID: 36509776

Ahmed, T. (2022). Organ-on-a-chip microengineering for bio-mimicking disease models and revolutionizing drug discovery. *Biosensors & Bioelectronics: X, 11*, 100194. DOI: 10.1016/j.biosx.2022.100194

Aihara, G., Clifton, K. P., Chen, M., Atta, L., Miller, B. F., & Fan, J. (2024). SEraster: a rasterization preprocessing framework for scalable spatial omics data analysis. *bioRxiv*. DOI: 10.1101/2024.02.01.578436

Akhtar, A., Khatun, R., Sengupta, S., & Bhattacharya, M. (2025). Role of Multi-Omics in Disease Biology. *Applied Sciences Research Periodicals., 3*(03), 02–21. https://doi.org/DOI: 10.63002/asrp.303.925

Akintunde, O., Tucker, T., & Carabetta, V. J. (2024). The evolution of next-generation sequencing technologies. In *High-throughput gene screening: Methods and protocols* (pp. 3–29). Springer.

Alagendran, S., Archunan, G., Prabhu, S. V., Orozco, B. E., & Guzman, R. G. (2010). Biochemical evaluation in human saliva with special reference to ovulation detection. *Indian Journal of Dental Research, 21*(2), 165–168. DOI: 10.4103/0970-9290.66625 PMID: 20657081

Alexandrov, T. (2020). Spatial metabolomics and imaging mass spectrometry in the age of artificial intelligence. *Annual Review of Biomedical Data Science*, *3*(1), 61–87. DOI: 10.1146/annurev-biodatasci-011420-031537 PMID: 34056560

Ali, K. A., Choudhuri, S., Bishal, A., Das, T., Nandy, S., & Hossain, C. M. (2025). Approaches of AI and Advanced Computational Biology for Combating HIV Infections. In J.-T. Chen, *AI and Precision Medicine in Infectious Disease Management* (1st ed., pp. 149–166). CRC Press. DOI: 10.1201/9781003615699-8

Ali, K. A., Choudhuri, S., Chakraborty, R., Pradhan, A., Kotal, A., & Shaharyar, M. A. (2025). AI Technology-Powered Precision Medicine and Smart Healthcare. In J.-T. Chen, *AI and Precision Medicine in Infectious Disease Management* (1st ed., pp. 51–70). CRC Press. DOI: 10.1201/9781003615699-3

Ali, H. (2023). Artificial intelligence in multi-omics data integration: Advancing precision medicine, biomarker discovery and genomic-driven disease interventions. *Int J Sci Res Arch*, *8*(1), 1012–1030. DOI: 10.30574/ijsra.2023.8.1.0189

Ali, H. (2023). Artificial intelligence in multi-omics data integration: Advancing precision medicine, biomarker discovery, and genomic-driven disease interventions. *International Journal of Scientific Research Archive*, *8*(1), 1012–1030.

Ali, K. A., Choudhuri, S., Roy, S. K., Reza, K. H., Mandal, D., Chakraborty, R., & Hossain, C. M. (2025). AI-Generated Nanoparticles in Immunological Research: Opportunities and Challenges. In Chen, J.-T. (Ed.), (pp. 295–322). Advances in Computational Intelligence and Robotics. IGI Global., DOI: 10.4018/979-8-3693-9725-1.ch011

Ali, K. A., Mohin, S., Mondal, P., Goswami, S., Ghosh, S., & Choudhuri, S. (2024). Influence of artificial intelligence in modern pharmaceutical formulation and drug development. *Future Journal of Pharmaceutical Sciences*, *10*(1), 53. DOI: 10.1186/s43094-024-00625-1

Allaband, C., Lingaraju, A., Flores Ramos, S., Kumar, T., Javaheri, H., Tiu, M. D., Dantas Machado, A. C., Richter, R. A., Elijah, E., Haddad, G. G., Leone, V. A., Dorrestein, P. C., Knight, R., & Zarrinpar, A. (2024). Time of sample collection is critical for the replicability of microbiome analyses. *Nature Metabolism*, *6*(7), 1282–1293. DOI: 10.1038/s42255-024-01064-1 PMID: 38951660

Alolga, R. N., Wang, S.-L., Qi, L.-W., Zang, H., & Huang, F.-Q. (2024). MALDI mass spectrometry imaging in targeted drug discovery and development: The pros, the cons, and prospects in global omics techniques. *Trends in Analytical Chemistry*, *178*, 117860. DOI: 10.1016/j.trac.2024.117860

Alshehri, S., Vitorino, R., Saleh, O., Al-Harthi, S., Alahmadi, A., Alotibi, R., Da Silva Rosa, S. C., Osama, A., Magedeldin, S., Alhattab, D., Emwas, A.-H., & Jaremko, M. (2025). Advances and applications of clinical proteomics in precision medicine. *Expert Review of Proteomics, 22*(10), 1–20. DOI: 10.1080/14789450.2025.2560919 PMID: 40963397

Al-Talib, M., Dimonte, S., & Humphreys, I. R. (2024). Mucosal T-cell responses to chronic viral infections: Implications for vaccine design. *Cellular & Molecular Immunology, 21*(9), 982–998. DOI: 10.1038/s41423-024-01140-2

Altschuler, S. J., & Wu, L. F. (2010). Cellular heterogeneity: Do differences make a difference? *Cell, 141*(4), 559–563. DOI: 10.1016/j.cell.2010.04.033

Alves, M. A., Lamichhane, S., Dickens, A., McGlinchey, A., Ribeiro, H. C., Sen, P., Wei, F., Hyötyläinen, T., & Orešič, M. (2021). Systems biology approaches to study lipidomes in health and disease. *Biochimica et Biophysica Acta (BBA)-. Biochimica et Biophysica Acta. Molecular and Cell Biology of Lipids, 1866*(2), 158857. DOI: 10.1016/j.bbalip.2020.158857

Amarasinghe, S. L., Su, S., Dong, X., Zappia, L., Ritchie, M. E., & Gouil, Q. (2020). Opportunities and challenges in long-read sequencing data analysis. *Genome Biology, 21*(1), 30. DOI: 10.1186/s13059-020-1935-5 PMID: 32033565

Ammons, D. T., Hopkins, L. S., Cronise, K. E., Kurihara, J., Regan, D. P., & Dow, S. (2024). Single-cell RNA sequencing reveals the cellular and molecular heterogeneity of treatment-naïve primary osteosarcoma in dogs. *Communications Biology, 7*(1), 496. DOI: 10.1038/s42003-024-06182-w PMID: 38658617

Ananthakumar, B., Nagaraja, B. G., Umesh, S. M., & Raghavan, R. (2025). Predictive Medicine. *Predictive Algorithms for Rehabilitation and Assistive Systems*, 291–314. DOI: 10.4018/979-8-3373-0194-5.ch011

Ananthakumar, B., Nanjundeshwara, N., Lakshmaiah, A., & Raghavan, R. (2024). Synthetic Gene Circuits as a Promising Approach in Cancer Immunotherapy. *Critical Developments in Cancer Immunotherapy*, 391–418. DOI: 10.4018/979-8-3693-3976-3.ch011

Ananthakumar, B., Nanjundeshwara, N., Lakshmaiah, A., & Raghavan, R. (2025). Nanocarriers for Delivering Nucleic Acids in Plants: Mechanisms and Applications. *Nanocarriers in Plant Science and Agriculture*, 207–234. DOI: 10.4018/979-8-3693-9869-2.ch008

Andreotti, F., Iervolino, A., Navarese, E. P., Maggioni, A. P., Crea, F., & Scambia, G. (2021). Precision phenomapping of acute coronary syndromes to improve patient outcomes. *Journal of Clinical Medicine*, *10*(8), 1755. DOI: 10.3390/jcm10081755 PMID: 33919478

Annink, M. E., Kraaijenhof, J. M., Stroes, E. S., & Kroon, J. (2024). Moving from lipids to leukocytes: Inflammation and immune cells in atherosclerosis. *Frontiers in Cell and Developmental Biology*, *12*, 1446758. DOI: 10.3389/fcell.2024.1446758 PMID: 39161593

Aoki, J., Kaya, C., Khalid, O., Kothari, T., Silberman, M. A., Skordis, C., Hughes, J., Hussong, J., & Salama, M. E. (2023). CKD progression prediction in a diverse US population: A machine-learning model. *Kidney Medicine*, *5*(3), 100692. DOI: 10.1016/j.xkme.2023.100692 PMID: 37637863

Aon, M. A., & Cortassa, S. (2015). Systems biology of the fluxome. *Processes (Basel, Switzerland)*, *3*(3), 607–618. DOI: 10.3390/pr3030607

Aoshi, T., Koyama, S., Kobiyama, K., Akira, S., & Ishii, K. J. (2011). Innate and adaptive immune responses to viral infection and vaccination. *Current Opinion in Virology*, *1*(4), 226–232. DOI: 10.1016/j.coviro.2011.07.002

Araújo, A. (2024). From artificial intelligence to semi-creative inorganic intelligence: A blockchain-based bioethical metamorphosis. *AI and Ethics*, •••, 1–6. DOI: 10.1007/s43681-024-00471-0

Argelaguet, R., Arnol, D., Bredikhin, D., Deloro, Y., Velten, B., Velten, B., Marioni, J. C., Marioni, J. C., Marioni, J. C., Stegle, O., & Stegle, O. (2020). MOFA+: A statistical framework for comprehensive integration of multi-modal single-cell data. *Genome Biology*, *21*(1), 1–17. DOI: 10.1186/s13059-020-02015-1

Argelaguet, R., Velten, B., Arnol, D., Dietrich, S., Zenz, T., Marioni, J. C., Buettner, F., Huber, W., & Stegle, O. (2018). Multi-omics factor analysis—A framework for unsupervised integration of multi-omics data sets. *Molecular Systems Biology*, *14*(6), e8124. DOI: 10.15252/msb.20178124 PMID: 29925568

Armand, T. P. T., Deji-Oloruntoba, O., Bhattacharjee, S., Nfor, K. A., & Kim, H.-C. (2024). Optimizing longevity: Integrating smart nutrition and digital technologies for personalized anti-aging healthcare. In *Proceedings of the 2024 International Conference on Artificial Intelligence in Information and Communication* (pp. 243–248). IEEE. DOI: 10.1109/ICAIIC60209.2024.10463262

Arneson, D., Shu, L., Tsai, B., Barrere-Cain, R., Sun, C., & Yang, X. (2017). Multidimensional integrative genomics approaches to dissecting cardiovascular disease. *Frontiers in Cardiovascular Medicine*, *4*, 8. DOI: 10.3389/fcvm.2017.00008 PMID: 28289683

Arumugam, V. A., Thangavelu, S., Fathah, Z., Ravindran, P., Sanjeev, A. M. A., Babu, S., Arun, M., Mohd. Iqbal, Y., Khan, S., Ruchi, T., Megha Katare, P., Ranjit, S., Chandra, R., & Dhama, K. (2020). COVID-19 and the world with co-morbidities of heart disease, hypertension and diabetes. *Journal of Pure & Applied Microbiology*, *14*(3), 1623–1638. DOI: 10.22207/JPAM.14.3.01

Ash, S. L., Orha, R., Mole, H., Dinesh-Kumar, M., Lee, S. P., Turrell, F. K., & Isacke, C. M. (2024). Targeting the activated microenvironment with endosialin (CD248)-directed CAR-T cells ablates perivascular cells to impair tumor growth and metastasis. *Journal for Immunotherapy of Cancer*, *12*(2), e008608. DOI: 10.1136/jitc-2023-008608

Asp, M., Bergenstråhle, J., & Lundeberg, J. (2020). Spatially Resolved Transcriptomes-Next Generation Tools for Tissue Exploration. *BioEssays*, *42*(10), 1900221. DOI: 10.1002/bies.201900221

Athanasopoulou, K., Boti, M. A., Adamopoulos, P. G., Skourou, P. C., & Scorilas, A. (2021). Third-generation sequencing: The spearhead towards the radical transformation of modern genomics. *Life (Chicago, Ill.)*, *12*(1), 30. DOI: 10.3390/life12010030 PMID: 35054423

Athanasopoulou, K., Michalopoulou, V.-I., Scorilas, A., & Adamopoulos, P. G. (2025). Integrating artificial intelligence in next-generation sequencing: Advances, challenges, and future directions. *Current Issues in Molecular Biology*, *47*(6), 470. DOI: 10.3390/cimb47060470 PMID: 40699869

Athaya, T., Ripan, R. C., Li, X., & Hu, H. (2023). Multimodal deep learning approaches for single-cell multi-omics data integration. *Briefings in Bioinformatics*, *24*(5), bbad313. Advance online publication. DOI: 10.1093/bib/bbad313

Augustyn, D. R., Wyciślik, Ł., & Mrozek, D. (2021). Perspectives of using Cloud computing in integrative analysis of multi-omics data. *Briefings in Functional Genomics*, *20*(4), 198–206. DOI: 10.1093/bfgp/elab007

Auslander, N., Gussow, A. B., & Koonin, E. V. (2021). Incorporating machine learning into established bioinformatics frameworks. *International Journal of Molecular Sciences*, *22*(6), 2903. DOI: 10.3390/ijms22062903 PMID: 33809353

Ayaz, A., Ali, A., Mohammed, I., & Ali, N. Alauddin, Singh, N., Jeelani, G., Mujtaba, H., & Bhat, K. A. (2024). *Multiomics for understanding neurodegenerative disorders* (pp. 95–125). Elsevier BV. https://doi.org/DOI: 10.1016/B978-0-443-23971-7.00005-5

Bai, X., Lau, B. T., Grimes, S. M., Sathe, A., & Ji, H. P. (2022). Single cell multi-omic mapping of subclonal architecture and pathway phenotype in primary gastric and metastatic colon cancers. *bioRxiv*. https://doi.org/DOI: 10.1101/2022.07.03.498616

Baker, L., Matthiopoulos, J., Müller, T., Freuling, C., & Hampson, K. (2020). Local rabies transmission and regional spatial coupling in European foxes. *PLoS One, 15*(5), e0220592. DOI: 10.1371/journal.pone.0220592 PMID: 32469961

Balakrishnan, G., Zhao, A., Sabuncu, M. R., Guttag, J., & Dalca, A. V. (2019). Voxelmorph: A learning framework for deformable medical image registration. *IEEE Transactions on Medical Imaging, 38*(8), 1788–1800. DOI: 10.1109/TMI.2019.2897538 PMID: 30716034

Balzer, M. S., Ma, Z., Zhou, J., Abedini, A., & Susztak, K. (2021). How to get started with single-cell RNA sequencing data analysis. *Journal of the American Society of Nephrology, 32*(6), 1279–1292. DOI: 10.1681/ASN.2020121742 PMID: 33722930

Bareche, Y., Kelly, D., Abbas-Aghababazadeh, F., Nakano, M., Esfahani, P., Tkachuk, D., Mohammad, H., Samstein, R., Lee, C.-H., Morris, L. G., Bedard, P. L., Haibe-Kains, B., & Stagg, J. (2022). Leveraging big data of immune checkpoint blockade response identifies novel potential targets. *Annals of Oncology : Official Journal of the European Society for Medical Oncology, 33*(12), 1304–1317. DOI: 10.1016/j.annonc.2022.08.084 PMID: 36055464

Barnett, J., Silverman, J., Wetzel, M., Rao, P., Sotudeh, N., & Wang, L. (2022). AtlasX-browser enables spatial multi-omics data analysis through the precise determination of the region of interest. *bioRxiv*. https://doi.org/DOI: 10.1101/2022.05.11.491526

Barsoum, I., Tawedrous, E., Faragalla, H., & Yousef, G. M. (2019). Histo-genomics: Digital pathology at the forefront of precision medicine. *Diagnosis (Berlin, Germany), 6*(3), 203–212. DOI: 10.1515/dx-2018-0064 PMID: 30827078

Barua, S., Fang, P., Sharma, A., Fujimoto, J., Wistuba, I., Rao, A. U. K., & Lin, S. H. (2018). Spatial interaction of tumor cells and regulatory T cells correlates with survival in non-small cell lung cancer. *Lung Cancer (Amsterdam, Netherlands), 117*, 73–79. DOI: 10.1016/j.lungcan.2018.01.022 PMID: 29409671

Barutta, F., Bellini, S., Canepa, S., Durazzo, M., & Gruden, G. (2021). Novel biomarkers of diabetic kidney disease: Current status and potential clinical application. *Acta Diabetologica*, *58*(7), 819–830. DOI: 10.1007/s00592-020-01656-9 PMID: 33528734

Bashiardes, S., Shapiro, H., Rozin, S., Shibolet, O., & Elinav, E. (2016). Nonalcoholic fatty liver and the gut microbiota. *Molecular Metabolism*, *5*(9), 782–794. DOI: 10.1016/j.molmet.2016.06.003

Bassez, A., Vos, H., Van Dyck, L., Floris, G., Arijs, I., Desmedt, C., Boeckx, B., Vanden Bempt, M., Nevelsteen, I., Lambein, K., Punie, K., Neven, P., Garg, A. D., Wildiers, H., Qian, J., Smeets, A., & Lambrechts, D. (2021). A single-cell map of intratumoral changes during anti-PD1 treatment of patients with breast cancer. *Nature Medicine*, *27*(5), 820–832. DOI: 10.1038/s41591-021-01323-8 PMID: 33958794

Basu, A., Carlson, J. J., & Veenstra, D. L. (2016). A framework for prioritizing research investments in precision medicine. *Medical Decision Making*, *36*(5), 567–580. DOI: 10.1177/0272989X15610780 PMID: 26502985

Battazza, A., Brasileiro, F. C. D. S., Tasaka, A. C., Bulla, C., Ximenes, P. P., Hosomi, J. E., da Silva, P. F., da Silva, L. F., de Moura, F. B. C., & Rocha, N. S. (2024). Integrating telepathology and digital pathology with artificial intelligence: An inevitable future. *Veterinary World*, *17*(8), 1667–1671. DOI: 10.14202/vetworld.2024.1667-1671 PMID: 39328444

Baumann, A. A., Buribayev, Z., Wolkenhauer, O., Salybekov, A. A., & Wolfien, M. (2025). Epigenomic echoes—Decoding genomic and epigenetic instability to distinguish lung cancer types and predict relapse. *Epigenomes*, *9*(1), 5. DOI: 10.3390/epigenomes9010005 PMID: 39982247

Bayaraa, O., Aksu, M., DeBose-Scarlett, E., Hocke, E., Jain, V., Wang, S.-H. J., Cruz, D., & Gregory, S. (2025). Multi-region spatial transcriptomics reveals region-specific differences in response to amyloid beta (Aβ) plaque-induced changes in Alzheimer's Disease (AD). *Genomics*. Advance online publication. DOI: 10.1186/s40246-025-00875-x PMID: 40766691

Bécavin, C., & Benecke, A. (2011). New dimensionality reduction methods for the representation of high dimensional "omics" data. *Expert Review of Molecular Diagnostics*, *11*(1), 27–34. DOI: 10.1586/erm.10.95

Beetz, M., Banerjee, A., Sang, Y., & Grau, V. (2022). Combined generation of electrocardiogram and cardiac anatomy models using multi-modal variational autoencoders. In *2022 IEEE 19th International Symposium on Biomedical Imaging (ISBI)* (pp. 1-4). IEEE. DOI: 10.1109/ISBI52829.2022.9761590

Behura, S. K., Tizioto, P. C., Kim, J., Grupioni, N. V., Seabury, C. M., Schnabel, R. D., Gershwin, L. J., Van Eenennaam, A. L., Toaff-Rosenstein, R., Neibergs, H. L., Regitano, L. C. A., & Taylor, J. F. (2017). Tissue tropism in host transcriptional response to members of the bovine respiratory disease complex. *Scientific Reports*, *7*(1), 17938. DOI: 10.1038/s41598-017-18205-0 PMID: 29263411

Benjamin, K., Bhandari, A., Shang, Z., Xing, Y., An, Y., Zhang, N., Hou, Y., Tillmann, U., Bull, K. R., & Harrington, H. A. (2022). Multiscale topology classifies and quantifies cell types in subcellular spatial transcriptomics. *arXiv*. https://arxiv.org/abs/2212.06505

Bera, K., Schalper, K. A., Rimm, D. L., Velcheti, V., & Madabhushi, A. (2019). Artificial intelligence in digital pathology—New tools for diagnosis and precision oncology. *Nature Reviews. Clinical Oncology*, *16*(11), 703–715. DOI: 10.1038/s41571-019-0252-y PMID: 31399699

Bhamidipati, K., & Wei, K. (2022). Precision medicine in rheumatoid arthritis. *Best Practice & Research. Clinical Rheumatology*, *36*(1), 101742. DOI: 10.1016/j.berh.2022.101742

Bhate, S. S., Barlow, G. L., Schürch, C. M., & Nolan, G. P. (2022). Tissue schematics map the specialization of immune tissue motifs and their appropriation by tumors. *Cell Systems*, *13*(2), 109–130.e6. DOI: 10.1016/j.cels.2021.09.012 PMID: 34653369

Bhotla, H. K., Balasubramanian, B., Meyyazhagan, A., Pushparaj, K., & Easwaran, M. (2021). Opportunistic mycoses in COVID-19 patients/survivors: Epidemic inside a pandemic. *Journal of Infection and Public Health*, *14*(11), 1720–1726. DOI: 10.1016/j.jiph.2021.10.010 PMID: 34700291

Bhuvana, G. N., Umesh, S. M., & Raghavan, R. (2025). Expanding the frontiers of Schistosomiasis Treatment: The promise of Nanoparticle-Based Therapies. *Journal of Vector Borne Diseases, 10*(4103/jvbd.jvbd_180_23). DOI: 10.4103/jvbd.jvbd_180_23

Biancalani, T., Scalia, G., Buffoni, L., Avasthi, R., Lu, Z., Sanger, A., Tokcan, N., Vanderburg, C. R., Segerstolpe, Å., Zhang, M., Avraham-Davidi, I., Vickovic, S., Nitzan, M., Ma, S., Subramanian, A., Lipinski, M., Buenrostro, J., Brown, N. B., Fanelli, D., & Regev, A. (2021). Deep learning and alignment of spatially resolved single-cell transcriptomes with tangram. *Nature Methods*, *18*(11), 1352–1362. DOI: 10.1038/s41592-021-01264-7 PMID: 34711971

Bianconi, I., Aschbacher, R., & Pagani, E. (2023). Current uses and future perspectives of genomic technologies in clinical microbiology. *Antibiotics (Basel, Switzerland)*, *12*(11), 1580. DOI: 10.3390/antibiotics12111580 PMID: 37998782

Binan, L., Jiang, A., Danquah, S. A., Valakh, V., Simonton, B., Bezney, J., Manguso, R. T., Yates, K. B., Nehme, R., Cleary, B., & Farhi, S. L. (2025). Simultaneous CRISPR screening and spatial transcriptomics reveal intracellular, intercellular, and functional transcriptional circuits. *Cell, 188*(8), 2141–2158.e18. DOI: 10.1016/j.cell.2025.02.012 PMID: 40081369

Biondi-Zoccai, G., D'Ascenzo, F., Giordano, S., Mirzoyev, U., Erol, Ç., Cenciarelli, S., Leone, P., & Versaci, F. (2025). Artificial Intelligence in Cardiology: General Perspectives and Focus on Interventional Cardiology. *The Anatolian Journal of Cardiology*, *29*(4), 152–163. DOI: 10.14744/AnatolJCardiol.2025.5237

Biradar, S., Suresh, C., Nanjundeshwara, N., & Raghavan, R. (2025). Spatially Variable Genes. *Spatially Variable Genes in Cancer*, 1–28. DOI: 10.4018/979-8-3693-7728-4.ch001

Biswas, A., Kumari, A., Gaikwad, D. S., & Pandey, D. K. (2023). Revolutionizing biological science: The synergy of genomics in health, bioinformatics, agriculture, and artificial intelligence. *OMICS: A Journal of Integrative Biology*, *27*(12), 550–569. DOI: 10.1089/omi.2023.0197 PMID: 38100404

Björkholm, B., Sjölund, M., Falk, P. G., Berg, O. G., Engstrand, L., & Andersson, D. I. (2001). Mutation frequency and biological cost of antibiotic resistance in Helicobacter pylori. *Proceedings of the National Academy of Sciences of the United States of America*, *98*(25), 14607–14612. DOI: 10.1073/pnas.241517298

Björnson, E., Borén, J., & Mardinoglu, A. (2016). Personalized cardiovascular disease prediction and treatment review of existing strategies and novel systems medicine tools. *Frontiers in Physiology*, *7*, 2. DOI: 10.3389/fphys.2016.00002 PMID: 26858650

Black, S., Phillips, D. J., Hickey, J. W., Kennedy-Darling, J., Venkataraaman, V. G., Samusik, N., Samusik, N., Goltsev, Y., Schürch, C. M., & Nolan, G. P. (2021). CODEX multiplexed tissue imaging with DNA-conjugated antibodies. *Nature Protocols*, *16*(8), 3802–3835. DOI: 10.1038/s41596-021-00556-8

Blader, I. J., & Saeij, J. P. (2009). Communication between *Toxoplasma gondii* and its host. *APMIS*, *117*(5-6), 458–476. DOI: 10.1111/j.1600-0463.2009.02453.x PMID: 19400868

Blakeley-Ruiz, J. A., Erickson, A. R., Cantarel, B. L., Xiong, W., Adams, R., Jansson, J. K., Fraser, C. M., & Hettich, R. L. (2019). Metaproteomics reveals persistent and phylum-redundant metabolic functional stability in adult human gut microbiomes of Crohn's remission patients despite temporal variations in microbial taxa, genomes, and proteomes. *Microbiome*, *7*(1), 18. DOI: 10.1186/s40168-019-0631-8

Blanco-Míguez, A., Beghini, F., Cumbo, F., McIver, L. J., Thompson, K. N., Zolfo, M., Manghi, P., Dubois, L., Huang, K. D., Thomas, A. M., Nickols, W. A., Piccinno, G., Piperni, E., Punčochář, M., Valles-Colomer, M., Tett, A., Giordano, F., Davies, R., Wolf, J., & Segata, N. (2023). Extending and improving metagenomic taxonomic profiling with uncharacterized species using MetaPhlAn 4. *Nature Biotechnology*, *41*(11), 1633–1644. DOI: 10.1038/s41587-023-01688-w

Blaser, M. J., & Atherton, J. C. (2004). Helicobacter pylori persistence: Biology and disease. *The Journal of Clinical Investigation*, *113*(3), 321–333. DOI: 10.1172/JCI20925

Bleckwehl, T., Maryam, S., Babler, A., Nyberg, M., Bosteen, M., Halder, M., & Hayat, S. (2023). Encompassing view of spatial and single-cell RNA-seq renews the role of the microvasculature in human atherosclerosis. bioRxiv, 2023-12.

Boe, R. H., Triandafillou, C. G., Lazcano, R., Wargo, J. A., & Raj, A. S. (2024). *Spatial transcriptomics reveals influence of microenvironment on intrinsic fates in melanoma therapy resistance*. https://doi.org/DOI: 10.1101/2024.06.30.601416

Böhmer, A., Schildgen, V., Lüsebrink, J., Ziegler, S., Tillmann, R. L., Kleines, M., & Schildgen, O. (2009). Novel application for isothermal nucleic acid sequence-based amplification (NASBA). *Journal of Virological Methods*, *158*(1–2), 199–201. DOI: 10.1016/j.jviromet.2009.02.010

Bond, D. R., Uddipto, K., Enjeti, A. K., Enjeti, A. K., & Lee, H. J. (2020). Single-cell epigenomics in cancer: Charting a course to clinical impact. *Epigenomics*, *12*(13), 1139–1151. DOI: 10.2217/epi-2020-0046

Border, S., Ferreira, R. M., Lucarelli, N., Kumar, S., Paul, A. S., Manthey, D., Barisoni, L., Strekalova, Y. A., Ray, J., Cheng, Y., Rosenberg, A. Z., Tomaszewski, J. E., Mimar, S., Hodgin, J. B., El-Achkar, T. M., Jain, S., Eadon, M. T., & Sarder, P. (2024). *FUSION: A web-based application for in-depth exploration of multi-omics data with brightfield histology*. https://doi.org/DOI: 10.1101/2024.07.09.602778

Boretti, G., Amirfallah, A., Edmunds, K. J., Hamzehpour, H., & Sigurjónsson, Ó. E. (2024). Advancing Cartilage Tissue Engineering: A Review of 3D Bioprinting Approaches and Bioink Properties. *Tissue Engineering. Part B, Reviews*, ten.teb.2024.0168. Advance online publication. DOI: 10.1089/ten.teb.2024.0168

Boukens, B. J., & Christoffels, V. M. (2012). Electrophysiological patterning of the heart. *Pediatric Cardiology*, *33*(6), 900–906. DOI: 10.1007/s00246-012-0237-4 PMID: 22367553

Bowers, R., Gonzalez-Pena, V., Wardhani, K., Goudeau, D., Blow, M. J., Udwary, D., Klein, D., Vill, A. C., Brito, I. L., Woyke, T., Malmstrom, R. R., & Gawad, C. (2024). scMicrobe PTA: Near Complete Genomes from Single Bacterial Cells. *bioRxiv*. https://doi.org/DOI: 10.1101/2024.01.30.577819

Branck, T., Hu, Z., Nickols, W. A., Walsh, A. M., Bhosle, A., Short, M. I., Nearing, J. T., Asnicar, F., McIver, L. J., Maharjan, S., Rahnavard, A., Louyakis, A. S., Badri, D. V., Brockel, C., Thompson, K. N., & Huttenhower, C. (2024). Comprehensive profile of the companion animal gut microbiome integrating reference-based and reference-free methods. *The ISME Journal, 18*(1), wrae201. Advance online publication. DOI: 10.1093/ismejo/wrae201 PMID: 39394961

Brandt, M. M., Meddens, C. A., Louzao-Martinez, L., Van Den Dungen, N. A. M., Lansu, N. R., Nieuwenhuis, E. E. S., Duncker, D. J., Verhaar, M. C., Joles, J. A., Mokry, M., & Cheng, C. (2018). Chromatin conformation links distal target genes to CKD loci. *Journal of the American Society of Nephrology*, *29*(2), 462–476. DOI: 10.1681/ASN.2016080875 PMID: 29093029

Bressan, D., Battistoni, G., & Hannon, G. J. (2023). The dawn of spatial omics. *Science*, *381*(6657), eabq4964. Advance online publication. DOI: 10.1126/science.abq4964

Brochu, E., Cora, V. M., & De Freitas, N. (2010). A tutorial on Bayesian optimization of expensive cost functions, with application to active user modeling and hierarchical reinforcement learning. *arXiv preprint arXiv:1012.2599.*

Bruynseels, K., Santoni de Sio, F., & Van den Hoven, J. (2018). Digital twins in health care: Ethical implications of an emerging engineering paradigm. *Frontiers in Genetics*, *9*, 31. DOI: 10.3389/fgene.2018.00031

Burke, W., Appelbaum, P., Dame, L., Marshall, P., Press, N., Pyeritz, R., Sharp, R., & Juengst, E. (2015). The translational potential of research on the ethical, legal, and social implications of genomics. *Genetics in Medicine*, *17*(1), 12–20. DOI: 10.1038/gim.2014.74 PMID: 24946153

Bush, C. L., Blumberg, J. B., El-Sohemy, A., Minich, D. M., Ordovás, J. M., Reed, D. G., & Behm, V. A. Y. (2020). Toward the definition of personalized nutrition. *Journal of the American College of Nutrition*, *39*(1), 5–15. DOI: 10.1080/07315724.2019.1685332

Byalahunasishetter, B. I., N., V., P., S., C., M., S., S., & Raghavan, R. (2025). Spatial Omics and Bioprinting. *Innovations in Precision Medicine and Genomics*, 85–110. DOI: 10.4018/979-8-3693-5787-3.ch003

Cai, G., Chen, F., Wen, K., Li, Y., & Ou-Yang, L. (2026). Identifying spatial domains from spatial multi-omics data using consistent and specific deep subspace learning. *Information Fusion*, *125*, 103428. DOI: 10.1016/j.inffus.2025.103428

Cao, J., Li, C., Cui, Z., Deng, S., Lei, T., Liu, W., Yang, H., & Chen, P. (2024a). Spatial Transcriptomics: A Powerful Tool in Disease Understanding and Drug Discovery. *Theranostics*, *14*(7), 2946–2968. DOI: 10.7150/thno.95908 PMID: 38773973

Cao, S., Liu, J., & Li, Y. (2025). Harnessing multi-omics and machine learning for predicting immune checkpoint blockade responses: Advances, challenges, and future directions. *Fundamental Research (Beijing)*, *S2667325825004613*. Advance online publication. DOI: 10.1016/j.fmre.2025.08.009

Caprioli, R. M., Farmer, T. B., & Gile, J. (1997). Molecular imaging of biological samples: Localization of peptides and proteins using MALDI-TOF MS. *Analytical Chemistry*, *69*(23), 4751–4760. DOI: 10.1021/ac970888i PMID: 9406525

Carmo-Fonseca, M. (2021). *Precision Medicine: Enabling Healthcare Progress in the Twenty-First Century*. Springer., DOI: 10.1007/978-3-030-70761-3_2

Carpenter, K. J. (2003). A short history of nutritional science: Part 1 (1785–1885). *The Journal of Nutrition*, *133*(3), 638–645. DOI: 10.1093/jn/133.3.638

Carrillo, C., Tulman, E. R., Delhon, G., Lu, Z., Carreno, A., Vagnozzi, A., Kutish, G. F., & Rock, D. L. (2005). Comparative genomics of foot-and-mouth disease virus. *Journal of Virology*, *79*(10), 6487–6504. DOI: 10.1128/JVI.79.10.6487-6504.2005 PMID: 15858032

Carstens, J. L., Correa De Sampaio, P., Yang, D., Barua, S., Wang, H., Rao, A., Allison, J. P., LeBleu, V. S., & Kalluri, R. (2017). Spatial computation of intratumoral T cells correlates with patient survival in pancreatic cancer. *Nature Communications*, *8*(1), 15095. DOI: 10.1038/ncomms15095 PMID: 28447602

Casamassimi, A., Federico, A., Rienzo, M., Esposito, S., & Ciccodicola, A. (2017). Transcriptome profiling in human diseases: New advances and perspectives. *International Journal of Molecular Sciences*, *18*(8), 1652. DOI: 10.3390/ijms18081652 PMID: 28758927

Cashaback, J. G. A., Allen, J. L., Chou, A. H.-Y., Lin, D. J., Price, M. A., Secerovic, N. K., Song, S., Zhang, H., & Miller, H. L. (2024). NSF DARE—transforming modeling in neurorehabilitation: A patient-in-the-loop framework. *Journal of Neuroengineering and Rehabilitation*, *21*(1), 23. DOI: 10.1186/s12984-024-01318-9

Caufield, J. H., Sigdel, D., Fu, J., Choi, H., Guevara-Gonzalez, V., Wang, D., & Ping, P. (2022). Cardiovascular informatics: Building a bridge to data harmony. *Cardiovascular Research, 118*(3), 732–745. DOI: 10.1093/cvr/cvab067 PMID: 33751044

Causer, A., Lu, T., & Fitzgerald, C., (2024). *SpaMTP: Integrative statistical analysis and visualisation of spatial metabolomics and transcriptomics data. bioRxiv*. https://doi.org/DOI: 10.1101/2024.10.31.621429

Cavarischia-Rega, C., Sharma, K., Fitzgerald, J. C., & Maček, B. (2025). Proteomic insights into the biology of dopaminergic neurons. *Frontiers in Molecular Neuroscience, 18*, 1642519. Advance online publication. DOI: 10.3389/fnmol.2025.1642519

Ceciliani, F., Eckersall, D., Burchmore, R., & Lecchi, C. (2014). Proteomics in veterinary medicine. *Veterinary Pathology, 51*(2), 351–362. DOI: 10.1177/0300985813502819 PMID: 24045891

Cedzyński, M., & Świerzko, A. S. (2023). Collectins and ficolins in neonatal health and disease. *Frontiers in Immunology, 14*, 1328658. DOI: 10.3389/fimmu.2023.1328658 PMID: 38193083

Cervilla, S., Grases, D., Perez, E., Real, F. X., Musulén, E., Esteller, M., & Porta-Pardo, E. (2024). Comparison of spatial transcriptomics technologies across six cancer types. *bioRxiv*. https://doi.org/DOI: 10.1101/2024.05.21.593407

Chambers, T. L., & Murach, K. A. (2025). A history of omics discoveries reveals the correlates and mechanisms of loading-induced hypertrophy in adult skeletal muscle. 2024 CaMPS young investigator award invited review. *American Journal of Physiology. Cell Physiology, 328*(5), C1535–C1557. DOI: 10.1152/ajpcell.00968.2024 PMID: 40172105

Chan, A. S. F., Greiner, J., Brennan, T., Agrawal, A., Hemmer, H., Sinning, K., & Grün, D. (2024). Spatio-temporal dynamics of the fibrotic niche in cardiac repair. bioRxiv, 2024-11.

Chang, Y. I. (2025). *A Survey: Potential Dimensionality Reduction Methods*. https://doi.org//arxiv.2502.11036DOI: 10.48550

Chang, H., Yang, X., You, K., Jiang, M., Cai, F., Zhang, Y., Liu, L., Liu, H., & Liu, X. (2020). Integrating multiple microarray dataset analysis and machine learning methods to reveal the key genes and regulatory mechanisms underlying human intervertebral disc degeneration. *PeerJ, 8*, e10120. DOI: 10.7717/peerj.10120 PMID: 33083145

Chan, K., & Li, X. (2021). Current epigenetic insights in kidney development. *Genes, 12*(8), 1281. DOI: 10.3390/genes12081281 PMID: 34440455

Chapman, E. A., Aballo, T. J., Melby, J. A., Zhou, T., Price, S. J., Rossler, K. J., & Ge, Y. (2023). Defining the sarcomeric proteoform landscape in ischemic cardiomyopathy by top-down proteomics. *Journal of Proteome Research*, *22*(3), 931–941. DOI: 10.1021/acs.jproteome.2c00729 PMID: 36800490

Chau, B. N., Xin, C., Hartner, J., Ren, S., Castano, A. P., Linn, G., Li, J., Tran, P. T., Kaimal, V., Huang, X., Chang, A. N., Li, S., Kalra, A., Grafals, M., Portilla, D., MacKenna, D. A., Orkin, S. H., & Duffield, J. S. (2012). MicroRNA-21 promotes fibrosis of the kidney by silencing metabolic pathways. *Science Translational Medicine*, *4*(121), 121ra18. Advance online publication. DOI: 10.1126/scitranslmed.3003205 PMID: 22344686

Chaurand, P., Schwartz, S. A., Reyzer, M. L., & Caprioli, R. M. (2005). Imaging mass spectrometry: Principles and potentials. *Toxicologic Pathology*, *33*(1), 92–101. DOI: 10.1080/01926230590881862 PMID: 15805060

Chebib, F. T., & Torres, V. E. (2016). Autosomal dominant polycystic kidney disease: Core curriculum 2016. *American Journal of Kidney Diseases*, *67*(5), 792–810. DOI: 10.1053/j.ajkd.2015.07.037 PMID: 26530876

Chen, C., Ge, Y., & Lu, L. (2023). Opportunities and challenges in the application of single-cell and spatial transcriptomics in plants. *Frontiers in Plant Science*, *14*, 1185377. Advance online publication. DOI: 10.3389/fpls.2023.1185377

Chen, C., Wang, J., Pan, D., Wang, X., Xu, Y., Yan, J., Wang, L., Yang, X., Yang, M., & Liu, G. (2023). Applications of multi-omics analysis in human diseases. *MedComm*, *4*(4), e315. DOI: 10.1002/mco2.315 PMID: 37533767

Cheng, K., & McCarrey, J. R. (2023). Profiling the Epigenetic Landscape of the Spermatogonial Stem Cell-Part 1: Epigenomics Assays. (Vol. 2656, pp. 71–108). https://doi.org/DOI: 10.1007/978-1-0716-3139-3_5

Cheng, X., Peng, T., Chu, T., Yang, Y., Liu, J., Gao, Q., Cao, C., & Wei, J. (2025). Application of single-cell and spatial omics in deciphering cellular hallmarks of cancer drug response and resistance. *Journal of Hematology & Oncology*, *18*(1), 70. DOI: 10.1186/s13045-025-01722-1 PMID: 40605007

Chen, J., & Xu, F. (2023). Application of nanopore sequencing in the diagnosis and treatment of pulmonary infections. *Molecular Diagnosis & Therapy*, *27*(6), 685–701. DOI: 10.1007/s40291-023-00669-8 PMID: 37563539

Chen, K. Y., Takeshima, Y., Zorzetto-Fernandes, A. L., Makino, K., Kibayashi, T., Hata, M., Ichiyama, K., Croteau, J., Taylor, K., & Sakaguchi, S. (2025). Joint single-cell measurements of surface proteins, intracellular proteins and gene expression with icCITE-seq. https://doi.org/DOI: 10.1101/2025.01.11.632564

Chen, L., Su, W., Chen, H., Chen, D.-Q., Wang, M., Guo, Y., & Zhao, Y.-Y. (2018). Proteomics for biomarker identification and clinical application in kidney disease. In Makowski, G. S. (Ed.), *Advances in Clinical Chemistry* (Vol. 85, pp. 91–113). Elsevier., DOI: 10.1016/bs.acc.2018.02.004

Chen, S., Zhu, B., Huang, S., Hickey, J. W., Lin, K. Z., Snyder, M., & Ma, Z. (2023). Integration of spatial and single-cell data across modalities with weak linkage. *BioRxiv*.

Chen, T.-Y., You, L., Hardillo, J. A. U., & Chien, M.-P. (2023). Spatial Transcriptomic Technologies. *Cells*, *12*(16), 2042. DOI: 10.3390/cells12162042 PMID: 37626852

Chen, W., Jin, B., Cheng, C., Peng, H., Zhang, X., Tan, W., Tang, R., Lian, X., Diao, H., Luo, N., Li, X., Fan, J., Shi, J., Yin, C., Wang, J., Peng, S., Yu, L., Li, J., Wu, R.-Q., & Jiang, X. (2024). Single-cell profiling reveals kidney CD163+ dendritic cell participation in human lupus nephritis. *Annals of the Rheumatic Diseases*, *83*(5), 608–623. DOI: 10.1136/ard-2023-224788 PMID: 38290829

Chen, W.-T., Lu, A., & Strohl, J. J.. (2022). Spatial transcriptomics and proteomics reveal cell-type-specific regulation in Alzheimer's disease. *Nature Neuroscience*, *25*(8), 1059–1071. DOI: 10.1038/s41593-022-01077-8 PMID: 35798980

Chen, W.-Y., Li, S., Kulkarni, A. S., Huang, L., Cao, J., Qian, K., & Wan, J. (2020). Single Cell Omics: From Assay Design to Biomedical Application. *Biotechnology Journal*, *15*(1), 1900262. DOI: 10.1002/biot.201900262

Chen, W., Zhang, P., Tran, T. N., Xiao, Y., Li, S., Shah, V. V., Cheng, H., Brannan, K. W., Youker, K., Lai, L., Fang, L., Yang, Y., Le, N.-T., Abe, J., Chen, S.-H., Ma, Q., Chen, K., Song, Q., Cooke, J. P., & Wang, G. (2025). A visual–omics foundation model to bridge histopathology with spatial transcriptomics. *Nature Methods*, *22*(7), 1568–1582. DOI: 10.1038/s41592-025-02707-1 PMID: 40442373

Chen, Y., (2025). Single-cell sequencing to multi-omics: Technologies and applications. *Molecular Systems Biology*, *21*(5).

Chen, Z., Du, C., Liu, S., Liu, J., Yang, Y., Dong, L., Zhao, W., Huang, W., & Lei, Y. (2024). Progress in biomaterials inspired by the extracellular matrix. *Giant (Oxford, England)*, *19*, 100323. DOI: 10.1016/j.giant.2024.100323

Cheung, E., Gakhar, R., & Sundaram, V. (2024). Xenium in situ analysis and reproducibility study of multiple carcinomas for clinical studies. *Cancer Research, 84*(6_Supplement), 6830. Advance online publication. DOI: 10.1158/1538-7445.AM2024-6830

Christodoulou, A. G., Shaw, J. L., Nguyen, C., Yang, Q., Xie, Y., Wang, N., & Li, D. (2018). Magnetic resonance multitasking for motion-resolved quantitative cardiovascular imaging. *Nature Biomedical Engineering, 2*(4), 215–226. DOI: 10.1038/s41551-018-0217-y PMID: 30237910

Chukwunweike, J. N., Praise, A., & Bashirat, B. A. (2024). *Harnessing Machine Learning for Cybersecurity: How Convolutional Neural Networks are Revolutionizing Threat Detection and Data Privacy.*

Cimenci, C. E. (2025). *Engineering the beating heart: cardiac organoid models for tissue regeneration.* IntechOpen.

Clark, D. J., Dhanasekaran, S. M., Petralia, F., Pan, J., Song, X., Hu, Y., Da Veiga Leprevost, F., Reva, B., Lih, T.-S. M., Chang, H.-Y., Ma, W., Huang, C., Ricketts, C. J., Chen, L., Krek, A., Li, Y., Rykunov, D., Li, Q. K., Chen, L. S., & Tu, Z. (2019). Integrated proteogenomic characterization of clear cell renal cell carcinoma. *Cell, 179*(4), 964–983.e31. DOI: 10.1016/j.cell.2019.10.007 PMID: 31675502

Clark, S. J., Argelaguet, R., Kapourani, C.-A., Stubbs, T. M., Lee, H. J., Krueger, F., Sanguinetti, G., Kelsey, G., Marioni, J. C., Stegle, O., & Reik, W. (2017). Joint Profiling Of Chromatin Accessibility, DNA Methylation And Transcription In Single Cells. *bioRxiv*, 138685. https://doi.org/DOI: 10.1101/138685

Clift, C. L., Blaser, M. C., Gerrits, W., Turner, M. E., Sonawane, A., Pham, T., Andresen, J. L., Fenton, O. S., Grolman, J. M., Campedelli, A., Buffolo, F., Schoen, F. J., Hjortnaes, J., Muehlschlegel, J. D., Mooney, D. J., Aikawa, M., Singh, S. A., Langer, R., & Aikawa, E. (2024). Intracellular proteomics and extracellular vesiculomics as a metric of disease recapitulation in 3D-bioprinted aortic valve arrays. *Science Advances, 10*(9), eadj9793. Advance online publication. DOI: 10.1126/sciadv.adj9793

Close, J. L., Long, B., & Zeng, H. (2021). Spatially resolved transcriptomics in neuroscience. *Nature Methods, 18*(1), 23–25. DOI: 10.1038/s41592-020-01040-z

Cockburn, J. G., Mariappan, V., Loke, M. F., Al-Maleki, A. R., Muttiah, B., Vellasamy, K. M., & Vadivelu, J. (2025). Integrative research: Current trends and considerations for biomarker discovery and precision medicine. *The Microbe, 7*, 100368. DOI: 10.1016/j.microb.2025.100368

Coletti, R., Carrilho, J., Martins, E. P., Gonçalves, C. S., Costa, B. M., & Lopes, M. B. (2024). A Novel Tool for Multi-Omics Network Integration and Visualization: A Study of Glioma Heterogeneity. *Bioinformatics (Oxford, England)*. Advance online publication. DOI: 10.1101/2024.03.18.585500

Conrad, T., & Altmüller, J. (2022). Single cell- and spatial 'Omics revolutionize physiology. *Acta Physiologica (Oxford, England), 235*(4), e13848. Advance online publication. DOI: 10.1111/apha.13848

Cooper, L., & Chen, J. (2022). Changes in Companion Diagnostic Labelling: Implementation of FDA's April 2020 Guidance for Industry for In Vitro CDx Labeling for Specific Oncology Therapeutic Groups. *Therapeutic Innovation & Regulatory Science, 56*(5), 689–697. DOI: 10.1007/s43441-022-00422-z PMID: 35689144

Cousins, P. D. (1999). Supply base rationalisation: myth or reality?. *European journal of purchasing & supply management, 5*(3-4), 143-155.

Cranley, J., Kanemaru, K., Bayraktar, S., Knight-Schrijver, V., Pett, J. P., Polanski, K., & Teichmann, S. A. (2024). Multiomic analysis reveals developmental dynamics of the human heart in health and disease. BioRxiv, 2024-04.

Cranley, J., Bayraktar, S., Kanemaru, K., Knight-Schrijver, V., Pett, J. P., Davaapil, H., & Teichmann, S. (2023). A spatially-resolved multiomic cell atlas reveals gene regulatory networks underlying cell specification in the developing human heart. *European Heart Journal, 44*(Supplement_2), ehad655–3042. DOI: 10.1093/eurheartj/ehad655.3042

Creswell, A., White, T., Dumoulin, V., Arulkumaran, K., Sengupta, B., & Bharath, A. A. (2018). Generative adversarial networks: An overview. *IEEE Signal Processing Magazine, 35*(1), 53–65. DOI: 10.1109/MSP.2017.2765202

Cs, C. (2021). Microscopic examination of spatial transcriptome using Seq-Scope. *Cell, 184*(13), 3559–3572.e22. Advance online publication. DOI: 10.1016/j.cell.2021.05.010

Cui, M., & Zhang, D. Y. (2021). Artificial intelligence and computational pathology. *Laboratory Investigation, 101*(4), 412–422. DOI: 10.1038/s41374-020-00514-0 PMID: 33454724

Cui, N., Xu, X., & Zhou, F. (2024). Single-cell technologies in psoriasis. *Clinical Immunology (Orlando, Fla.), 110242*, 110242. Advance online publication. DOI: 10.1016/j.clim.2024.110242

Cui, S., Wu, Q., West, J., & Bai, J. (2019). Machine learning-based microarray analyses indicate low-expression genes might collectively influence PAH disease. *PLoS Computational Biology*, *15*(8), e1007264. DOI: 10.1371/journal.pcbi.1007264 PMID: 31404060

Czarnewski, P., Mahfouz, A., Calogero, R. A., Palagi, P. M., Portell-Silva, L., Gonzalez-Uriarte, A., Soneson, C., Burdett, T., Szomolay, B., Videm, P., Hotz, H.-R., Papatheodorou, I., Hancock, J. M., Grüning, B., Haerty, W., Krause, R., Capella-Gutierrez, S., Leskošek, B., Alessandri, L., & Korpelainen, E. (2022). Community-driven ELIXIR activities in single-cell omics. *F1000 Research*, *11*, 869. DOI: 10.12688/f1000research.122312.1

D'Souza, R. F., Mathew, M., Mishra, V., & Surapaneni, K. M. (2024). Twelve tips for addressing ethical concerns in the implementation of artificial intelligence in medical education. *Medical Education Online*, *29*(1), 2330250. DOI: 10.1080/10872981.2024.2330250

D'Urso, F., & Broccolo, F. (2024). Applications of artificial intelligence in microbiome analysis and probiotic interventions—An overview and perspective based on the current state of the art. *Applied Sciences (Basel, Switzerland)*, *14*(19), 8627. DOI: 10.3390/app14198627

Dai, W., Qiao, X., Fang, Y., Guo, R., Bai, P., Liu, S., Li, T., Jiang, Y., Wei, S., Na, Z., Xiao, X., & Li, D. (2024). Epigenetics-targeted drugs: Current paradigms and future challenges. *Signal Transduction and Targeted Therapy*, *9*(1), 332. DOI: 10.1038/s41392-024-02039-0 PMID: 39592582

Das, K., Tay, M. L. I., Yong, E. Y., & Chuah, K. L. (2025). A targeted next-generation sequencing panel for identification of clinically relevant mutation profiles in solid tumours. *Scientific Reports*, *15*(1), 20740. DOI: 10.1038/s41598-025-08039-6 PMID: 40593313

Datlinger, P., Rendeiro, A. F., Boenke, T., Senekowitsch, M., Krausgruber, T., Barreca, D., & Bock, C. (2021). Ultra- high-throughput single- cell RNA sequencing and perturbation screening with combinatorial fluidic indexing. *Nature Methods*, *18*(6), 635–642. DOI: 10.1038/s41592-021-01153-z

Davenport, T. H. (2018). From analytics to artificial intelligence. *Journal of Business Analytics*, *1*(2), 73–80. DOI: 10.1080/2573234X.2018.1543535

Dave, R., Pandey, K., Patel, R., Gour, N., & Bhatia, D. (2025). Leveraging 3D cell culture and AI technologies for next-generation drug discovery. *Cell Biomaterials*, *1*(3), 100050. DOI: 10.1016/j.celbio.2025.100050

Davies, R. (2018). The metabolomic quest for a biomarker in chronic kidney disease. *Clinical Kidney Journal, 11*(6), 694–703. DOI: 10.1093/ckj/sfy037 PMID: 30288265

Dawson, T. (2024). *Sinai SCENT TMC - Single Cell Assay for Transposase Accessible Chromatin (ATAC-seq) v1*. https://doi.org/DOI: 10.17504/protocols.io.j8nlkoqo1v5r/v1

De Falco, M., Colella, S., Abdelkarim, J. A. Y., Antonacci, A., Scognamiglio, V., & De Felice, M. (2025). Helicase-dependent isothermal amplification turns biosensing into modern diagnostic technologies. *Microchemical Journal, 216*, 114772. DOI: 10.1016/j.microc.2025.114772

De Souza, M. L. M., Lopes, G. A., Branco, A. C., Fairley, J. K., & Fraga, L. A. D. O. (2021). Leprosy screening based on artificial intelligence: Development of a cross-platform app. *JMIR mHealth and uHealth, 9*(4), e23718. DOI: 10.2196/23718

de Toro-Martín, J., Arsenault, B. J., Després, J. P., & Vohl, M. C. (2017). Precision nutrition: A review of personalized nutritional approaches for the prevention and management of metabolic syndrome. *Nutrients, 9*(8), 913. DOI: 10.3390/nu9080913

Debbadi, R. K., & Boateng, O. (2025). Optimizing end-to-end business processes by integrating machine learning models with UiPath for predictive analytics and decision automation. *Int J Sci Res Arch, 14*(2), 778–796. DOI: 10.30574/ijsra.2025.14.2.0448

Debbadi, R. K., & Boateng, O. (2025a). Developing intelligent automation workflows in Microsoft Power Automate by embedding deep learning algorithms for real-time process adaptation. *Int J Sci Res Arch, 14*(2), 802–820. DOI: 10.30574/ijsra.2025.14.2.0449

DeBlasio, D., Kim, K., & Kingsford, C. (2020). More accurate transcript assembly via parameter advising. *Journal of Computational Biology, 27*(8), 1181–1189. DOI: 10.1089/cmb.2019.0286 PMID: 32315544

Deehan, E. C., Zhang, Z., Riva, A., Armet, A. M., Perez-Muñoz, M. E., Nguyen, N. K., Krysa, J. A., Seethaler, B., Zhao, Y.-Y., Cole, J., Li, F., Hausmann, B., Spittler, A., Nazare, J.-A., Delzenne, N. M., Curtis, J. M., Wismer, W. V., Proctor, S. D., Bakal, J. A., & Walter, J. (2022). Elucidating the role of the gut microbiota in the physiological effects of dietary fiber. *Microbiome, 10*(1), 77. DOI: 10.1186/s40168-022-01248-5

DelGiudice, M., Peirone, S., Perrone, S., Priante, F., Varese, F., Tirtei, E., Fagioli, F., & Cereda, M. (2021). Artificial Intelligence in Bulk and Single-Cell RNA-Sequencing Data to Foster Precision Oncology. *International Journal of Molecular Sciences, 22*(9), 4563. DOI: 10.3390/ijms22094563 PMID: 33925407

Deng, Y., Bartosovic, M., Ma, S., Zhang, D., Liu, Y., Qin, X., & Fan, R. (2021). Spatial-ATAC-seq: spatially resolved chromatin accessibility profiling of tissues at genome scale and cellular level. bioRxiv, 2021-06.

Deng, Y., Bai, Z., & Fan, R. (2023). Microtechnologies for single-cell and spatial multi-omics. *Nature Reviews Bioengineering*, *1*(10), 769–784. DOI: 10.1038/s44222-023-00084-y

Deng, Y., Bartosovic, M., Ma, S., Zhang, D., Kukanja, P., Xiao, Y., Su, G., Liu, Y., Qin, X., Rosoklija, G. B., Dwork, A. J., Mann, J. J., Xu, M. L., Halene, S., Craft, J. E., Leong, K. W., Boldrini, M., Castelo-Branco, G., & Fan, R. (2022). Spatial profiling of chromatin accessibility in mouse and human tissues. *Nature*, *609*(7926), 375–383. DOI: 10.1038/s41586-022-05094-1 PMID: 35978191

Derks, S., Derks, S., & Diosdado, B. (2015). Personalized cancer medicine: Next steps in the genomic era. *Cellular Oncology (Dordrecht)*, *38*(1), 1–2. DOI: 10.1007/s13402-015-0221-0

Devuyst, O., & Bochud, M. (2015). Uromodulin, kidney function, cardiovascular disease, and mortality. *Kidney International*, *88*(5), 944–946. DOI: 10.1038/ki.2015.267 PMID: 26579681

Dezem, F., Morosini, N. S., Arjumand, W., DuBose, H., & Plummer, J. (2024). Spatially Resolved Single-Cell Omics: Methods, Challenges, and Future Perspectives. *Annual Review of Biomedical Data Science*, *7*(1), 131–153. Advance online publication. DOI: 10.1146/annurev-biodatasci-102523-103640

Dhar, R., Kumar, A., & Karmakar, S. (2025). The future of immunohistochemistry: Advancing precision medicine through artificial intelligence-driven spatial proteomics and multiplexed diagnostic technologies. *Asian Journal of Medical Sciences*, *16*(8), 1–2. DOI: 10.71152/ajms.v16i8.4674

Dhas, Y., Arshad, N., Biswas, N., Jones, L., & Ashili, S. (2023). MicroRNA-21 silencing in diabetic nephropathy: Insights on therapeutic strategies. *Biomedicines*, *11*(9), 2583. DOI: 10.3390/biomedicines11092583 PMID: 37761024

Dhiflaoui, A., & Almawi, W. Y. (2025). Mechanisms and clinical implications of microRNA associations and signaling pathways in B-cell acute lymphoblastic leukemia. *Gene*, *967*, 149730. DOI: 10.1016/j.gene.2025.149730

Díaz-Menéndez, M., Esteban, E. T., Ujiie, M., Calleri, G., Rothe, C., Malvy, D., Nicastri, E., Bissinger, A. L., Grandadam, M., Alpern, J. D., Gobbi, F., Schlagenhauf, P., Duvignaud, A., Javelle, E., Nakamoto, T., Antinori, S., & Hamer, D. H. (2020). Travel-associated chikungunya acquired in Myanmar in 2019. *Euro Surveillance : Bulletin Europeen Sur Les Maladies Transmissibles = European. Eurosurveillance, 25*(1). Advance online publication. DOI: 10.2807/1560-7917.ES.2020.25.1.1900721

Dieter, C., Lemos, N. E., Girardi, E., Ramos, D. T., Corrêa, N. R. D. F., Canani, L. H., Bauer, A. C., Assmann, T. S., & Crispim, D. (2023). The lncRNA MALAT1 is upregulated in urine of type 1 diabetes mellitus patients with diabetic kidney disease. *Genetics and Molecular Biology, 46*(2), e20220291. DOI: 10.1590/1678-4685-gmb-2022-0291 PMID: 37272835

Dimitriu, M.-A., Lazar-Contes, I., Roszkowski, M., & Mansuy, I. M. (2022). Single-Cell Multiomics Techniques: From Conception to Applications. *Frontiers in Cell and Developmental Biology, 10*, 854317. Advance online publication. DOI: 10.3389/fcell.2022.854317

Ding, L., Li, X., Zhu, H., & Luo, H. (2022). Single-Cell Sequencing in Rheumatic Diseases: New Insights from the Perspective of the Cell Type. *Aging and Disease, 13*(6), 1633–1651. DOI: 10.14336/AD.2022.0323

Ding, Y., Almasian, M., Zhang, X., Kadam, V., & Yuan, J. (2023). Abstract P1180: Multi-scale imaging and analysis to elucidate cardiac structure and contractile function. *Circulation Research, 133*(Suppl_1), AP1180–AP1180. DOI: 10.1161/res.133.suppl_1.P1180

Dinu, M., Bita, B., Parau, A. C., Ristoscu, C., & Negut, I. (2025). Bioactive Glass and Melittin Thin Films Deposited by MAPLE for Titanium Implant Functionalization. *Materials (Basel), 18*(10), 2410. DOI: 10.3390/ma18102410

Dlamini, Z., Francies, F. Z., Hull, R., & Marima, R. (2020). Artificial intelligence (AI) and big data in cancer and precision oncology. *Computational and Structural Biotechnology Journal, 18*, 2300–2311. https://doi.org/https://doi.org/10.1016/j.csbj.2020.08.019. DOI: 10.1016/j.csbj.2020.08.019

Dong, M., Wang, L., Hu, N., Rao, Y., Wang, Z., & Zhang, Y. (2025). Integration of multi-omics approaches in exploring intra-tumoral heterogeneity. *Cancer Cell International, 25*(1), 317. Advance online publication. DOI: 10.1186/s12935-025-03944-2

Doran, S., Arif, M., Lam, S., Bayraktar, A., Turkez, H., Uhlen, M., & Mardinoglu, A. (2021). Multi-omics approaches for revealing the complexity of cardiovascular disease. *Briefings in Bioinformatics*, *22*(5), bbab061. DOI: 10.1093/bib/bbab061 PMID: 33725119

Doshi, M. D., Gordon, E. J., Freedman, B. I., Glover, C., Locke, J. E., & Thomas, C. P. (2021). Integrating APOL1 kidney-risk variant testing in live kidney donor evaluation: An expert panel opinion. *Transplantation*, *105*(10), 2132–2134. DOI: 10.1097/TP.0000000000003641 PMID: 33534524

Dries, R., Chen, J., Del Rossi, N., Khan, M. M., Sistig, A., & Yuan, G.-C. (2021). Advances in spatial transcriptomic data analysis. *Genome Research*, *31*(10), 1706–1718. DOI: 10.1101/gr.275224.121 PMID: 34599004

Dries, R., Zhu, Q., & Eng, C.-H. L.. (2021). Giotto: A toolbox for integrative analysis and visualization of spatial expression data. *Genome Biology*, *22*(1), 78. DOI: 10.1186/s13059-021-02286-2 PMID: 33685491

Duan, A., Wang, H., Zhu, Y., Wang, Q., Zhang, J., Hou, Q., Xing, Y., Shi, J., Hou, J., Qin, Z., Chen, Z., Liu, Z., & Yang, J. (2021). Chromatin architecture reveals cell type-specific target genes for kidney disease risk variants. *BMC Biology*, *19*(1), 38. DOI: 10.1186/s12915-021-00977-7 PMID: 33627123

Du, B., Mao, K., Bateni, S. M., Meng, F., Wang, X.-M., Guo, Z., Jun, C., & Du, G. (2022). A novel fully coupled physical–statistical–deep learning method for retrieving near-surface air temperature from multisource data. *Remote Sensing (Basel)*, *14*(22), 5812. DOI: 10.3390/rs14225812

Dubin, R. F., & Rhee, E. P. (2020). Proteomics and metabolomics in kidney disease, including insights into etiology, treatment, and prevention. *Clinical Journal of the American Society of Nephrology; CJASN*, *15*(3), 404–411. DOI: 10.2215/CJN.07420619 PMID: 31636087

Duncan, K. D., Fyrestam, J., & Lanekoff, I. (2019). Advances in mass spectrometry based single-cell metabolomics. *Analyst*, *144*(3), 782–793. DOI: 10.1039/C8AN01581C

Du, P., Fan, R., Zhang, N., Wu, C., & Zhang, Y. (2024a). Advances in Integrated Multi-omics Analysis for Drug-Target Identification. *Biomolecules*, *14*(6), 692. DOI: 10.3390/biom14060692 PMID: 38927095

Du, S., Zhai, L., Ye, S., Wang, L., Liu, M., & Tan, M. (2023). In-depth urinary and exosome proteome profiling analysis identifies novel biomarkers for diabetic kidney disease. *Science China. Life Sciences*, *66*(12), 2587–2603. DOI: 10.1007/s11427-022-2348-0 PMID: 37405567

Duvallet, C., Gibbons, S. M., Gurry, T., Irizarry, R. A., & Alm, E. J. (2017). Meta-analysis of gut microbiome studies identifies disease-specific and shared responses. *Nature Communications*, *8*(1), 1784. DOI: 10.1038/s41467-017-01973-8

Du, Y., Ding, X., & Ye, Y. (2024). The spatial multi-omics revolution in cancer therapy: Precision redefined. *Cell Reports Medicine*, *5*(9), 101740. DOI: 10.1016/j.xcrm.2024.101740

Efremova, M., Vento-Tormo, M., Teichmann, S. A., & Vento-Tormo, R. (2020). CellPhoneDB: Inferring cell–cell communication from combined expression of multi-subunit ligand–receptor complexes. *Nature Protocols*, *15*(4), 1484–1506. DOI: 10.1038/s41596-020-0292-x PMID: 32103204

Egbuna, O., Zimmerman, B., Manos, G., Fortier, A., Chirieac, M. C., Dakin, L. A., Friedman, D. J., Bramham, K., Campbell, K., Knebelmann, B., Barisoni, L., Falk, R. J., Gipson, D. S., Lipkowitz, M. S., Ojo, A., Bunnage, M. E., Pollak, M. R., Altshuler, D., & Chertow, G. M. (2023). Inaxaplin for proteinuric kidney disease in persons with two APOL1 variants. *The New England Journal of Medicine*, *388*(11), 969–979. DOI: 10.1056/NEJMoa2202396 PMID: 36920755

Elangovan, A., Li, Y., Pires, D. E. V., Davis, M. J., & Verspoor, K. (2022). Large-scale protein-protein post-translational modification extraction with distant supervision and confidence calibrated BioBERT. *BMC Bioinformatics*, *23*(1), 4. DOI: 10.1186/s12859-021-04504-x PMID: 34983371

ElRakaiby, M., Dutilh, B. E., Rizkallah, M. R., Boleij, A., Cole, J. N., & Aziz, R. K. (2014). Pharmacomicrobiomics: The impact of human microbiome variations on systems pharmacology and personalized therapeutics. *OMICS: A Journal of Integrative Biology*, *18*(7), 402–414. DOI: 10.1089/omi.2014.0018

Elsamanoudy, A. Z., Neamat-Allah, M. A. M., Mohammad, F. A. H., Hassanien, M., & Nada, H. A. (2016). The role of nutrition related genes and nutrigenetics in understanding the pathogenesis of cancer. *Journal of Microscopy and Ultrastructure*, *4*(3), 115–122. DOI: 10.1016/j.jmau.2016.02.002

El-Serag, H. B., & Sonnenberg, A. (1998). Opposing time trends of peptic ulcer and reflux disease. *Gut*, *43*(3), 327–333. DOI: 10.1136/gut.43.3.327

Emanuel, G., & He, J. (2021). MERFISH Enables Mapping of Cellular Diversity with Spatial Context at Subcellular Resolution. *Microscopy Today, 29*(4), 30–33. DOI: 10.1017/S1551929521000894

Emons, M., Gunz, S., Crowell, H. L., Mallona, I., Kuehl, M., Furrer, R., & Robinson, M. D. (2025). Harnessing the potential of spatial statistics for spatial omics data with pasta. *Nucleic Acids Research, 53*(17), gkaf870. Advance online publication. DOI: 10.1093/nar/gkaf870

Eng, C. H. L., Lawson, M., Zhu, Q., Dries, R., Koulena, N., Takei, Y., & Cai, L. (2019). Transcriptome-scale super-resolved imaging in tissues by RNA seqFISH+. *Nature, 568*(7751), 235–239. DOI: 10.1038/s41586-019-1049-y PMID: 30911168

Eraslan, G., Avsec, Ž., Gagneur, J., & Theis, F. J. (2019). Deep learning: New computational modelling techniques for genomics. *Nature Reviews. Genetics, 20*(7), 389–403. DOI: 10.1038/s41576-019-0122-6 PMID: 30971806

Erdoğan, D. E., (2025). ATAC-seq in emerging model organisms: Challenges and strategies. *Journal of Experimental Zoology. Part B, Molecular and Developmental Evolution*. PMID: 40452285

Esteva, A., Robicquet, A., Ramsundar, B., Kuleshov, V., DePristo, M., Chou, K., & Dean, J. (2019). A guide to deep learning in healthcare. *Nature Medicine, 25*(1), 24–29. DOI: 10.1038/s41591-018-0316-z

Fakruddin, M. D., Mazumdar, R. M., Chowdhury, A., & Mannan, K. S. Bin. (2012). Nucleic acid sequence based amplification (NASBA)-prospects and applications. *International Journal of Life Science and Pharma Research, 2*(1), 106–121.

Fan, R., & Lin, F. (2023). Spatial Delivery. [editorial]. *GEN Biotechnology, 2*(5), 331–332. Advance online publication. DOI: 10.1089/genbio.2023.29117.editorial

Farhud, D. D., & Yeganeh, M. Z. (2010). Nutrigenomics and nutrigenetics. *Iranian Journal of Public Health, 39*(4), 1.

Farhud, D. D., & Zokaei, S. (2021). Ethical issues of artificial intelligence in medicine and healthcare. *Iranian Journal of Public Health, 50*(11), i. DOI: 10.18502/ijph.v50i11.7600

Farnham, P. J. (2009). Insights from genomic profiling of transcription factors. *Nature Reviews. Genetics, 10*(9), 605–616. DOI: 10.1038/nrg2636

Farrugia, G., Harper, R., Krstevski, C., Lelliott, P., Hsu, I., Dona, M., & Pinto, A. (2024). Abstract P355: Multiomic analyses reveal sex-specific mechanisms governing cardiac fibrosis. *Hypertension, 81*(Suppl_1), AP355–AP355. DOI: 10.1161/hyp.81.suppl_1.P355

Feng, Y., Cai, L., Hong, W., Zhang, C., Tan, N., Wang, M., & Jiang, L. (2022). Rewiring of 3D chromatin topology orchestrates transcriptional reprogramming and the development of human dilated cardiomyopathy. *Circulation, 145*(22), 1663–1683. DOI: 10.1161/CIRCULATIONAHA.121.055781 PMID: 35400201

Fernández-Moya, S. M., Ganesh, A. J., & Plass, M. (2024). Neural cell diversity in the light of single-cell transcriptomics. *Transcription*, 1–19. DOI: 10.1080/21541264.2023.2295044

Ferrari, V., Mograbi, B., Gal, J., & Milano, G. (2024). Companion tests and personalized cancer therapy: Reaching a glass ceiling. *International Journal of Molecular Sciences, 25*(18), 9991. DOI: 10.3390/ijms25189991 PMID: 39337479

Ferrocino, I., Rantsiou, K., McClure, R., Kostic, T., De Souza, R. S. C., Lange, L., FitzGerald, J., Kriaa, A., Cotter, P., Maguin, E., Schelkle, B., Schloter, M., Berg, G., Sessitsch, A., & Cocolin, L. (2023). The need for an integrated multi-OMICs approach in microbiome science in the food system. *Comprehensive Reviews in Food Science and Food Safety, 22*(2), 1082–1103. DOI: 10.1111/1541-4337.13103

Fiehn, O., & Kind, T. (2007). Metabolite profiling in blood plasma. In *Metabolomics: Methods and protocols* (pp. 3–17). Springer. DOI: 10.1007/978-1-59745-244-1_1

Figgett, W. A., Monaghan, K., Ng, M., Alhamdoosh, M., Maraskovsky, E., Wilson, N. J., & Mackay, F. (2019). Machine learning applied to whole-blood RNA-sequencing data uncovers distinct subsets of patients with systemic lupus erythematosus. *Clinical & Translational Immunology, 8*(12), e01093. DOI: 10.1002/cti2.1093 PMID: 31921420

Firdous, A., Firdous, S., Ali, A., Nisar, B., Ahmad, H., Iqbal, G., & Ahmad, S. B. (2024). *Machine learning approaches for multiomics data analysis*. Elsevier BV., DOI: 10.1016/B978-0-443-23971-7.00013-4

Fontecha-Barriuso, M., Martin-Sanchez, D., Ruiz-Andres, O., Poveda, J., Sanchez-Niño, M. D., Valiño-Rivas, L., Ruiz-Ortega, M., Ortiz, A., & Sanz, A. B. (2018). Targeting epigenetic DNA and histone modifications to treat kidney disease. *Nephrology, Dialysis, Transplantation, 33*(11), 1875–1886. DOI: 10.1093/ndt/gfy009 PMID: 29534238

Forbes, T. A., Howden, S. E., Lawlor, K., Phipson, B., Maksimovic, J., Hale, L., Wilson, S., Quinlan, C., Ho, G., Holman, K., Bennetts, B., Crawford, J., Trnka, P., Oshlack, A., Patel, C., Mallett, A., Simons, C., & Little, M. H. (2018). Patient-iPSC-derived kidney organoids show functional validation of a ciliopathic renal phenotype and reveal underlying pathogenetic mechanisms. *American Journal of Human Genetics*, *102*(5), 816–831. DOI: 10.1016/j.ajhg.2018.03.014 PMID: 29706353

Forestier, H., Grenet, M., Borel, A., & Celiberti, V. (2017). Les productions lithiques de l'Archipel indonésien. *Journal of Lithic Studies*, *4*(2), 231–303. DOI: 10.2218/jls.v4i2.2544

Forte, E., McLellan, M. A., Skelly, D. A., & Rosenthal, N. A. (2021). Ex uno, plures–from one tissue to many cells: A review of single-cell transcriptomics in cardiovascular biology. *International Journal of Molecular Sciences*, *22*(4), 2071. DOI: 10.3390/ijms22042071 PMID: 33669808

Fox, A. G., Buonodono, K. E., Jones, A. R., Thomas, M., Schepers, L. E., Earl, C. C., & Crouch, A. C. (2025). Multimodal study of murine cardiovascular remodeling: Four-dimensional ultrasound and mass spectrometry imaging. *Journal of Visualized Experiments*, (215), e67347. DOI: 10.3791/67347 PMID: 39878950

Francisco, K. K. Y., Apuhin, A. E. C., Tan, M. J. T., Byers, M. C., Maravilla, N. M. A. T., Karim, H. A., & AlDahoul, N. (2024). Can personalized medicine coexist with health equity? Examining the cost barrier and ethical implications. *arXiv preprint arXiv:2411.02307.*

Francis, L. P. (2014). Genomic knowledge sharing: A review of the ethical and legal issues. *Applied & Translational Genomics*, *3*(4), 111–115. DOI: 10.1016/j.atg.2014.09.003 PMID: 27294025

Franzosa, E. A., Hsu, T., Sirota-Madi, A., Shafquat, A., Abu-Ali, G., Morgan, X. C., & Huttenhower, C. (2015). Sequencing and beyond: Integrating molecular'omics' for microbial community profiling. *Nature Reviews. Microbiology*, *13*(6), 360–372. DOI: 10.1038/nrmicro3451

Freedman, B. I., Kopp, J. B., Sampson, M. G., & Susztak, K. (2021). APOL1 at 10 years: Progress and next steps. *Kidney International*, *99*(6), 1296–1302. DOI: 10.1016/j.kint.2021.03.013 PMID: 33794228

Freedman, B. I., Limou, S., Ma, L., & Kopp, J. B. (2018). APOL1-associated nephropathy: A key contributor to racial disparities in CKD. *American Journal of Kidney Diseases*, *72*(5, Suppl 1), S8–S16. DOI: 10.1053/j.ajkd.2018.06.020 PMID: 30343724

Freudenheim, J. L., & Gower, E. (2017). Interaction of genetic factors with nutrition in cancer. In *Nutrition in the Prevention and Treatment of Disease* (pp. 733–747). Academic Press. DOI: 10.1016/B978-0-12-802928-2.00033-3

Frey, N., & Olson, E. N. (2000). Making matters worse for a broken heart. *The Journal of Clinical Investigation*, *106*(12), 1437–1439. DOI: 10.1172/JCI11733 PMID: 11120749

Fu, X., & Chen, Y. (2024). Pix2Path: integrating spatial transcriptomics and digital pathology with deep learning to score pathological risk and link gene expression to disease mechanisms. bioRxiv, 2024-08.

Fujimoto, M., & Tanaka, T. (2025). Advancing liver metabolic zonation with single-cell and spatial omics. *Endocrine Journal*, *72*(10), 1069–1078. Advance online publication. DOI: 10.1507/endocrj.EJ25-0140

Gajjala, P. R., Fliser, D., Speer, T., Jankowski, V., & Jankowski, J. (2019). Proteomic biomarkers of diabetic kidney disease: Progress and challenges. *Current Diabetes Reports*, *19*(9), 87. DOI: 10.1007/s11892-019-1198-0

Gao, M.-Y., Wang, J.-Q., He, J., Gao, R., Zhang, Y., & Li, X. (2023). Single-Cell RNA-Sequencing in Astrocyte Development, Heterogeneity, and Disease. *Cellular and Molecular Neurobiology*, *43*(7), 3449–3464. DOI: 10.1007/s10571-023-01397-7

Gao, R., Yang, L., Zhu, Y., Zhang, T., Gu, W., & Yang, L. (2024). Advancing precision medicine through single-cell sequencing: Insights and implications. *Clinical and Translational Discovery*, *4*(2), e293. Advance online publication. DOI: 10.1002/ctd2.293

Gao, S., Zhao, J., Guan, Y., Tang, Y., Li, Y., & Liu, L. (2022). Mass Spectrometry Imaging technology in metabolomics: A systematic review. *Biomedical Chromatography*, *37*(7), e5494. DOI: 10.1002/bmc.5494

Gato-Diaz, U., Concheiro, A., Alvarez-Lorenzo, C., & Blanco-Fernandez, B. (2025). 4D printed hydrogels for precision delivery of bioactive molecules in cancer. *Expert Opinion on Drug Delivery*, *0*(0), 1–17. DOI: 10.1080/17425247.2025.2539959

Genomics. (2021). *Visium Spatial Gene Expression User Guide*. 10x Genomics. https://www.10xgenomics.com (Access on 22nd October 2025)

Ge, S., Sun, S., Xu, H., Cheng, Q., & Ren, Z. (2025). Deep learning in single-cell and spatial transcriptomics data analysis: Advances and challenges from a data science perspective. *Briefings in Bioinformatics*, *26*(2), bbaf136. Advance online publication. DOI: 10.1093/bib/bbaf136

Getz, G. S., & Reardon, C. A. (2014). The mutual interplay of lipid metabolism and the cells of the immune system in relation to atherosclerosis. *Clinical Lipidology, 9*(6), 657–671. DOI: 10.2217/clp.14.50 PMID: 25705263

Ghamsari, M. S. (2025). Quantum computing applications in biology. *Discover Computing, 28*(1), 276. DOI: 10.1007/s10791-025-09803-y

Ghram, A., & Lavie, C. J. (2025). Integrating omics to optimize precision cardiac rehabilitation. *Metabolism and Target Organ Damage, 5*(3), N-A.

Giaimo, B. D., Ferrante, F., Herchenröther, A., Hake, S. B., & Borggrefe, T. (2019). The histone variant H2A.Z in gene regulation. *Epigenetics & Chromatin, 12*(1), 37. DOI: 10.1186/s13072-019-0274-9 PMID: 31200754

Giani, A. M., Gallo, G. R., Gianfranceschi, L., & Formenti, G. (2020). Long walk to genomics: History and current approaches to genome sequencing and assembly. *Computational and Structural Biotechnology Journal, 18*, 9–19. DOI: 10.1016/j.csbj.2019.11.002 PMID: 31890139

Giesen, C., Wang, H. A., Schapiro, D., Zivanovic, N., Jacobs, A., Hattendorf, B., Schüffler, P. J., Grolimund, D., Buhmann, J. M., Brandt, S., Varga, Z., Wild, P. J., Günther, D., & Bodenmiller, B. (2014). Highly multiplexed imaging of tumor tissues with subcellular resolution by mass cytometry. *Nature Methods, 11*(4), 417–422. DOI: 10.1038/nmeth.2869 PMID: 24584193

Gilyazova, I., Ivanova, E., Izmailov, A., Sharifgaliev, I., Karunas, A., Pudova, E., Kobelyatskaya, A., Gilyazova, G., Izmailova, A., Pavlov, V., & Khusnutdinova, E. (2023). MicroRNA expression signatures in clear cell renal cell carcinoma: High-throughput searching for key miRNA markers in patients from the Volga-Ural region of Eurasian continent. *International Journal of Molecular Sciences, 24*(13), 6909. DOI: 10.3390/ijms24086909 PMID: 37108073

Gioacchino, E., Vandelannoote, K., Ruberto, A. A., Popovici, J., & Cantaert, T. (2024). Unraveling the intricacies of host-pathogen interaction through single-cell genomics. *Microbes and Infection, 26*(7), 105313. DOI: 10.1016/j.micinf.2024.105313 PMID: 38369008

Giudicelli, G. C., De Souza, C. M. B., Veronese, F. V., Pereira, L. V., Hünemeier, T., & Vianna, F. S. L. (2022). Precision medicine implementation challenges for APOL1 testing in chronic kidney disease in admixed populations. *Frontiers in Genetics, 13*, 1016341. DOI: 10.3389/fgene.2022.1016341 PMID: 36588788

Giuffrida, M. C., & Spoto, G. (2017). Integration of isothermal amplification methods in microfluidic devices: Recent advances. *Biosensors & Bioelectronics*, *90*, 174–186. DOI: 10.1016/j.bios.2016.11.045

Giuntini, E. B., Sardá, F. A. H., & de Menezes, E. W. (2022). The effects of soluble dietary fibers on glycemic response: An overview and futures perspectives. *Foods*, *11*(23), 3934. DOI: 10.3390/foods11233934

Gogoi, P., & Valan, J. A. (2024). Machine learning approaches for predicting and diagnosing chronic kidney disease: Current trends, challenges, solutions, and future directions. *International Urology and Nephrology*, *57*(4), 1245–1268. Advance online publication. DOI: 10.1007/s11255-024-04281-5 PMID: 39560857

Goker, M., Derici, U. S., Gokyer, S., Parmaksiz, M. G., Kaya, B., Can, A., & Yilgor, P. (2024). Spatial Growth Factor Delivery for 3D Bioprinting of Vascularized Bone with Adipose-Derived Stem/Stromal Cells as a Single Cell Source. *ACS Biomaterials Science & Engineering*, *10*(3), 1607–1619. DOI: 10.1021/acsbiomaterials.3c01222

Gokhale, T. A., Asfour, H., Verma, S., Bursac, N., & Henriquez, C. S. (2018). Microheterogeneity-induced conduction slowing and wavefront collisions govern macroscopic conduction behavior: A computational and experimental study. *PLoS Computational Biology*, *14*(7), e1006276. DOI: 10.1371/journal.pcbi.1006276 PMID: 30011279

Goltsev, Y., Samusik, N., Kennedy-Darling, J., Bhate, S. S., Hale, M., Vasquez, G., & Nolan, G. P. (2018). Deep Profiling of Mouse Splenic Architecture with CODEX Multiplexed Imaging. *Social Science Research Network*. https://doi.org/ DOI: 10.2139/SSRN.3155542

Gómez, A., Tacheau, A., Le Floc'h, S., Pettigrew, R. I., Cloutier, G., Finet, G., & Ohayon, J. (2021). Intravascular ultrasound imaging of human coronary atherosclerotic plaque: novel morpho-elastic biomarkers of instability. In *Biomechanics of coronary atherosclerotic plaque* (pp. 465–489). Academic Press. DOI: 10.1016/B978-0-12-817195-0.00020-2

Gong, D., Arbesfeld-Qiu, J. M., Perrault, E., Bae, J. W., & Hwang, W. L. (2024). Spatial oncology: Translating contextual biology to the clinic. *Cancer Cell*, *42*(10), 1653–1675. Advance online publication. DOI: 10.1016/j.ccell.2024.09.001

Goodfellow, I. (2016). Deep learning.

Goodwin, R. J. A., Platz, S. J., Reis-Filho, J. S., & Barry, S. T. (2024). Accelerating Drug Development Using Spatial Multi-omics. *Cancer Discovery*, *14*(4), 620–624. DOI: 10.1158/2159-8290.CD-24-0101 PMID: 38571424

Goodwin, S., McPherson, J. D., & McCombie, W. R. (2016). Coming of age: Ten years of next-generation sequencing technologies. *Nature Reviews. Genetics*, *17*(6), 333–351. DOI: 10.1038/nrg.2016.49 PMID: 27184599

Gough, A., Stern, A. M., Maier, J., Lezon, T., Shun, T.-Y., Chennubhotla, C., Schurdak, M. E., Haney, S. A., & Taylor, D. L. (2017). Biologically Relevant Heterogeneity: Metrics and Practical Insights. *SLAS Discovery*, *22*(3), 213–237. DOI: 10.1177/2472555216682725 PMID: 28231035

Govek, K. W., Yamajala, V. S., & Camara, P. G. (2019). Clustering-independent analysis of genomic data using spectral simplicial theory. *PLoS Computational Biology*, *15*(11), e1007509. DOI: 10.1371/journal.pcbi.1007509 PMID: 31756191

Graham, B. S. (2020). Rapid COVID-19 vaccine development. *Science*, *368*(6494), 945–946. DOI: 10.1126/science.abb8923

Green, E. D., Gunter, C., Biesecker, L. G., Di Francesco, V., Easter, C. L., Feingold, E. A., & Manolio, T. A. (2020). Strategic vision for improving human health at the forefront of genomics. *Nature*, *586*(7831), 683–692. DOI: 10.1038/s41586-020-2817-4

Greenstreet, L., Afanassiev, A., Kijima, Y., Heitz, M., Ishiguro, S., King, S., Yachie, N., & Schiebinger, G. (2023). DNA-GPS: A theoretical framework for optics-free spatial genomics and synthesis of current methods. *Cell Systems*, *14*(10), 844–859.e4. Advance online publication. DOI: 10.1016/j.cels.2023.08.005

Groopman, E. E., Marasa, M., Cameron-Christie, S., Petrovski, S., Aggarwal, V. S., Milo-Rasouly, H., Li, Y., Zhang, J., Nestor, J., Krithivasan, P., Lam, W. Y., Mitrotti, A., Piva, S., Kil, B. H., Chatterjee, D., Reingold, R., Bradbury, D., DiVecchia, M., Snyder, H., & Gharavi, A. G. (2019). Diagnostic utility of exome sequencing for kidney disease. *The New England Journal of Medicine*, *380*(2), 142–151. DOI: 10.1056/NEJMoa1806891 PMID: 30586318

Grubaugh, N. D., Ladner, J. T., Kraemer, M. U. G., Dudas, G., Tan, A. L., Gangavarapu, K., Wiley, M. R., White, S., Thézé, J., Magnani, D. M., Prieto, K., Reyes, D., Bingham, A. M., Paul, L. M., Robles-Sikisaka, R., Oliveira, G., Pronty, D., Barcellona, C. M., Metsky, H. C., & Andersen, K. G. (2017). Genomic epidemiology reveals multiple introductions of Zika virus into the United States. *Nature*, *546*(7658), 401–405. DOI: 10.1038/nature22400

Grzeski, M., Jensen, P. M., Hempel, B.-F., Thiele, H., Lellmann, J., Schallenberg, S., Budach, V., Keilholz, U., Tinhofer, I., & Klein, O. (2025). Integrating MALDI-MSI-Based Spatial Proteomics and Machine Learning to Predict Chemoradiotherapy Outcomes in Head and Neck Cancer. *International Journal of Molecular Sciences*, *26*(18), 9084. DOI: 10.3390/ijms26189084 PMID: 41009656

Gundappa, M., Arumugam, V. A., Hsieh, H. L., Balasubramanian, B., & Shanmugam, V. (2022). Expression of tissue factor and TF-mediated integrin regulation in HTR-8/SVneo trophoblast cells. *Journal of Reproductive Immunology*, *150*, 103473. DOI: 10.1016/j.jri.2022.103473 PMID: 35030354

Guo, P., Mao, L., Chen, Y., & Zhang, W. (2023). Spatial FFPE-ATAC-seq enables epigenomic profiling of archived tissue sections. *Nature Communications*, *14*(1), 1529. PMID: 36934097

Guo, X., Zhang, Y., Zheng, L., Zheng, C., Song, J., Zhang, Q., Kang, B., Liu, Z., Jin, L., Xing, R., Gao, R., Zhang, L., Dong, M., Hu, X., Ren, X., Kirchhoff, D., Roider, H. G., Yan, T., & Zhang, Z. (2018). Global characterization of T cells in non-small-cell lung cancer by single-cell sequencing. *Nature Medicine*, *24*(7), 978–985. DOI: 10.1038/s41591-018-0045-3 PMID: 29942094

Gupta, A., Madhavan, M. V., Sehgal, K., Nair, N., Mahajan, S., Sehrawat, T. S., Bikdeli, B., Ahluwalia, N., Ausiello, J. C., Wan, E. Y., Freedberg, D. E., Kirtane, A. J., Parikh, S. A., Maurer, M. S., Nordvig, A. S., Accili, D., Bathon, J. M., Mohan, S., Bauer, K. A., & Landry, D. W. (2020). Extrapulmonary manifestations of COVID-19. *Nature Medicine*, *26*(7), 1017–1032. DOI: 10.1038/s41591-020-0968-3

Gupta, N., Maurya, S., Verma, H., & Verma, V. K. (2019). Unraveling the factors and mechanism involved in persistence: Host-pathogen interactions in Helicobacter pylori. *Journal of Cellular Biochemistry*, *120*(11), 18572–18587. DOI: 10.1002/jcb.29201

Gupta, R. K., & Kuznicki, J. (2020). Biological and medical importance of cellular heterogeneity deciphered by single-cell RNA sequencing. *Cells*, *9*(8), 1751. DOI: 10.3390/cells9081751

Gupta, R. M., Schnitzler, G. R., Fang, S., Lee-Kim, V. S., & Barry, A. (2023). Multiomic analysis and CRISPR perturbation screens identify endothelial cell programs and novel therapeutic targets for coronary artery disease. *Arteriosclerosis, Thrombosis, and Vascular Biology*, *43*(5), 600–608. DOI: 10.1161/ATVBAHA.123.318328 PMID: 36994731

Gupta, S., Gupta, M. K., Shabaz, M., & Sharma, A. (2022). Deep learning techniques for cancer classification using microarray gene expression data. *Frontiers in Physiology, 13*, 952709. DOI: 10.3389/fphys.2022.952709 PMID: 36246115

Gupta, T., Aggarwal, S. K., & Mathur, T. (2025). Integrative imaging and molecular profiling: unveiling the nexus of genomics and precision medicine. In *Computer-Assisted Analysis for Digital Medicinal Imagery* (pp. 183–208). IGI Global.

Guryanov, I., Fiorucci, S., & Tennikova, T. (2016). Receptor-ligand inter actions: Advanced biomedical applications. *Materials Science and Engineering C, 68*, 890–903. DOI: 10.1016/j.msec.2016.07.072 PMID: 27524092

Gutierrez Reyes, C. D., Alejo-Jacuinde, G., Perez Sanchez, B., Chavez Reyes, J., Onigbinde, S., Mogut, D., Hernández-Jasso, I., Calderón-Vallejo, D., Quintanar, J. L., & Mechref, Y. (2024). Multi-omics applications in biological systems. *Current Issues in Molecular Biology, 46*(6), 5777–5793. DOI: 10.3390/cimb46060345 PMID: 38921016

Gu, Y.-Y., Lu, F.-H., Huang, X.-R., Zhang, L., Mao, W., Yu, X.-Q., Liu, X.-S., & Lan, H.-Y. (2021). Non-coding RNAs as biomarkers and therapeutic targets for diabetic kidney disease. *Frontiers in Pharmacology, 11*, 583528. DOI: 10.3389/fphar.2020.583528 PMID: 33574750

Hadfield, J., Megill, C., Bell, S. M., Huddleston, J., Potter, B., Callender, C., Sagulenko, P., Bedford, T., & Neher, R. A. (2018). Nextstrain: Real-time tracking of pathogen evolution. *Bioinformatics (Oxford, England), 34*(23), 4121–4123. DOI: 10.1093/bioinformatics/bty407

Haider, S., & Pal, R. (2013). Integrated Analysis of Transcriptomic and Proteomic Data. *Current Genomics, 14*(2), 91–110. DOI: 10.2174/1389202911314020003

Hajdarpašić, A., Tukker, M., Rijdt, W. T., Mohamedhoesein, S., Meijers, W. C., & Caliskan, K. (2025). Epigenetics of cardiomyopathies: The next frontier. *Heart Failure Reviews, 30*(1), 257–270. DOI: 10.1007/s10741-024-10460-4 PMID: 39586986

Hall, G. (2024). Genetic insights into the mechanisms of proliferative glomerulonephritis. *The Journal of Clinical Investigation, 134*(2), e183090. DOI: 10.1172/JCI183090 PMID: 39087471

Hamburg, M. A., & Collins, F. S. (2010). The path to personalized medicine. *The New England Journal of Medicine, 363*(4), 301–304. DOI: 10.1056/NEJMp1006304

Hamdani, N., Kooij, V., Van Dijk, S., Merkus, D., Paulus, W. J., Remedios, C. D., & van der Velden, J. (2008). Sarcomeric dysfunction in heart failure. *Cardiovascular Research, 77*(4), 649–658. DOI: 10.1093/cvr/cvm079 PMID: 18055579

Han, S., Xu, Q., Du, Y., Tang, C., Cui, H., Xia, X., & Shang, H. (2024). Single-cell spatial transcriptomics in cardiovascular development, disease, and medicine. *Genes & Diseases, 11*(6), 101163. DOI: 10.1016/j.gendis.2023.101163 PMID: 39224111

Hansson, F. G., Madsen, N. G., Hansen, L. G., Jakočiūnas, T., Lengger, B., Keasling, J. D., Jensen, M. K., Acevedo-Rocha, C. G., & Jensen, E. D. (2025). Labels as a feature: Network homophily for systematically annotating human GPCR drug-target interactions. *Nature Communications, 16*(1), 4121. DOI: 10.1038/s41467-025-59418-6 PMID: 40316519

Hao, P., Guo, T., & Sze, S. K. (2011). Simultaneous analysis of proteome, phospho- and glycoproteome of rat kidney tissue with electrostatic repulsion hydrophilic interaction chromatography. *PLoS One, 6*(2), e16884. DOI: 10.1371/journal.pone.0016884 PMID: 21373199

Hasan, G. M., Mohammad, T., Shamsi, A., Sohal, S. S., & Hassan, M. I. (2025). Applications of genome sequencing in infectious diseases: From pathogen identification to precision medicine. *Pharmaceuticals (Basel, Switzerland), 18*(11), 1687. DOI: 10.3390/ph18111687 PMID: 41304932

Hasanzadeh, A., Hajiramezanali, E., Duffield, N., & Qian, X. (2022). MoReL: Multi-omics relational learning. *arXiv*. https://arxiv.org/abs/2203.08149

Hasanzad, M., Sarhangi, N., Naghavi, A., Ghavimehr, E., Khatami, F., Ehsani Chimeh, S., Larijani, B., & Aghaei Meybodi, H. R. (2022). Genomic medicine on the frontier of precision medicine. *Journal of Diabetes and Metabolic Disorders, 21*(1), 853–861. DOI: 10.1007/s40200-021-00880-6 PMID: 35673457

Hasegawa, S., & Inagi, R. (2021). Harnessing metabolomics to describe the pathophysiology underlying progression in diabetic kidney disease. *Current Diabetes Reports, 21*(7), 21. DOI: 10.1007/s11892-021-01390-8 PMID: 33974145

Haskins, G., Kruger, U., & Yan, P. (2020). Deep learning in medical image registration: A survey. *Machine Vision and Applications, 31*(1), 8. DOI: 10.1007/s00138-020-01060-x

Ha, T., Kaiser, C., Myong, S., Wu, B., & Xiao, J. (2022). Next generation single-molecule techniques: Imaging, labeling, and manipulation <em>in vitro</em> and <em>in cellulo</em>. *Molecular Cell, 82*(2), 304–314. DOI: 10.1016/j.molcel.2021.12.019

Hayashi, H. (2010). Does remodelling of gap junctions and connexin expression contribute to arrhythmogenesis?–study in an immobilization rat model–. *Circulation Journal, 74*(12), 2558–2559. DOI: 10.1253/circj.CJ-10-0983 PMID: 21088334

He, Z., Liu, L., Han, D., Gao, K., Dong, L., Bu, D., Huo, P., Wang, Z., Deng, W., Liu, J., Guo, J., Zhao, Y., & Wu, Y. (2025). *POINT: A web-based platform for pharmacological investigation enhanced by multi-omics networks and knowledge graphs* (Version 1). arXiv. https://doi.org/DOI: 10.48550/ARXIV.2503.07203

Heather, J. M., & Chain, B. (2016). The sequence of sequencers: The history of sequencing DNA. *Genomics*, *107*(1), 1–8. DOI: 10.1016/j.ygeno.2015.11.003 PMID: 26554401

Hegenbarth, J. C., Lezzoche, G., De Windt, L. J., & Stoll, M. (2022). Perspectives on bulk-tissue RNA sequencing and single-cell RNA sequencing for cardiac transcriptomics. *Frontiers in Molecular Medicine*, *2*, 839338. DOI: 10.3389/fmmed.2022.839338 PMID: 39086967

Helmink, B. A., Reddy, S. M., Gao, J., Zhang, S., Basar, R., Thakur, R., Yizhak, K., Sade-Feldman, M., Blando, J., Han, G., Gopalakrishnan, V., Xi, Y., Zhao, H., Amaria, R. N., Tawbi, H. A., Cogdill, A. P., Liu, W., LeBleu, V. S., Kugeratski, F. G., & Wargo, J. A. (2020). B cells and tertiary lymphoid structures promote immunotherapy response. *Nature*, *577*(7791), 549–555. DOI: 10.1038/s41586-019-1922-8 PMID: 31942075

He, M. J., Pu, W., Wang, X., Zhang, W., Tang, D., & Dai, Y. (2022). Comparing DESI-MSI and MALDI-MSI Mediated Spatial Metabolomics and Their Applications in Cancer Studies. *Frontiers in Oncology*, *12*, 891018. DOI: 10.3389/fonc.2022.891018 PMID: 35924152

Hendra, C., Pratanwanich, P. N., Wan, Y. K., Goh, W. S., Thiery, A., & Göke, J. (2022). Detection of m6A from direct RNA sequencing using a multiple instance learning framework. *Nature Methods*, *19*(12), 1590–1598. DOI: 10.1038/s41592-022-01666-1 PMID: 36357692

He, Q. (2016). Tumor heterogeneity and drug resistance of targeted antitumor agents. *Yao Xue Xue Bao*, *51*(2), 197–201. https://pubmed.ncbi.nlm.nih.gov/29856199/

Herrero-Fernandez, B., Gomez-Bris, R., Somovilla-Crespo, B., & Gonzalez-Granado, J. M. (2019). Immunobiology of atherosclerosis: A complex net of interactions. *International Journal of Molecular Sciences*, *20*(21), 5293. DOI: 10.3390/ijms20215293 PMID: 31653058

Hesse, K., Khanji, M. Y., Aung, N., Dabbagh, G. S., Petersen, S. E., & Chahal, C. A. A. (2024). Assessing heterogeneity on cardiovascular magnetic resonance imaging: A novel approach to diagnosis and risk stratification in cardiac diseases. *European Heart Journal Cardiovascular Imaging*, *25*(4), 437–445. DOI: 10.1093/ehjci/jead285 PMID: 37982176

He, Y., Lu, W., Zhou, X., Mu, J., & Shen, W. (2024). Unraveling Alzheimer's disease: Insights from single-cell sequencing and spatial transcriptomic. *Frontiers in Neurology, 15*, 1515981. Advance online publication. DOI: 10.3389/fneur.2024.1515981

Heydari, A. A., & Sindi, S. S. (2023). Deep learning in spatial transcriptomics: Learning from the next next-generation sequencing. *Biophysics Reviews*, *4*(1), 011306. DOI: 10.1063/5.0091135 PMID: 38505815

Heynen-Genel, S., Pache, L., Chanda, S. K., & Rosen, J. (2012). Functional genomic and high-content screening for target discovery and deconvolution. *Expert Opinion on Drug Discovery*, *7*(10), 955–968. DOI: 10.1517/17460441.2012.711311 PMID: 22860749

Hilafu, H., & Safo, S. E. (2022). Sparse sliced inverse regression for high dimensional data analysis. *BMC Bioinformatics*, *23*(1), 168. DOI: 10.1186/s12859-022-04700-3 PMID: 35525975

Hippen, A. A., Falco, M. M., Weber, L. M., Erkan, E. P., Zhang, K., Doherty, J. A., Vähärautio, A., Greene, C. S., & Hicks, S. C. (2021). miQC: An adaptive probabilistic framework for quality control of single-cell RNA-sequencing data. *PLoS Computational Biology*, *17*(8), 1009290. DOI: 10.1371/journal.pcbi.1009290

Hodgin, J. B., Smith, C., & Kretzler, M. (2024). Multi-omics data integration shines a light on the renal medulla. *Kidney International*, *105*(2), 242–244. DOI: 10.1016/j.kint.2023.11.014 PMID: 38245213

Horton, R., & Lucassen, A. (2023). Ethical considerations in research with genomic data. *The New Bioethics : a Multidisciplinary Journal of Biotechnology and the Body*, *29*(1), 37–51. DOI: 10.1080/20502877.2022.2060590 PMID: 35484929

Horvath, P., Aulner, N., Bickle, M., Davies, A. M., Del Nery, E., Ebner, D., Montoya, M. C., Östling, P., Pietiäinen, V., Price, L. S., Shorte, S. L., Turcatti, G., von Schantz, C., & Carragher, N. O. (2016). Screening out irrelevant cell-based models of disease. *Nature Reviews. Drug Discovery*, *15*(11), 751–769. DOI: 10.1038/nrd.2016.175

Hosseinpour, M., Ardalani, F., Mohseni, M., Beheshtian, M., Arzhangi, S., Ossareh, S., Najmabadi, H., Nobakht, A., Kahrizi, K., & Broumand, B. (2022). Targeted next generation sequencing revealed novel variants in the PKD1 and PKD2 genes of Iranian patients with autosomal dominant polycystic kidney disease. *Archives of Iranian Medicine*, *25*(9), 600–608. DOI: 10.34172/aim.2022.95 PMID: 37543885

Houseini, S. T., Mahmanzar, M., Rahimian, K., Bayat, S., Gholamzad, A., Mollapour Sisakht, M., Farhadi, A., Kuehu, D. L., & Deng, Y. (2025). Spatial Transcriptomics and Single Cell-RNASeq Reveals Cellular Heterogeneity of SARS-CoV-2 in Lung Tissues and Global Mutational Patterns in COVID-19 Patients. *Journal of Medical Virology, 97*(9), e70586. Advance online publication. DOI: 10.1002/jmv.70586

Houston, A. (2024). *10x Genomics Xenium for FFPE User Guides v1*. https://doi.org/DOI: 10.17504/protocols.io.81wgbxjonlpk/v1

Huan, C., Li, J., Li, Y., Zhao, S., Yang, Q., Zhang, Z., & Zhou, L. (2025). Spatially resolved multiomics: data analysis from monoomics to multiomics. *BME frontiers, 6*, 0084.

Huang, C., Wang, Y., Li, X., Ren, L., Zhao, J., Hu, Y., Zhang, L., Fan, G., Xu, J., Gu, X., Cheng, Z., Yu, T., Xia, J., Wei, Y., Wu, W., Xie, X., Yin, W., Li, H., Liu, M., & Cao, B. (2020). Clinical features of patients infected with 2019 novel coronavirus in Wuhan, China. *Lancet, 395*(10223), 497–506. DOI: 10.1016/S0140-6736(20)30183-5

Huang, L., Ma, F., Chapman, A., Lu, S., & Xie, X. S. (2015). Single-Cell Whole-Genome Amplification and Sequencing: Methodology and Applications. *Annual Review of Genomics and Human Genetics, 16*(1), 79–102. DOI: 10.1146/annurev-genom-090413-025352

Huang, Q., Li, Y., Huang, Y., Wu, J., Bao, W., Xue, C., Li, X., Dong, S., Dong, Z., & Hu, S. (2025). Advances in molecular pathology and therapy of non-small cell lung cancer. *Signal Transduction and Targeted Therapy, 10*(1), 186. DOI: 10.1038/s41392-025-02243-6

Huang, S., Ouyang, L., Tang, J., Qian, K., Chen, X., Zhang, X., Ming, J., & Xi, R. (2024). Spatial transcriptomics: A new frontier in cancer research. *Clinical Cancer Bulletin, 3*(1), 13. Advance online publication. DOI: 10.1007/s44272-024-00018-8

Hu, J., Schroeder, A., Coleman, K., Chen, C., Auerbach, B. J., & Li, M. (2021). Statistical and machine learning methods for spatially resolved transcriptomics with histology. *Computational and Structural Biotechnology Journal, 19*, 3829–3841. DOI: 10.1016/j.csbj.2021.06.052

Hu, K., Harman, A., & Baharlou, H. (2024). Imaging mass cytometry for in situ immune profiling. In *Flow Cytometry Protocols* (pp. 407–423). Springer US. DOI: 10.1007/978-1-0716-3738-8_19

Hunter, M. V., Moncada, R., Weiss, J. M., Yanai, I., & White, R. M. (2021). Spatially resolved transcriptomics reveals the architecture of the tumor-microenvironment interface. *Nature Communications*, *12*(1), 6278. DOI: 10.1038/s41467-021-26614-z PMID: 34725363

Hu, W., Zhang, Y., Mei, J., & Fang, X. (2023). Spatial transcriptomics in human biomedical research and clinical application. *Current Medicine (Cham, Switzerland)*, *2*(1), 6. Advance online publication. DOI: 10.1007/s44194-023-00023-4

Hwang, T. J., Carpenter, D., Lauffenburger, J. C., Wang, B., Franklin, J. M., & Kesselheim, A. S. (2016). Failure of Investigational Drugs in Late-Stage Clinical Development and Publication of Trial Results. *JAMA Internal Medicine*, *176*(12), 1826. DOI: 10.1001/jamainternmed.2016.6008 PMID: 27723879

Iafrate, A., Onozato, M., Elliot, H., & Yapp, C. (2020). Highly multiplexed fluorescence in situ hybridization (fish) platform for gene copy number evaluation.

Ilicic, T., Ilicic, T., Kim, J. K., Kolodziejczyk, A. A., Kolodziejczyk, A. A., Bagger, F. O., Bagger, F. O., Bagger, F. O., McCarthy, D. J., McCarthy, D. J., Marioni, J. C., Marioni, J. C., Marioni, J. C., Teichmann, S. A., Teichmann, S. A., & Teichmann, S. A. (2016). Classification of low quality cells from single-cell RNA-seq data. *Genome Biology*, *17*(1), 29. DOI: 10.1186/s13059-016-0888-1

Inocencio, I., Rai, A., Donner, D., & Greening, D. W. (2025). The proteomic landscape of the coronary accessible heart cell surfaceome. *Proteomics*, *25*(7), e202400320. DOI: 10.1002/pmic.202400320 PMID: 39790063

Isaic, A., Motofelea, N., Hoinoiu, T., Motofelea, A. C., Leancu, I. C., Stan, E., Gheorghe, S. R., Dutu, A. G., & Crintea, A. (2025). Next-generation sequencing: A review of its transformative impact on cancer diagnosis, treatment, and resistance management. *Diagnostics (Basel)*, *15*(19), 2425. DOI: 10.3390/diagnostics15192425 PMID: 41095644

Isaza-Ruget, M. A., Yomayusa, N., González, C. A., De Oro, V. F. A., Cely, A., Murcia, J., Gonzalez-Velez, A., Robayo, A., Colmenares Mejía, C. C., & Castillo, A.. (2024). Predicting chronic kidney disease progression with artificial intelligence. *BMC Nephrology*, *25*(1), 148. DOI: 10.1186/s12882-024-03545-7 PMID: 38671349

Isnard, P., Li, D., Xuanyuan, Q., Wu, H., & Humphreys, B. D. (2024). Histopathological-based analysis of human kidney spatial transcriptomics data. *American Journal of Pathology*. Advance online publication. DOI: 10.1016/j.ajpath.2023.12.001

Itoku, A., Isaac, J., Wilson, S., Reidy, K., & Kaskel, F. (2024). APOL1 nephropathy risk variants through the life course: A review. *American Journal of Kidney Diseases, 84*(1), 102–110. DOI: 10.1053/j.ajkd.2023.12.014 PMID: 38341125

Ivanisevic, T., & Sewduth, R. N. (2023). Multi-Omics Integration for the Design of Novel Therapies and the Identification of Novel Biomarkers. *Proteomes, 11*(4), 34. Advance online publication. DOI: 10.3390/proteomes11040034

Jackson, H. W., Fischer, J. R., Zanotelli, V. R. T., Ali, H. R., Mechera, R., Soysal, S. D., Moch, H., Muenst, S., Varga, Z., Weber, W. P., & Bodenmiller, B. (2020). The single-cell pathology landscape of breast cancer. *Nature, 578*(7796), 615–620. DOI: 10.1038/s41586-019-1876-x PMID: 31959985

Jacobs, R. T., Nare, B., Wring, S. A., Orr, M. D., Chen, D., Sligar, J. M., Jenks, M. X., Noe, R. A., Bowling, T. S., Mercer, L. T., Rewerts, C., Gaukel, E., Owens, J., Parham, R., Randolph, R., Beaudet, B., Bacchi, C. J., Yarlett, N., Plattner, J. J., & Don, R. (2011). SCYX-7158, an orally-active benzoxaborole for the treatment of stage 2 human African trypanosomiasis. *PLoS Neglected Tropical Diseases, 5*(6), e1151. DOI: 10.1371/journal.pntd.0001151

Jager, R. (2023). Special Issue on Whole Genome Amplification. *International Journal of Molecular Sciences, 24*(11), 9626. DOI: 10.3390/ijms24119626

Jain, S., & Eadon, M. T. (2024). Spatial transcriptomics in health and disease. *Nature Reviews. Nephrology, 20*(10), 659–671. DOI: 10.1038/s41581-024-00841-1 PMID: 38719971

Jamalinia, M., & Weiskirchen, R. (2025). Advances in personalized medicine: Translating genomic insights into targeted therapies for cancer treatment. *Annals of Translational Medicine, 13*(2), 18. DOI: 10.21037/atm-25-34 PMID: 40438512

Jean Beltran, P. M., Federspiel, J. D., Sheng, X., & Cristea, I. M. (2017). Proteomics and integrative omic approaches for understanding host–pathogen interactions and infectious diseases. *Molecular Systems Biology, 13*(3), 922. DOI: 10.15252/msb.20167062

Jiang, P., Gu, S., Pan, D., Fu, J., Sahu, A., Hu, X., Li, Z., Traugh, N., Bu, X., Li, B., Liu, J., Freeman, G. J., Brown, M. A., Wucherpfennig, K. W., & Liu, X. S. (2018). Signatures of T cell dysfunction and exclusion predict cancer immunotherapy response. *Nature Medicine, 24*(10), 1550–1558. DOI: 10.1038/s41591-018-0136-1 PMID: 30127393

Jiang, W., Ye, W., Tan, X., & Bao, Y.-J. (2025). Network-based multi-omics integrative analysis methods in drug discovery: A systematic review. *BioData Mining*, *18*(1), 27. DOI: 10.1186/s13040-025-00442-z PMID: 40155979

Jiang, X., Liu, B., Nie, Z., Duan, L., Xiong, Q., Jin, Z., Yang, C., & Chen, Y. (2021). The role of m6A modification in the biological functions and diseases. *Signal Transduction and Targeted Therapy*, *6*(1), 74. DOI: 10.1038/s41392-020-00450-x PMID: 33611339

Jimenez-Carvelo, A. M., & Cuadros-Rodríguez, L. (2021). Data mining/machine learning methods in foodomics. *Current Opinion in Food Science*, *37*, 76–82. DOI: 10.1016/j.cofs.2020.09.008

Jiménez-Gracia, L., Marchese, D., Nieto, J. C., Caratu, G., Melón-Ardanaz, E., Gudiño, V., Wise, K., Ryan, N. S., Hernando-Momblona, X., Bernardes, J. P., Tran, F., Sievers, L. K., Schreiber, S., van den Berge, M., Kole, T. M. T. M., Nawijn, M. C., Rosenstiel, P., Batlle, E., Parish, I. A., & Martelotto, L. G. (2023). FixNCut: Single-cell genomics through reversible tissue fixation and dissociation. *bioRxiv*. DOI: 10.1101/2023.06.16.545221

Jin, F., Li, J., Guo, J., Doeppner, T. R., Hermann, D. M., Yao, G., & Dai, Y. (2021). Targeting epigenetic modifiers to reprogramme macrophages in non-resolving inflammation-driven atherosclerosis. *European Heart Journal Open*, *1*(2), oeab022. DOI: 10.1093/ehjopen/oeab022 PMID: 35919269

Johnson-Laird, P. N. (1983). *Mental models: Towards a cognitive science of language, inference, and consciousness* (No. 6). Harvard University Press.

Jonjić, S. (2024). Tissue-specific antiviral immunity. *Cellular & Molecular Immunology*, *21*(9), 941–942. DOI: 10.1038/s41423-024-01200-7

Jordan, M. I., & Mitchell, T. M. (2015). Machine learning: Trends, perspectives, and prospects. *Science*, *349*(6245), 255–260. DOI: 10.1126/science.aaa8415

Jørgensen, J. T. (2021). Companion and complementary diagnostics: A key to precision medicine. In *Precision Medicine and Artificial Intelligence* (pp. 97–107). Elsevier., DOI: 10.1016/B978-0-12-820239-5.00003-6

Joshi, A., Rienks, M., Theofilatos, K., & Mayr, M. (2021). Systems biology in cardiovascular disease: A multiomics approach. *Nature Reviews. Cardiology*, *18*(5), 313–330. DOI: 10.1038/s41569-020-00477-1 PMID: 33340009

Jung, N., & Kim, T. K. (2023). Spatial transcriptomics in neuroscience. *Experimental & Molecular Medicine*, *55*(10), 2105–2115. DOI: 10.1038/s12276-023-01093-y

Jung, S. H., Hwang, B. H., Shin, S., Park, E. H., Park, S. H., Kim, C. W., & Chung, Y. J. (2022). Spatiotemporal dynamics of macrophage heterogeneity and a potential function of Trem2hi macrophages in infarcted hearts. *Nature Communications*, *13*(1), 4580. DOI: 10.1038/s41467-022-32284-2 PMID: 35933399

Jyothsna, V., Sandhya, E., Pydala, B., Swetha, N. N., & Sree, T. S. D. (2024). *Visual data analysis and inference through dimensionality reduction techniques*. 21–68. https://doi.org/DOI: 10.1201/9781003542735-2

Kafle, A., & Ojha, S. C. (2024). Advancing vaccine development against Opisthorchis viverrini: A synergistic integration of omics technologies and advanced computational tools. *Frontiers in Pharmacology*, *15*, 1410453. DOI: 10.3389/fphar.2024.1410453

Kalikar, S., Jain, C., Vasimuddin, M., & Misra, S. (2022). Accelerating minimap2 for long-read sequencing applications on modern CPUs. *Nature Computational Science*, *2*(2), 78–83. DOI: 10.1038/s43588-022-00201-8 PMID: 38177520

Kalim, S., & Rhee, E. P. (2017). An overview of renal metabolomics. *Kidney International*, *91*(1), 61–69. DOI: 10.1016/j.kint.2016.08.021 PMID: 27692817

Kalisky, T., & Quake, S. R. (2011). Single-cell genomics. *Nature Methods*, *8*(4), 311–314. DOI: 10.1038/nmeth0411-311

Kamat, S., Kumari, M., & Jayabaskaran, C. (2021). Nano-engineered tools in the diagnosis, therapeutics, prevention, and mitigation of SARS-CoV-2. *Journal of Controlled Release*, *338*, 813–836. DOI: 10.1016/j.jconrel.2021.08.046 PMID: 34478750

Kamsalem, A. S., Ubah, O. M., Olatokun, T. P., Oluwaniyi, R. O., Philips, O. A., Ojo-omoniyi, D. S., & Oluwasakinju, O. L. (2025). Unraveling tumor microenvironments with spatial genomics: Implications for precision breast cancer therapy. *International Journal of Life Science Research Archive*, *9*(1), 043–063. https://doi.org/DOI: 10.53771/ijlsra.2025.9.1.0045

Kanehisa, M., & Goto, S. (2000). KEGG: Kyoto encyclopedia of genes and genomes. *Nucleic Acids Research*, *28*(1), 27–30. DOI: 10.1093/nar/28.1.27

Kanemaru, K., Cranley, J., Muraro, D., Miranda, A. M., Ho, S. Y., Wilbrey-Clark, A., & Teichmann, S. A. (2023). Spatially resolved multiomics of human cardiac niches. *Nature*, *619*(7971), 801–810. DOI: 10.1038/s41586-023-06311-1 PMID: 37438528

Kang, J., Guo, X., Peng, H., Deng, Y., Lai, J., Tang, L., Aoieong, C., Tou, T., Tsai, T., & Liu, X. (2023). Metabolic implications of amino acid metabolites in chronic kidney disease progression: A metabolomics analysis using OPLS-DA and MBRole2.0 database. *International Urology and Nephrology*, *56*(6), 1173–1184. DOI: 10.1007/s11255-023-03779-8 PMID: 37728808

Kang, S. L., Woo, J. H., Kim, N. H., Kwon, J. Y., & Kim, S. M. (2023). Necessity of strengthening the current clinical regulatory framework for companion diagnostics: An institutional comparison of the FDA, EMA, and MFDS. *Molecular Therapy. Methods & Clinical Development*, *30*, 447–458. DOI: 10.1016/j.omtm.2023.08.008 PMID: 37663648

Kanneganti, T.-D. (2020). Intracellular innate immune receptors: Life inside the cell. *Immunological Reviews*, *297*(1), 5–12. DOI: 10.1111/imr.12912

Kantaros, A., Petrescu, F. I. T., & Ganetsos, T. (2025). From Stents to Smart Implants Employing Biomimetic Materials: The Impact of 4D Printing on Modern Healthcare. *Biomimetics (Basel, Switzerland)*, *10*(2), 2. Advance online publication. DOI: 10.3390/biomimetics10020125

Kaput, J. (2008). Nutrigenomics research for personalized nutrition and medicine. *Current Opinion in Biotechnology*, *19*(2), 110–120. DOI: 10.1016/j.copbio.2008.02.005

Kary, A. D., Noelle, H., & Magin, C. M. (2025). Tissue-Informed Biomaterial Innovations Advance Pulmonary Regenerative Engineering. *ACS Macro Letters*, *14*(4), 434–447. DOI: 10.1021/acsmacrolett.5c00075

Kasemeier-Kulesa, J. C., Morrison, J. A., McKinney, S., Li, H., Gogol, M., Hall, K., Chen, S., Wang, Y., Perera, A., McLennan, R., & Kulesa, P. M. (2023). Cell-type profiling of the sympathetic nervous system using spatial transcriptomics and spatial mapping of mRNA. *Developmental Dynamics*, *252*(8), 1130–1142. DOI: 10.1002/dvdy.577

Kashtan, C. E., Ding, J., Garosi, G., Heidet, L., Massella, L., Nakanishi, K., Nozu, K., Renieri, A., Rheault, M., Wang, F., & Gross, O. (2018). Alport syndrome: A unified classification of genetic disorders of collagen IV A345: A position paper of the Alport Syndrome Classification Working Group. *Kidney International*, *93*(5), 1045–1051. DOI: 10.1016/j.kint.2017.12.018 PMID: 29551517

Kashyap, B. K., Singh, V. V., Solanki, M. K., Kumar, A., Ruokolainen, J., & Kesari, K. K. (2023). Smart Nanomaterials in Cancer Theranostics: Challenges and Opportunities. *ACS Omega*, *8*(16), 14290–14320. DOI: 10.1021/acsomega.2c07840

Kaur, E., Lopos, L. C., & Bilichak, A. (2025). Mitigating Heat Stress Response in CRISPR/Cas-Mediated Edited Crops by Altering the Expression Pattern of Non-coding DNA. *Genome and Epigenome Editing for Stress-Tolerant Crops*, 1-31.

Kaur, J., Thakran, A., & Naqvi, S. (2025). Recent advances in cell membrane-based biomimetic delivery systems for Parkinson's disease: Perspectives and challenges. *Asian Journal of Pharmaceutical Sciences*, *20*(4), 101060. DOI: 10.1016/j.ajps.2025.101060

Kaur, P., Singh, A., & Chana, I. (2021). Computational Techniques and Tools for Omics Data Analysis: State-of-the-Art, Challenges, and Future Directions. *Archives of Computational Methods in Engineering*, *28*(7), 1–37. DOI: 10.1007/s11831-021-09547-0

Kaur, R. P., Ludhiadch, A., & Munshi, A. (2019). *Single-Cell Genomics: Technology and Applications*. Academic Press., DOI: 10.1016/B978-0-12-814919-5.00009-9

Keilberg, D., & Ottemann, K. M. (2016). How H elicobacter pylori senses, targets and interacts with the gastric epithelium. *Environmental Microbiology*, *18*(3), 791–806. DOI: 10.1111/1462-2920.13222

Kennedy-Darling, J., Nolan, G. P., Goltsev, Y., & Samusik, N. (2016). *Abstract A089: Multiparametric immunofluorescence analysis of the tumor microenvironment using CODEX. 4*. https://doi.org/DOI: 10.1158/2326-6066.IMM2016-A089

Keren, L., Bosse, M., Thompson, S., Risom, T., Vijayaragavan, K., McCaffrey, E., Marquez, D., Angoshtari, R., Greenwald, N. F., Fienberg, H., Wang, J., Kambham, N., Kirkwood, D., Nolan, G., Montine, T. J., Galli, S. J., West, R., Bendall, S. C., & Angelo, M. (2019). MIBI-TOF: A multiplexed imaging platform relates cellular phenotypes and tissue structure. *Science Advances*, *5*(10), eaax5851. Advance online publication. DOI: 10.1126/sciadv.aax5851 PMID: 31633026

Keutzer, L., Wicha, S. G., & Simonsson, U. S. H. (2020). Mobile health apps for improvement of tuberculosis treatment: Descriptive review. *JMIR mHealth and uHealth*, *8*(4), e17246. DOI: 10.2196/17246

Khairnar, M., & Marda, S. (2024). Integrative Multi-Omics in Precision Medicine: From Molecular Interconnectivity to Single-Cell Resolution. *International Journal of Scientific Research & Technology*, *2*(7).

Khan, A., Barapatre, A. R., Babar, N., Doshi, J., Ghaly, M., Patel, K. G., Nawaz, S., Hasana, U., Khatri, S. P., Pathange, S., Pesaru, A. R., Puvvada, C. S., Billoo, M., & Jamil, U. (2025). Genomic medicine and personalized treatment: A narrative review. *Annals of Medicine and Surgery (London)*, *87*(3), 1406–1414. DOI: 10.1097/MS9.0000000000002965 PMID: 40213198

Khan, N., Jameel, N., & Khan, M. (2019). Functional genomics: Linking genotype with phenotype on a genome-wide scale. *International Journal of Applied Pharmaceutical Sciences and Research*, *4*(1), 4–12. DOI: 10.21477/ijapsr.4.1.2

Khare, K., & Pandey, R. (2022). Cellular heterogeneity in disease severity and clinical outcome: Granular understanding of immune response is key. *Frontiers in Immunology, Volume 13-2022*. DOI: 10.3389/fimmu.2022.973070

Khatri, I., & Bhasin, M. K. (2020). A transcriptomics-based meta-analysis combined with machine learning identifies a secretory biomarker panel for diagnosis of pancreatic adenocarcinoma. *Frontiers in Genetics*, *11*, 572284. DOI: 10.3389/fgene.2020.572284 PMID: 33133160

Khobragade, S. S., Deshmukh, M., Vyas, U., & Ingle, R. G. (2025). Innovative Approaches in Bone Tissue Engineering: Strategies for Cancer Treatment and Recovery. *International Journal of Molecular Sciences*, *26*(9), 3937. DOI: 10.3390/ijms26093937

Khodadadian, A., Darzi, S., Haghi-Daredeh, S., sadat Eshaghi, F., Babakhanzadeh, E., Mirabutalebi, S. H., & Nazari, M. (2020). Genomics and transcriptomics: Powerful technologies in precision medicine. *International Journal of General Medicine*, *13*, 627–640. DOI: 10.2147/IJGM.S249970 PMID: 32982380

Khoury, M. J., Iademarco, M. F., & Riley, W. T. (2015). Precision public health for the era of precision medicine. *American Journal of Preventive Medicine*, *50*(3), 398–401. DOI: 10.1016/j.amepre.2015.08.031

Kiessling, P., & Kuppe, C. (2024). Spatial multi-omics: Novel tools to study the complexity of cardiovascular diseases. *Genome Medicine*, *16*(1), 14. DOI: 10.1186/s13073-024-01282-y PMID: 38238823

Kim, H. K., Pham, M. H. C., Ko, K. S., Rhee, B. D., & Han, J. (2018). Alternative splicing isoforms in health and disease. *Pflügers Archiv*, *470*(7), 995–1016. DOI: 10.1007/s00424-018-2136-x PMID: 29536164

Kim, H., Choi, S., Heo, H., Cho, S. H., Lee, Y., Kim, D., & Rhee, S. (2025). Applications of single-cell omics technologies for induced pluripotent stem cell-based cardiovascular research. *International Journal of Stem Cells*, *18*(1), 37–48. DOI: 10.15283/ijsc23183 PMID: 39129179

Kim, H., Kim, H.-H., Chang, C. L., Song, S. H., & Kim, N. (2021). Novel PKD1 mutations in patients with autosomal dominant polycystic kidney disease. *Laboratory Medicine*, *52*(2), 174–180. DOI: 10.1093/labmed/lmaa047 PMID: 32816041

Kim, J., Ha, S., Son, M., Kim, D., Kim, M.-J., Kim, B., Kim, D., Chung, H. Y., & Chung, K. W. (2023). TLR7 activation by miR-21 promotes renal fibrosis by activating the pro-inflammatory signaling pathway in tubule epithelial cells. *Cell Communication and Signaling : CCS*, *21*(1), 215. DOI: 10.1186/s12964-023-01234-w PMID: 37596656

Kim, Y.-S., Choi, J., & Lee, S. H. (2023). Single-cell and spatial sequencing application in pathology. *Journal of Pathology and Translational Medicine*, *57*(1), 43–51. DOI: 10.4132/jptm.2022.12.12

Kingma, D. P., & Ba, J. (2014). Adam: A method for stochastic optimization. *arXiv preprint arXiv:1412.6980*.

Kirchhof, P., Sipido, K. R., Cowie, M. R., Eschenhagen, T., Fox, K. A., Katus, H., Schroeder, S., Schunkert, H., Priori, S., Alonso, A., Chezaubernard, C., Doevendans, P., Eschenhagen, T., Fox, K., Katus, H., Khder, Y., Kirchhof, P., Kramer, F., Kristensen, S., & Herttuala, S. Y. (2014). The continuum of personalized cardiovascular medicine: A position paper of the european society of cardiology. *European Heart Journal*, *35*(46), 3250–3257. DOI: 10.1093/eurheartj/ehu312 PMID: 25148837

Knepper, M. A., & Pisitkun, T. (2007). Exosomes in urine: Who would have thought...? *Kidney International*, *72*(9), 1043–1045. DOI: 10.1038/sj.ki.5002510 PMID: 17943150

Kocaturk, N. M., & Gozuacik, D. (2018). Crosstalk between mammalian autophagy and the ubiquitin-proteasome system. *Frontiers in Cell and Developmental Biology*, *6*, 128. DOI: 10.3389/fcell.2018.00128 PMID: 30333975

Koelwyn, G. J., Corr, E. M., Erbay, E., & Moore, K. J. (2018). Regulation of macrophage immunometabolism in atherosclerosis. *Nature Immunology*, *19*(6), 526–537. DOI: 10.1038/s41590-018-0113-3 PMID: 29777212

Konuma, T., & Okada, Y. (2021). Statistical genetics and polygenic risk score for precision medicine. *Inflammation and Regeneration*, *41*(1), 18. DOI: 10.1186/s41232-021-00172-9 PMID: 34140035

Korir, P. K., Geeleher, P., & Seoighe, C. (2015). Seq-ing improved gene expression estimates from microarrays using machine learning. *BMC Bioinformatics*, *16*(1), 1–11. DOI: 10.1186/s12859-015-0712-z PMID: 26338512

Kosaraju, C., Chiruguri, S., Mohammad, N., Jetty, B. T., & Gali, G. (2022). A model for analysis of diseases based on nutrition deficiency using random forest. In *Proceedings of the 7th International Conference on Communication and Electronics Systems* (pp. 1527–1531). IEEE. DOI: 10.1109/ICCES54183.2022.9835805

Kostas, J. C., Brainard, C. S., & Cristea, I. M. (2025). A Primer on Proteomic Characterization of Intercellular Communication in a Virus Microenvironment. *Molecular & Cellular Proteomics*, *24*(3), 100913. DOI: 10.1016/j.mcpro.2025.100913 PMID: 39862905

Köttgen, A., Cornec-Le Gall, E., Halbritter, J., Kiryluk, K., Mallett, A. J., Parekh, R. S., Rasouly, H. M., Sampson, M. G., Tin, A., Antignac, C., Ars, E., Bergmann, C., Bleyer, A. J., Bockenhauer, D., Devuyst, O., Florez, J. C., Fowler, K. J., Franceschini, N., Fukagawa, M., & Gharavi, A. G. (2022). Genetics in chronic kidney disease: Conclusions from a Kidney Disease: Improving Global Outcomes (KDIGO) controversies conference. *Kidney International*, *101*(6), 1126–1141. DOI: 10.1016/j.kint.2022.03.019 PMID: 35460632

Krassowski, M., Das, V., Sahu, S. K., & Misra, B. B. (2020). State of the field in multi-omics research: From computational needs to data mining and sharing. *Frontiers in Genetics*, *11*, 610798. DOI: 10.3389/fgene.2020.610798 PMID: 33362867

Kukurba, K. R., & Montgomery, S. B. (2015). RNA sequencing and analysis. *Cold Spring Harbor Protocols*, *2015*(11), 951–969. DOI: 10.1101/pdb.top084970 PMID: 25870306

Kulasinghe, A., Berrell, N., Donovan, M. L., & Nilges, B. S. (2025). *Spatial-Omics Methods and Applications*. https://doi.org/DOI: 10.1007/978-1-0716-4276-4_5

Kulkarni, A., Anderson, A. G., Merullo, D. P., & Konopka, G. (2019). Beyond bulk: A review of single cell transcriptomics methodologies and applications. *Current Opinion in Biotechnology*, *58*, 129–136. https://doi.org/https://doi.org/10.1016/j.copbio.2019.03.001. DOI: 10.1016/j.copbio.2019.03.001

Kulkarni, S., Arumugam, T., Chuturgoon, A., An, P., & Ramsuran, V. (2022). Epigenetics of infectious diseases. [). Frontiers Media SA.]. *Frontiers in Immunology*, *13*, 1054151. DOI: 10.3389/fimmu.2022.1054151

Kuppe, C., Ramirez Flores, R. O., Li, Z., Hayat, S., Levinson, R. T., Liao, X., & Kramann, R. (2022). Spatial multi-omic map of human myocardial infarction. *Nature*, *608*(7924), 766–777. DOI: 10.1038/s41586-022-05060-x PMID: 35948637

Kuswanto, W., Nolan, G., & Lu, G. (2023). Highly multiplexed spatial profiling with CODEX: Bioinformatic analysis and application in human disease. [). Berlin/Heidelberg: Springer Berlin Heidelberg.]. *Seminars in Immunopathology*, *45*(1), 145–157. DOI: 10.1007/s00281-022-00974-0 PMID: 36414691

Kuta Alam, T. P. J. (2022). *Visualizing the Interactions Shaping the Imaging of the Microenvironment in Human Cancers* (pp. 67–79). *Methods in Molecular Biology (Clifton, N.J.)*. Advance online publication. DOI: 10.1007/978-1-0716-2703-7_5

Lac, L., Leung, C. K., & Hu, P. (2024). Computational frameworks integrating deep learning and statistical models in mining multimodal omics data. *Journal of Biomedical Informatics, 152*, 104629. DOI: 10.1016/j.jbi.2024.104629

LaFave, L. M., Savage, R. E., & Buenrostro, J. D. (2022). Single-Cell Epigenomics Reveals Mechanisms of Cancer Progression. *Annual Review of Cancer Biology, 6*(1), 167–185. DOI: 10.1146/annurev-cancerbio-070620-094453

Lai, H. M., & Chan, S. L. (2024). A scalable 3D spatial multi-omics method for high-plex morpho-molecular imaging of millimeter-thick tissues. *Journal of Clinical Oncology, 42*(23), 162–162. DOI: 10.1200/JCO.2024.42.23_suppl.162

Laird, J., Perera, G., Batorsky, R., Wang, H., Arkun, K., & Chin, M. T. (2023). Spatial transcriptomic analysis of focal and normal areas of myocyte disarray in human hypertrophic cardiomyopathy. *International Journal of Molecular Sciences, 24*(16), 12625. DOI: 10.3390/ijms241612625 PMID: 37628806

Lai, Y., Xue, J., Liu, C. W., Gao, B., Chi, L., Tu, P., Lu, K., & Ru, H. (2019). Serum metabolomics identifies altered bioenergetics, signaling cascades in parallel with exposome markers in Crohn's disease. *Molecules (Basel, Switzerland), 24*(3), 449. DOI: 10.3390/molecules24030449

Lakshmaiah, A., Korrapati, C., Suresh, C., & Raghavan, R. (2025). Synthetic Biology and AI. *Spatially Variable Genes in Cancer*, 265–290. DOI: 10.4018/979-8-3693-7728-4.ch010

Lamichhane, J. R., Osdaghi, E., Behlau, F., Köhl, J., Jones, J. B., & Aubertot, J. N. (2018). Thirteen decades of antimicrobial copper compounds applied in agriculture. A review. *Agronomy for Sustainable Development, 38*(3), 28. DOI: 10.1007/s13593-018-0503-9

Lammers, W. J., Schalij, M. J., Kirchhof, C. J., & Allessie, M. A. (1990). Quantification of spatial inhomogeneity in conduction and initiation of reentrant atrial arrhythmias. *American Journal of Physiology. Heart and Circulatory Physiology, 259*(4), H1254–H1263. DOI: 10.1152/ajpheart.1990.259.4.H1254 PMID: 1699438

Langbein, T., Weber, W. A., & Eiber, M. (2019). Future of theranostics: An outlook on precision oncology in nuclear medicine. *Journal of Nuclear Medicine, 60*(Supplement 2), 13S–19S. DOI: 10.2967/jnumed.118.220566

Langlieb, J., Sachdev, N. S., Balderrama, K. S., Nadaf, N. M., Raj, M., Murray, E., Webber, J. T., Vanderburg, C., Gazestani, V., Tward, D., Mezias, C., Li, X., Flowers, K., Cable, D. M., Norton, T., Mitra, P., Chen, F., & Macosko, E. Z. (2023). The molecular cytoarchitecture of the adult mouse brain. *Nature*, *624*(7991), 333–342. DOI: 10.1038/s41586-023-06818-7

Lareau, C. A., Duarte, F. M., Chew, J. G., Kartha, V. K., Burkett, Z. D., Kohlway, A. S., Pokholok, D., Aryee, M. J., Steemers, F. J., Lebofsky, R., & Buenrostro, J. D. (2019). Droplet- based combinatorial indexing for massive- scale single- cell chromatin accessibility. *Nature Biotechnology*, *37*(8), 916–924. DOI: 10.1038/s41587-019-0147-6

Lázár, E., Mauron, R., Andrusivová, Ž., Foyer, J., He, M., Larsson, L., & Lundeberg, J. (2024). Spatial dynamics of the developing human heart. bioRxiv, 2024-03.

Lazarova, M., Labaj, J., Kovacikova, Z., & Slamenova, D. (2004). Diet containing fungal (1 3)—D-glucan derivative exhibits protective effects against DNA lesions induced in freshly isolated rat cells. *Neoplasma*, *51*, 431–435.

LeCun, Y., Bengio, Y., & Hinton, G. (2015). Deep learning. *nature, 521*(7553), 436-444.

LeCun, Y., Bengio, Y., & Hinton, G. (2015). Deep learning. *nature, 521*(7553), 436-444. DOI: 10.1038/nature14539

Lee, D., Jung, Y., Lee, J. E., Lim, M., Lee, D., Na, K. J., & Choi, H. (2025). Abstract 6276: Comparative analysis and integration of image-based and high-resolution sequence-based spatial transcriptomics platforms for tumor microenvironment analysis. *Cancer Research*, *85*(8, Supplement_1), 6276. DOI: 10.1158/1538-7445.AM2025-6276

Lee, J. A., Carragher, N. O., & Berg, E. L. (2017). Empirical drug discovery: A view from the proteome. *Drug Discovery Today. Technologies*, *23*, 1–5. DOI: 10.1016/j.ddtec.2017.05.003

Lee, J. H. (2025). From single-cell maps to diagnostics: Enabling biomarker discovery in precision medicine. *Academia Molecular Biology and Genomics.*, *2*(3). Advance online publication. DOI: 10.20935/AcadMolBioGen7859

Lee, J. S., & Yoon, S. (2017). Integrative analysis of multi-omics data for cancer research: Methods and applications. *BMC Bioinformatics*, *18*(Suppl 2), 83.

Lee, J., Yoo, M., & Choi, J. (2022). Recent advances in spatially resolved transcriptomics: Challenges and opportunities. *BMB Reports*, *55*(3), 113–124. DOI: 10.5483/BMBRep.2022.55.3.014

Lee, R. Y., Ng, C. W., Rajapakse, M., Ang, N., Yeong, J. P. S., & Lau, M. C. (2023). The promise and challenge of spatial omics in dissecting tumour microenvironment and the role of AI. *Frontiers in Oncology*, *13*, 1172314. Advance online publication. DOI: 10.3389/fonc.2023.1172314

Lee, T. H., Liu, P. S., Wang, S. J., Tsai, M. M., Shanmugam, V., & Hsieh, H. L. (2021). Bradykinin, as a reprogramming factor, induces transdifferentiation of brain astrocytes into neuron-like cells. *Biomedicines*, *9*(8), 923. DOI: 10.3390/biomedicines9080923 PMID: 34440126

Lee, Y., Lee, M., Shin, Y., Kim, K., & Kim, T. (2025a). Spatial Omics in Clinical Research: A Comprehensive Review of Technologies and Guidelines for Applications. *International Journal of Molecular Sciences*, *26*(9), 3949. DOI: 10.3390/ijms26093949 PMID: 40362187

Le, J., Dian, Y., Zhao, D., Guo, Z., Luo, Z., Chen, X., Zeng, F., & Deng, G. (2025). Single-cell multi-omics in cancer immunotherapy: From tumor heterogeneity to personalized precision treatment. *Molecular Cancer*, *24*(1), 221. Advance online publication. DOI: 10.1186/s12943-025-02426-3

Le-Khac, P. H., Healy, G., & Smeaton, A. F. (2020). Contrastive representation learning: A framework and review. *IEEE Access : Practical Innovations, Open Solutions*, *8*, 193907–193934. DOI: 10.1109/ACCESS.2020.3031549

Leopold, J. A. (2022). Personalizing treatments for patients based on cardiovascular phenotyping. *Expert Review of Precision Medicine and Drug Development*, *7*(1), 4–16. DOI: 10.1080/23808993.2022.2028548 PMID: 36778892

Lerch, J. P. (2024). *eLife Assessment: Three-dimensional single-cell transcriptome imaging of thick tissues*. https://doi.org/DOI: 10.7554/elife.90029.3.sa0

Le, T. T., Fu, W., & Moore, J. H. (2020). Scaling tree-based automated machine learning to biomedical big data with a feature set selector. *Bioinformatics (Oxford, England)*, *36*(1), 250–256. DOI: 10.1093/bioinformatics/btz470 PMID: 31165141

Levy, B., Burnside, R. D., & Akkari, Y. (2025). Optical genome mapping: A new tool for cytogenomic analysis. *Genes*, *16*(8), 924. DOI: 10.3390/genes16080924 PMID: 40869972

Lewis, S. M., Asselin-Labat, M.-L., Nguyen, Q., Berthelet, J., Tan, X., Wimmer, V. C., Merino, D., Rogers, K. L., & Naik, S. H. (2021). Spatial omics and multiplexed imaging to explore cancer biology. *Nature Methods*, *18*(9), 997–1012. DOI: 10.1038/s41592-021-01203-6 PMID: 34341583

Liang, A., Kong, Y., Chen, Z., Qiu, Y., Wu, Y., Zhu, X., & Li, Z. (2024). Advancements and applications of single-cell multi-omics techniques in cancer research: Unveiling heterogeneity and paving the way for precision therapeutics. *Biochemistry and Biophysics Reports*, *37*, 101589. DOI: 10.1016/j.bbrep.2023.101589 PMID: 38074997

Liao, H., Lyon, C. J., Ying, B., & Hu, T. (2024). Climate change, its impact on emerging infectious diseases and new technologies to combat the challenge. *Emerging Microbes & Infections*, *13*(1), 2356143. DOI: 10.1080/22221751.2024.2356143

Liau, B., Zhang, D., & Bursac, N. (2012). Functional cardiac tissue engineering. *Regenerative Medicine*, *7*(2), 187–206. DOI: 10.2217/rme.11.122 PMID: 22397609

Libbrecht, M. W., & Noble, W. S. (2015). Machine learning applications in genetics and genomics. *Nature Reviews. Genetics*, *16*(6), 321–332. DOI: 10.1038/nrg3920 PMID: 25948244

Li, F., Tang, Z., Zheng, Y., Jiang, T., Deng, L., Dai, W., Zhao, Y., Zheng, N., Liu, S., Fan, Y., Lu, S., Chen, Y., Liu, G., Zhang, Y., & Xiong, Y. (2026). Responsive nanomedicine strategies achieve pancreatic cancer precise theranostics. *Bioactive Materials*, *55*, 334–375. DOI: 10.1016/j.bioactmat.2025.08.012

Li, H. (2018). Minimap2: Pairwise alignment for nucleotide sequences. *Bioinformatics (Oxford, England)*, *34*(18), 3094–3100. DOI: 10.1093/bioinformatics/bty191 PMID: 29750242

Li, H., Calder, C. A., & Cressie, N. (2007). Beyond moran's i: Testing for spatial dependence based on the spatial autoregressive model. *Geographical Analysis*, *39*(4), 357–375. DOI: 10.1111/j.1538-4632.2007.00708.x

Li, K. W., Ganz, A. B., & Smit, A. B. (2019). Proteomics of neurodegenerative diseases: Analysis of human post-mortem brain. *Journal of Neurochemistry*, *151*(4), 435–445. DOI: 10.1111/jnc.14603 PMID: 30289976

Lillis, L., Siverson, J., Lee, A., Cantera, J., Parker, M., Piepenburg, O., Lehman, D. A., & Boyle, D. S. (2016). Factors influencing recombinase polymerase amplification (RPA) assay outcomes at point of care. *Molecular and Cellular Probes*, *30*(2), 74–78. DOI: 10.1016/j.mcp.2016.01.009

Li, M., Guo, M., Peng, Y., Jiang, Q., Xia, L., Zhong, S., Qiu, Y., Su, X., Zhang, S., Yang, C., Mijiti, P., Mao, Q., Takiff, H., Li, F., Chen, C., & Gao, Q. (2022). High proportion of tuberculosis transmission among social contacts in rural China: A 12-year prospective population-based genomic epidemiological study. *Emerging Microbes & Infections*, *11*(1), 2102–2111. DOI: 10.1080/22221751.2022.2112912

Lindström, N. O., De Sena Brandine, G., Ransick, A., & McMahon, A. P. (2019). Single-cell RNA sequencing of the adult mouse kidney: From molecular cataloging of cell types to disease-associated predictions. *American Journal of Kidney Diseases, 73*(1), 140–142. DOI: 10.1053/j.ajkd.2018.07.002 PMID: 30241960

Ling, X. C., & Kuo, K.-L. (2018). Oxidative stress in chronic kidney disease. *Renal Replacement Therapy, 4*(1), 53. DOI: 10.1186/s41100-018-0195-2

Lin, W. N., Tay, M. Z., Lu, R., Liu, Y., Chen, C.-H., & Cheow, L. F. (2020). The Role of Single-Cell Technology in the Study and Control of Infectious Diseases. *Cells, 9*(6), 1440. DOI: 10.3390/cells9061440

Lipworth, S., Hammond, R. J. H., Baron, V. O., Hu, Y., Coates, A., & Gillespie, S. H. (2016). Defining dormancy in mycobacterial disease. *Tuberculosis (Edinburgh, Scotland), 99*, 131–142. https://doi.org/https://doi.org/10.1016/j.tube.2016.05.006. DOI: 10.1016/j.tube.2016.05.006

Li, T., Gu, Z., & Zhou, G. (2025). Next-Generation Barcoding for Single-Cell Omics. *Analytical Chemistry, 97*(31), 16708–16713. Advance online publication. DOI: 10.1021/acs.analchem.5c03587

Litviňuková, M., Talavera-López, C., Maatz, H., Reichart, D., Worth, C. L., Lindberg, E. L., & Teichmann, S. A. (2020). Cells of the adult human heart. *Nature, 588*(7838), 466–472. DOI: 10.1038/s41586-020-2797-4 PMID: 32971526

Liu, B., Wu, F.-X., & Zou, X. (2020). scASK: A novel ensemble framework for classifying cell types based on single-cell RNA-seq data. *bioRxiv*. https://doi.org/ DOI: 10.1101/2020.06.07.138271

Liu, F., Chen, J., Li, Z., & Meng, X. (2022). Recent advances in epigenetics of age-related kidney diseases. *Genes, 13*(5), 796. DOI: 10.3390/genes13050796 PMID: 35627181

Liu, H., Begik, O., Lucas, M. C., Ramirez, J. M., Mason, C. E., Wiener, D., Schwartz, S., Mattick, J. S., Smith, M. A., & Novoa, E. M. (2019). Accurate detection of m6A RNA modifications in native RNA sequences. *Nature Communications, 10*(1), 4079. DOI: 10.1038/s41467-019-11713-9 PMID: 31501426

Liu, L., Chen, A., Li, Y., Mulder, J., Heyn, H., & Xu, X. (2024). Spatiotemporal omics for biology and medicine. *Cell, 187*(17), 4488–4519. DOI: 10.1016/j.cell.2024.07.040

Liu, Q., Martínez-Jarquín, S., & Zenobi, R. (2022). Recent Advances in Single-Cell Metabolomics Based on Mass Spectrometry. *CCS Chemistry, 5*(2), 310–324. DOI: 10.31635/ccschem.022.202202333

Liu, T., Li, X., Pan, X., Zhang, L., Liao, J., & Yin, M. (2023a). STGMVA: clustering, imputation, and integration for spatial resolved transcriptomics using spatio-temporal gaussian mixture variational autoencoder. DOI: 10.21203/rs.3.pex-2322/v1

Liu, X., (2024). Spatial metabolomics reveals energy-metabolic dysregulation in neurodegeneration. *Translational Neurodegeneration*, *13*(7), e00397. DOI: 10.1186/s40035-024-00397-x PMID: 38254235

Liu, X., Peng, T., Xu, M., Lin, S., Hu, B., Chu, T., Liu, B., Xu, Y., Ding, W., Li, L., Cao, C., & Wu, P. (2024a). Spatial multi-omics: Deciphering the technological landscape of integration of multi-omics and its applications. *Journal of Hematology & Oncology*, *17*(1), 72. DOI: 10.1186/s13045-024-01596-9 PMID: 39182134

Liu, X., Shi, J., Jiao, Y., An, J., Tian, J., Yang, Y., & Zhuo, L. (2024). Integrated multi-omics with machine learning to uncover the intricacies of kidney disease. *Briefings in Bioinformatics*, *25*(5), bbae364. Advance online publication. DOI: 10.1093/bib/bbae364 PMID: 39082652

Liu, X., Wang, L., Wang, Y., Qiao, X., Chen, N., Liu, F., & Shen, H. (2024b). Myocardial infarction complexity: A multi-omics approach. *Clinica Chimica Acta*, *552*, 117680. DOI: 10.1016/j.cca.2023.117680 PMID: 38008153

Liu, X., Yin, K., Chen, L., Chen, W., Li, W., Zhang, T., & Zhou, Z. (2023b). Lineage-specific regulatory changes in hypertrophic cardiomyopathy unraveled by single-nucleus RNA-seq and spatial transcriptomics. *Cell Discovery*, *9*(1), 6. DOI: 10.1038/s41421-022-00490-3 PMID: 36646705

Liu, Y., Li, L., & Zhang, X. (2020). Slide-seq2: Scalable spatially resolved transcriptomics at near single-cell resolution. *Nature Methods*, *17*(7), 755–758. DOI: 10.1038/s41592-020-0909-8

Liu, Y., Zhang, X., Yang, S., Zhou, Z., Tian, L., Li, W., & Wang, Z. (2023c). Integrated mass spectrometry imaging reveals spatial-metabolic alteration in diabetic cardiomyopathy and the intervention effects of ferulic acid. *Journal of Pharmaceutical Analysis*, *13*(12), 1496–1509. DOI: 10.1016/j.jpha.2023.08.011 PMID: 38223449

Liu, Z., Zhang, S., James, B. T., Galani, K., Mangan, R. J., Fass, S. B., Liang, C., Wagle, M. M., Boix, C. A., Tanigawa, Y., Yun, S., Sung, Y., Xiong, X., Sun, N., Hou, L., Wohlwend, M., Qiu, M., Han, X., Xiong, L., & Kellis, M. (2025). Single-cell multiregion epigenomic rewiring in Alzheimer's disease progression and cognitive resilience. *Cell*, *188*(18), 4980–5002.e29. Advance online publication. DOI: 10.1016/j.cell.2025.06.031

Li, W., Chen, Z., Wu, A., Qin, F. X.-F., & Zhang, L. (2020). *Systems immunology meets epigenetics*. Academic Press., DOI: 10.1016/B978-0-12-817964-2.00010-1

Li, X., & Hou, J. (2018). A richer understanding of intratumoral heterogeneity: Single-cell genomics put it within reach. *Journal of Thoracic Disease*, *10*(3), 1178–1182. DOI: 10.21037/jtd.2018.03.17

Li, Y., & Luo, Y. (2023). Spatial transcriptomic cell-type deconvolution using graph neural networks. *bioRxiv*.

Liyanage, S. H., & Guo, J. (2024). *Single-cell spatial proteomics*. 18861. https://doi.org/DOI: 10.14670/hh-18-861

Llorens-Bobadilla, E., Zamboni, M., Marklund, M., Bhalla, N., Chen, X., Hartman, J., & Ståhl, P. L. (2022). Chromatin accessibility profiling in tissue sections by spatial ATAC. bioRxiv, 2022-07.

Lloyd-Price, J., Abu-Ali, G., & Huttenhower, C. (2016). The healthy human microbiome. *Genome Medicine*, *8*(1), 51. DOI: 10.1186/s13073-016-0307-y

Lo, P.-K., & Zhou, Q. (2018). Emerging techniques in single-cell epigenomics and their applications to cancer research. 1(1), 1–8. https://doi.org/DOI: 10.4172/JCG.1000103

Logsdon, G. A., Vollger, M. R., & Eichler, E. E. (2020). Long-read human genome sequencing and its applications. *Nature Reviews. Genetics*, *21*(10), 597–614. DOI: 10.1038/s41576-020-0236-x PMID: 32504078

Lomakin, A., Svedlund, J., Strell, C., Gataric, M., Shmatko, A., Rukhovich, G. D., Park, J. S., Ju, Y. S., Dentro, S. C., Kleshchevnikov, V., Vaskivskyi, V., Li, T., Bayraktar, O. A., Pinder, S. L., Richardson, A. L., Santagata, S., Campbell, P. J., Russnes, H. G., Gerstung, M., & Yates, L. R. (2022). Spatial genomics maps the structure, nature and evolution of cancer clones. *Nature*, *611*(7936), 594–602. DOI: 10.1038/s41586-022-05425-2

Lombaert, H., Peyrat, J. M., Croisille, P., Rapacchi, S., Fanton, L., Cheriet, F., & Ayache, N. (2012). Human atlas of the cardiac fiber architecture: Study on a healthy population. *IEEE Transactions on Medical Imaging*, *31*(7), 1436–1447. DOI: 10.1109/TMI.2012.2192743 PMID: 22481815

Long, Y., Ang, K. S., Liao, S., Sethi, R., Heng, Y., Zhong, C., & Chen, J. (2023b). Integrated analysis of spatial multi-omics with SpatialGlue. bioRxiv, 2023-04.

Long, X., Yuan, X., & Du, J. (2023a). Single-cell and spatial transcriptomics: Advances in heart development and disease applications. *Computational and Structural Biotechnology Journal, 21*, 2717–2731. DOI: 10.1016/j.csbj.2023.04.007 PMID: 37181659

Long, Y., Ang, K. S., Li, M., Chong, K. L. K., Sethi, R., Zhong, C., Xu, H., Ong, Z., Sachaphibulkij, K., Chen, A., Zeng, L., Fu, H., Wu, M., Lim, L. H. K., Liu, L., & Chen, J. (2023). Spatially informed clustering, integration, and deconvolution of spatial transcriptomics with graphst. *Nature Communications, 14*(1), 1155. DOI: 10.1038/s41467-023-36796-3 PMID: 36859400

Loomba, R., Seguritan, V., Li, W., Long, T., Klitgord, N., Bhatt, A., Dulai, P. S., Caussy, C., Bettencourt, R., Highlander, S. K., Jones, M. B., Sirlin, C. B., Schnabl, B., Brinkac, L., Schork, N., Chen, C.-H., Brenner, D. A., Biggs, W., Yooseph, S., & Nelson, K. E. (2017). Gut microbiome-based metagenomic signature for non-invasive detection of advanced fibrosis in human nonalcoholic fatty liver disease. *Cell Metabolism, 25*(5), 1054–1062. DOI: 10.1016/j.cmet.2017.04.001

Loscalzo, J. (2024). Multi-Omics and Single-Cell Omics: New Tools in Drug Target Discovery. *Arteriosclerosis, Thrombosis, and Vascular Biology, 44*(4), 759–762. DOI: 10.1161/ATVBAHA.124.320686

Lowe, R., Shirley, N., Bleackley, M., Dolan, S., & Shafee, T. (2017). Transcriptomics technologies. *PLoS Computational Biology, 13*(5), e1005457. DOI: 10.1371/journal.pcbi.1005457 PMID: 28545146

Lucas, M. C., & Novoa, E. M. (2023). Long-read sequencing in the era of epigenomics and epitranscriptomics. *nature methods, 20*(1), 25-29. DOI: 10.1038/s41592-022-01724-8

Lu, H.-C., Dai, W.-N., & He, L.-Y. (2021). Epigenetic histone modifications in the pathogenesis of diabetic kidney disease. *Diabetes, Metabolic Syndrome and Obesity, 14*, 329–344. DOI: 10.2147/DMSO.S288500 PMID: 33519221

Lu, N., Qiao, Y., Lu, Z., & Tu, J. (2023). Chimera: The spoiler in multiple displacement amplification. *Computational and Structural Biotechnology Journal, 21*, 1688–1696. DOI: 10.1016/j.csbj.2023.02.034

Luo, Y., Zhao, C., & Chen, F. (2024). Multiomics research: Principles and challenges in integrated analysis. *BioDesign Research, 6*, 0059.

Luo, L., & Leeper, N. J. (2023). Spatial metabolomics and the vulnerable atherosclerotic plaque. *Arteriosclerosis, Thrombosis, and Vascular Biology, 43*(9), 1636–1638. DOI: 10.1161/ATVBAHA.123.319739 PMID: 37534466

Luo, T., Rosasco, M. G., Hoeman, C. M., Khare, S., Gorski, D. J., Rand, T., Klinghoffer, R. A., Guinney, J., & Ho, C.-S. (2025). Abstract 2085: Visium HD combined with deep-learning-based cell segmentation on H&E images yield accurate cell annotation at single-cell resolution. *Cancer Research, 85*(8, Supplement_1), 2085. DOI: 10.1158/1538-7445.AM2025-2085

Lurie, N., Saville, M., Hatchett, R., & Halton, J. (2020). Developing Covid-19 vaccines at pandemic speed. *The New England Journal of Medicine, 382*(21), 1969–1973. DOI: 10.1056/NEJMp2005630

Lu, S., Hua, J., Xu, J., Wei, M., Liang, C., Meng, Q., Liu, J., Zhang, B., Wang, W., Yu, X., & Shi, S. (2022). Turning towards nonimmunoreactive tumors: Evaluation of cancer-associated fibroblasts enables prediction of the immune microenvironment and treatment sensitivity in pancreatic cancer. *Computational and Structural Biotechnology Journal, 20*, 3911–3923. DOI: 10.1016/j.csbj.2022.07.029 PMID: 35950187

Lu, Y.-Q., & Wang, Y. (2024). Multi-omic analysis reveals genetic determinants and therapeutic targets of chronic kidney disease and kidney function. *International Journal of Molecular Sciences, 25*(11), 6033. DOI: 10.3390/ijms25116033 PMID: 38892221

Lv, H., Sun, F., & Chen, Y. (2025). Decoding congenital heart disease: A multi-omic framework for cardiac lineage and regulatory dysfunction. *Frontiers in Cell and Developmental Biology, 13*, 1659884. DOI: 10.3389/fcell.2025.1659884 PMID: 41049289

MacDonald, M. L., & Westwick, J. K. (2006). Exploiting Network Biology to Improve Drug Discovery. In G. A. Kenneth, H. R. Jeffrey, & T. D. Lansing, *High Content Screening* (Vol. 356, pp. 221–232). Humana Press. DOI: 10.1385/1-59745-217-3:221

Maes, F., Collignon, A., Vandermeulen, D., Marchal, G., & Suetens, P. (2002). Multimodality image registration by maximization of mutual information. *IEEE Transactions on Medical Imaging, 16*(2), 187–198. DOI: 10.1109/42.563664 PMID: 9101328

Magnano, C. S., & Gitter, A. (2021). Automating parameter selection to avoid implausible biological pathway models. *NPJ Systems Biology and Applications, 7*(1), 12. DOI: 10.1038/s41540-020-00167-1 PMID: 33623016

Maintz, J. A., & Viergever, M. A. (1998). A survey of medical image registration. *Medical Image Analysis, 2*(1), 1–36. DOI: 10.1016/S1361-8415(01)80026-8 PMID: 10638851

Maiorino, E., & Loscalzo, J. (2023). Phenomics and robust multiomics data for cardiovascular disease subtyping. *Arteriosclerosis, Thrombosis, and Vascular Biology, 43*(7), 1111–1123. DOI: 10.1161/ATVBAHA.122.318892 PMID: 37226730

Mallikarjun, V., Yin, B., Caggiano, L. R., Blimbaum, S., Pavelec, C. M., Holmes, J. W., & Ewald, S. E. (2024). Automated spatially targeted optical micro proteomics identifies fibroblast-and macrophage-specific regulation of myocardial infarction scar maturation in rats. *Journal of Molecular and Cellular Cardiology, 186*, 1–15. DOI: 10.1016/j.yjmcc.2023.10.005 PMID: 37951204

Malone, A. F. (2021). APOL1 risk variants in kidney transplantation: A modulation of immune cell function. *The Journal of Clinical Investigation, 131*(9), e154676. DOI: 10.1172/JCI154676 PMID: 34779415

Manabe, O., Oyama-Manabe, N., & Tamaki, N. (2020). Positron emission tomography/ MRI for cardiac diseases assessment. *The British Journal of Radiology, 93*(1113), 20190836. DOI: 10.1259/bjr.20190836 PMID: 32023123

Mani, S., Lalani, S. R., & Pammi, M. (2025). Genomics and multiomics in the age of precision medicine. *Pediatric Research, 97*(4), 1–12. DOI: 10.1038/s41390-025-04021-0 PMID: 40185865

Manzoni, C., Kia, D. A., Vandrovcova, J., Hardy, J., Wood, N. W., Lewis, P. A., Lewis, P. A., & Ferrari, R. (2018). Genome, transcriptome and proteome: The rise of omics data and their integration in biomedical sciences. *Briefings in Bioinformatics, 19*(2), 286–302. DOI: 10.1093/bib/bbw114

Marco, A., Berkenkamp, F., Hennig, P., Schoellig, A. P., Krause, A., Schaal, S., & Trimpe, S. (2017, May). Virtual vs. real: Trading off simulations and physical experiments in reinforcement learning with Bayesian optimization. In *2017 IEEE International Conference on Robotics and Automation (ICRA)* (pp. 1557-1563). IEEE. DOI: DOI: 10.1109/ICRA.2017.7989186

Marconato, L., Palla, G., Yamauchi, K. A., Virshup, I. H., Heidari, E., Treis, T., Vöhringer, H., Huber, W., Gerstung, M., Moore, J., Theis, F. J., & Stegle, O. (2023). SpatialData: an open and universal data framework for spatial omics. *bioRxiv*. https://doi.org/DOI: 10.1101/2023.05.05.539647

Marques, L., Costa, B., Pereira, M., Silva, A., Santos, J., Saldanha, L., Silva, I., Magalhães, P., Schmidt, S., & Vale, N. (2024). Advancing precision medicine: A review of innovative in silico approaches for drug development, clinical pharmacology, and personalized healthcare. *Pharmaceutics, 16*(3), 332. DOI: 10.3390/pharmaceutics16030332 PMID: 38543226

Marshall, B., & Warren, J. R. (1984). UNIDENTIFIED CURVED BACILLI IN THE STOMACH OF PATIENTS WITH GASTRITIS AND PEPTIC ULCERATION. *Lancet, 323*(8390), 1311–1315. https://doi.org/https://doi.org/10.1016/S0140-6736(84)91816-6. DOI: 10.1016/S0140-6736(84)91816-6

Marshall, C., Farrow, M. A., Djambazova, K. V., & Spraggins, J. M. (2023). Untangling Alzheimer's disease with spatial multi-omics: A brief review. *Frontiers in Aging Neuroscience*, *15*, 1150512. Advance online publication. DOI: 10.3389/fnagi.2023.1150512

Marshall, J. L., Noel, T., Wang, Q. S., Chen, H., Murray, E., Subramanian, A., Vernon, K. A., Bazua-Valenti, S., Liguori, K., Keller, K., Stickels, R. R., McBean, B., Heneghan, R. M., Weins, A., Macosko, E. Z., Chen, F., & Greka, A. (2022). High-resolution Slide-seqV2 spatial transcriptomics enables the discovery of disease-specific cell neighborhoods and pathways. *iScience*, *25*(4), 104097. DOI: 10.1016/j.isci.2022.104097 PMID: 35372810

Martinelli, A. L., & Rapsomaniki, M. A. (2022). ATHENA: Analysis of tumor heterogeneity from spatial omics measurements. *Bioinformatics (Oxford, England)*, *38*(11), 3151–3153. DOI: 10.1093/bioinformatics/btac303 PMID: 35485743

Martinelli, G., & Foreman, N. (2015). Advancing precision medicine through multi-omics: An integrated approach to tumor profiling. *Science*, *349*(6253), 1246. DOI: 10.1126/science.349.6253.1246-c

Martire, S., & Banaszynski, L. A. (2020). The roles of histone variants in fine-tuning chromatin organization and function. *Nature Reviews. Molecular Cell Biology*, *21*(9), 522–541. DOI: 10.1038/s41580-020-0262-8 PMID: 32665685

Marumo, T., Yamasaki, Y., Ueda, S., & Kawachi, H. (2020). Epigenetic regulation of RASAL1 promotes kidney fibrosis. *Nature Communications*, *11*(1), 4469. DOI: 10.1038/s41467-020-18241-9 PMID: 32901013

Marx, V. (2021). Method of the Year: Spatially resolved transcriptomics. *Nature Methods*, *18*(1), 9–14. DOI: 10.1038/s41592-020-01033-y PMID: 33408395

Ma, S.-X., & Lim, S. B. (2021). Single-Cell RNA Sequencing in Parkinson's Disease. *Biomedicines*, *9*(4), 368. DOI: 10.3390/biomedicines9040368

Ma, S., Zhang, B., LaFave, L. M., & Fan, R. (2023). Spatial co-profiling of epigenome and transcriptome reveals developmental regulatory programs. *Nature Biotechnology*, *41*(2), 250–261.

Mason, C. E., Porter, S. G., & Smith, T. M. (2013). Characterizing multi-omic data in systems biology. *Systems Analysis of Human Multigene Disorders*, 15–38.

Mathys, H., Davila-Velderrain, J., Peng, Z., Gao, F., Mohammadi, S., Young, J. Z., Menon, M., He, L., Abdurrob, F., Jiang, X., Martorell, A. J., Ransohoff, R. M., Hafler, B. P., Bennett, D. A., Kellis, M., & Tsai, L.-H. (2019). Single-cell transcriptomic analysis of Alzheimer's disease. *Nature*, *570*(7761), 332–337. DOI: 10.1038/s41586-019-1195-2 PMID: 31042697

Matsubayashi, J., & Takano, T. (2025). Proximity labeling uncovers the synaptic proteome under physiological and pathological conditions. *Frontiers in Cellular Neuroscience*, *19*, 1638627. Advance online publication. DOI: 10.3389/fncel.2025.1638627

Ma, Y., Shi, W., Dong, Y., Sun, Y., & Jin, Q.-G. (2024). Spatial Multi-Omics in Alzheimer's Disease: A Multi-Dimensional Approach to Understanding Pathology and Progression. *Current Issues in Molecular Biology*, *46*(5), 4968–4990. DOI: 10.3390/cimb46050298

McAloon, C. J., Osman, F., Glennon, P., Lim, P. B., & Hayat, S. A. (2016). Global epidemiology and incidence of cardiovascular disease. In *Cardiovascular Diseases* (pp. 57–96). Academic Press. DOI: 10.1016/B978-0-12-803312-8.00004-5

McBride, D. A., Jones, R. M., Bottini, N., & Shah, N. J. (2024). The therapeutic potential of immunoengineering for systemic autoimmunity. *Nature Reviews. Rheumatology*, *20*(4), 203–215. DOI: 10.1038/s41584-024-01084-x

McClellan, J., & King, M.-C. (2010). Genetic heterogeneity in human disease. *Cell*, *141*(2), 210–217. DOI: 10.1016/j.cell.2010.03.032

McClymont, H., Lambert, S. B., Barr, I., Vardoulakis, S., Bambrick, H., & Hu, W. (2024). Internet-based surveillance systems and infectious diseases prediction: An updated review of the last 10 years and lessons from the COVID-19 pandemic. *Journal of Epidemiology and Global Health*, *14*(3), 645–657. DOI: 10.1007/s44197-024-00272-y

McCoy, D. C., Cuartas, J., Behrman, J., Cappa, C., Heymann, J., López Bóo, F., Lu, C., Raikes, A., Richter, L., Stein, A., & Fink, G. (2021). Global estimates of the implications of COVID-19-related preprimary school closures for children's instructional access, development, learning, and economic wellbeing. *Child Development*, *92*(5), e883–e899. DOI: 10.1111/cdev.13658

Mei, Y., Xiao, W., Hu, H., Lu, G., Chen, L., Sun, Z., Lü, M., Ma, W., Jiang, T., Gao, Y., Li, L., Chen, G., Wang, Z., Li, H., Wu, D., Zhou, P., Leng, Q., & Jia, G. (2021). Single-cell analyses reveal suppressive tumor microenvironment of human colorectal cancer. *Clinical and Translational Medicine*, *11*(6), e422. DOI: 10.1002/ctm2.422 PMID: 34185431

Melms, J. C., Biermann, J., Huang, H., Wang, Y., Nair, A., Tagore, S., Katsyv, I., Rendeiro, A. F., Amin, A. D., Schapiro, D., Frangieh, C. J., Luoma, A. M., Filliol, A., Fang, Y., Ravichandran, H., Clausi, M. G., Alba, G. A., Rogava, M., Chen, S. W., & Izar, B. (2021). A molecular single-cell lung atlas of lethal COVID-19. *Nature*, *595*(7865), 114–119. DOI: 10.1038/s41586-021-03569-1

Meng, Z., Dong, J., Guo, L., Su, F., Liu, J., Wang, G., & Zhao, Z. (2026). Now and future of artificial intelligence-based signet ring cell diagnosis: A survey. *Expert Systems with Applications*, *296*, 129188. DOI: 10.1016/j.eswa.2025.129188

Mensah, G. A., Moran, A. E., Roth, G. A., & Narula, J. (2014). The global burden of cardiovascular diseases, 1990–2010. *Global Heart*, *9*(1), 183. DOI: 10.1016/j.gheart.2014.01.008 PMID: 25432128

Merritt, C. R., Ong, G. T., Church, S. E., Barker, K., Danaher, P., Geiss, G., Hoang, M., Jung, J., Liang, Y., McKay-Fleisch, J., Nguyen, K., Norgaard, Z., Sorg, K., Sprague, I., Warren, C., Warren, S., Webster, P. J., Zhou, Z., Zollinger, D. R., & Beechem, J. M. (2020). Multiplex digital spatial profiling of proteins and RNA in fixed tissue. *Nature Biotechnology*, *38*(8), 1000–1008. DOI: 10.1038/s41587-020-0472-9 PMID: 32393914

Meyer-Schwesinger, C. (2019). The ubiquitin–proteasome system in kidney physiology and disease. *Nature Reviews. Nephrology*, *15*(7), 393–411. DOI: 10.1038/s41581-019-0148-1 PMID: 31036905

Meyyazhagan, A., Balasubramanian, B., Easwaran, M., Alagamuthu, K. K., Shanmugam, S., Kuchi Bhotla, H., Pappusamy, M., Arumugam, V. A., Thangaraj, A., Kaul, T., Keshavarao, S., & Cacabelos, R. (2020). Biomarker study of the biological parameter and neurotransmitter levels in autistics. *Molecular and Cellular Biochemistry*, *474*(1-2), 277–284. DOI: 10.1007/s11010-020-03851-2 PMID: 32740790

Meyyazhagan, A., Pushparaj, K., Balasubramanian, B., Kuchi Bhotla, H., Pappusamy, M., Arumugam, V. A., Easwaran, M., Pottail, L., Mani, P., Tsibizova, V., & Di Renzo, G. C. (2022). COVID-19 in pregnant women and children: Insights on clinical manifestations, complexities, and pathogenesis. *International Journal of Gynaecology and Obstetrics: the Official Organ of the International Federation of Gynaecology and Obstetrics*, *156*(2), 216–224. DOI: 10.1002/ijgo.14007 PMID: 34735717

Michot, J., Bigenwald, C., Champiat, S., Collins, M., Carbonnel, F., Postel-Vinay, S., Berdelou, A., Varga, A., Bahleda, R., Hollebecque, A., Massard, C., Fuerea, A., Ribrag, V., Gazzah, A., Armand, J. P., Amellal, N., Angevin, E., Noel, N., Boutros, C., & Lambotte, O. (2016). Immune-related adverse events with immune checkpoint blockade: A comprehensive review. *European Journal of Cancer (Oxford, England)*, *54*, 139–148. DOI: 10.1016/j.ejca.2015.11.016 PMID: 26765102

Mihalić, A., Železnjak, J., Lisnić, B., Jonjić, S., Juranić Lisnić, V., & Brizić, I. (2024). Immune surveillance of cytomegalovirus in tissues. *Cellular & Molecular Immunology*, *21*(9), 959–981. DOI: 10.1038/s41423-024-01186-2

Milošević, V. (2023). Different approaches to Imaging Mass Cytometry data analysis. *Bioinformatics Advances*, *3*(1), vbad046. Advance online publication. DOI: 10.1093/bioadv/vbad046

Min, S., Lee, B., & Yoon, S. (2017). Deep learning in bioinformatics. *Briefings in Bioinformatics*, *18*(5), 851–869. DOI: 10.1093/bib/bbw068 PMID: 27473064

Misra, B. B., Langefeld, C., Olivier, M., & Cox, L. A. (2019). Integrated omics: Tools, advances and future approaches. *Journal of Molecular Endocrinology*, *62*(1), R21–R45. DOI: 10.1530/JME-18-0055

Mitre, L., Burkard, A., Hansen, S., Li, X., Ehret, F., Heitlinger, S., & Westermann, D. (2023). Changes in the immune cell landscape in the infarcted mouse heart. *European Heart Journal*, *44*(Supplement_2), ehad655–3212. DOI: 10.1093/eurheartj/ehad655.3212

Miyamoto, A. T., Shimagami, H., Kumanogoh, A., & Nishide, M. (2025). Spatial transcriptomics in autoimmune rheumatic disease: Potential clinical applications and perspectives. *Inflammation and Regeneration*, *45*(1), 6. DOI: 10.1186/s41232-025-00369-2 PMID: 39980019

Moffat, J. G., Vincent, F., Lee, J. A., Eder, J., & Prunotto, M. (2017). Opportunities and challenges in phenotypic drug discovery: An industry perspective. *Nature Reviews. Drug Discovery*, *16*(8), 531–543. DOI: 10.1038/nrd.2017.111

Moffitt, J. R., Hao, J., Wang, G., Chen, K. H., Babcock, H. P., & Zhuang, X. (2016). High-throughput single-cell gene-expression profiling with multiplexed error-robust fluorescence in situ hybridization. *Proceedings of the National Academy of Sciences of the United States of America*, *113*(39), 11046–11051. DOI: 10.1073/pnas.1612826113

Mohd, Y., Balasubramanian, B., Meyyazhagan, A., Kuchi Bhotla, H., Shanmugam, S.K., Ramesh Kumar, M.K., Pappusamy, M., Alagamuthu, K.K., Keshavarao, S., & Arumugam, V.A. (2021). Extricating the association between the prognostic factors of colorectal cancer. *J Gastrointest Cancer, 52*(3), 022-1028.

Mohd, Y., Kumar, P., Kuchi Bhotla, H., Meyyazhagan, A., Balasubramanian, B., Ramesh Kumar, M. K., Pappusamy, M., Alagamuthu, K. K., Orlacchio, A., Keshavarao, S., Sampathkumar, P., & Arumugam, V. A. (2021). Transmission jeopardy of adenomatosis polyposis coli and methylenetetrahydrofolate reductase in colorectal cancer. *Journal of the Renin-Angiotensin-Aldosterone System : JRAAS, 2021*, 2021. DOI: 10.1155/2021/7010706 PMID: 34956401

Mohenska, M., Tan, N. M., Tokolyi, A., Furtado, M. B., Costa, M. W., Perry, A. J., & Ramialison, M. (2022). 3D-cardiomics: A spatial transcriptional atlas of the mammalian heart. *Journal of Molecular and Cellular Cardiology, 163*, 20–32. DOI: 10.1016/j.yjmcc.2021.09.011 PMID: 34624332

Moiz, B., Li, A., Padmanabhan, S., Sriram, G., & Clyne, A. M. (2022). Isotope-assisted metabolic flux analysis: A powerful technique to gain new insights into the human metabolome in health and disease. *Metabolites, 12*(11), 1066. DOI: 10.3390/metabo12111066 PMID: 36355149

Montesinos López, O. A., Montesinos López, A., & Crossa, J. (2022). *Multivariate statistical machine learning methods for genomic prediction*. Springer Nature., DOI: 10.1007/978-3-030-89010-0

Moon, K. R., Van Dijk, D., Wang, Z., Gigante, S., Burkhardt, D. B., Chen, W. S., Yim, K., Elzen, A. V. D., Hirn, M. J., Coifman, R. R., Ivanova, N. B., Wolf, G., & Krishnaswamy, S. (2019). Visualizing structure and transitions in high-dimensional biological data. *Nature Biotechnology, 37*(12), 1482–1492. DOI: 10.1038/s41587-019-0336-3 PMID: 31796933

Morevati, M., Fang, E. F., Mace, M. L., Kanbay, M., Gravesen, E., Nordholm, A., Egstrand, S., & Hornum, M. (2022). Roles of NAD+ in acute and chronic kidney diseases. *International Journal of Molecular Sciences, 24*(1), 137. DOI: 10.3390/ijms24010137 PMID: 36613582

Morse, D. B., Michalowski, A. M., Ceribelli, M., De Jonghe, J., Vias, M., Riley, D., Davies-Hill, T., Voss, T. C., Pittaluga, S., Muus, C., Liu, J., Weitz, D. A., Brenton, J. D., Buenrostro, J. D., Knowles, T. P. J., & Thomas, C. J. (2023). Positional influence on cellular transcriptional identity revealed through spatially segmented single-cell transcriptomics. Cell Systems, 14 6(6), 464-481.e7. https://doi.org/DOI: 10.1016/j.cels.2023.05.003

Moss, S. F. (2017). The Clinical Evidence Linking Helicobacter pylori to Gastric Cancer. *Cellular and Molecular Gastroenterology and Hepatology*, *3*(2), 183–191. https://doi.org/https://doi.org/10.1016/j.jcmgh.2016.12.001. DOI: 10.1016/j.jcmgh.2016.12.001

Mozaffarian, D., Rosenberg, I., & Uauy, R. (2018). History of modern nutrition science—Implications for current research, dietary guidelines, and food policy. *BMJ (Clinical Research Ed.)*, *361*, k2392. DOI: 10.1136/bmj.k2392

Murphy, K. P. (2022). *Probabilistic machine learning: an introduction*. MIT press.

Mutch, D. M., Wahli, W., & Williamson, G. (2005). Nutrigenomics and nutrigenetics: The emerging faces of nutrition. *The FASEB Journal*, *19*(12), 1602–1616. DOI: 10.1096/fj.05-3911rev

Myers, R. (2022). scNMT-seq v2 v1. https://doi.org/DOI: 10.17504/protocols.io.q26g7yrz3gwz/v1

Myers, R. J., Tretter, Z. M., Daffron, A. G., Fritschi, E., Santos, W. T., Foster, M., Klotz, M. G., Stafford, K. M., Kasch, C., Taylor, T. J., Tellefson, L. C., Hartman, T. P. V., Hackler, D., Specque, F., Barre, A., Nikolski, M., & Chalopin, D. (2025). SeuratIntegrate: An R package to facilitate the use of integration methods with Seurat. *Bioinformatics (Oxford, England)*, *41*(6), btaf358. Advance online publication. DOI: 10.1093/bioinformatics/btaf358

Nadaf, N., & Rayan, N. (2025). *Slide-Tags using Tissue-Tape v1*. https://doi.org/ DOI: 10.17504/protocols.io.36wgqdpdkvk5/v1

Nagashri, N., Archana, L., & Raghavan, R. (2024). Microbial Volatile Compounds (MVOCs) in Food Industries and Food Safety Applications. *Innovations, Securities, and Case Studies Across Healthcare, Business, and Technology*, 75–96. DOI: 10.4018/979-8-3693-1906-2.ch005

Nair, D. G., & Weiskirchen, R. (2024). Advanced In Vitro Models for Preclinical Drug Safety: Recent Progress and Prospects. *Current Issues in Molecular Biology*, *47*(1), 7. DOI: 10.3390/cimb47010007 PMID: 39852122

Nanjundeshwara, N., Lakshmaiah, A., & Raghavan, R. (2025). Aflatoxigenic Fungi. In A. Sankara Narayanan, S. Gomathi, V. Ambikapathy, & A. Panneerselvam, *Aflatoxigenic Fungi* (1st ed., pp. 225–239). CRC Press. DOI: 10.1201/9781003487043-29

Nassar, S. F., Raddassi, K., & Wu, T. (2021). Single-Cell Multiomics Analysis for Drug Discovery. *Metabolites*, *11*(11), 729. DOI: 10.3390/metabo11110729 PMID: 34822387

Nekolla, S. G., Rischpler, C., & Kunze, K. P. (2015). PET/MRI for cardiac imaging: Technical considerations and potential applications. In *Molecular and Multimodality Imaging in Cardiovascular Disease* (pp. 29–48). Springer International Publishing. DOI: 10.1007/978-3-319-19611-4_3

Neumann, E. K. (2025). Spatial Multiomics Toward Understanding Neurological Systems. *Journal of Mass Spectrometry : JMS*, *60*(6), e5143. DOI: 10.1002/jms.5143 PMID: 40360168

Nguyen, N., Souza, T., Verheijen, M. C., Gmuender, H., Selevsek, N., Schlapbach, R., & Jennen, D. (2021). Translational proteomics analysis of anthracycline-induced cardiotoxicity from cardiac microtissues to human heart biopsies. *Frontiers in Genetics*, *12*, 695625. DOI: 10.3389/fgene.2021.695625 PMID: 34211507

Nicholson, R. J., Pezzolesi, M. G., & Summers, S. A. (2021). Rotten to the cortex: Ceramide-mediated lipotoxicity in diabetic kidney disease. *Frontiers in Endocrinology*, *11*, 622692. DOI: 10.3389/fendo.2020.622692 PMID: 33584550

Niehaus, M., Soltwisch, J., Belov, M. E., & Dreisewerd, K. (2019). Transmission-mode MALDI-2 mass spectrometry imaging of cells and tissues at subcellular resolution. *Nature Methods*, *16*(9), 925–931. DOI: 10.1038/s41592-019-0536-2 PMID: 31451764

Ning, Z., & Xu, G. (2022). Can spatially resolved metabolomics uncover weak points in tumors? *Expert Review of Proteomics*, *19*(4–6), 227–230. DOI: 10.1080/14789450.2023.2176754

Nirmal, A. J., Maliga, Z., Vallius, T., Quattrochi, B., Chen, A. A., Jacobson, C. A., Pelletier, R. J., Yapp, C., Arias-Camison, R., Chen, Y.-A., Lian, C. G., Murphy, G. F., Santagata, S., & Sorger, P. K. (2022). The Spatial Landscape of Progression and Immunoediting in Primary Melanoma at Single-Cell Resolution. *Cancer Discovery*, *12*(6), 1518–1541. DOI: 10.1158/2159-8290.CD-21-1357 PMID: 35404441

Nkuipou-Kenfack, E., Zürbig, P., & Mischak, H. (2014). The use of urine proteomic classifiers in diagnosis and prediction of kidney disease. *Nature Reviews. Nephrology*, *10*(3), 153–162. DOI: 10.1038/nrneph.2013.282

Nomura, A., Takeji, Y., Shimojima, M., & Takamura, M. (2025). Digitalomics: Towards artificial intelligence/machine learning-based precision cardiovascular medicine. *Circulation Journal*, •••, CJ-24. DOI: 10.1253/circj.CJ-24-0865 PMID: 39894532

Nunes, F., Behrens, L. M. P., Weimer, R. D., Gonçalves, G. F., Fernandes, G., & Dorn, M. (2025). Deep learning methods and applications in single-cell multimodal data integration. *Molecular Omics*, *21*(6), 545–565. Advance online publication. DOI: 10.1039/D5MO00062A

Nunez, M. D., Fernandez, K., Srinivasan, R., & Vandekerckhove, J. (2024). A tutorial on fitting joint models of M/EEG and behavior to understand cognition. *Behavior Research Methods*, *56*(6), 1–31. DOI: 10.3758/s13428-023-02331-x

Nurminen, V., Ravindran, A., Valkonen, M., Örd, T., Mikkola, L., Lönnberg, T., & Kaikkonen, M. (2023). Disease-associated cell states in atherosclerosis defined by spatial transcriptomics. *Atherosclerosis*, *379*, S15. DOI: 10.1016/j.atherosclerosis.2023.06.725

O'Hara, A. M., & Shanahan, F. (2006). The gut flora as a forgotten organ. *EMBO Reports*, *7*(7), 688–693. DOI: 10.1038/sj.embor.7400731

Obande, G. A., & Banga Singh, K. K. (2020). Current and future perspectives on isothermal nucleic acid amplification technologies for diagnosing infections. *Infection and Drug Resistance*, *13*, 455–483. DOI: 10.2147/IDR.S217571

Obermeyer, Z., Powers, B., Vogeli, C., & Mullainathan, S. (2019). Dissecting racial bias in an algorithm used to manage the health of populations. *Science*, *366*(6464), 447–453. DOI: 10.1126/science.aax2342

Oe, Y. (2024). Editorial for the special issue: Pathophysiology of chronic kidney disease and its complications. *Biomedicines*, *12*(416), 416. Advance online publication. DOI: 10.3390/biomedicines12020416 PMID: 38398018

Ogidi, O. I. (2024). Recent advances in anticancer activity and Bioinformatics Approach from Potential Plants. In: Computational Approaches in Biomaterials, Bioinformatics and Biomedical Engineering Applications. Published *CRC Press, Taylor and Francis Group, Florida, USA*. DOI: 10.1201/9781003354437-3

Ogidi, O. I., & Ajoko, I. (2024). Neuroprotective Potentials of Phytochemicals. In: M.S. Rajput, T. Sarachana, P.D. Sarkar, M. Raval (eds) NeuroPhytomedicine. Published by *CRC Press, Routledge Taylor and Francis Group.*

Ogidi, O. I., & Emaikwu, N. G. (2024). Plant Phenolic compound isolation and its Bioinformatics Approaches of Molecular Mechanisms in Antimicrobial activities and Resistance. In: Computational Approaches in Biomaterials, Bioinformatics and Biomedical Engineering Applications. Publishered *CRC Press, Taylor and Francis Group, Florida, USA*. DOI: 10.1201/9781003354437-8

Ogidi, O. I., Chukwunonso, O. A., Afakwu-Adimoha, H. C., & Barwant, M. M. (2025b). Ethical Considerations in Generative Artificial Intelligence in Healthcare. In: Generative Artificial Intelligence in Healthcare: Current Practice and Future Development. *CRC Press, Taylor and Francis Group, Florida, USA.* https://doi.org/ DOI: 10.1201/9781003488255-5

Ogidi, O. I., Orlu, H. A., & Poripo, B. E. (2025h). Evaluation of Hepatoprotective Activities of *Bryophyllum pinnatum* Leaf Extract in Paracetamol Induced Toxicity In Wistar Rats. *Trends in Pharmaceutical Sciences, 11*(1), -. DOI: 10.30476/tips.2025.105055.1272

Ogidi, O. I. (2025). Natural Products Modulating Mitophagy: Implications for Cancer Therapy. In Pandurangan, A. (Ed.), *Roles of Mitophagy in Cancer Regulation* (pp. 325–358). IGI Global Scientific Publishing., DOI: 10.4018/979-8-3693-3836-0.ch012

Ogidi, O. I., Alfred-Ugbenbo, D., & Ebitimitula, O. (2025f). Drug Discovery and Development with Generative Artificial Intelligence. In Gupta, S., Gupta, U., Sharma, M., & Malik, K. (Eds.), *Generative Intelligence in Healthcare: Transforming Patient Care with AI Creativity* (1st ed.). CRC Press., DOI: 10.1201/9781003539483-1

Ogidi, O. I., Alfred-Ugbenbo, D., & Poripo, B. E. (2025d). AI-Driven Drug Discovery, Design, and Development in Immunological Disorders. In Chen, J. (Ed.), *AI-Assisted Computational Approaches for Immunological Disorders* (pp. 143–180). IGI Global Scientific Publishing., DOI: 10.4018/979-8-3693-9725-1.ch005

Ogidi, O. I., Aye, T., & Felagha, I. (2025c). Antiviral Therapeutics for Emerging Viruses. In Aslam, M., Yingngam, B., Javed, T., & Riaz, S. (Eds.), *Global Perspectives on Antiviral Drug Development* (pp. 217–260). IGI Global Scientific Publishing., DOI: 10.4018/979-8-3693-9276-8.ch006

Ogidi, O. I., Barwant, M. M., & Okara, E. (2024a). Biomarker Discovery: Protein Phosphorylation State. In *Spatially Variable Genes in Cancer: Development, Progression, and Treatment Response*. IGI Global., DOI: 10.4018/979-8-3693-7728-4.ch005

Ogidi, O. I., & Bunu, E. M. (2025). Blockchain Technology and Genomic Research. In Chaudhary, A., Singh, R., & Agarwal, G. (Eds.), *Blockchain in Health.* John Wiley & Sons, Inc., DOI: 10.1002/9781394336043.ch19

Ogidi, O. I., Christopher, E., & Orlu, H. A. (2025a). Environmental DNA and RNA Metabarcoding: Advancing Insights in Cancer Research. In Raghavan, R. (Ed.), *Cancer Exposomics and Environmental Influences on Carcinogenesis* (pp. 279–322). IGI Global Scientific Publishing., DOI: 10.4018/979-8-3373-2165-3.ch007

Ogidi, O. I., Erho, S. A., & Omu, O. (2025e). Reactive Oxygen Species in Cancer. In Prabhakar, P. (Ed.), *The Role of Reactive Oxygen Species in Human Health and Disease* (pp. 339–376). IGI Global Scientific Publishing., DOI: 10.4018/979-8-3693-7919-6.ch010

Ogidi, O. I., Ijere, N. D., & Sogoi, P. B. (2025g). Neural and Endocrine Regulation of Immunity. In Reddy, K. (Ed.), *Research Methodologies and Practical Applications in Psychoneuroimmunology* (pp. 17–42). IGI Global Scientific Publishing., DOI: 10.4018/979-8-3693-7432-0.ch002

Ogidi, O. I., Ogregade, I., & Joshua, S. (2024b). Investigation of Renal Protective Effects of Bryophyllum pinnatum (Lam) Extract in Acetaminophen Induced Toxicity in Albino Rats. *Medical and Pharmaceutical Journal*, *3*(2), 82–93. DOI: 10.55940/medphar202485

Ogidi, O. I., & Tawariowei, A. M. (2023). Endocrine Disrupting Chemicals and their Role in Cancer-A review. *Molecular Sciences And Applications.*, *3*, 5–14. DOI: 10.37394/232023.2023.3.2

Ohtsuki, S. (2012). Pharmacoproteomic approach by quantitative targeted proteomics. *Yakugaku Zasshi*, *132*(4), 479–487. DOI: 10.1248/yakushi.132.479 PMID: 22465926

Oliveira Chaves, L., Gomes Domingos, A. L., Louzada Fernandes, D., Ribeiro Cerqueira, F., Siqueira-Batista, R., & Bressan, J. (2023). Applicability of machine learning techniques in food intake assessment: A systematic review. *Critical Reviews in Food Science and Nutrition*, *63*(7), 902–919. DOI: 10.1080/10408398.2021.1956425

Ondov, B. D., Treangen, T. J., Melsted, P., Mallonee, A. B., Bergman, N. H., Koren, S., & Phillippy, A. M. (2016). Mash: Fast genome and metagenome distance estimation using MinHash. *Genome Biology*, *17*(1), 1–14. DOI: 10.1186/s13059-016-0997-x PMID: 27323842

Ordovas, J. M., & Corella, D. (2004). Nutritional genomics. *Annual Review of Genomics and Human Genetics*, *5*(1), 71–118. DOI: 10.1146/annurev.genom.5.061903.180008

Ospina, O. E., Soupir, A., & Fridley, B. L. (2023). *A Primer on Preprocessing* (Vol. 2629). Visualization, Clustering, and Phenotyping of Barcode-Based Spatial Transcriptomics Data., DOI: 10.1007/978-1-0716-2986-4_7

Ospina, O. E., Wilson, C. M., Soupir, A. C., Berglund, A., Smalley, I., Tsai, K. Y., & Fridley, B. L. (2022). spatialGE: Quantification and visualization of the tumor microenvironment heterogeneity using spatial transcriptomics. *Bioinformatics (Oxford, England), 38*(9), 2645–2647. DOI: 10.1093/bioinformatics/btac145 PMID: 35258565

Otsuji, R., Fujioka, Y., Hata, N., Kuga, D., Hatae, R., Sangatsuda, Y., Nakamizo, A., Mizoguchi, M., & Yoshimoto, K. (2024). Liquid Biopsy for Glioma Using Cell-Free DNA in Cerebrospinal Fluid. *Cancers (Basel), 16*(5), 1009. DOI: 10.3390/cancers16051009

Pagel, O., Loroch, S., Sickmann, A., & Zahedi, R. P. (2015). Current strategies and findings in clinically relevant post-translational modification-specific proteomics. *Expert Review of Proteomics, 12*(3), 235–253. DOI: 10.1586/14789450.2015.1042867 PMID: 25955281

Pagès, F., Mlecnik, B., Marliot, F., Bindea, G., Ou, F.-S., Bifulco, C., Lugli, A., Zlobec, I., Rau, T. T., Berger, M. D., Nagtegaal, I. D., Vink-Börger, E., Hartmann, A., Geppert, C., Kolwelter, J., Merkel, S., Grützmann, R., Van Den Eynde, M., Jouret-Mourin, A., & Galon, J. (2018). International validation of the consensus Immunoscore for the classification of colon cancer: A prognostic and accuracy study. *Lancet, 391*(10135), 2128–2139. DOI: 10.1016/S0140-6736(18)30789-X PMID: 29754777

Pahnke, A., Montgomery, M., & Radisic, M. (2014). Spatial and electrical factors regulating cardiac regeneration and assembly. *Biomaterials for Cardiac Regeneration*, 71-92.

Palla, G., Spitzer, H., Klein, M., Fischer, D., Schaar, A. C., Kuemmerle, L. B., Rybakov, S., Ibarra, I. L., Holmberg, O., Virshup, I., Lotfollahi, M., Richter, S., & Theis, F. J. (2022). Squidpy: A scalable framework for spatial omics analysis. *Nature Methods, 19*(2), 171–178. DOI: 10.1038/s41592-021-01358-2 PMID: 35102346

Palmer, J. A., Rosenthal, N., Teichmann, S. A., & Litvinukova, M. (2024). Revisiting cardiac biology in the era of single cell and spatial omics. *Circulation Research, 134*(12), 1681–1702. DOI: 10.1161/CIRCRESAHA.124.323672 PMID: 38843288

Pandey, G., Pandey, O. P., Rogers, A. J., Ahsen, M. E., Hoffman, G. E., Raby, B. A., Weiss, S. T., Schadt, E. E., & Bunyavanich, S. (2018). A nasal brush-based classifier of asthma identified by machine learning analysis of nasal RNA sequence data. *Scientific Reports, 8*(1), 8826. DOI: 10.1038/s41598-018-27189-4 PMID: 29891868

Pang, M., Su, K., & Li, M. (2021). Leveraging information in spatial tran scriptomics to predict super-resolution gene expression from histology images in tumors. *BioRxiv* pp. 2021–11. DOI: 10.1101/2021.11.28.470212

Pang, Z., & Ye, L. (2022). Illuminating spatial pharmacology with in situ drug imaging. *Clinical and Translational Medicine*, *12*(8), e1024. DOI: 10.1002/ctm2.1024 PMID: 36030504

Panizo, S., Martínez-Arias, L., Alonso-Montes, C., Cannata, P., Martín-Carro, B., Fernández-Martín, J. L., Naves-Díaz, M., Carrillo López, N., & Cannata-Andía, J. B. (2021). Fibrosis in chronic kidney disease: Pathogenesis and consequences. *International Journal of Molecular Sciences*, *22*(8), 408. DOI: 10.3390/ijms22010408 PMID: 33401711

Papetti, D. M., Spolaor, S., Nazari, I., Tirelli, A., Leonardi, T., Caprioli, C., Besozzi, D., Vlachou, T., Pelicci, P. G., Cazzaniga, P., & Nobile, M. S. (2023). Barcode demultiplexing of nanopore sequencing raw signals by unsupervised machine learning. *Frontiers in Bioinformatics*, *3*, 1067113. DOI: 10.3389/fbinf.2023.1067113 PMID: 37181486

Paranitharan, N., Kataria, S., Arumugam, V. A., Hsieh, H., Muthukrishnan, S., & Velayuthaprabhu, S. (2025). Integrin α1 upregulation by tf:fviia complex promotes cervical cancer migration through par2-dependent mek1/2 activation. *Biochemical and Biophysical Research Communications*, *742*, 151151. DOI: 10.1016/j.bbrc.2024.151151 PMID: 39657349

Parasa, N. A., Namgiri, J. V., Mohanty, S. N., & Dash, J. K. (2021). Introduction to unsupervised learning in bioinformatics. *Data analytics in bioinformatics: A machine learning perspective*, 35-49. DOI: 10.1002/9781119785620.ch2

Park, J. (2023). *Multi-omics modeling for the recovery of disrupted biological systems*. Weill Medical College of Cornell University.

Park, J., Foox, J., Hether, T., Danko, D., Warren, S., Kim, Y., Reeves, J., Butler, D., Mozsary, C., Rosiene, J., Shaiber, A., Afshin, E. E., MacKay, M., Rendeiro, A. F., Bram, Y., Chandar, V., Geiger, H., Craney, A., Velu, P., & Mason, C. E. (2022). System-wide transcriptome damage and tissue identity loss in COVID-19 patients. *Cell Reports Medicine*, *3*(2), 100522. DOI: 10.1016/j.xcrm.2022.100522

Park, J., Kim, H., Kim, J., & Cheon, M. (2020). A practical application of generative adversarial networks for RNA-seq analysis to predict the molecular progress of Alzheimer's disease. *PLoS Computational Biology*, *16*(7), e1008099. DOI: 10.1371/journal.pcbi.1008099 PMID: 32706788

Park, J., Shrestha, R., Qiu, C., Kondo, A., Huang, S., Werth, M., Li, M., Barasch, J., & Susztak, K. (2018). Single-cell transcriptomics of the mouse kidney reveals potential cellular targets of kidney disease. *Science*, *360*(6390), 758–763. DOI: 10.1126/science.aar2131 PMID: 29622724

Park, P. J. (2009). ChIP-seq: Advantages and challenges of a maturing technology. *Nature Reviews. Genetics*, *10*(10), 669–680. DOI: 10.1038/nrg2641 PMID: 19736561

Parra, E. R., Uraoka, N., Jiang, M., Cook, P., Gibbons, D., Forget, M.-A., Bernatchez, C., Haymaker, C., Wistuba, I. I., & Rodriguez-Canales, J. (2017). Validation of multiplex immunofluorescence panels using multispectral microscopy for immune-profiling of formalin-fixed and paraffin-embedded human tumor tissues. *Scientific Reports*, *7*(1), 13380. DOI: 10.1038/s41598-017-13942-8 PMID: 29042640

Pascucci, M., Royer, G., Adamek, J., Al Asmar, M., Aristizabal, D., Blanche, L., Bezzarga, A., Boniface-Chang, G., Brunner, A., Curel, C., Dulac-Arnold, G., Fakhri, R. M., Malou, N., Nordon, C., Runge, V., Samson, F., Sebastian, E., Soukieh, D., Vert, J.-P., & Madoui, M.-A. (2021). AI-based mobile application to fight antibiotic resistance. *Nature Communications*, *12*(1), 1173. DOI: 10.1038/s41467-021-21187-3

Pasolli, E., Truong, D. T., Malik, F., Waldron, L., & Segata, N. (2016). Machine learning meta-analysis of large metagenomic datasets: Tools and biological insights. *PLoS Computational Biology*, *12*(7), e1004977. DOI: 10.1371/journal.pcbi.1004977

Patel, S. K., Surve, J., Parmar, J., Ahmed, K., Bui, F. M., & Al-Zahrani, F. A. (2022). Recent advances in biosensors for detection of COVID-19 and other viruses. *IEEE Reviews in Biomedical Engineering*, *16*, 22–37. DOI: 10.1109/RBME.2022.3212038

Patera, F., Gatticchi, L., Cellini, B., Chiasserini, D., & Reboldi, G. (2024). Kidney fibrosis and oxidative stress: From molecular pathways to new pharmacological opportunities. *Biomolecules*, *14*(1), 137. DOI: 10.3390/biom14010137 PMID: 38275766

Patterson, N. H., & Caprioli, R. M. (2022). Integrated molecular imaging of kidney tissue by mass spectrometry. *Nature Reviews. Nephrology*, *18*(1), 23–38. DOI: 10.1038/s41581-021-00487-2 PMID: 36253509

Pavlou, A., Mulenge, F., Gern, O. L., Busker, L. M., Greimel, E., Waltl, I., & Kalinke, U. (2024). Orchestration of antiviral responses within the infected central nervous system. *Cellular & Molecular Immunology*, *21*(9), 943–958. DOI: 10.1038/s41423-024-01181-7

Peisker, F., Halder, M., Nagai, J., Ziegler, S., Kaesler, N., Hoeft, K., & Kramann, R. (2022). Mapping the cardiac vascular niche in heart failure. *Nature Communications*, *13*(1), 3027. DOI: 10.1038/s41467-022-30682-0 PMID: 35641541

Pei, X. M., Yeung, M. H. Y., Wong, A. N. N., Tsang, H. F., Yu, A. C. S., Yim, A. K. Y., & Wong, S. C. C. (2023). Targeted sequencing approach and its clinical applications for the molecular diagnosis of human diseases. *Cells*, *12*(3), 493. DOI: 10.3390/cells12030493 PMID: 36766834

Penaranda, C., & Hung, D. T. (2019). Single-cell RNA sequencing to understand host–pathogen interactions. *ACS Infectious Diseases*, *5*(3), 336–344. DOI: 10.1021/acsinfecdis.8b00369

Peng, N., Song, S. J., Hui, V. W.-K., Lai, J. C.-T., Wong, G. L.-H., Wong, V. W.-S., & Yip, T. C.-F. (2026). Foundations of Artificial Intelligence in Hepatology: What a Clinician Needs to Know. *Journal of Clinical and Experimental Hepatology*, *16*(1), 103183. Advance online publication. DOI: 10.1016/j.jceh.2025.103183

Pereira, P. R., Carrageta, D. F., Oliveira, P. F., Rodrigues, A., Alves, M. G., & Monteiro, M. P. (2022). Metabolomics as a tool for the early diagnosis and prognosis of diabetic kidney disease. *Medicinal Research Reviews*, *42*(3), 1518–1544. DOI: 10.1002/med.21883 PMID: 35274315

Pertea, M., Pertea, G. M., Antonescu, C. M., Chang, T. C., Mendell, J. T., & Salzberg, S. L. (2015). StringTie enables improved reconstruction of a transcriptome from RNA-seq reads. *Nature Biotechnology*, *33*(3), 290–295. DOI: 10.1038/nbt.3122 PMID: 25690850

Petti, M., & Farina, L. (2023). Network medicine for patients' stratification: From single-layer to multi-omics. *WIREs Mechanisms of Disease*, *15*(6), e1623. DOI: 10.1002/wsbm.1623 PMID: 37323106

Pham, D., Tan, X., Balderson, B., Xu, J., Grice, L. F., Yoon, S., Willis, E. F., Tran, M., Lam, P. Y., Raghubar, A., Kalita-de Croft, P., Lakhani, S., Vukovic, J., Ruitenberg, M. J., & Nguyen, Q. H. (2023). Robust mapping of spa tiotemporal trajectories and cell–cell interactions in healthy and diseased tissues. *Nature Communications*, *14*(1), 7739. DOI: 10.1038/s41467-023-43120-6 PMID: 38007580

Picard, M., Scott-Boyer, M. P., Bodein, A., Périn, O., & Droit, A. (2021). Integration strategies of multi-omics data for machine learning analysis. *Computational and Structural Biotechnology Journal*, *19*, 3735–3746. DOI: 10.1016/j.csbj.2021.06.030 PMID: 34285775

Pierantoni, L., Reis, R. L., Silva-Correia, J., Oliveira, J. M., & Heavey, S. (2023). Spatial -omics technologies: The new enterprise in 3D breast cancer models. *Trends in Biotechnology*, *41*(12), 1488–1500. DOI: 10.1016/j.tibtech.2023.07.003

Pirillo, A., Bonacina, F., Norata, G. D., & Catapano, A. L. (2018). The interplay of lipids, lipoproteins, and immunity in atherosclerosis. *Current Atherosclerosis Reports*, *20*(3), 12. DOI: 10.1007/s11883-018-0715-0 PMID: 29445885

Pleasance, E., Titmuss, E., Williamson, L., Kwan, H., Culibrk, L., Zhao, E. Y., Dixon, K., Fan, K., Bowlby, R., Jones, M. R., Shen, Y., Grewal, J. K., Ashkani, J., Wee, K., Grisdale, C. J., Thibodeau, M. L., Bozoky, Z., Pearson, H., Majounie, E., & Marra, M. A. (2020). Pan-cancer analysis of advanced patient tumors reveals interactions between therapy and genomic landscapes. *Nature Cancer*, *1*(4), 452–468. DOI: 10.1038/s43018-020-0050-6

Plemel, J. R., Stratton, J. A., Michaels, N. J., Rawji, K. S., Zhang, E., Sinha, S., Baaklini, C. S., Dong, Y., Ho, M., Thorburn, K., Friedman, T. N., Jawad, S., Silva, C., Caprariello, A. V., Hoghooghi, V., Yue, J., Jaffer, A., Lee, K., Kerr, B. J., & Yong, V. W. (2020). Microglia response following acute demyelination is heterogeneous and limits infiltrating macrophage dispersion. *Science Advances*, *6*(3), eaay6324. DOI: 10.1126/sciadv.aay6324

Plottel, C. S., & Blaser, M. J. (2011). Microbiome and malignancy. *Cell Host & Microbe*, *10*(4), 324–335. DOI: 10.1016/j.chom.2011.10.003

Pluim, J. P., Maintz, J. A., & Viergever, M. A. (2003). Mutual-information-based registration of medical images: A survey. *IEEE Transactions on Medical Imaging*, *22*(8), 986–1004. DOI: 10.1109/TMI.2003.815867 PMID: 12906253

Pollard, M. O., Gurdasani, D., Mentzer, A. J., Porter, T., & Sandhu, M. S. (2018). Long reads: Their purpose and place. *Human Molecular Genetics*, *27*(R2), R234–R241. DOI: 10.1093/hmg/ddy177 PMID: 29767702

Polnaszek, B., Gilmore-Bykovskyi, A., Hovanes, M., Roiland, R., Ferguson, P., Brown, R., & Kind, A. J. H. (2016). Overcoming the challenges of unstructured data in multisite, electronic medical record-based abstraction. *Medical Care*, *54*(10), e65–e72. DOI: 10.1097/MLR.0000000000000108

Popova, L., & Carabetta, V. J. (2024). The use of next-generation sequencing in personalized medicine. *High Throughput Gene Screening: Methods and Protocols*, 287-315.

Pozojevic, J., & Spielmann, M. (2023). Single-Cell Sequencing in Neurodegenerative Disorders. *Molecular Diagnosis & Therapy, 27*(5), 553–561. DOI: 10.1007/s40291-023-00668-9

Prasad, A., & Tsimikas, S. (2008). Candidate biomarkers for the detection of coronary plaque destabilization and rupture. *Current Atherosclerosis Reports, 10*(4), 309–317. DOI: 10.1007/s11883-008-0048-5 PMID: 18618976

Pratanwanich, P. N., Yao, F., Chen, Y., Koh, C. W., Wan, Y. K., Hendra, C., Poon, P., Goh, Y. T., Yap, P. M., Chooi, J. Y., Chng, W. J., Ng, S. B., Thiery, A., Goh, W. S. S., & Göke, J. (2021). Identification of differential RNA modifications from nanopore direct RNA sequencing with xPore. *Nature Biotechnology, 39*(11), 1394–1402. DOI: 10.1038/s41587-021-00949-w PMID: 34282325

Pribluda, A., de la Cruz, C., & Jackson, E. L. (2015). Intratumoral Heterogeneity: From Diversity Comes Resistance. *Clinical Cancer Research, 21*(13), 2916–2923. DOI: 10.1158/1078-0432.CCR-14-1213

Prideaux, B., & Stoeckli, M. (2012). Mass spectrometry imaging for drug distribution studies. *Journal of Proteomics, 75*(16), 4999–5013. DOI: 10.1016/j.jprot.2012.07.028 PMID: 22842290

Priyadharshini, V. S., & Teran, L. M. (2016). Personalized medicine in respiratory disease: Role of proteomics. *Advances in Protein Chemistry and Structural Biology, 102*, 115–146. DOI: 10.1016/bs.apcsb.2015.11.008

Priya, S., Boyd, A. S., Torres-Bautista, A., Sawadkar, P., Mosahebi, A., Narayan Rath, S., & Kalaskar, D. M. (2024). Accelerating vascular graft development: Adipose-derived stem cells and PODS® (Polyhedrin delivery system with tissue-specific growth factors) – Enhanced 3D bioprinting for functional blood vessels. *Chemical Engineering Journal, 494*, 152934. DOI: 10.1016/j.cej.2024.152934

Ptacek, J., Geyer, F. C., Mignault, A., Deeds, J., Giedt, J., Gu, J., McLaughlin, M. E., Sigal, Y., Tarolli, J. G., Aksoy, M., Zhang, Y., Hav, M., Finck, R., & Finn, J. (2020). 48 Advances in multiplexed ion beam imaging (MIBI) for immune profiling of the tumor microenvironment. *Journal for Immunotherapy of Cancer, 8*, A30.1–A30. Advance online publication. DOI: 10.1136/jitc-2020-SITC2020.0048

Pucciarelli, D. M., Lu, B. Y., Zlobec, I., & DiStasio, M. (2023). Beyond the lab and into the hospital: An outlook on the clinical utility of spatial omics technologies. *GEN Biotechnology, 2*(5), 360–371. DOI: 10.1089/genbio.2023.0030

Pulley, J. M., Rhoads, J. P., Jerome, R. N., Challa, A. P., Erreger, K. B., Joly, M. M., Lavieri, R. R., Perry, K. E., Zaleski, N. M., Shirey-Rice, J. K., & Aronoff, D. M. (2020). Using what we already have: Uncovering new drug repurposing strategies in existing omics data. *Annual Review of Pharmacology and Toxicology*, *60*(1), 333–352. DOI: 10.1146/annurev-pharmtox-010919-023537 PMID: 31337270

Pushkaran, A. C., & Arabi, A. A. (2024). From understanding diseases to drug design: Can artificial intelligence bridge the gap? *Artificial Intelligence Review*, *57*(4), 86. DOI: 10.1007/s10462-024-10714-5

Pushparaj, K., Kuchi Bhotla, H., Arumugam, V. A., Pappusamy, M., Easwaran, M., Liu, W. C., Issara, U., Rengasamy, K. R. R., Meyyazhagan, A., & Balasubramanian, B. (2022). Mucormycosis (black fungus) ensuing COVID-19 and comorbidity meets - magnifying global pandemic grieve and catastrophe begins. *The Science of the Total Environment*, *805*, 150355. DOI: 10.1016/j.scitotenv.2021.150355 PMID: 34818767

Pw, F. (2023). Longitudinal scRNA-seq analysis in mouse and human informs optimization of rapid mouse astrocyte differentiation protocols. *Nature Neuroscience*, *26*(10), 1726–1738. Advance online publication. DOI: 10.1038/s41593-023-01424-2

Qahwaji, R., Ashankyty, I., Sannan, N. S., Hazzazi, M. S., Basabrain, A. A., & Mobashir, M. (2024). Pharmacogenomics: A genetic approach to drug development and therapy. *Pharmaceuticals (Basel, Switzerland)*, *17*(7), 940. DOI: 10.3390/ph17070940 PMID: 39065790

Qi, J., Wang, K., Han, Y., Huang, X., Liu, X., Xu, J., Duan, R., Xu, J., Chen, J., Wang, J., Chen, G., Kang, M., Chen, Y., Li, Q., Wang, S., & Chen, J. (2026). Computed Tomography and Whole Slide Image-Based Prediction of Pathological Responses to Neoadjuvant Chemotherapy in Locally Advanced Gastric Cancer: A Multi-Omics Study. *Biomedical Signal Processing and Control*, *112*, 108489. DOI: 10.1016/j.bspc.2025.108489

Qin, J., Li, R., Raes, J., Arumugam, M., Burgdorf, K. S., Manichanh, C., ... & Wang, J. (2010). A human gut microbial gene catalogue established by metagenomic sequencing. *nature, 464*(7285), 59-65.

Qin, D. (2019). Next-generation sequencing and its clinical application. *Cancer Biology & Medicine*, *16*(1), 4–10. DOI: 10.20892/j.issn.2095-3941.2018.0055 PMID: 31119042

Qin, S., Miao, D., Zhang, Y., & Bai, Y. (2023). Methods developments of mass spectrometry based single cell metabolomics. *Trends in Analytical Chemistry*, *117086*, 117086. Advance online publication. DOI: 10.1016/j.trac.2023.117086

Qiu, X., Wu, Z., Xu, Q., Sheng, C., & Jiao, Z. (2020). Pharmacogenomics of immunosuppressants. In Cai, W., Liu, Z., Miao, L., & Xiang, X. (Eds.), *Pharmacogenomics in precision medicine* (pp. 83–106). Springer., DOI: 10.1007/978-981-15-3895-7_5

Qu, H., Kao, C., & Hakonarson, H. (2023). Single-Cell RNA Sequencing Technology Landscape in 2023. *Stem Cells (Dayton, Ohio)*. Advance online publication. DOI: 10.1093/stmcls/sxad077

Qu, Y., He, F., & Chen, Y. (2010). Different effects of the probe summarization algorithms PLIER and RMA on high-level analysis of Affymetrix exon arrays. *BMC Bioinformatics*, *11*(1), 1–13. DOI: 10.1186/1471-2105-11-211 PMID: 20426803

Qu, Z., Hanna, P., Ajijola, O. A., Garfinkel, A., & Shivkumar, K. (2025). Ultrastructure and cardiac impulse propagation: Scaling up from microscopic to macroscopic conduction. *The Journal of Physiology*, *603*(7), 1887–1901. DOI: 10.1113/JP287632 PMID: 39612369

Rademacher, A., Huseynov, A., Bortolomeazzi, M., Wille, S. J., Schumacher, S., Sant, P., Keitel, D., Okonechnikov, K., Ghasemi, D. R., Pajtler, K. W., Mallm, J., & Rippe, K. (2025). Comparison of spatial transcriptomics technologies using tumor cryosections. *Genome Biology*, *26*(1), 176. Advance online publication. DOI: 10.1186/s13059-025-03624-4

Radke, J., Ishaque, N., Koll, R., Gu, Z., Schumann, E., Sieverling, L., Uhrig, S., Hübschmann, D., Toprak, U. H., López, C., Hostench, X. P., Borgoni, S., Juraeva, D., Pritsch, F., Paramasivam, N., Balasubramanian, G. P., Schlesner, M., Sahay, S., Weniger, M., & Heppner, F. L. (2022). The genomic and transcriptional landscape of primary central nervous system lymphoma. *Nature Communications*, *13*(1), 2558. DOI: 10.1038/s41467-022-30050-y

Raghavan, R., Moger, S. S., Umesh, S., & Bhuvana, G. N. (2025). Revolutionizing Pain Management. *Unveiling Technological Advancements and Interdisciplinary Solutions for Pain Care*, 129–164. DOI: 10.4018/979-8-3693-9501-1.ch005

Raghubar, A. M., Pham, D. T., Tan, X., Grice, L. F., Crawford, J., Lam, P. Y., Andersen, S. B., Yoon, S., Teoh, S. M., Matigian, N. A., Stewart, A., Francis, L., Ng, M. S. Y., Healy, H. G., Combes, A. N., Kassianos, A. J., Nguyen, Q., & Mallett, A. J. (2022). Spatially resolved transcriptomes of mammalian kidneys illustrate the molecular complexity and interactions of functional nephron segments. *Frontiers in Medicine*, *9*, 873923. DOI: 10.3389/fmed.2022.873923 PMID: 35872784

Rahate, K. P., Sengar, M. S., & Patel, S. (2024). Artificial Intelligence in Medical Imaging by Machine Learning and Deep Learning. In *Approaches to Human-Centered AI in Healthcare* (pp. 121–159). IGI Global. DOI: 10.4018/979-8-3693-2238-3.ch006

Rahimzadeh, N., Srinivasan, S. S., Zhang, J., & Swarup, V. (2024). Gene networks and systems biology in Alzheimer's disease: Insights from multi-omics approaches. *Alzheimer's & Dementia*, *20*(5), 3587–3605. DOI: 10.1002/alz.13790

Raimundo, F., Papaxanthos, L., Vallot, C., Vert, J.-P., & Vert, J.-P. (2021). Machine learning for single cell genomics data analysis. *bioRxiv*. https://doi.org/DOI: 10.1101/2021.02.04.429763

Ramírez Medina, C. R., Ali, I., Baricevic-Jones, I., Odudu, A., Saleem, M. A., Whetton, A. D., Kalra, P. A., & Geifman, N. (2023). Proteomic signature associated with chronic kidney disease (CKD) progression identified by data-independent acquisition mass spectrometry. *Clinical Proteomics*, *20*(1), 19. DOI: 10.1186/s12014-023-09405-0 PMID: 37076799

Ramnauth, A. D., Tippani, M., Divecha, H. R., Papariello, A. R., Miller, R. A., Pattie, E. A., Kleinman, J. E., Maynard, K. R., Collado-Torres, L., Hyde, T. M., Martinowich, K., Hicks, S. C., & Page, S. C. (2023). Spatially-resolved transcriptomics of human dentate gyrus across postnatal lifespan reveals heterogeneity in markers for proliferation, extracellular matrix, and neuroinflammation. *bioRxiv, 4*(12–01), 2023.11.20.567883. DOI: 10.1101/2023.11.20.567883

Rampton, V., Böhmer, M., & Winkler, A. (2022). Medical technologies past and present: How history helps to understand the digital era. *The Journal of Medical Humanities*, *43*(2), 343–364. DOI: 10.1007/s10912-021-09699-x

Ramya, S., Poornima, P., Jananisri, A., Geofferina, I. P., Bavyataa, V., Divya, M., Priyanga, P., Vadivukarasi, J., Sujitha, S., Elamathi, S., Anand, A. V., & Balamuralikrishnan, B. (2023). Role of hormones and the potential impact of multiple stresses on infertility. *Stresses*, *3*(2), 454–474. DOI: 10.3390/stresses3020033

Rana, D., Rangel, V. R., Padmanaban, P., Trikalitis, V. D., Kandar, A., Kim, H.-W., & Rouwkema, J. (2025). Bioprinting of Aptamer-Based Programmable Bioinks to Modulate Multiscale Microvascular Morphogenesis in 4D. *Advanced Healthcare Materials*, *14*(1), 2402302. DOI: 10.1002/adhm.202402302

Ranek, J. S., Stanley, N., & Purvis, J. E. (2022). Integrating temporal single-cell gene expression modalities for trajectory inference and disease prediction. *Genome Biology*, *23*(1), 186. DOI: 10.1186/s13059-022-02749-0 PMID: 36064614

Rangisetti, L. S., Shaik, F., Bhagavatula, A., & Kommareddy, L. S. (2024). Heart disease detection using graph neural networks (GNNs). In *2024 5th International Conference on Electronics and Sustainable Communication Systems (ICESC)* (pp. 1436-1441). IEEE.

Rao, M. S., Van Vleet, T. R., Ciurlionis, R., Buck, W. R., Mittelstadt, S. W., Blomme, E. A., & Liguori, M. J. (2019). Comparison of RNA-Seq and microarray gene expression platforms for the toxicogenomic evaluation of liver from short-term rat toxicity studies. *Frontiers in Genetics*, *9*, 636. DOI: 10.3389/fgene.2018.00636 PMID: 30723492

Rappez, L., Stadler, M., Triana, S., Gathungu, R. M., Ovchinnikova, K., Phapale, P., Heikenwalder, M., & Alexandrov, T. (2021). SpaceM reveals metabolic states of single cells. *Nature Methods*, *18*(7), 799–805. DOI: 10.1038/s41592-021-01198-0 PMID: 34226721

Rappoport, N., & Shamir, R. (2018). Multi-omic and multi-view clustering algorithms: Review and cancer benchmark. *Nucleic Acids Research*, *46*(20), 10546–10562. DOI: 10.1093/nar/gky889 PMID: 30295871

Raufaste-Cazavieille, V., Santiago, R., & Droit, A. (2022). Multi-omics analysis: Paving the path toward achieving precision medicine in cancer treatment and immuno-oncology. *Frontiers in Molecular Biosciences*, *9*, 962743. Advance online publication. DOI: 10.3389/fmolb.2022.962743

Rawat, S. (2023). Spatial Omics Spotlights the Players in the Tumor Microenvironment. *GEN Biotechnology*, *2*(5), 342–346. DOI: 10.1089/genbio.2023.29115.sro

Recanatini, M., & Cabrelle, C. (2020). Drug Research Meets Network Science: Where Are We? *Journal of Medicinal Chemistry*, *63*(16), 8653–8666. DOI: 10.1021/acs.jmedchem.9b01989 PMID: 32338900

Reisqs, J. B., Qu, Y. S., & Boutjdir, M. (2024). Ion channel trafficking implications in heart failure. *Frontiers in Cardiovascular Medicine*, *11*, 1351496. DOI: 10.3389/fcvm.2024.1351496 PMID: 38420267

Reitz, C. J., Kuzmanov, U., & Gramolini, A. O. (2023). Multi-omic analyses and network biology in cardiovascular disease. *Proteomics*, *23*(21-22), 2200289. DOI: 10.1002/pmic.202200289 PMID: 37691071

Rendeiro, A. F., Ravichandran, H., Bram, Y., Kim, J., Meydan, C., Park, J., Foox, J., Hether, T., Warren, S., Kim, Y., Reeves, J., Salvatore, S., Mason, C. E., Swanson, E., Borczuk, A., Elemento, O., & Schwartz, R. E. (2021). *The spatial landscape of lung pathology during COVID-19 progression - targeted spatial transcriptomics data*. https://doi.org/DOI: 10.5281/zenodo.4635285

Ren, P., Lu, L., Cai, S., Chen, J., Lin, W., & Han, F. (2021). Alternative splicing: A new cause and potential therapeutic target in autoimmune disease. *Frontiers in Immunology*, *12*, 713540. DOI: 10.3389/fimmu.2021.713540 PMID: 34484216

Ren, Z., Zeng, X., Lao, Y., Zheng, H., You, Z., Xiang, H., & Zou, Q. (2024). A spatial hierarchical network learning framework for drug repositioning, allowing interpretation from macro to micro scale. *Communications Biology*, *7*(1), 1413. DOI: 10.1038/s42003-024-07107-3 PMID: 39478146

Reznichenko, A., Nair, V., Eddy, S., Fermin, D., Tomilo, M., Slidel, T., Ju, W., Henry, I., Badal, S. S., Wesley, J. D., Liles, J. T., Moosmang, S., Williams, J. M., Quinn, C. M., Bitzer, M., Hodgin, J. B., Barisoni, L., Karihaloo, A., Breyer, M. D., & Kretzler, M. (2024). Unbiased kidney-centric molecular categorization of chronic kidney disease as a step towards precision medicine. *Kidney International*, *105*(6), 1263–1278. DOI: 10.1016/j.kint.2024.01.012 PMID: 38286178

Rivas-Ortiz, C. I., Lopez-Vidal, Y., Arredondo-Hernandez, L. J. R., & Castillo-Rojas, G. (2017). Genetic alterations in gastric cancer associated with Helicobacter pylori infection. *Frontiers in Medicine*, *4*, 47. DOI: 10.3389/fmed.2017.00047

Rocque, B., Guion, K., Singh, P., Bangerth, S., Pickard, L., Bhattacharjee, J., Eguizabal, S., Weaver, C., Chopra, S., Zhou, S., Kohli, R., Sher, L., Akbari, O., Ekser, B., & Emamaullee, J. A. (2024). Technical optimization of spatially resolved single-cell transcriptomic datasets to study clinical liver disease. *Scientific Reports*, *14*(1), 3612. Advance online publication. DOI: 10.1038/s41598-024-53993-2

Rodriques, S. G., Stickels, R. R., Goeva, A., Martin, C. A., Murray, E., Vanderburg, C. R., Welch, J., Chen, L. M., Chen, F., & Macosko, E. Z. (2019). Slide-seq: A scalable technology for measuring genome-wide expression at high spatial resolution. *Science*, *363*(6434), 1463–1467. DOI: 10.1126/science.aaw1219 PMID: 30923225

Roess, A. A., Winch, P. J., Akhter, A., Afroz, D., Ali, N. A., Shah, R., Begum, N., Seraji, H. R., El Arifeen, S., Darmstadt, G. L., & Baqui, A. H.Bangladesh Projahnmo Study Group. (2015). Household animal and human medicine use and animal husbandry practices in rural Bangladesh: Risk factors for emerging zoonotic disease and antibiotic resistance. *Zoonoses and Public Health*, *62*(7), 569–578. DOI: 10.1111/zph.12186 PMID: 25787116

Rouskas, K., Bocher, O., Simistiras, A., Emmanouil, C., Mantas, P., Skoulakis, A., ... & Dimas, A. S. (2025). Periodic dietary restriction of animal products induces metabolic reprogramming in humans with effects on cardiometabolic health. *npj Metabolic Health and Disease, 3*(1), 14.

Ruan, X., Hu, K., & Zhang, X. (2023). PIE-seq: Identifying RNA-binding protein targets by dual RNA-deaminase editing and sequencing. *Nature Communications*, *14*(1), 3275. DOI: 10.1038/s41467-023-39054-8

Russell, S. K., Harrison, J. K., Olson, B. S., Lee, H. J., O'Brien, V. P., Xing, X., Livny, J., Yu, L., Roberson, E. D. O., Bomjan, R., Fan, C., Sha, M., Estfanous, S., Amer, A. O., Colonna, M., Stappenbeck, T. S., Wang, T., Hannan, T. J., & Hultgren, S. J. (2023). Uropathogenic Escherichia coli infection-induced epithelial trained immunity impacts urinary tract disease outcome. *Nature Microbiology*, *8*(5), 875–888. DOI: 10.1038/s41564-023-01346-6

Ryu, Y., Han, G. H., Jung, E., & Hwang, D. (2023). Integration of single-cell RNA-seq datasets: A review of computational methods. *Molecules and Cells*, *46*(2), 106–119. DOI: 10.14348/molcells.2023.0009 PMID: 36859475

Sabit, H., Arneth, B., Pawlik, T. M., Abdel-Ghany, S., Ghazy, A., Abdelazeem, R. M., Alqosaibi, A., Al-Dhuayan, I. S., Almulhim, J., Alrabiah, N. A., & Hashash, A. (2025). Leveraging Single-Cell Multi-Omics to Decode Tumor Microenvironment Diversity and Therapeutic Resistance. *Pharmaceuticals (Basel, Switzerland)*, *18*(1), 75. DOI: 10.3390/ph18010075

Sadraei, A., Naghib, Seyed Morteza, & and Rabiee, N. (2025). 4D printing biological stimuli-responsive hydrogels for tissue engineering and localized drug delivery applications – part 1. *Expert Opinion on Drug Delivery, 22*(4), 471–490. DOI: 10.1080/17425247.2025.2466772

Saez-Rodriguez, J., Rinschen, M. M., Floege, J., & Kramann, R. (2019). Big science and big data in nephrology. *Kidney International*, *95*(6), 1326–1337. DOI: 10.1016/j.kint.2018.11.048 PMID: 30982672

Sager, M., Yeat, N. C., Pajaro-Van der Stadt, S., Lin, C., Ren, Q., & Lin, J. (2015). Transcriptomics in cancer diagnostics: Developments in technology, clinical research and commercialization. *Expert Review of Molecular Diagnostics*, *15*(12), 1589–1603. DOI: 10.1586/14737159.2015.1105133 PMID: 26565429

Sahl, S. J., & Moerner, W. E. (2013). Super-resolution fluorescence imaging with single molecules. *Current Opinion in Structural Biology*, *23*(5), 778–787. https://doi.org/https://doi.org/10.1016/j.sbi.2013.07.010. DOI: 10.1016/j.sbi.2013.07.010

Saiselet, M., Rodrigues-Vitória, J., Tourneur, A., Craciun, L., Spinette, A., Larsimont, D., Andry, G., Lundeberg, J., Maenhaut, C., & Detours, V. (2020). Transcrip tional output, cell-type densities, and normalization in spatial transcriptomics. *Journal of Molecular Cell Biology*, *12*(11), 906–908. DOI: 10.1093/jmcb/mjaa028 PMID: 32573704

Saliba, A., Du, Y., Feng, T., & Garmire, L. (2024). Multi-Omics Integration in Nephrology: Advances, Challenges, and Future Directions. *Seminars in Nephrology*, *44*(6), 151584. DOI: 10.1016/j.semnephrol.2025.151584 PMID: 40216576

Salman, R. A. S., Frantzias, J., Lee, R. J., Lyden, P. D., Battey, T. W., Ayres, A. M., & Lanfermann, H. (2018). Absolute risk and predictors of the growth of acute spontaneous intracerebral haemorrhage: A systematic review and meta-analysis of individual patient data. *Lancet Neurology*, *17*(10), 885–894. DOI: 10.1016/S1474-4422(18)30253-9

Samek, W., Wiegand, T., & Müller, K. R. (2017). Explainable artificial intelligence: Understanding, visualizing and interpreting deep learning models. *arXiv preprint arXiv:1708.08296*.

Sanches, P. H. G., De Melo, N. C., Porcari, A. M., & De Carvalho, L. M. (2024). Integrating Molecular Perspectives: Strategies for Comprehensive Multi-Omics Integrative Data Analysis and Machine Learning Applications in Transcriptomics, Proteomics, and Metabolomics. *Biology (Basel)*, *13*(11), 848. DOI: 10.3390/biology13110848 PMID: 39596803

Sangeetha, T., Anand, A. V., & Begum, T. N. (2022). Assessment of inter-relationship between anemia and COPD in accordance with altitude. *The Open Respiratory Medicine Journal*, *16*(1), e187430642206270. DOI: 10.2174/18743064-v16-e2206270 PMID: 37273961

Sangeetha, T., Nargis Begum, T., Balamuralikrishnan, B., Arun, M., Rengasamy, K. R. R., Senthilkumar, N., Velayuthaprabhu, S., Saradhadevi, M., Sampathkumar, P., & Vijaya Anand, A. (2022). Influence of SERPINA1 gene polymorphisms on anemia and chronic obstructive pulmonary disease. *Journal of the Renin-Angiotensin-Aldosterone System : JRAAS*, *2022*, 2022. DOI: 10.1155/2022/2238320 PMID: 36320441

Saratkar, S. Y., Langote, M., Kumar, P., Gote, P., Weerarathna, I. N., & Mishra, G. V. (2025). Digital twin for personalized medicine development. *Frontiers in Digital Health*, *7*, 1583466. DOI: 10.3389/fdgth.2025.1583466 PMID: 40851640

Sarkar, D. I. (2024). The evolution of bioinformatics: From gene sequencing to personalized medicine. *Innovation and Integrative Research Center Journal*, *2*(6), 307–337.

Sarker, I. H. (2021). Machine learning: Algorithms, real-world applications and research directions. *SN Computer Science*, *2*(3), 160. DOI: 10.1007/s42979-021-00592-x PMID: 33778771

Satam, H., (2024). Correction: Next-generation sequencing technology: Current trends and advancements. *Biology (Basel)*, *13*(5), 286. DOI: 10.3390/biology13050286 PMID: 38785841

Saul, D., & Kosinsky, R. L. (2023). Spatial transcriptomics herald a new era of transcriptome research. *Clinical and Translational Medicine, 13*(5), e1264. Advance online publication. DOI: 10.1002/ctm2.1264

Schena, F. P., Chiurlia, S., Abbrescia, D. I., & Cox, S. N. (2024). Kidney and urine cell transcriptomics in IgA nephropathy and lupus nephritis: A narrative review. *Clinical Kidney Journal, 17*(1), sfad121. Advance online publication. DOI: 10.1093/ckj/sfad121 PMID: 38186900

Schlecht, A., Boneva, S., Salié, H., Killmer, S., Wolf, J., Hajdu, R. I., Hajdu, R. I., Auw-Haedrich, C., Agostini, H., Reinhard, T., Schlunck, G., Bengsch, B., & Lange, C. (2021). Imaging mass cytometry for high-dimensional tissue profiling in the eye. *BMC Ophthalmology, 21*(1), 338. DOI: 10.1186/s12886-021-02099-8

Schonrock, N., Callis, T., Hatchell, K., Esplin, E., Morales, A., Garcia, J., & Nussbaum, R. (2022). Precision Medicine Opportunities for Familial Arrhythmias and Cardiomyopathies Identified When Cost of Genetic Testing is Removed as a Barrier. *Heart Lung and Circulation, 31*, S19. DOI: 10.1016/j.hlc.2022.04.041

Schulte-Schrepping, J., Ferreira, H. J., Saglam, A., Hinkley, E. R., Schultze, J. L., & Schultze, J. L. (2020). *Advances in single-cell epigenomics of the immune system*. Academic Press., DOI: 10.1016/B978-0-12-817964-2.00008-3

Schulz, E., Speekenbrink, M., & Krause, A. (2018). A tutorial on Gaussian process regression: Modelling, exploring, and exploiting functions. *Journal of Mathematical Psychology, 85*, 1–16. DOI: 10.1016/j.jmp.2018.03.001

Schumacher, D., & Kramann, R. (2023). Multiomic spatial mapping of myocardial infarction and implications for personalized therapy. *Arteriosclerosis, Thrombosis, and Vascular Biology, 43*(2), 192–202. DOI: 10.1161/ATVBAHA.122.318333 PMID: 36579644

Schupack, D. A., Mars, R. A., Voelker, D. H., Abeykoon, J. P., & Kashyap, P. C. (2022). The promise of the gut microbiome as part of individualized treatment strategies. *Nature Reviews. Gastroenterology & Hepatology, 19*(1), 7–25. DOI: 10.1038/s41575-021-00499-1

Schürch, C. M., Bhate, S. S., Barlow, G. L., Phillips, D. J., Noti, L., Zlobec, I., Chu, P., Black, S., Demeter, J., McIlwain, D. R., Kinoshita, S., Samusik, N., Goltsev, Y., & Nolan, G. P. (2020). Coordinated Cellular Neighborhoods Orchestrate Antitumoral Immunity at the Invasive Front of Colorectal Cancer. *Cell, 182*(5), 1341–1359.e19. DOI: 10.1016/j.cell.2020.07.005 PMID: 32763154

Schwalm, S., Manaila, R., Oftring, A., Schaefer, L., Von Gunten, S., & Pfeilschifter, J. (2024). The contribution of the sphingosine 1-phosphate signaling pathway to chronic kidney diseases: Recent findings and new perspectives. *Pflügers Archiv*, *476*(10), 1845–1861. DOI: 10.1007/s00424-024-03029-5 PMID: 39384640

Schwarze, K., Buchanan, J., Fermont, J. M., Dreau, H., Tilley, M. W., Taylor, J. M., Antoniou, P., Knight, S. J., Camps, C., Pentony, M. M., Kvikstad, E. M., Harris, S., Popitsch, N., Pagnamenta, A. T., Schuh, A., Taylor, J. C., & Wordsworth, S. (2020). The complete costs of genome sequencing: A microcosting study in cancer and rare diseases from a single center in the United Kingdom. *Genetics in Medicine*, *22*(1), 85–94. DOI: 10.1038/s41436-019-0618-7 PMID: 31358947

Seeley, E. H., Liu, Z., Yuan, S., Stroope, C., Cockerham, E., Rashdan, N. A., & Yurdagul, A.Jr. (2023). Spatially resolved metabolites in stable and unstable human atherosclerotic plaques identified by mass spectrometry imaging. *Arteriosclerosis, Thrombosis, and Vascular Biology*, *43*(9), 1626–1635. DOI: 10.1161/ATVBA-HA.122.318684 PMID: 37381983

Seers, T., Rothe, C., Hamer, D. H., Denny, S., Spindler, R., Schwartz, E., & Johnston, V. (2023). Zika virus infection in European travellers returning from Thailand in 2022: A GeoSentinel case series. *Tropical Medicine & International Health*, *28*(7), 576–579. DOI: 10.1111/tmi.13900

Seidel, K., Piskol, R., Thao Nguyen, T. T., Shelton, A., Cottonham, C., de la Cruz, C. C., . . . Siebel, C. (2024). Notch signaling maintains a progenitor-like subclass of hepatocellular carcinoma. bioRxiv, 2024-12.

Seo, G., Lee, G., Kim, M. J., Baek, S.-H., Choi, M., Ku, K. B., Lee, C.-S., Jun, S., Park, D., Kim, H. G., Kim, S.-J., Lee, J.-O., Kim, B. T., Park, E. C., & Kim, S. (2020). Rapid Detection of COVID-19 Causative Virus (SARS-CoV-2) in Human Nasopharyngeal Swab Specimens Using Field-Effect Transistor-Based Biosensor. *ACS Nano*, *14*(4), 5135–5142. DOI: 10.1021/acsnano.0c02823

Severs, N. J. (2001). Gap junction remodelling and cardiac arrhythmogenesis: Cause or coincidence? *Journal of Cellular and Molecular Medicine*, *5*(4), 355–366. DOI: 10.1111/j.1582-4934.2001.tb00170.x PMID: 12067469

Severs, N. J. (2003). Communicating junctions, connexins and the cardiomyocyte: from cell biology to cardiology. In *Cardiac Remodelling and Failure* (pp. 417–434). Springer US. DOI: 10.1007/978-1-4419-9262-8_29

Shabani, M., & Borry, P. (2018). Rules for processing genetic data for research purposes in view of the new EU General Data Protection Regulation. *European Journal of Human Genetics : EJHG*, *26*(2), 149–156. DOI: 10.1038/s41431-017-0045-7

Shah, P. K. (2014). Biomarkers of plaque instability. *Current Cardiology Reports, 16*(12), 547. DOI: 10.1007/s11886-014-0547-7 PMID: 25326730

Shah, S. H., & Gerszten, R. E. (2024). Sifting for Gold in Terabytes of Data: Illuminating Cardiovascular Biology in the 'Omics Age. *Circulation, 149*(17), 1323–1325. DOI: 10.1161/CIRCULATIONAHA.123.066988 PMID: 38648274

Shah, S., Lubeck, E., Zhou, W., & Cai, L. (2016). In situ transcription profiling of single cells reveals spatial organization of cells in the mouse hippocampus. *Neuron, 92*(2), 342–357. DOI: 10.1016/j.neuron.2016.10.001 PMID: 27764670

Shang, Y., Wang, Z., Xi, L., Wang, Y., Liu, M., Feng, Y., Wang, S., Wu, Q., Xiang, X., Chen, M., & Ding, Y. (2024). Droplet-based single-cell sequencing: Strategies and applications. *Biotechnology Advances, 77*, 108454. DOI: 10.1016/j.biotechadv.2024.108454

Shao, M., & Kingsford, C. (2017). Accurate assembly of transcripts through phase-preserving graph decomposition. *Nature Biotechnology, 35*(12), 1167–1169. DOI: 10.1038/nbt.4020 PMID: 29131147

Sharma, A., & DasGupta, R. (2019). Tracking tumor evolution one-cell-at-a-time. *Molecular & Cellular Oncology, 6*(3), 1590089. DOI: 10.1080/23723556.2019.1590089

Sharpe, J. D., & Kamara, M. T. (2018). A systematic evaluation of mobile apps to improve the uptake of and adherence to HIV pre-exposure prophylaxis. *Sexual Health, 15*(6), 587–594. DOI: 10.1071/SH18120

Shendure, J., & Ji, H. (2008). Next-generation DNA sequencing. *Nature Biotechnology, 26*(10), 1135–1145. DOI: 10.1038/nbt1486

Shengquan, C., Boheng, Z., Xiaoyang, C., Xuegong, Z., & Rui, J. (2021). STPLUS: A reference based method for the accurate enhancement of spatial tran scriptomics. *Bioinformatics (Oxford, England), 37*(1, Supplement_1), i299–i307. DOI: 10.1093/bioinformatics/btab298 PMID: 34252941

Sheng, X., Qiu, C., Liu, H., Gluck, C., Hsu, J. Y., He, J., Hsu, C., Sha, D., Weir, M. R., Isakova, T., Raj, D., Rincon-Choles, H., Feldman, H. I., Townsend, R., Li, H., & Susztak, K. (2020). Systematic integrated analysis of genetic and epigenetic variation in diabetic kidney disease. *Proceedings of the National Academy of Sciences of the United States of America, 117*(47), 29013–29024. DOI: 10.1073/pnas.2005905117 PMID: 33144501

Shen, L., Chen, W., Ding, J., Shu, G., Chen, M., Zhao, Z., & Ji, J. (2023). The role of metabolic reprogramming of oxygen-induced macrophages in the dynamic changes of atherosclerotic plaques. *The FASEB Journal*, *37*(3), e22791. DOI: 10.1096/fj.202201486R PMID: 36723768

Sherif, Z. A., Ogunwobi, O. O., & Ressom, H. W. (2025). Mechanisms and technologies in cancer epigenetics. *Frontiers in Oncology*, *14*, 1513654. Advance online publication. DOI: 10.3389/fonc.2024.1513654

Sheu, Y. H. (2020). Illuminating the black box: Interpreting deep neural network models for psychiatric research. *Frontiers in Psychiatry*, *11*, 551299. DOI: 10.3389/fpsyt.2020.551299 PMID: 33192663

Sheynkman, G. M., Shortreed, M. R., Cesnik, A. J., & Smith, L. M. (2016). Proteogenomics: Integrating next-generation sequencing and mass spectrometry to characterize human proteomic variation. *Annual Review of Analytical Chemistry (Palo Alto, Calif.)*, *9*(1), 521–545. DOI: 10.1146/annurev-anchem-071015-041722 PMID: 27049631

Shi, H., El Kazzi, M., Liu, Y., Gao, A., Schroder, A. L., Vuong, S., & Witting, P. K. (2022). Multiplex analysis of mass imaging data: Application to the pathology of experimental myocardial infarction. *Acta Physiologica (Oxford, England)*, *235*(2), e13790. DOI: 10.1111/apha.13790 PMID: 35080155

Shimagami, H., Nishimura, K., Matsushita, H., Metsugi, S., Kato, Y., Kawasaki, T., Tsujimoto, K., Edahiro, R., Itotagawa, E., Naito, M., Kawada, S., Nakatsubo, D., Matsukawa, K., Namba-Hamano, T., Inoue, K., Takahashi, A., Mizui, M., Kato, S., Hikita, H., & Nishide, M. (2025). Single-cell analysis reveals immune cell abnormalities underlying the clinical heterogeneity of patients with systemic sclerosis. *Nature Communications*, *16*(1), 4949. Advance online publication. DOI: 10.1038/s41467-025-60034-7

Shi, X., Wei, F., Mehrpouyan, M., Chen, Y., D'Cruz, L. M., Widmann, S., & Tyznik, A. J. (2023). Flow cytometry analysis of protein expression using antibody-derived tags followed by CITE-Seq. *Cytometry. Part A*. Advance online publication. DOI: 10.1002/cyto.a.24792

Shi, Y., Huang, L., Dong, H., Yang, M., Ding, W., Zhou, X., Lu, T., Liu, Z., Zhou, X., Wang, M., Zeng, B., Sun, Y., Zhong, S., Wang, B., Wang, W., Yin, C., Wang, X., & Wu, Q. (2024). Decoding the spatiotemporal regulation of transcription factors during human spinal cord development. *Cell Research*, *34*(3), 193–213. DOI: 10.1038/s41422-023-00897-x

Shi, Y., Johnson, J., Wang, B., Chen, B., Fisher, G. L., Urabe, G., & Li, L. (2019). Mass spectrometric imaging reveals temporal and spatial dynamics of bioactive lipids in arteries undergoing restenosis. *Journal of Proteome Research*, *18*(4), 1669–1678. DOI: 10.1021/acs.jproteome.8b00941 PMID: 30784274

Shu, C., Wang, Q., Yan, X., & Wang, J. (2018). Whole-genome expression microarray combined with machine learning to identify prognostic biomarkers for high-grade glioma. *Journal of Molecular Neuroscience*, *64*(4), 491–500. DOI: 10.1007/s12031-018-1049-7 PMID: 29502292

Shu, H., Chen, J., Xu, C., Hu, J., Wang, Y., Peng, J., Jiang, Q., Shang, X., & Wang, T. (2024). Efficient integration of spatial omics data for joint domain detection, matching, and alignment with stMSA. *Bioinformatics (Oxford, England)*. Advance online publication. DOI: 10.1101/2024.07.29.605604

Sielaff, M., Kuharev, J., Bohn, T., Hahlbrock, J., Bopp, T., Tenzer, S., & Distler, U. (2017). Evaluation of FASP, SP3, and iST protocols for proteomic sample preparation in the low microgram range. *Journal of Proteome Research*, *16*(11), 4060–4072. DOI: 10.1021/acs.jproteome.7b00433

Singh, P. R., Dadireddy, V., Udupa, S., Kalladi, S. M., Shee, S., Khosla, S., Rajmani, R. S., Singh, A., Ramakumar, S., & Nagaraja, V. (2023). The Mycobacterium tuberculosis methyltransferase Rv2067c manipulates host epigenetic programming to promote its own survival. *Nature Communications*, *14*(1), 8497. DOI: 10.1038/s41467-023-43940-6

Singh, V. (2025). *Single-Cell Proteomics: Recent Advancement and Prospects*. https://doi.org/DOI: 10.20944/preprints202505.0303.v1

Si, S., Liu, H., Xu, L., & Zhan, S. (2024). Identification of novel therapeutic targets for chronic kidney disease and kidney function by integrating multi-omics proteome with transcriptome. *Genome Medicine*, *16*(1), 84. DOI: 10.1186/s13073-024-01356-x PMID: 38898508

Slijkhuis, N., Towers, M., Claude, E., & Van Soest, G. (2025). MALDI versus DESI mass spectrometry imaging of lipids in atherosclerotic plaque. *Rapid Communications in Mass Spectrometry*, *39*(1), e9927. DOI: 10.1002/rcm.9927 PMID: 39435741

Soja, E., Bhargava, S., Sivajothi, S., Flynn, W. F., & Courtois, E. T. (2025). KAPP-Sen TMC: Xenium - Cell DIVE - H&amp. *E (Norwalk, Conn.)*, *v1*, •••. DOI: 10.17504/protocols.io.n92ld6zj7g5b/v1

Soliman, H., Tung, L. W., & Rossi, F. M. (2021). Fibroblast and myofibroblast subtypes: single cell sequencing. In *Myofibroblasts: Methods and Protocols* (pp. 49–84). Springer US. DOI: 10.1007/978-1-0716-1382-5_4

Song, Q., & Su, J. (2021). Dstg: Deconvoluting spatial transcriptomics data through graph-based artificial intelligence. *Briefings in Bioinformatics*, *22*(5), bbaa414. Advance online publication. DOI: 10.1093/bib/bbaa414 PMID: 33480403

Song, X., Liang, H., Nan, F., Chen, W., Li, J., He, L., Cun, Y., Li, Z., Zhang, W., & Zhang, D. (2025). Autoimmune Diseases: Molecular Pathogenesis and Therapeutic Targets. *MedComm*, *6*(7), e70262. DOI: 10.1002/mco2.70262 PMID: 40529617

Sotiras, A., Davatzikos, C., & Paragios, N. (2013). Deformable medical image registration: A survey. *IEEE Transactions on Medical Imaging*, *32*(7), 1153–1190. DOI: 10.1109/TMI.2013.2265603 PMID: 23739795

Spiegel, B. M. R., Rizzo, A., Persky, S., Liran, O., Wiederhold, B., Woods, S., Donovan, K., Sarkar, K., Xiang, H., Joo, S., Jotwani, R., Lang, M., Paul, M., Senter-Zapata, M., Widmeier, K., & Zhang, H. (2024). What Is Medical Extended Reality? A Taxonomy Defining the Current Breadth and Depth of an Evolving Field. *Journal of Medical Extended Reality*, *1*(1), 4–12. DOI: 10.1089/jmxr.2023.0012

Spitzenberger, F., Patel, J., Gebuhr, I., Kruttwig, K., Safi, A., & Meisel, C. (2022). Laboratory-Developed Tests: Design of a Regulatory Strategy in Compliance with the International State-of-the-Art and the Regulation (EU) 2017/746 (EU IVDR [In Vitro Diagnostic Medical Device Regulation]). *Therapeutic Innovation & Regulatory Science*, *56*(1), 47–64. DOI: 10.1007/s43441-021-00323-7 PMID: 34291407

Spurbeck, R. R. (2024). Altered epigenetic landscape as infectious disease diagnostics. *Epigenomics*, *16*(19–20), 1269–1272. DOI: 10.1080/17501911.2024.2415282

Ståhl, P. L., Salmén, F., Vickovic, S., Lundmark, A., Navarro, J. F., Magnusson, J., Giacomello, S., Asp, M., Westholm, J. O., Huss, M., Mollbrink, A., Linnarsson, S., Codeluppi, S., Borg, Å., Pontén, F., Costea, P. I., Sahlén, P., Mulder, J., Bergmann, O., & Frisén, J. (2016). Visualization and analysis of gene expression in tissue sections by spatial transcriptomics. *Science*, *353*(6294), 78–82. DOI: 10.1126/science.aaf2403 PMID: 27365449

Stambas, J., Lu, C., & Tripp, R. A. (2020). Innate and adaptive immune responses in respiratory virus infection: Implications for the clinic. *Expert Review of Respiratory Medicine*, *14*(11), 1141–1147. DOI: 10.1080/17476348.2020.1807945

Steenbergen, E. J., Troyanov, S., Rovin, B. H., & Parikh, S. V. (2021). Multi-omics integration in lupus nephritis identifies molecular subgroups and therapeutic targets. *Kidney International, 100*(5), 1013–1026. DOI: 10.1016/j.kint.2021.05.012

Stegger, L., Schülke, C., Wenning, C., Rahbar, K., Kies, P., Schober, O., & Schäfers, M. (2013). Cardiac PET/MRI. *Current Cardiovascular Imaging Reports, 6*(2), 169–178. DOI: 10.1007/s12410-012-9189-6

Steinbrenner, I., Schultheiss, U. T., Kotsis, F., Schlosser, P., Stockmann, H., Mohney, R. P., Schmid, M., Oefner, P. J., Eckardt, K.-U., Köttgen, A., Sekula, P., Eckardt, K.-U., Meiselbach, H., Schneider, M. P., Schiffer, M., Prokosch, H.-U., Bärthlein, B., Beck, A., Reis, A., & Nadal, J. (2021). Urine metabolite levels, adverse kidney outcomes, and mortality in CKD patients: A metabolome-wide association study. *American Journal of Kidney Diseases, 78*(5), 669–677.e1. DOI: 10.1053/j.ajkd.2021.01.018 PMID: 33839201

Stephen, P., & Sumner, L. W. (2025). Spatially Resolved Plant Metabolomics. *Metabolites, 15*(8), 539. DOI: 10.3390/metabo15080539

Stern, P. L. (2020). Key steps in vaccine development. *Annals of Allergy, Asthma & Immunology, 125*(1), 17–27. DOI: 10.1016/j.anai.2020.01.025

Stickels, R. R., Murray, E., Kumar, P., Li, J., Marshall, J. L., Di Bella, D. J., & Macosko, E. Z. (2021). Highly sensitive spatial transcriptomics at near-cellular resolution with Slide-seqV2. *Nature Biotechnology, 39*(3), 313–319. DOI: 10.1038/s41587-020-0739-1 PMID: 33288904

Stoeckius, M., Hafemeister, C., Stephenson, W., Houck-Loomis, B., Chattopadhyay, P. K., Swerdlow, H., Satija, R., & Smibert, P. (2017). Simultaneous epitope and transcriptome measurement in single cells. *Nature Methods, 14*(9), 865–868. DOI: 10.1038/nmeth.4380

Stoykova, Z., & Koeva–Dimitrova, L. (2019). History and development of personalized medicine. *Health Economics and Management, 19*(1).

Subhalakshmi, R. T., Nivaashini, M., Ganesh, G. R., & Sathishkumar, A. (2025). Integrating AI into PET/CT and PET/MRI: A paradigm shift in hybrid imaging. In *AI Insights on Nuclear Medicine* (pp. 311-332). IGI Global Scientific Publishing.

Subrahmanyam, P. B., & Maecker, H. T. (2017). CyTOF Measurement of Immunocompetence Across Major Immune Cell Types. *Current Protocols in Cytometry, 82*(1). Advance online publication. DOI: 10.1002/cpcy.27 PMID: 28967988

Subramanian, I., Verma, S., Kumar, S., Jere, A., & Anamika, K. (2020). Multi-omics data integration, interpretation, and its application. *Bioinformatics and Biology Insights, 14*, 1177932219899051. DOI: 10.1177/1177932219899051 PMID: 32076369

Suchkov, S., Scherman, D., Wu, A., Borisova, A., Murphy, S., Smith, D., & Abe, H. (2024). Personalized and precision medicine (ppm) through the view of biomarker-driven applications to support their global impact in the upgraded healthcare-related services. *Journal of Biomedical and Engineering Research*, *2*(2), 1–12. DOI: 10.64030/3065-8780.02.02.04

Sugimoto, K., Moriyasu, F., Oshiro, H., Takeuchi, H., Abe, M., Yoshimasu, Y., Kasai, Y., Sakamaki, K., Hara, T., & Itoi, T. (2020). The role of multiparametric US of the liver for the evaluation of nonalcoholic steatohepatitis. *Radiology*, *296*(3), 532–540. DOI: 10.1148/radiol.2020192665

Sun, S., Zhu, J., & Zhou, X. (2021). Statistical analysis of spatial expres sion patterns for spatially resolved transcriptomic studies. *Nature Methods*, *17*(2), 193–200. DOI: 10.1038/s41592-019-0701-7 PMID: 31988518

Sun, Y., Piya, S., Ouyang, Z., Chen, Y., Gagnon, J., Cao, S., Zhang, H., Song, B., Zhu, J., Chandratre, K., Yu, H., Hu, W., Ryals, M., Casey, F., Huh, D., & Zhang, B. (2025). *SpaceSequest: A unified pipeline for spatial transcriptomics data analysis*. https://doi.org/DOI: 10.1101/2025.09.15.676389

Suo, Y., Thimme, R., & Bengsch, B. (2025). Spatial single-cell omics: new insights into liver diseases. *Gut*, gutjnl-332105. https://doi.org/DOI: 10.1136/gutjnl-2024-332105

Suphavilai, C., Chia, S., Sharma, A., Tu, L., Da Silva, R. P., Mongia, A., DasGupta, R., & Nagarajan, N. (2021). Predicting heterogeneity in clone-specific therapeutic vulnerabilities using single-cell transcriptomic signatures. *Genome Medicine*, *13*(1), 189. DOI: 10.1186/s13073-021-01000-y PMID: 34915921

Suvà, M. L., & Tirosh, I. (2019). Single-Cell RNA Sequencing in Cancer: Lessons Learned and Emerging Challenges. *Molecular Cell*, *75*(1), 7–12. DOI: 10.1016/j.molcel.2019.05.003 PMID: 31299208

Swain, S. S., Sharma, D., Hussain, T., & Pati, S. (2020). Molecular mechanisms of underlying genetic factors and associated mutations for drug resistance in Mycobacterium tuberculosis. *Emerging Microbes & Infections*, *9*(1), 1651–1663. DOI: 10.1080/22221751.2020.1785334

Swanson, E., Lord, C., Reading, J., Heubeck, A. T., Genge, P. C., Thomson, Z., Da Weiss, M., Li, X., Savage, A. K., Green, R., Green, R., Torgerson, T. R., Torgerson, T. R., Bumol, T. F., Graybuck, L. T., & Skene, P. J. (2021). Simultaneous trimodal single-cell measurement of transcripts, epitopes, and chromatin accessibility using TEA-seq. *eLife*, *10*, e63632. Advance online publication. DOI: 10.7554/eLife.63632

Swinney, D. C., & Anthony, J. (2011). How were new medicines discovered? *Nature Reviews. Drug Discovery*, *10*(7), 507–519. DOI: 10.1038/nrd3480

Szabo de Edelenyi, F., Goumidi, L., Bertrais, S., Phillips, C., MacManus, R., Roche, H., & Lairon, D. (2008). Prediction of metabolic syndrome status based on dietary and genetic parameters using random forest. *Genes & Nutrition*, *3*(3-4), 173–176. DOI: 10.1007/s12263-008-0097-y

Tadros, R., Francis, C., Xu, X., Vermeer, A. M., Harper, A. R., Huurman, R., & Bezzina, C. R. (2021). Shared genetic pathways contribute to risk of hypertrophic and dilated cardiomyopathies with opposite directions of effect. *Nature Genetics*, *53*(2), 128–134. DOI: 10.1038/s41588-020-00762-2 PMID: 33495596

Takabatake, Y., Kimura, T., Takahashi, A., & Isaka, Y. (2014). Autophagy and the kidney: Health and disease. *Nephrology, Dialysis, Transplantation*, *29*(9), 1639–1647. DOI: 10.1093/ndt/gft535 PMID: 24520117

Tanevski, J., Flores, R. O. R., Gabor, A., Schapiro, D., & Saez-Rodriguez, J. (2020). Explainable multi-view framework for dissecting intercellular signaling from highly multiplexed spatial data. *BioRxiv*, pp. 2020–05. DOI: 10.1101/2020.05.08.084145

Tang, F., Barbacioru, C., Wang, Y., Nordman, E., Lee, C., Xu, N., Wang, X., Bodeau, J., Tuch, B. B., Siddiqui, A., Lao, K., & Surani, M. A. (2009). mRNA-Seq whole-transcriptome analysis of a single cell. *Nature Methods*, *6*(5), 377–382. DOI: 10.1038/nmeth.1315 PMID: 19349980

Tang, Z., Alrumaihi, F., Alwanian, W. M., Alharbi, H. O., Allemailem, K. S., Alissa, M., & Chen, X. (2025). The future of cardiology: Integrating single-cell transcriptomics with multi-omics for enhanced cardiac disease insights. *Current Problems in Cardiology*, *50*(4), 103005. DOI: 10.1016/j.cpcardiol.2025.103005 PMID: 39894239

Tan, S., Yang, J., Hu, S., & Lei, W. (2024). Cell-cell interactions in the heart: Advanced cardiac models and omics technologies. *Stem Cell Research & Therapy*, *15*(1), 362. DOI: 10.1186/s13287-024-03982-z PMID: 39396018

Tan, W. C. C., Nerurkar, S. N., Cai, H. Y., Ng, H. H. M., Wu, D., Wee, Y. T. F., Lim, J. C. T., Yeong, J., & Lim, T. K. H. (2020). Overview of multiplex immunohistochemistry/immunofluorescence techniques in the era of cancer immunotherapy. *Cancer Communications*, *40*(4), 135–153. DOI: 10.1002/cac2.12023 PMID: 32301585

Tavallaee, G., & Orouji, E. (2025). Mapping the 3D genome architecture. *Computational and Structural Biotechnology Journal*, *27*, 89–101. DOI: 10.1016/j.csbj.2024.12.018 PMID: 39816913

Taye, M. M. (2023). Understanding machine learning with deep learning: Architectures, workflow, applications and future directions. *Computers*, *12*(5), 91. DOI: 10.3390/computers12050091

Tedesco, M., Giannese, F., Lazarevic, D., Giansanti, V., Rosano, D., Monzani, S., Catalano, I., Grassi, E., Zanella, E. R., Botrugno, O. A., Morelli, L., Panina Bordignon, P., Caravagna, G., Bertotti, A., Martino, G., Aldrighetti, L., Pasqualato, S., Trusolino, L., Cittaro, D., & Tonon, G. (2021). Chromatin Velocity reveals epigenetic dynamics by single-cell profiling of heterochromatin and euchromatin. *Nature Biotechnology*, 1–10. DOI: 10.1038/s41587-021-01031-1

Tekpınar, L., & Erdem, R. (2019). Kişiselleştirilmiş Tip Ve Genom Araştirmalarinin Sağlik Çiktilari Bağlaminda Değerlendirilmesi. *Hacettepe Sağlık İdaresi Dergisi*, *22*(4), 843–862.

Tezval, H., Merseburger, A. S., Matuschek, I., Machtens, S., Kuczyk, M. A., & Serth, J. (2008). RASSF1A protein expression and correlation with clinicopathological parameters in renal cell carcinoma. *BMC Urology*, *8*(1), 12. DOI: 10.1186/1471-2490-8-12 PMID: 18822131

Theodore Armand, T. P., Kim, H. C., & Kim, J. I. (2024). Digital anti-aging healthcare: Applications of digital technologies in diet management. *Journal of Personalized Medicine*, *14*(3), 254. DOI: 10.3390/jpm14030254

Theodosiou, A. A., & Read, R. C. (2023). Artificial intelligence, machine learning and deep learning: Potential resources for the infection clinician. *The Journal of Infection*, *87*(4), 287–294. https://doi.org/https://doi.org/10.1016/j.jinf.2023.07.006. DOI: 10.1016/j.jinf.2023.07.006

Thul, P. J., Åkesson, L., Wiking, M., Mahdessian, D., Geladaki, A., Ait Blal, H., Alm, T., Asplund, A., Björk, L., Breckels, L. M., Bäckström, A., Danielsson, F., Fagerberg, L., Fall, J., Gatto, L., Gnann, C., Hober, S., Hjelmare, M., Johansson, F., & Lundberg, E. (2017). A subcellular map of the human proteome. *Science*, *356*(6340), eaal3321. DOI: 10.1126/science.aal3321 PMID: 28495876

Tian, L., Su, S., Dong, X., Amann-Zalcenstein, D., Biben, C., Seidi, A., Hilton, D. J., Naik, S. H., & Ritchie, M. E. (2018). scPipe: a flexible R/Bioconductor preprocessing pipeline for single-cell RNA-sequencing data. *bioRxiv*, 175927. https://doi.org/DOI: 10.1101/175927

Tirgar, P., Vikstrom, A., Sepúlveda, J. M. R., Srivastava, L. K., Amini, A., Tabata, T., & Ehrlicher, A. (2025). Heart-on-a-Miniscope: A miniaturized solution for electrophysiological drug screening in cardiac organoids. *Small*, *21*(6), 2409571. DOI: 10.1002/smll.202409571 PMID: 39937454

Topçuoğlu, M. A. (2022). Stroke epidemiology and near future projection in Turkey: Analysis of Turkey data from the Global Burden of Disease Study. *Turkish Journal of Nephrology*, *28*(4), 200–211.

Topol, E. (2019). *Deep medicine: How artificial intelligence can make healthcare human again*. Hachette UK.

Topol, E. J. (2019). High-performance medicine: The convergence of human and artificial intelligence. *Nature Medicine*, *25*(1), 44–56. DOI: 10.1038/s41591-018-0300-7 PMID: 30617339

Trabulus, S., Zor, M. S., Alagoz, S., Dincer, M. T., Meşe, M., Yilmaz, E., Tahir Turanli, E., & Seyahi, N. (2024). Profiling of five urinary exosomal miRNAs for the differential diagnosis of patients with diabetic kidney disease and focal segmental glomerulosclerosis. *PLoS One*, *19*(1), e0312470. DOI: 10.1371/journal.pone.0312470 PMID: 39471136

Tran, T., Park, J., Kim, O., Kim, B., Choi, D., Lee, J., Kim, K., Oh, B., & Lee, H. (2013). Combined phospho- and glycoproteome enrichment in nephrocalcinosis tissues of phytate-fed rats. *Rapid Communications in Mass Spectrometry*, *27*(23), 2767–2776. DOI: 10.1002/rcm.6742 PMID: 24214862

Troulé, K., Petryszak, R., Cakir, B., Cranley, J., Harasty, A., Prete, M., & Vento-Tormo, R. (2025). CellPhoneDB v5: Inferring cell–cell communication from single-cell multiomics data. *Nature Protocols*, *20*(12), 1–29. DOI: 10.1038/s41596-024-01137-1 PMID: 40133495

Turanli, B., Karagoz, K., Gulfidan, G., Sinha, R., Mardinoglu, A., & Arga, K. Y. (2018). A network-based cancer drug discovery: From integrated multi-omics approaches to precision medicine. *Current Pharmaceutical Design*, *24*(32), 3778–3790. DOI: 10.2174/1381612824666181106095959

Ungvari, Z., Tarantini, S., Kiss, T., Wren, J. D., Giles, C. B., Griffin, C. T., & Csiszar, A. (2018). Endothelial dysfunction and angiogenesis impairment in the ageing vasculature. *Nature Reviews. Cardiology*, *15*(9), 555–565. DOI: 10.1038/s41569-018-0030-z PMID: 29795441

Vabret, N., Britton, G. J., Gruber, C., Hegde, S., Kim, J., Kuksin, M., Levantovsky, R., Malle, L., Moreira, A., Park, M. D., Pia, L., Risson, E., Saffern, M., Salomé, B., Esai Selvan, M., Spindler, M. P., Tan, J., van der Heide, V., Gregory, J. K., & Laserson, U. (2020). Immunology of COVID-19: Current state of the science. *Immunity*, *52*(6), 910–941. DOI: 10.1016/j.immuni.2020.05.002

Valcz, G., Buzás, E. I., Gatenby, R. A., Újvári, B., & Molnár, B. (2024). Small extracellular vesicles from surviving cancer cells as multiparametric monitoring tools of measurable residual disease and therapeutic efficiency. *Biochimica et Biophysica Acta (BBA) -. Biochimica et Biophysica Acta. Reviews on Cancer*, *1879*(2), 189088. DOI: 10.1016/j.bbcan.2024.189088

Valous, N. A., Popp, F., Zörnig, I., Jäger, D., & Charoentong, P. (2024). Graph machine learning for integrated multi-omics analysis. *British Journal of Cancer*, *131*(2), 205–211. DOI: 10.1038/s41416-024-02706-7 PMID: 38729996

Vandereyken, K., Sifrim, A., Thienpont, B., & Voet, T. (2023). Methods and applications for single-cell and spatial multi-omics. *Nature Reviews. Genetics*, *24*(8), 1–22. DOI: 10.1038/s41576-023-00580-2

Varghese, M. S., Liu, C. L., & Kazi, D. S. (2021). The price of progress: Cost, access, and adoption of novel cardiovascular drugs in clinical practice. *Current Cardiology Reports*, *23*(11), 163. DOI: 10.1007/s11886-021-01598-w PMID: 34599393

Vaswani, A., Shazeer, N., Parmar, N., Uszkoreit, J., Jones, L., Gomez, A. N., Kaiser, Ł., & Polosukhin, I. (2017). Attention is all you need. In von Luxburg, U. (Ed.), Advances in neural information processing systems: Vol. 30. *NIPS 2017). Curran Associates Inc* (pp. 5998–6008).

Vayena, E., Blasimme, A., & Cohen, I. G. (2018). Machine learning in medicine: Addressing ethical challenges. *PLoS Medicine*, *15*(11), e1002689. DOI: 10.1371/journal.pmed.1002689

Velayuthaprabhu, S., & Archunan, G. (2005). Evaluation of anticardiolipin antibodies and antiphosphatidylserine antibodies in women with recurrent abortion. *Indian Journal of Medical Sciences*, *59*(8), 347–352. DOI: 10.4103/0019-5359.16651 PMID: 16129928

Velayuthaprabhu, S., Archunan, G., & Balakrishnan, K. (2007). Placental thrombosis in experimental anticardiolipin antibodies-mediated intrauterine fetal death. *American Journal of Reproductive Immunology (New York, N.Y.), 57*(4), 270–276. DOI: 10.1111/j.1600-0897.2007.00474.x PMID: 17362388

Velayuthaprabhu, S., Matsubayashi, H., & Archunan, G. (2016). Beta-2 GPI induced tissue factor and placental apoptosis for the pathophysiology of pregnancy loss in antiphospholipid syndrome. *International Journal of Research in Medical Sciences, 4*(8), 3109–3113. DOI: 10.18203/2320-6012.ijrms20162267

Velayuthaprabhu, S., Matsubayashi, H., Sugi, T., Nakamura, M., Ohnishi, Y., Ogura, T., & Archunan, G. (2013). Expression of apoptosis in placenta of experimental antiphospholipid syndrome (APS) mouse. *American Journal of Reproductive Immunology (New York, N.Y.), 69*(5), 486–494. DOI: 10.1111/aji.12084 PMID: 23398188

Velayuthaprabhu, S., Matsubayashi, H., Sugi, T., Nakamura, M., Ohnishi, Y., Ogura, T., Tomiyama, T., & Archunan, G. (2011). A unique preliminary study on fetal resorption and placental apoptosis in mice with passive immunization of anti-phosphatidylethanolamine antibodies and anti-Factor XII antibodies. *American Journal of Reproductive Immunology (New York, N.Y.), 66*(5), 373–384. DOI: 10.1111/j.1600-0897.2011.01008.x PMID: 21623987

Velázquez, O. C., Seto, R. W., & Rombeau, J. L. (1996). The scientific rationale and clinical application of short-chain fatty acids and medium-chain triacylglycerols. *The Proceedings of the Nutrition Society, 55*(1B), 49–78. DOI: 10.1079/PNS19960011

Venerito, M., Vasapolli, R., Rokkas, T., Delchier, J., & Malfertheiner, P. (2017). Helicobacter pylori, gastric cancer and other gastrointestinal malignancies. *Helicobacter, 22*(S1), e12413. DOI: 10.1111/hel.12413

Verbeke, F., Siwy, J., Van Biesen, W., Mischak, H., Pletinck, A., Schepers, E., Neirynck, N., Magalhães, P., Pejchinovski, M., & Pontillo, C.. (2021). The urinary proteomics classifier CKD273 predicts cardiovascular outcome in patients with chronic kidney disease. *Nephrology, Dialysis, Transplantation, 36*(5), 811–818. DOI: 10.1093/ndt/gfz242 PMID: 31837226

Verrier, R. L., Nearing, B. D., & Zimetbaum, P. J. (2022). *U.S. Patent Application No. 17/607,711.*

Verstegen, M. M. A., Coppes, R. P., Beghin, A., De Coppi, P., Gerli, M. F. M., de Graeff, N., Pan, Q., Saito, Y., Shi, S., Zadpoor, A. A., & van der Laan, L. J. W. (2025). Clinical applications of human organoids. *Nature Medicine, 31*(2), 409–421. DOI: 10.1038/s41591-024-03489-3

Větrovský, T., & Baldrian, P. (2013). The variability of the 16S rRNA gene in bacterial genomes and its consequences for bacterial community analyses. *PLoS One*, *8*(2), e57923. DOI: 10.1371/journal.pone.0057923

Vetter, V., Denizer, G., Friedland, L. R., Krishnan, J., & Shapiro, M. (2018). Understanding modern-day vaccines: What you need to know. *Annals of Medicine*, *50*(2), 110–120. DOI: 10.1080/07853890.2017.1407035

Vincent, F., Loria, P., Pregel, M., Stanton, R., Kitching, L., Nocka, K., Doyonnas, R., Steppan, C., Gilbert, A., & Schroeter, T. (2015). Developing predictive assays: the phenotypic screening "rule of 3." *Science Translational Medicine, 7*(293), 293ps15-293ps15.

Vinkšel, M., Writzl, K., Maver, A., & Peterlin, B. (2021). Improving diagnostics of rare genetic diseases with NGS approaches. *Journal of Community Genetics*, *12*(2), 247–256. DOI: 10.1007/s12687-020-00500-5 PMID: 33452619

Visvikis-Siest, S., Theodoridou, D., Kontoe, M. S., Kumar, S., & Marschler, M. (2020). Milestones in personalized medicine. *Frontiers in Genetics*, *11*, 569175. DOI: 10.3389/fgene.2020.569175

Vitorino, R. (2024). Transforming clinical research: The power of high-throughput omics integration. *Proteomes*, *12*(3), 25. DOI: 10.3390/proteomes12030025 PMID: 39311198

Vollert, I., Eder, A., Hansen, A., & Eschenhagen, T. (2015). Engineering cardiovascular regeneration. *Current Stem Cell Reports*, *1*(2), 67–78. DOI: 10.1007/s40778-015-0010-8

Vujic, A., Robinson, E., Ito, M., Haider, S., Skepper, J., Methner, C., & Foo, R. (2012). An altered epigenomic mark predisposes to dilated cardiomyopathy through alternative splicing of sarcomeric gene expression in mice. *Circulation Research*, *111*(suppl_1), A100–A100. DOI: 10.1161/res.111.suppl_1.A100

Vyas, R., & D. M. S. (2024). Harnessing AI and machine learning for advanced biological data analysis. *African Journal of Biomedical Research*, *27*(1S), 1694–1701.

Waarts, M. R., Stonestrom, A. J., Park, Y. C., & Levine, R. L. (2022). Targeting mutations in cancer. *The Journal of Clinical Investigation*, *132*(8), e154943. DOI: 10.1172/JCI154943 PMID: 35426374

Waldman, S. A., & Terzic, A. (2011). Widening the path to personalized medicine. *Clinical and Translational Science*, *4*(5), 392–394. DOI: 10.1111/j.1752-8062.2011.00345.x

Walker, H. J. (2021). *Metabolite imaging by mass spectrometry: A new discovery tool* (Vol. 98). Academic Press., DOI: 10.1016/bs.abr.2020.09.022

Walling, M. A., & Shepard, J. R. E. (2011). Cellular heterogeneity and live cell arrays. *Chemical Society Reviews*, *40*(7), 4049–4076. DOI: 10.1039/c0cs00212g

Wang, B., Mezlini, A. M., Demir, F., Fiume, M., Tu, Z., Brudno, M., Haibe-Kains, B., & Goldenberg, A. (2014). Similarity network fusion for aggregating data types on a genomic scale. *Nature Methods*, *11*(3), 333–337. DOI: 10.1038/nmeth.2810 PMID: 24464287

Wang, D., & Bodovitz, S. (2010). Single cell analysis: The new frontier in 'omics.'. *Trends in Biotechnology*, *28*(6), 281–290. DOI: 10.1016/j.tibtech.2010.03.002

Wang, D., Pan, M., Li, H., Li, M., Li, P., Xiong, F., & Xiao, H. (2024). Four novel mutations identified in the COL4A3, COL4A4 and COL4A5 genes in 10 families with Alport syndrome. *BMC Medical Genomics*, *17*(1), 181. DOI: 10.1186/s12920-024-01953-0 PMID: 38978054

Wang, G., Moffitt, J. R., & Zhuang, X. (2017). Multiplexed imaging of high density libraries of RNAs with MERFISH and expansion microscopy. *bioRxiv*, 238899. https://doi.org/DOI: 10.1101/238899

Wang, L., Yang, Y., Ma, H., Xie, Y., Xu, J., Near, D., & Qian, L. (2022). Single-cell dual-omics reveals the transcriptomic and epigenomic diversity of cardiac non-myocytes. *Cardiovascular Research*, *118*(6), 1548–1563. DOI: 10.1093/cvr/cvab134 PMID: 33839759

Wang, M., Gu, M., Liu, L., Liu, Y., & Tian, L. (2021). Single-cell RNA sequencing (scRNA-seq) in cardiac tissue: Applications and limitations. *Vascular Health and Risk Management*, *17*, 641–657. DOI: 10.2147/VHRM.S288090 PMID: 34629873

Wang, N., & He, Q. (2024). *Artificial Intelligence and Bioinformatics Applications in Precision Medicine and Future Implications*. Elsevier BV., DOI: 10.1016/B978-0-12-824010-6.00058-7

Wang, Q., Zhu, H.-Y., Deng, L., Xu, S., Xie, W., Li, M., Wang, R., Liang, T., Zhan, L., & Yu, G. (2025). Spatial Transcriptomics: Biotechnologies, Computational Tools, and Neuroscience Applications. *Small Methods*, *9*(5), 2401107. Advance online publication. DOI: 10.1002/smtd.202401107

Wang, R. S., Maron, B. A., & Loscalzo, J. (2023). Multiomics network medicine approaches to precision medicine and therapeutics in cardiovascular diseases. *Arteriosclerosis, Thrombosis, and Vascular Biology*, *43*(4), 493–503. DOI: 10.1161/ATVBAHA.122.318731 PMID: 36794589

Wang, X., & Connolly, T. M. (2010). Biomarkers of vulnerable atheromatous plaques: Translational medicine perspectives. *Advances in Clinical Chemistry, 50*, 1–22. DOI: 10.1016/S0065-2423(10)50001-5 PMID: 20521438

Wang, Y., Tang, S., Ma, R., Zamit, I., Wei, Y., & Pan, Y. (2022). Multi-modal intermediate integrative methods in neuropsychiatric disorders: A review. *Computational and Structural Biotechnology Journal, 20*, 6149–6162. DOI: 10.1016/j.csbj.2022.11.008 PMID: 36420153

Wang, Y., Wang, Y., Ma, A.-J., Li, D.-X., Luo, L.-J., Liu, D.-X., Jin, D., Liu, K., & Ye, C.-Y. (2015). Rapid and sensitive isothermal detection of nucleic-acid sequence by multiple cross displacement amplification. *Scientific Reports, 5*(1), 11902. DOI: 10.1038/srep11902

Wang, Y., Wei, J., Zhang, W., Doherty, M., Zhang, Y., Xie, H., Li, W., Wang, N., Lei, G., & Zeng, C. (2022). Gut dysbiosis in rheumatic diseases: A systematic review and meta-analysis of 92 observational studies. *EBioMedicine, 80*, 104055. DOI: 10.1016/j.ebiom.2022.104055 PMID: 35594658

Wanichthanarak, K., Fahrmann, J. F., & Grapov, D. (2015). Genomic, Proteomic, and Metabolomic Data Integration Strategies. *Biomarker Insights, 10s4. Biomarker Insights, S29511*, BMI.S29511. Advance online publication. DOI: 10.4137/BMI.S29511 PMID: 26396492

Wanner, N., & Bechtel-Walz, W. (2017). Epigenetics of kidney disease. *Cell and Tissue Research, 369*(1), 75–92. DOI: 10.1007/s00441-017-2588-x PMID: 28286899

Want, E. J., Cravatt, B. F., & Siuzdak, G. (2005). The expanding role of mass spectrometry in metabolite profiling and characterization. *ChemBioChem, 6*(11), 1941–1951. DOI: 10.1002/cbic.200500151

Ward, R. A., Aghaeepour, N., Bhattacharyya, R. P., Clish, C. B., Gaudillière, B., Hacohcn, N., Mansour, M. K., Mudd, P. A., Pasupneti, S., Presti, R. M., Rhee, E. P., Sen, P., Spec, A., Tam, J. M., Villani, A.-C., Woolley, A. E., Hsu, J. L., & Vyas, J. M. (2021). Harnessing the potential of multiomics studies for precision medicine in infectious disease. *Open Forum Infectious Diseases, 8*(11), ofab483. DOI: 10.1093/ofid/ofab483

Warnat-Herresthal, S., Perrakis, K., Taschler, B., Becker, M., Baßler, K., Beyer, M., & Schultze, J. L. (2020). Scalable prediction of acute myeloid leukemia using high-dimensional machine learning and blood transcriptomics. *iScience, 23*(1), 100780. Advance online publication. DOI: 10.1016/j.isci.2019.100780 PMID: 31918046

Warr, A., Robert, C., Hume, D., Archibald, A., Deeb, N., & Watson, M. (2015). Exome sequencing: Current and future perspectives. *G3 (Bethesda, Md.)*, *5*(8), 1543–1550. DOI: 10.1534/g3.115.018564 PMID: 26139844

Watson, J. A., Strub-Wourgraft, N., Tarral, A., Ribeiro, I., Tarning, J., & White, N. J. (2019). Pharmacokinetic-pharmacodynamic assessment of the hepatic and bone marrow toxicities of the new trypanoside fexinidazole. *Antimicrobial Agents and Chemotherapy*, *63*(4), 10–1128. DOI: 10.1128/AAC.02515-18

Wayne, L. G., & Hayes, L. G. (1996). An in vitro model for sequential study of shift-down of Mycobacterium tuberculosis through two stages of nonreplicating persistence. *Infection and Immunity*, *64*(6), 2062–2069. DOI: 10.1128/iai.64.6.2062-2069.1996

Wei, M., Wang, H., Zeng, Y., Ye, D., Wang, G., & Shi, L. (2025). Multi-Modal Spatial Metabolomics with Mid-Infrared Microscope Guided Mass Spectrometry Imaging. *Journal of Visualized Experiments*, *221*(221). Advance online publication. DOI: 10.3791/68709

Wei, S., Shen, H., Zhang, Y., Liu, C., Li, S., Yao, J., Jin, Z., & Yu, H. (2024). Integrative analysis of single-cell and bulk transcriptome data reveal the significant role of macrophages in lupus nephritis. *Arthritis Research & Therapy*, *26*(1), 84. DOI: 10.1186/s13075-024-03311-y PMID: 38610007

Wei, Z., Wang, X., Feng, H., Ji, F., Bai, D., Dong, X., & Huang, W. (2023). Isothermal nucleic acid amplification technology for rapid detection of virus. *Critical Reviews in Biotechnology*, *43*(3), 415–432. DOI: 10.1080/07388551.2022.2030295

Wen, Y., Ma, L., & Ju, C. (2023). Recent insights into the pathogenesis and therapeutic targets of chronic liver diseases. *eGastroenterology*, *1*(2), e100020. DOI: 10.1136/egastro-2023-100020

Wess, M., Andersen, M. K., Midtbust, E., Guillem, J. C. C., Viset, T., Størkersen, Ø., Krossa, S., Rye, M. B., & Tessem, M.-B. (2025). Spatial integration of multi-omics data from serial sections using the novel Multi-Omics Imaging Integration Toolset. *GigaScience*, *14*, giaf035. Advance online publication. DOI: 10.1093/gigascience/giaf035

Wick, R. R., Judd, L. M., & Holt, K. E. (2019). Performance of neural network basecalling tools for Oxford Nanopore sequencing. *Genome Biology*, *20*(1), 1–10. DOI: 10.1186/s13059-019-1727-y PMID: 31234903

Williams, C. G., Lee, H. J., Asatsuma, T., Vento-Tormo, R., & Haque, A. (2022). An introduction to spatial transcriptomics for biomedical research. *Genome Medicine*, *14*(1), 68. Advance online publication. DOI: 10.1186/s13073-022-01075-1

Wilson, P. C., Ledru, N., & Humphreys, B. D. (2020). Epigenomics and the kidney. *Current Opinion in Nephrology and Hypertension*, *29*(3), 280–285. DOI: 10.1097/MNH.0000000000000602 PMID: 32235270

Wirbel, J., Pyl, P. T., Kartal, E., Zych, K., Kashani, A., Milanese, A., Fleck, J. S., Voigt, A. Y., Palleja, A., Ponnudurai, R., Sunagawa, S., Coelho, L. P., Schrotz-King, P., Vogtmann, E., Habermann, N., Niméus, E., Thomas, A. M., Manghi, P., Gandini, S., & Zeller, G. (2019). Meta-analysis of fecal metagenomes reveals global microbial signatures that are specific for colorectal cancer. *Nature Medicine*, *25*(4), 679–689. DOI: 10.1038/s41591-019-0406-6

Workman, R. E., Tang, A. D., Tang, P. S., Jain, M., Tyson, J. R., Razaghi, R., Zuzarte, P. C., Gilpatrick, T., Payne, A., Quick, J., Sadowski, N., Holmes, N., de Jesus, J. G., Jones, K. L., Soulette, C. M., Snutch, T. P., Loman, N., Paten, B., Loose, M., & Timp, W. (2019). Nanopore native RNA sequencing of a human poly (A) transcriptome. *Nature Methods*, *16*(12), 1297–1305. DOI: 10.1038/s41592-019-0617-2 PMID: 31740818

World Health Organization. (2025). *Brucellosis*. https://www.who.int/news-room/fact-sheets/detail/brucellosis

Wu, H., Malone, A. F., Donnelly, E. L., Kirita, Y., Uchimura, K., Ramakrishnan, S. M., Gaut, J. P., & Humphreys, B. D. (2018). Single-cell transcriptomics of a human kidney allograft biopsy specimen defines a diverse inflammatory response. *Journal of the American Society of Nephrology*, *29*(8), 2069–2080. DOI: 10.1681/ASN.2018020125 PMID: 29980650

Wünnemann, F., Sicklinger, F., Bestak, K., Nimo, J., Thiemann, T., Amrute, J., & Schapiro, D. (2024). Spatial omics of acute myocardial infarction reveals a novel mode of immune cell infiltration. BioRxiv, 2024-05.

Wu, S. Z., Al-Eryani, G., Roden, D. L., Junankar, S., Harvey, K., Andersson, A., Thennavan, A., Wang, C., Torpy, J. R., Bartonicek, N., Wang, T., Larsson, L., Kaczorowski, D., Weisenfeld, N. I., Uytingco, C. R., Chew, J. G., Bent, Z. W., Chan, C.-L., Gnanasambandapillai, V., & Swarbrick, A. (2021). A single-cell and spatially resolved atlas of human breast cancers. *Nature Genetics*, *53*(9), 1334–1347. DOI: 10.1038/s41588-021-00911-1 PMID: 34493872

Wu, X., Xu, W., Lin, D., Sun, L., Loo, L.-H., Dai, J., & Cao, G. (2025). The application and prospects of spatial omics technologies in clinical medical research and molecular diagnostics. *Journal of Genetics and Genomics*. Advance online publication. DOI: 10.1016/j.jgg.2025.09.003

Wu, X., Yang, X., Dai, Y., Zhao, Z., Zhu, J., Guo, H., & Yang, R. (2024). Single-cell sequencing to multi-omics: Technologies and applications. *Biomarker Research, 12*(1), 110. Advance online publication. DOI: 10.1186/s40364-024-00643-4

Wu, Y. C., & Feng, J. W. (2018). Development and application of artificial neural network. *Wireless Personal Communications, 102*(2), 1645–1656. DOI: 10.1007/s11277-017-5224-x

Wu, Y., Cheng, Y., Wang, X., Fan, J., & Gao, Q. (2022). Spatial omics: Navigating to the golden era of cancer research. *Clinical and Translational Medicine, 12*(1), e696. DOI: 10.1002/ctm2.696

Wu, Y., & Xie, L. (2025). AI-driven multi-omics integration for multi-scale predictive modeling of genotype–environment–phenotype relationships. *Computational and Structural Biotechnology Journal, 27*, 265–277. DOI: 10.1016/j.csbj.2024.12.030

Wu, Z., Boen, J., Jindal, S., Basu, S., Bieniosek, M., He, S., LaPelusa, M., Mayer, A. T., Kaseb, A. O., Zou, J., Sharma, P., & Trevino, A. E. (2025). *Spatial multi-omics and deep learning reveal fingerprints of immunotherapy response and resistance in hepatocellular carcinoma.* Cancer Biology. DOI: 10.1101/2025.06.11.656869

Xia, C., Fan, J., Emanuel, G., Hao, J., & Zhuang, X. (2019). Spatial transcriptome profiling by MERFISH reveals subcellular RNA compartmentalization and cell cycle-dependent gene expression. *Nature Methods, 16*(10), 989–993. DOI: 10.1038/s41592-019-0548-y PMID: 31501331

Xia, C., Fan, J., Emanuel, G., Hao, J., & Zhuang, X. (2019). Spatial transcriptome profiling by MERFISH reveals subcellular RNA compartmentalization and cell cycle-dependent gene expression. *Proceedings of the National Academy of Sciences of the United States of America, 116*(39), 19490–19499. DOI: 10.1073/pnas.1912459116 PMID: 31501331

Xia, M., Liu, Q., Zhang, W., Ge, J., & Mei, Z. (2025). Spatiotemporal Dynamics of Central Nervous System Diseases: Advancing Translational Neuropathology via Single-Cell and Spatial Multiomics. *MedComm, 6*(9), e70328. DOI: 10.1002/mco2.70328 PMID: 40843131

Xiao, Q., Hou, R., Xie, L., Niu, M., Pan, X., & Zhu, X. (2023). Macrophage metabolic reprogramming and atherosclerotic plaque microenvironment: Fostering each other? *Clinical and Translational Medicine, 13*(5), e1257. DOI: 10.1002/ctm2.1257 PMID: 37151159

Xia, T., Hu, L., Zuo, L., Cao, L., Zhang, Y., Xu, M., & Fang, S. (2024). ST-GEARS: Advancing 3D downstream research through accurate spatial information recovery. *Nature Communications*, *15*(1), 7806. DOI: 10.1038/s41467-024-51935-0 PMID: 39242563

Xie, C., Wang, J., Diao, X., Wang, X., & Cai, Z. (2024). *Single-Cell Resolution DESI Mass Spectrometry Imaging through 10-Fold Sample Expansion*. DOI: 10.1101/2024.10.21.619369

Xie, H., Zhang, B., Xie, M., & Li, T. (2023). Circulating metabolic signatures of heart failure in precision cardiology. *Precision Clinical Medicine*, *6*(1), pbad005. DOI: 10.1093/pcmedi/pbad005 PMID: 37020642

Xie, Y., Ruan, F., Li, Y., & Chen, Y. (2024). Spatial chromatin accessibility sequencing resolves high-order spatial interactions of epigenomic markers. *eLife*, *13*, e87868. DOI: 10.7554/eLife.87868 PMID: 38236718

Xu, C., & Jackson, S. A. (2019). Machine learning and complex biological data. *Genome Biology*, *20*(1), 1–4. DOI: 10.1186/s13059-019-1689-0 PMID: 30992073

Xu, C., Jin, X., Wei, S., Wang, P., Luo, M., Xu, Z., Yang, W., Cai, Y., Xiao, L., Lin, X., Liu, H., Cheng, R., Pang, F., Chen, R., Su, X., Hu, Y., Wang, G., & Jiang, Q. (2022). DeepST: Identifying spatial domains in spatial transcriptomics by deep learning. *Nucleic Acids Research*, *50*(22), e131–e131. DOI: 10.1093/nar/gkac901 PMID: 36250636

Xue, Q., Hu, H., Wang, R., Wu, F., & Xiong, H. (2025). The applications of single-cell multiomics in drug screening. *Pharmaceutical Science Advances*, *3*, 100090. DOI: 10.1016/j.pscia.2025.100090

Xu, H., Wang, S., Fang, M., Luo, S., Chen, C., Wan, S., & Qu, K. (2023). SPACEL: Deep learning-based characterization of spatial transcriptome architectures. *Nature Communications*, *14*(1), 7603. DOI: 10.1038/s41467-023-43220-3 PMID: 37990022

Xu, J., Jiang, M., Koga, S., Zhang, N., & Huang, Z. (2025). *SpatialFinder: A Human-in-the-Loop Vision-Language Framework for Prioritizing High-Value Regions in Spatial Transcriptomics*. https://doi.org/DOI: 10.1101/2025.08.16.670684

Xu, P., Peng, J., Yuan, T., Chen, Z., He, H., Wu, Z., Li, T., Li, X., Wang, L., Gao, L., Yan, J., Wei, W., Li, C. T., Luo, Z.-G., & Chen, Y. (2024). High-throughput mapping of single-neuron projection and molecular features by retrograde barcoded labeling. *eLife*, *13*, e85419. DOI: 10.7554/eLife.85419

Xu, T., Cai, H., Liu, J., Mao, X., Chen, Y., Chen, M., & Wang, Y. (2025). Exploring the role of unconventional T cells in rheumatoid arthritis. *Frontiers in Immunology, 16*, 1656994. Advance online publication. DOI: 10.3389/fimmu.2025.1656994

Xu, X., Su, J., Zhu, R., Li, K., Zhao, X., Fan, J., & Mao, F. (2025). From morphology to single-cell molecules: High-resolution 3D histology in biomedicine. *Molecular Cancer, 24*(1), 63. DOI: 10.1186/s12943-025-02240-x PMID: 40033282

Xu, Y., & McCord, R. P. (2021). Costa: Unsupervised convolutional neural network learning for spatial transcriptomics analysis. *BMC Bioinformatics, 22*(1), 397. DOI: 10.1186/s12859-021-04314-1 PMID: 34372758

Yamada, S., Ko, T., Hatsuse, S., Nomura, S., Zhang, B., Dai, Z., & Komuro, I. (2022). Spatiotemporal transcriptome analysis reveals critical roles for mechano-sensing genes at the border zone in remodeling after myocardial infarction. *Nature Cardiovascular Research, 1*(11), 1072–1083. DOI: 10.1038/s44161-022-00140-7 PMID: 39195917

Yan, X., Ang, K. S., van Olst, L., Edwards, A., Watson, T., Zheng, R., & Chen, J. (2024). Mosaic integration of spatial multi-omics with SpaMosaic. BioRxiv, 2024-10.

Yang, F., Zhao, Z., Zhang, D., Xiong, Y., Dong, X., Wang, Y., Yang, M., Pan, T., Liu, C., Liu, K., Lin, Y., Liu, Y., Tu, Q., Dang, Y., Xia, M., Mi, D., Zhou, W., & Xu, Z. (2024). Single-cell multi-omics analysis of lineage development and spatial organization in the human fetal cerebellum. *Cell Discovery, 10*(1), 22. DOI: 10.1038/s41421-024-00656-1

Yang, P., Jin, L., Liao, J., Jin, K., Shao, X., Li, C., Qian, J., Cheng, J., Yu, D., Guo, R., Xu, X., Lu, X., & Fan, X. (2023). Revealing spatial multimodal heterogeneity in tissues with SpaTrio. *Cell Genomics, 3*(12), 100446. DOI: 10.1016/j.xgen.2023.100446 PMID: 38116121

Yang, P., Rong, X., Gao, Z., Wang, J., & Liu, Z. (2025). Metabolic and epigenetic regulation of macrophage polarization in atherosclerosis: Molecular mechanisms and targeted therapies. *Pharmacological Research, 212*, 107588. DOI: 10.1016/j.phrs.2025.107588 PMID: 39778637

Yang, Z., Song, D., Wang, Y., & Tang, L. (2022). lncRNA MALAT1 promotes diabetic nephropathy progression via miR-15b-5p/TLR4 signaling axis. *Journal of Immunology Research, 8098001*, 1–13. Advance online publication. DOI: 10.1155/2022/8098001 PMID: 35910856

Yan, H., Shi, J., Dai, Y., Li, X., Wu, Y., Zhang, J., Gu, Z., Zhang, C., & Leng, J. (2022). Technique integration of single-cell RNA sequencing with spatially resolved transcriptomics in the tumor microenvironment. *Cancer Cell International, 22*(1), 155. Advance online publication. DOI: 10.1186/s12935-022-02580-4

Yanni, J., Tellez, J. O., Mączewski, M., Mackiewicz, U., Beresewicz, A., Billeter, R., & Boyett, M. R. (2011). Changes in ion channel gene expression underlying heart failure-induced sinoatrial node dysfunction. *Circulation: Heart Failure, 4*(4), 496–508. DOI: 10.1161/CIRCHEARTFAILURE.110.957647 PMID: 21565973

Yao, L., He, F., Zhao, Q., Li, D., Fu, S., Zhang, M., & Wang, L. (2023). Spatial multiplexed protein profiling of cardiac ischemia-reperfusion injury. *Circulation Research, 133*(1), 86–103. DOI: 10.1161/CIRCRESAHA.123.322620 PMID: 37249015

Yao, Z., van Velthoven, C. T. J., Kunst, M., Zhang, M., McMillen, D., Lee, C., Jung, W., Goldy, J., Abdelhak, A., Aitken, M., Baker, K., Baker, P., Barkan, E., Bertagnolli, D., Bhandiwad, A., Bielstein, C., Bishwakarma, P., Campos, J., Carey, D., & Zeng, H. (2023). A high-resolution transcriptomic and spatial atlas of cell types in the whole mouse brain. *Nature, 624*(7991), 317–332. DOI: 10.1038/s41586-023-06812-z

Yeo, Y. Y., Chang, Y., Qiu, H., Yiu, S. P. T., Michel, H. A., Wu, W., Jin, X., Kure, S., Parmelee, L., Luo, S., Cramer, P., Lee, J. L., Wang, Y., Yeung, J., El Ahmar, N., Simsek, B., Mohanna, R. A., Van Orden, M., Lu, W. S., & Jiang, S. (2024). *Same-Slide Spatial Multi-Omics Integration Reveals Tumor Virus-Linked Spatial Reorganization of the Tumor Microenvironment*. https://doi.org/DOI: 10.1101/2024.12.20.629650

Ye, Z., Lai, Z., Zheng, S., & Chen, Y. (2023). Spatial-Live: A lightweight and versatile tool for single-cell spatial-omics data visualization. *Bioinformatics (Oxford, England)*. Advance online publication. DOI: 10.1101/2023.09.24.559173 PMID: 37873441

Yin, Q., Zhong, L., Song, Y., Bai, L., Wang, Z., Li, C., & Yang, X. (2025). A decision support system in precision medicine: Contrastive multimodal learning for patient stratification. *Annals of Operations Research, 348*(1), 579–607. DOI: 10.1007/s10479-023-05545-6

You, Y.-H., Quach, T., Saito, R., Pham, J., & Sharma, K. (2016). Metabolomics reveals a key role for fumarate in mediating the effects of NADPH oxidase 4 in diabetic kidney disease. *Journal of the American Society of Nephrology, 27*(2), 466–481. DOI: 10.1681/ASN.2015030302 PMID: 26203118

Yu, Z. B., & Sheng, J. J. (2011). Remodelling of cardiac gap junctions and arrhythmias. *Sheng li xue bao:[Acta physiologica Sinica], 63*(6), 586-592.

Yuan, G.-C., Cai, L., Elowitz, M. B., Enver, T., Fan, G., Guo, G., Irizarry, R. A., Kharchenko, P. V., Kim, J., Orkin, S. H., Orkin, S. H., Orkin, S. H., Quackenbush, J., Saadatpour, A., Schroeder, T., Shivdasani, R. A., & Tirosh, I. (2017). Challenges and emerging directions in single-cell analysis. *Genome Biology, 18*(1), 84. DOI: 10.1186/s13059-017-1218-y

Yuan, H., Reddy, M. A., Sun, G., Lanting, L., Wang, M., Kato, M., & Natarajan, R. (2013). Involvement of P300/CBP and epigenetic histone acetylation in TGF-β1-mediated gene transcription in mesangial cells. *American Journal of Physiology. Renal Physiology, 304*(5), F601–F613. DOI: 10.1152/ajprenal.00523.2012 PMID: 23235480

Yuan, X., Liu, C., & Wang, Y. (2023). 4D myocardium reconstruction with decoupled motion and shape model. In *Proceedings of the IEEE/CVF International Conference on Computer Vision* (pp. 21252-21262). DOI: 10.1109/ICCV51070.2023.01943

Yurdagul, A.Jr, & Aikawa, E. (2023). Discovering a rare smooth muscle cell population specific to men in ascending aortic aneurysm using spatial transcriptomics. *Arteriosclerosis, Thrombosis, and Vascular Biology, 43*(12), 2298–2300. DOI: 10.1161/ATVBAHA.123.320235 PMID: 37916413

Yusuf, A. A., Govender, M. A., Brandenburg, J.-T., & Winkler, C. A. (2021). Kidney disease and APOL1. *Human Molecular Genetics, 30*(R1, R2), R129–R137. DOI: 10.1093/hmg/ddab024 PMID: 33744923

Zacharias, H. U., Altenbuchinger, M., Solbrig, S., Schäfer, A., Buyukozkan, M., Schultheiß, U. T., Kotsis, F., Köttgen, A., Krumsiek, J., & Theis, F. J., (2018). Fully integrative data analysis of NMR metabolic fingerprints with comprehensive patient data: A case report based on the German Chronic Kidney Disease (GCKD) study. *arXiv*. https://arxiv.org/abs/1810.04281

Zaidan, M., Burtin, M., Zhang, J. D., Blanc, T., Barre, P., Garbay, S., Nguyen, C., Vasseur, F., Yammine, L., Germano, S., Badi, L., Gubler, M.-C., Gallazzini, M., Friedlander, G., Pontoglio, M., & Terzi, F. (2020). Signaling pathways predisposing to chronic kidney disease progression. *JCI Insight, 5*(10), e126183. DOI: 10.1172/jci.insight.126183 PMID: 32376805

Zeevi, D., Korem, T., Zmora, N., Israeli, D., Rothschild, D., Weinberger, A., & Segal, E. (2015). Personalized nutrition by prediction of glycemic responses. *Cell, 163*(5), 1079–1094. DOI: 10.1016/j.cell.2015.11.001

Zeng, Z., Gu, S. S., Wong, C. J., Yang, L., Ouardaoui, N., Li, D., Zhang, W., Brown, M., & Liu, X. S. (2022). Machine learning on syngeneic mouse tumor profiles to model clinical immunotherapy response. *Science Advances*, *8*(41), eabm8564. Advance online publication. DOI: 10.1126/sciadv.abm8564 PMID: 36240281

Zhan, C., Tang, T., Wu, E., Zhang, Y., He, M., Wu, R., & Shen, B. (2023). From multi-omics approaches to personalized medicine in myocardial infarction. *Frontiers in Cardiovascular Medicine*, *10*, 1250340. DOI: 10.3389/fcvm.2023.1250340 PMID: 37965091

Zhang, C., Feng, H., Zhu, J., & Hong, W. (2025). Application of spatial and single-cell omics in tumor immunotherapy biomarkers. *LabMed Discovery.*, *2*(2), 100076. DOI: 10.1016/j.lmd.2025.100076

Zhang, H., Liu, W., Chang, S., Wang, H., He, J., & Huang, Q. (2023a). ST-ReGE: A novel spatial-temporal residual graph convolutional network for CVD. *IEEE Journal of Biomedical and Health Informatics*, *28*(1), 216–227. DOI: 10.1109/JBHI.2023.3327025 PMID: 37871091

Zhang, K., Feng, W., & Wang, P. (2022). Identification of spatially variable genes with graph cuts. *Nature Communications*, *13*(1), 5488. DOI: 10.1038/s41467-022-33182-3 PMID: 36123336

Zhang, Q., Li, T., Xu, M., Islam, B., & Wang, J. (2024). Application of Optogenetics in Neurodegenerative Diseases. *Cellular and Molecular Neurobiology*, *44*(1), 57. DOI: 10.1007/s10571-024-01486-1

Zhang, R., Li, Y., Fu, H., Zhao, C., Li, X., Wang, Y., Sun, Y., & Li, Y. (2025). Nanomedicine strategies for cuproptosis: Metabolic reprogramming and tumor immunotherapy. *Acta Pharmaceutica Sinica. B*, *15*(9), 4582–4613. Advance online publication. DOI: 10.1016/j.apsb.2025.07.007

Zhang, S., Deshpande, A., Verma, B. K., Wang, H., Mi, H., Yuan, L., Ho, W. J., Jaffee, E. M., Zhu, Q., Anders, R. A., Yarchoan, M., Kagohara, L. T., Fertig, E. J., & Popel, A. S. (2024). Integration of Clinical Trial Spatial Multiomics Analysis and Virtual Clinical Trials Enables Immunotherapy Response Prediction and Biomarker Discovery. *Cancer Research*, *84*(16), 2734–2748. DOI: 10.1158/0008-5472.CAN-24-0943

Zhang, X., Chen, Y., Hua, K., Xu, S., You, R., Hao, M., & Jiang, R. (2023b). uniHEART: an ensemble atlas of cardiac cells provides multifaceted portraits of the human heart. DOI: 10.21203/rs.3.rs-3215038/v1

Zhang, X., Marand, A. P., Yan, H., & Schmitz, R. J. (2023). Massive-scale single-cell chromatin accessibility sequencing using combinatorial fluidic indexing. *bioRxiv*. https://doi.org/DOI: 10.1101/2023.09.17.558155

Zhang, Y. (2022). An omic and multidimensional spatial atlas from serial biopsies of an evolving metastatic breast cancer. *Cell Reports Medicine*, *3*(2), 100525. DOI: 10.1016/j.xcrm.2022.100525

Zhang, Y., (2023). Spatial transcriptomic landscape of Alzheimer's and Parkinson's disease brains. *Nature Neuroscience*, *26*(12), 1895–1909. DOI: 10.1038/s41593-023-01347-8 PMID: 37814027

Zhang, Y., Yang, C., Chen, X., Wu, L., Yuan, Z., Zhang, F., & Qian, B. (2025). Cancer therapy resistance from a spatial-omics perspective. *Clinical and Translational Medicine*, *15*(7), e70396. DOI: 10.1002/ctm2.70396 PMID: 40677104

Zhao, T., Chiang, Z. D., Morriss, J. W., LaFave, L. M., Murray, E. M., Del Priore, I., Meli, K., Lareau, C. A., Nadaf, N. M., Li, J., Earl, A. S., Macosko, E. Z., Jacks, T., Buenrostro, J. D., & Chen, F. (n.d.). *Spatial Genomics Enables Multi-Modal Study of Clonal Heterogeneity in Tissues*. https://doi.org/DOI: 10.1038/s41586-021-04217-4

Zhao, L., Li, T., Dang, M., Li, Y., Fan, H., Hao, Q., & Zhang, G. (2023). Association of methylenetetrahydrofolate reductase (MTHFR) rs1801133 (677C>T) gene polymorphism with ischemic stroke risk in different populations: An updated meta-analysis. *Frontiers in Genetics*, *13*, 1021423. DOI: 10.3389/fgene.2022.1021423

Zheng, X. (2023). Spatial transcriptomic characterization of COVID-19 pneumonitis identifies immune circuits related to tissue injury. *JCI Insight*, *8*(2), e157837. Advance online publication. DOI: 10.1172/jci.insight.157837

Zhou, R., Yang, G., Zhang, Y., & Wang, Y. (2023a). Spatial transcriptomics in development and disease. *Molecular Biomedicine*, *4*(1), 32. DOI: 10.1186/s43556-023-00144-0 PMID: 37806992

Zhou, S., Wang, Y., Wakhloo, D., Dikshit, A., Wang, L.-C., & Srinivasan, M. (2025). Abstract 6471: Spatial multiomics reveals T-cell activation and exhaustion states in the tumor microenvironment. *Cancer Research*, *85*(8, Supplement_1), 6471. DOI: 10.1158/1538-7445.AM2025-6471

Zhou, X., Xu, Y., Zhu, L., Su, Z., Han, X., Zhang, Z., Huang, Y., & Liu, Q. (2020). Comparison of Multiple Displacement Amplification (MDA) and Multiple Annealing and Looping-Based Amplification Cycles (MALBAC) in Limited DNA Sequencing Based on Tube and Droplet. *Micromachines*, *11*(7), 645. DOI: 10.3390/mi11070645

Zhou, Y., Xiao, X., Dong, L., Tang, C., Xiao, G., & Xu, L. (2025). Cooperative integration of spatially resolved multi-omics data with COSMOS. *Nature Communications, 16*(1), 27. DOI: 10.1038/s41467-024-55204-y PMID: 39747840

Zhuang, Q., Chen, Z., Shen, J., Fan, M., Xue, D., Lu, H., Xu, R., & He, X. (2018). RASSF1A promoter methylation correlates with development, progression, and poor cancer-specific survival of renal cell carcinoma: Trial sequential analysis. *OncoTargets and Therapy, 12*, 119–134. DOI: 10.2147/OTT.S183142 PMID: 30588036

Zhuang, S., & Liu, N. (2014). EGFR signaling in renal fibrosis. *Kidney International Supplements, 4*(1), 70–74. DOI: 10.1038/kisup.2014.13 PMID: 26312153

Zong, C., Lu, S., Chapman, A. R., & Xie, X. S. (2012). Genome-wide detection of singlenucleotide and copy-number variations of a single human cell. Science. 338:1622–1626. https://doi.org/. 1.DOI: 10.1126/science.1229164

Zong, Y., Yu, T., Wang, X., Wang, Y., Hu, Z., & Li, Y. (2022). conST:an interpretable multi-modal contrastive learning framework for spatial transcriptomics. bioRxiv 2022, 2022–01. DOI: 10.1101/2022.01.14.476408

Zou, Z., Jin, Y., Nevalainen, P., Huan, Y., Heikkonen, J., & Westerlund, T. (2019, March). Edge and fog computing enabled AI for IoT-an overview. In *2019 IEEE International Conference on Artificial Intelligence Circuits and Systems (AICAS)* (pp. 51-56). IEEE. DOI: 10.1109/AICAS.2019.8771621

Zubair, A., Chapple, R., Natarajan, S., Wright, W. C., Pan, M., Lee, H. M., Tillman, H., Easton, J., & Geeleher, P. (2022). Jointly leveraging spatial transcriptomics and deep learning models for image annotation achieves better-than-pathologist performance in cell type identification in tumors. *Cancer Research, 82*(12), 456–456. DOI: 10.1158/1538-7445.AM2022-456

About the Contributors

Jen-Tsung Chen is a professor of cell biology at the National University of Kaohsiung in Taiwan. He teaches cell biology, genomics, proteomics, plant physiology, and plant biotechnology. Dr. Chen's research interests include bioactive compounds, chromatography techniques, plant molecular biology, plant biotechnology, bioinformatics, and systems pharmacology. He is an active editor of academic books and international journals to advance the exploration of multidisciplinary knowledge involving plant physiology, plant biotechnology, nanotechnology, materials science, ethnopharmacology, and systems biology. He serves as an associate editor, editorial board member, and guest editor in reputed journals. Dr. Chen published books in collaboration with international publishers, and he is handling book projects on diverse topics such as AI technologies, drug discovery, drug development, herbal medicine, medicinal biotechnology, nanotechnology, bioengineering, plant functional genomics, plant speed breeding, epigenetics, functional RNAs, and CRISPR-based genome editing. Dr. Chen is a productive author in academic publications and was recognized as one of the World's Top 2% Scientists 2023, 2024, and 2025 by Elsevier and Stanford University. Dr. Chen received the Springer Nature Editor of Distinction Award in 2025.

***

Deghinmotei Alfred-Ugbenbo is a Pharmacist and Researcher, whose formal education spans from Africa(Nigeria) to eastern Europe (M.Pharm in Belarus and PhD in Ukraine). Currently, he is a Lecturer in the Department of Pharmaceutical Chemistry, Faculty of Pharmaceutical Sciences, Bayelsa Medical University and an Associate Lecturer in the Department of Chemistry, Faculty of Sciences, Bayelsa Medical University, Bayelsa State, Nigeria. His interests cut across chemical, pharmaceutical, environmental and medical sciences.

Kazi Asraf Ali is an accomplished pharmaceutical researcher and educator, currently serving as Assistant Professor in the Department of Pharmaceutical Technology at Maulana Abul Kalam Azad University of Technology (MAKAUT), West Bengal. He holds M. Pharm (2009) and Ph.D. (2014) degrees from Jadavpur University. With over 11 years of teaching experience, Dr. Ali has guided more than 40 M. Pharm students and one Ph.D. scholar. He has published extensively in peer-reviewed journals and contributed book chapters of national and international repute. His research focuses on novel drug delivery systems, pharmaceutical formulations, nanomaterials, and the use of artificial intelligence in pharmaceutical development. Dr. Ali has authored 38 publications, and 530 citations on Google Scholar with h-index 11. His recent studies explore AI-generated nanoparticles, buoyant drug formulations, and nanocomposite applications in dermatology. He actively collaborates with researchers across institutions and has served as a question paper setter, moderator, evaluator, head examiner, and RAC member in various universities. Through his dedication to research and education, Dr. Ali continues to contribute significantly to advancements in pharmaceutical sciences.

Adriana Elena Ani ă, DVM, MVSc, PhD, is a graduate of the Faculty of Veterinary Medicine (class of 2004) at the "Ion Ionescu de la Brad" University of Agricultural Sciences and Veterinary Medicine Ia i, where she also completed a master's program in Epidemiology and Veterinary Legislation (2004–2006). She earned her PhD in 2010 in Veterinary Medicine, followed by postdoctoral training (2011–2013) within a POSDRU project. The research experience includes advanced training stages in virology and laboratory diagnostics in France and Switzerland, participation in over ten national and two international research grant. With over 80 scientific publications and contributions to major Treatises of Infectious Diseases, she has established expertise in zoonotic and emerging infections, virology, and molecular diagnostics. Currently, she serves as an associate professor and course coordinator in Virology, demonstrating extensive experience in academic leadership, project management, and interdisciplinary research on zoonotic and foodborne pathogens.

Drago Anita is a veterinarian, academic, and researcher specialized in epidemiology and molecular biology, with a strong focus on pathogen detection and One Health approaches. He serves as Scientific Vice-Dean of the Faculty of Veterinary Medicine and conducts advanced research at the ROVETEMERG platform, including in vivo imaging and infectious disease investigations. His expertise spans molecular diagnostics, bioinformatics, antimicrobial resistance, and translational applications in public and animal health.

Ebimobotei Bunu is currently a lecturer at Bayelsa Medical University in Nigeria. A medical doctor since 2012, with an MD from Belarus State Medical University and a Master of Science in Human Physiology from the University of Port Harcourt, Nigeria. I was born on August 28, 1980, in Korokorosei, Bayelsa State of Nigeria.

Rishi Chhabra is currently pursuing M.Pharm in Pharmaceutical Quality Assurance department from ISF College of Pharmacy, Moga, Punjab. My expertise are in Qulatiy control, Quality assurance, Qbd handling, Ocular disease, novel drug development

Ranjay Kumar Choudhary is a seasoned academician and researcher with over 20 years of experience in Clinical Laboratory Sciences. He holds a PhD from Manav Rachna International University, India, where his research focused on enhancing hepatitis B diagnostics using nucleic acid technology. His academic qualifications include a Master's and Bachelor's degree, along with a diploma in Medical Laboratory Technology. Dr. Choudhary is a registered laboratory specialist with both the Saudi Commission for Health Specialties and the Uttarakhand Paramedical Council, India. His current academic and research interests lie in interdisciplinary advancements in Medical Laboratory Sciences, particularly in integrating Artificial Intelligence, robotics, and Next-Generation Sequencing (NGS) into diagnostic applications.

Sourav Dutta is an M.Pharm (Pharmaceutics) graduate from MAKAUT with hands-on experience in novel drug delivery systems and formulation development. My research includes oral thin films and sustained-release formulations, along with presentations at national and international seminars. I have also gained industry exposure in production and quality control. I aim to contribute to academic teaching and research as an Assistant Professor

Syed Nazimul Haque completed M. Pharm (Pharmaceutics) course from Maulana Abul Kalam Azad University of Technology. Currently is working in production jr executive at Bharat Biotech International Limited.

Sai Priyanka is a Research Scholar in Sri Sathya Sai University for Human Excellence, India.

Mokesh Kadhirvel, a Cardiac care technologist with over Four years of hands on experience in Cardiac diagnostic procedures and patient care. My expertise includes performing 2D echocardiogram,, TMT, Holter, ECG, Non invasive cardiac tests and I can assist to manage patients in cardiac ICU.

Ebenezar Kesavarao Kumar is currently an administrator at SRMIST, Trichy. I have completed my UG, PG and PhD in Biochemistry and pursued PDF in Neonatal Physiology at UofC, Canada. Have around 20+ yrs of teaching and research experience.

Manish Kumar is a highly accomplished and research-driven Professor of Pharmaceutics with over 20 years of academic experience and 15 years in pharmaceutical research, currently serving as Professor & Head at ISF College of Pharmacy. Holding a Ph.D. in Pharmacy, I bring proven expertise in Drug Delivery Systems, Biopharmaceutics, and Pharmaceutical Technology, along with a strong record of academic leadership and curriculum development. I have guided numerous Ph.D. and M.Pharm research projects, authored 161 research publications, contributed 12 book chapters, and filed/granted 32 patents. With a Google Scholar H-index of 22 and extensive involvement in NAAC, NBA, and NIRF accreditation processes, I combine scholarly excellence with administrative acumen. Passionate about creating meaningful academic impact, I seek to contribute to a progressive institution where I can advance pharmaceutical education, mentor future scientists, and drive innovative research.

Rahul Kumar is an Assistant Professor at the Faculty of Pharmacy, Vidya University, Meerut. He holds an M.Pharm in Pharmaceutical Chemistry from Galgotias University and a B.Pharm from Swami Vivekanand Subharti University, Meerut. His research interests include Computer-Aided Drug Design (CADD), pharmaceutical chemistry, and molecular modeling. He has authored research articles, book chapters with international publishers (Springer Nature, Elsevier, CRC Press, IGI Global), and is an inventor on a design patent for a MUC-1 based SPR biosensor for Triple-Negative Breast Cancer diagnosis. He has actively participated in national and international conferences, contributing to discussions on drug discovery, computational approaches, and pharmaceutical innovation. Rahul has been recognized with several awards for research and innovation, including top positions at Anveshan 2025, Health-A-Thon 2.0, Pharmathon 2024, and Idea Hackathon 2024. His ongoing work emphasizes integrating computational and experimental strategies to advance drug discovery and therapeutic development.

Dipika Mandal is working as an Assistant Professor in the Department of Pharmaceutical Technology, Maulana Abul Kalam Azad University of Technology, West Bengal, India. Dr. Dipika has nearly 9 years of teaching and 7 years of research experience in the field of Pharmacy. She completed her B. Pharmacy (2006) and M. Pharmacy (2009) from Jadavpur University (JU), India. She has also completed her Ph.D. from JU (2019), India. She has been a recipient of Research Fellowship from

UGC, New Delhi, India. Dr. Mandal is working in the field of novel drug delivery systems. She has published several research articles, book chapters and review article in various reputed journals.

Manoj Mani is a distinguished MSc graduate with a specialized background in human genetics and a profound interest in animal research. I have an exceptional research acumen, authoring over ten scientific publications and earning several awards for my outstanding presentation skills at academic and scientific forums. My research interests also extend to phytotherapeutics and molecular biology, where I explore the molecular underpinnings of genetic and reproductive processes.

Kamalesh Marudhachalam is Project Associate - I, CSIR-IICB, Kolkata. I completed my graduation form Bharathiar university. I am interested in the study of Immunology and its therapeutics

Prem Shankar Mishra is working as a Professor & Dean of faculty of Pharmacy, vidya University, Meerut, Uttar Pradesh, India. He has more than 16 years of teaching and research experience. He has published more than 37 national and international research articles in indexed journals, more than 15 book chapters, and authored two textbooks. He has credited total 11 patents with three granted Indian patents (one for design and two for process), two overseas patents, and 08 published Indian patents. He also has been received 15.36 lakh grant from Council of Science & Technology, Uttar Pradesh, India, grant from an Indian Government fund of 0.5 lacs from DRDO, New Delhi and 0.5 lacs from the Council of Science & Technology U.P., India for organizing National Conferences in the years 2023 and 2024. He received the International Academic Excellence Award 2022 at Global Edu-Conclave on 25 Dec 2022. He has supervised 02 PhD*, 10 PG and 30 UG students for their projects. His research areas of interest are Drug design, Synthesis, herbal formulations, and anticancer agents.

Srusti Moger is a graduate students at Sri Sathya Sai University for Human Excellence, India.

Odangowei Inetiminebi Ogidi is a Lecturer, Acting Head, Department of Biochemistry and Sub-Dean, Faculty of Basic Medical Sciences, Bayelsa Medical University, Yenagoa, Bayelsa State, Nigeria. He holds a Bachelor of Science degree in Biochemistry, Master of Science and Doctor of Philosophy degrees in Biotechnology. He has also served as a Lecturer, Head of Departments of Science Laboratory Technology and Biochemistry, and Deputy Director, Centre for Continuing Education in Federal Polytechnic Ekowe, Bayelsa State. He has authored over 80 peer-reviewed academic articles in reputable local and international journals, published over 50 book chapters and presented over 40 conference papers. He is a member of various Journal Editorial Boards. He is indexed in World Scientist and University Ranking (AD Scientific Index ID https://www.adscientificindex.com/scientist/ogidi-odangowei-inetiminebi/4306173) amongst other international bodies. He have supervised over 30 ND and HND students and currently supervising over 20 undergraduate students. His research interest has been Medical/Pharmaceutical and Environmental Biotechnology/Biochemistry, Phytomedicine, Phytochemistry.

Ramya Raghavan, Research interests focus on epigenetic factors contributing to intellectual disability. Further investigated neural markers, molecular mechanisms associated with severe speech impairment. Her recent editorial work on a book dedicated to advancing cancer treatment through spatial omics approaches underscores her commitment to innovation. As an experienced editor specializing in medical and scientific publications, Dr. Raghavan possesses a unique ability to bridge complex scientific concepts with broader societal impact. Her expertise extends to the critical intersection of the exposome and Social Determinants of Health (SDOH). Dr. Raghavan's work underscores how a holistic understanding of environmental exposures, from the molecular level to population-wide influences, is crucial for addressing health disparities and advancing a more equitable approach to disease prevention. With a robust background in biology and a dedication to science communication, Dr. Raghavan ensures the clarity, accuracy, and accessibility of cutting-edge medical advancements, particularly in the evolving fields of omics, exposomics, and their implications for public health.

Ashok Kumar Sah is an Assistant Professor in the Department of Medical Laboratory Science, College of Applied Health Sciences, A' Sharqiyah University, Oman. A certified Medical Laboratory Scientist (ASCPi), he earned his PhD in Medical Laboratory Science from Glocal University, India, following his Bachelor's and Master's degrees from HNB Garhwal University. His expertise spans Molecular Pathology, Clinical Biochemistry, Microbiology, Immunology, Clinical Pathology, Transplant Technology, and Cancer Biomarkers. Dr. Sah has published 55 peer-reviewed papers, five books, and three book chapters, with contributions to leading journals including The Lancet, NPJ Clean Water, Biomarker Research, and Molecular and Cellular Biology. With over 15 years of academic and research experience across India, Nepal, and Oman, he actively presents in national and international conferences. Beyond academia, he enjoys reading, traveling, and engaging with diverse cultures, and is committed to global public health and scientific advancement.

Basudev Sahana, have been engaged as a faculty member at Department of Pharmaceutics, Institute of Pharmacy, Jalpaiguri, Govt. of West Bengal, working in the fields of Pharmaceutical formulation design, Nanoparticulate formulation development, Mucoadhesive formulation design and development by using adhesive polymers obtained from edible plant origin, and some other fields of health care services.

Gowri Subramaniam serves as Professor and Head at Dr. N.G.P. Arts and Science College. Her academic work bridges biochemistry and bioinformatics, focusing on areas such as phytoremediation and cytogenetic molecular analysis. Her professional contributions are documented under ResearcherID Y-8899-2018. Since joining the Department of Biochemistry in 2007, her work has centered on the intersection of traditional medicinal knowledge and modern biochemical techniques.

Jayasree Talukdar is an Assistant Professor, Department of Biotechnology, Pandu College, Guwahati. PHD RESEARCH WORK PhD from Department of Bioengineering & Technology, Gauhati University, Guwahati, Assam (June 2014 – December 2020) on the topic "Association of expressions profile of genes in the transforming growth factor-β signaling pathway with life style and food habits for risk of esophageal Cancer". Guide: Dr Subhash Medhi, Assistant Professor, Department of Bioengineering & Technology, Gauhati University. PUBLICATIONS: 6 BOOK CHAPTER: 2

Index

A

B

C

D

E

F

G

H

I

M

N

O

P

R

S

T

V